T0314291

Feedback Systems

Feedback Systems

An Introduction for Scientists and Engineers

SECOND EDITION

Karl Johan Åström

Richard M. Murray

PRINCETON UNIVERSITY PRESS

PRINCETON AND OXFORD

Requests for permission to reproduce material from this work should be sent to permissions@press.princeton.edu

Published by Princeton University Press
41 William Street, Princeton, New Jersey 08540
6 Oxford Street, Woodstock, Oxfordshire OX20 1TR

press.princeton.edu

All Rights Reserved

Library of Congress Control Number: 2020949670

ISBN 9780691193984
ISBN (e-book) 9780691213477

British Library Cataloging-in-Publication Data is available

Editorial: Susannah Shoemaker and Kristen Hop
Production Editorial: Nathan Carr
Production: Jacquie Poirier
Copyeditor: Lor Campbell Gehret

This book has been composed in LaTeX

Printed on acid-free paper. ∞

Printed in the United States of America

10 9 8 7 6 5 4 3 2 1

Contents

Preface to the Second Edition

THE SECOND EDITION of *Feedback Systems* contains a variety of changes that are based on feedback on the first edition, particularly in its use for introductory courses in control. One of the primary comments from users of the text was that the use of control tools for design purposes occurred only after several chapters of analytical tools, leaving the instructor having to try to convince students that the techniques would soon be useful. In our own teaching, we find that we often use design examples in the first few weeks of the class and use this to motivate the various techniques that follow. This approach has been particularly useful in engineering courses, where students are often eager to apply the tools to examples as part of gaining insight into the methods. We also found that universities that have a laboratory component attached to their controls class need to introduce some basic design techniques early, so that students can be implementing control laws in the laboratory in the early weeks of the course.

To help emphasize this more design-oriented flow, we have added a new chapter on "Feedback Principles" that illustrates some simple design principles and tools that can be used to show students what types of problems can be solved using feedback. This new chapter uses simple models, simulations, and elementary analysis techniques, so that it should be accessible to students from a variety of engineering and scientific backgrounds. For courses in which students have already been exposed to the basic ideas of feedback, perhaps in an earlier discipline-specific course, this new chapter can easily be skipped without any loss of continuity.

We have also rearranged some of the material in the final chapters of the book, moving material on fundamental limits from the chapters on frequency domain design (Chapter 11 in the original text, now Chapter 12) and robust performance (Chapter 12 in the original text, now Chapter 13) into a separate chapter on fundamental limits (Chapter 14). This new chapter also contains some additional material on techniques for robust pole placement as well as on limits imposed by nonlinearities.

In addition to these relatively large changes, we have made many other smaller changes based on the feedback we have received from early adopters of the text. We have added some material on the Routh–Hurwitz criterion and root locus plots, to at least serve as "hooks" for instructors who wish to cover that material in more detail. We have also made some notational changes throughout, most notably changing the symbols for disturbance and noise signals to v and w, respectively. The notation in the biological examples has also been updated to match the notation used in the textbook by Del Vecchio and Murray [70].

Overall, we have tried to maintain the style and organization of the book in a manner that is consistent with our goals for the first edition. In particular, we have targeted the material toward a wide range of audiences rather than any specific discipline. One consequence is that instructors who are teaching department-specific courses may find there are other texts that are better suited to these audiences.

Books written over the past few years that are tuned to non-traditional audiences, include Janert [131] (computer science), Del Vecchio and Murray [70] (biology), and Bechhoefer [31] (physics). In addition, the textbook *Feedback Control for Everyone* by Albertos and Mareels [7] provides a readable introduction requiring minimal mathematical background.

Finally, we are indebted to numerous individuals who have taught out of the text and sent us feedback on changes that would better serve their needs. In addition to the many individuals listed in the preface to the first edition, we would like to thank Kalle Åström, Bo Bernhardsson, Karl Berntorp, Constantine Caramanis, Shuo Han, Björn Olofsson, Noah Olsman, Richard Pates, Jason Rolfe, Clancy Rowley, and André Tits for their feedback, insights, and contributions. Vickie Kearn, our recently-retired editor at Princeton University Press, has continued to serve as an enthusiastic advocate for our efforts and we particularly appreciate her support over the years in our vision for the book and for her advocacy of making the material available for free download. We also thank Nathan Carr for his relentless attention to detail in the final stages of production.

Karl Johan Åström   Richard M. Murray
Lund, Sweden   Pasadena, California

Preface to the First Edition

THIS BOOK PROVIDES an introduction to the basic principles and tools for the design and analysis of feedback systems. It is intended to serve a diverse audience of scientists and engineers who are interested in understanding and utilizing feedback in physical, biological, information, and social systems. We have attempted to keep the mathematical prerequisites to a minimum while being careful not to sacrifice rigor in the process. We have also attempted to make use of examples from a variety of disciplines, illustrating the generality of many of the tools while at the same time showing how they can be applied in specific application domains.

A major goal of this book is to present a concise and insightful view of the current knowledge in feedback and control systems. The field of control started by teaching everything that was known at the time and, as new knowledge was acquired, additional courses were developed to cover new techniques. A consequence of this evolution is that introductory courses have remained the same for many years, and it is often necessary to take many individual courses in order to obtain a good perspective on the field. In developing this book, we have attempted to condense the current knowledge by emphasizing fundamental concepts. We believe that it is important to understand why feedback is useful, to know the language and basic mathematics of control and to grasp the key paradigms that have been developed over the past half century. It is also important to be able to solve simple feedback problems using back-of-the-envelope techniques, to recognize fundamental limitations and difficult control problems and to have a feel for available design methods.

This book was originally developed for use in an experimental course at Caltech involving students from a wide set of backgrounds. The course was offered to undergraduates at the junior and senior levels in traditional engineering disciplines, as well as first- and second-year graduate students in engineering and science. This latter group included graduate students in biology, computer science, and physics. Over the course of several years, the text has been classroom tested at Caltech and at Lund University, and the feedback from many students and colleagues has been incorporated to help improve the readability and accessibility of the material.

Because of its intended audience, this book is organized in a slightly unusual fashion compared to many other books on feedback and control. In particular, we introduce a number of concepts in the text that are normally reserved for second-year courses on control and hence often not available to students who are not control systems majors. This has been done at the expense of certain traditional topics, which we felt that the astute student could learn independently and are often explored through the exercises. Examples of topics that we have included are nonlinear dynamics, Lyapunov stability analysis, the matrix exponential, reachability and observability, and fundamental limits of performance and robustness. Topics that we have de-emphasized include root locus techniques, lead/lag compensation, and detailed rules for generating Bode and Nyquist plots by hand.

Several features of the book are designed to facilitate its dual function as a basic engineering text and as an introduction for researchers in natural, information, and social sciences. The bulk of the material is intended to be used regardless of the audience and covers the core principles and tools in the analysis and design of feedback systems. Advanced sections, marked by the "dangerous bend" symbol shown here, contain material that requires a slightly more technical background, of the sort that would be expected of senior undergraduates in engineering. A few sections are marked by two dangerous bend symbols and are intended for readers with more specialized backgrounds, identified at the beginning of the section. To limit the length of the text, several standard results and extensions are given in the exercises, with appropriate hints toward their solutions.

To further augment the printed material contained here, a companion web site has been developed and is available from the publisher's web page:

https://press.princeton.edu/books/hardcover/9780691193984/feedback-systems

The web site contains a database of frequently asked questions, supplemental examples and exercises, and lecture material for courses based on this text. The material is organized by chapter and includes a summary of the major points in the text as well as links to external resources. The web site also contains the source code for many examples in the book, as well as utilities to implement the techniques described in the text. Most of the code was originally written using MATLAB M-files but was also tested with LabView MathScript to ensure compatibility with both packages. Many files can also be run using other scripting languages such as Octave, SciLab, SysQuake, and Xmath.

The first half of the book focuses almost exclusively on state space control systems. We begin in Chapter 3[*] with a description of modeling of physical, biological, and information systems using ordinary differential equations and difference equations.Chapter 4 presents a number of examples in some detail, primarily as a reference for problems that will be used throughout the text. Following this, Chapter 5 looks at the dynamic behavior of models, including definitions of stability and more complicated nonlinear behavior. We provide advanced sections in this chapter on Lyapunov stability analysis because we find that it is useful in a broad array of applications and is frequently a topic that is not introduced until later in one's studies.

The remaining three chapters of the first half of the book focus on linear systems, beginning with a description of input/output behavior in Chapter 6. In Chapter 7, we formally introduce feedback systems by demonstrating how state space control laws can be designed. This is followed in Chapter 8 by material on output feedback and estimators. Chapters 7 and 8 introduce the key concepts of reachability and observability, which give tremendous insight into the choice of actuators and sensors, whether for engineered or natural systems.

The second half of the book presents material that is often considered to be from the field of "classical control." This includes the transfer function, introduced in Chapter 9, which is a fundamental tool for understanding feedback systems. Using transfer functions, one can begin to analyze the stability of feedback systems using frequency domain analysis, including the ability to reason about the closed loop behavior of a system from its open loop characteristics. This is the subject of Chapter 10, which revolves around the Nyquist stability criterion.

---

[*]Chapter numbers reflect those in the second edition.

In Chapters 11 and 12, we again look at the design problem, focusing first on proportional-integral-derivative (PID) controllers and then on the more general process of loop shaping. PID control is by far the most common design technique in control systems and a useful tool for any student. The chapter on frequency domain design introduces many of the ideas of modern control theory, including the sensitivity function. In Chapter 13, we combine the results from the second half of the book to analyze some of the fundamental trade-offs between robustness and performance. This is also a key chapter illustrating the power of the techniques that have been developed and serving as an introduction for more advanced studies.

The book is designed for use in a 10- to 15-week course in feedback systems that provides many of the key concepts needed in a variety of disciplines. For a 10-week course, Chapters 1–3, 5–7, and 9–12 can each be covered in a week's time, with the omission of some topics from the final chapters. A more leisurely course, spread out over 14–15 weeks, could cover the entire book, with 2 weeks on modeling (Chapters 3 and 2)—particularly for students without much background in ordinary differential equations—and 2 weeks on robust performance (Chapter 13).

The mathematical prerequisites for the book are modest and in keeping with our goal of providing an introduction that serves a broad audience. We assume familiarity with the basic tools of linear algebra, including matrices, vectors, and eigenvalues. These are typically covered in a sophomore-level course on the subject, and the textbooks by Apostol [12], Arnold [15], and Strang [233] can serve as good references. Similarly, we assume basic knowledge of differential equations, including the concepts of homogeneous and particular solutions for linear ordinary differential equations in one variable. Apostol [12] and Boyce and DiPrima [53] cover this material well. Finally, we also make use of complex numbers and functions and, in some of the advanced sections, more detailed concepts in complex variables that are typically covered in a junior-level engineering or physics course in mathematical methods. Apostol [11] or Stewart [232] can be used for the basic material, with Ahlfors [6], Marsden and Hoffman [177], or Saff and Snider [212] being good references for the more advanced material. We have chosen not to include appendices summarizing these various topics since there are a number of good books available.

One additional choice that we felt was important was the decision not to rely on a knowledge of Laplace transforms in the book. While their use is by far the most common approach to teaching feedback systems in engineering, many students in the natural and information sciences may lack the necessary mathematical background. Since Laplace transforms are not required in any essential way, we have included them only in an advanced section intended to tie things together for students with that background. Of course, we make tremendous use of *transfer functions*, which we introduce through the notion of response to exponential inputs, an approach we feel is more accessible to a broad array of scientists and engineers. For classes in which students have already had Laplace transforms, it should be quite natural to build on this background in the appropriate sections of the text.

Acknowledgments

The authors would like to thank the many people who helped during the preparation of this book. The idea for writing this book came in part from a report on future directions in control [187] to which Stephen Boyd, Roger Brockett, John Doyle and Gunter Stein were major contributors. Kristi Morgansen and Hideo Mabuchi helped teach early versions of the course at Caltech on which much of the text is based, and Steve Waydo served as the head TA for the course taught at Caltech in 2003–2004 and provided numerous comments and corrections. Charlotta Johnsson and Anton Cervin taught from early versions of the manuscript in Lund in 2003–2007 and gave very useful feedback. Other colleagues and students who provided feedback and advice include Leif Andersson, John Carson, K. Mani Chandy, Michel Charpentier, Domitilla Del Vecchio, Kate Galloway, Per Hagander, Toivo Hennings-son Perby, Joseph Hellerstein, George Hines, Tore Hägglund, Cole Lepine, Anders Rantzer, Anders Robertsson, Dawn Tilbury and Francisco Zabala. The reviewers for Princeton University Press and Tom Robbins at NI Press also provided valu-able comments that significantly improved the organization, layout and focus of the book. Our editor, Vickie Kearn, was a great source of encouragement and help throughout the publishing process. Finally, we would like to thank Caltech, Lund University, and the University of California at Santa Barbara for providing many resources, stimulating colleagues and students, and pleasant working environments that greatly aided in the writing of this book.

Karl Johan Åström   Richard M. Murray
Lund, Sweden   Pasadena, California
Santa Barbara, California

Feedback Systems

Chapter One

Introduction

Feedback is a central feature of life. The process of feedback governs how we grow, respond to stress and challenge, and regulate factors such as body temperature, blood pressure, and cholesterol level. The mechanisms operate at every level, from the interaction of proteins in cells to the interaction of organisms in complex ecologies.

— M. B. Hoagland and B. Dodson, *The Way Life Works*, 1995 [119].

In this chapter we provide an introduction to the basic concept of *feedback* and the related engineering discipline of *control*. We focus on both historical and current examples, with the intention of providing the context for current tools in feedback and control.

1.1 WHAT IS FEEDBACK?

A *dynamical system* is a system whose behavior changes over time, often in response to external stimulation or forcing. The term *feedback* refers to a situation in which two (or more) dynamical systems are connected together such that each system influences the other and their dynamics are thus strongly coupled. Simple causal reasoning about a feedback system is difficult because the first system influences the second and the second system influences the first, leading to a circular argument. A consequence of this is that the behavior of feedback systems is often counterintuitive, and it is therefore necessary to resort to formal methods to understand them.

Figure 1.1 illustrates in block diagram form the idea of feedback. We often use the terms *open loop* and *closed loop* when referring to such systems. A system is said to be a closed loop system if the systems are interconnected in a cycle, as shown in Figure 1.1a. If we break the interconnection, we refer to the configuration as an open loop system, as shown in Figure 1.1b. Note that since the system is in a feedback loop, the choice of system 1 versus system 2 is somewhat arbitrary. It just depends where you want to start describing how the system works.

As the quote at the beginning of this chapter illustrates, a major source of examples of feedback systems is biology. Biological systems make use of feedback in an extraordinary number of ways, on scales ranging from molecules to cells to organisms to ecosystems. One example is the regulation of glucose in the bloodstream through the production of insulin and glucagon by the pancreas. The body attempts to maintain a constant concentration of glucose, which is used by the body's cells to produce energy. When glucose levels rise (after eating a meal, for example), the hormone insulin is released and causes the body to store excess glucose in the liver.

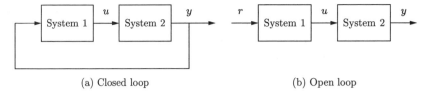

(a) Closed loop   (b) Open loop

Figure 1.1: Open and closed loop systems. (a) The output of system 1 is used as the input of system 2, and the output of system 2 becomes the input of system 1, creating a closed loop system. (b) The interconnection between system 2 and system 1 is removed, and the system is said to be open loop.

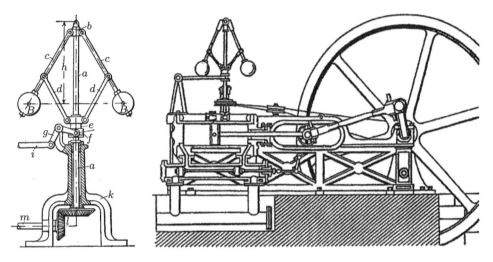

Figure 1.2: The centrifugal governor and the steam engine. The centrifugal governor on the left consists of a set of flyballs that spread apart as the speed of the engine increases. The steam engine on the right uses a centrifugal governor (above and to the left of the flywheel) to regulate its speed. (Credit: Machine a Vapeur Horizontale de Philip Taylor [1828].)

When glucose levels are low, the pancreas secretes the hormone glucagon, which has the opposite effect. Referring to Figure 1.1, we can view the liver as system 1 and the pancreas as system 2. The output from the liver is the glucose concentration in the blood, and the output from the pancreas is the amount of insulin or glucagon produced. The interplay between insulin and glucagon secretions throughout the day helps to keep the blood-glucose concentration constant, at about 90 mg per 100 mL of blood.

An early engineering example of a feedback system is a centrifugal governor, in which the shaft of a steam engine is connected to a flyball mechanism that is itself connected to the throttle of the steam engine, as illustrated in Figure 1.2. The system is designed so that as the speed of the engine increases (perhaps because of a lessening of the load on the engine), the flyballs spread apart and a linkage causes the throttle on the steam engine to be closed. This in turn slows down the engine, which causes the flyballs to come back together. We can model this system as a closed loop system by taking system 1 as the steam engine and system 2 as the governor. When properly designed, the flyball governor maintains a constant speed of the engine, roughly independent of the loading conditions. The centrifugal governor was an enabler of the successful Watt steam engine, which fueled the industrial revolution.

The examples given so far all deal with *negative feedback*, in which we attempt to react to disturbances in such a way that their effects decrease. *Positive feedback* is the opposite, where the increase in some variable or signal leads to a situation in which that quantity is further increased through feedback. This has a destabilizing effect and is usually accompanied by a saturation that limits the growth of the quantity. Although often considered undesirable, this behavior is used in biological (and engineering) systems to obtain a very fast response to a condition or signal. Encouragement is a type of positive feedback that is very useful in both industry and academia. Another common use of positive feedback is in the design of systems with oscillatory dynamics.

Feedback has many interesting properties that can be exploited in designing systems. As in the case of glucose regulation or the flyball governor, feedback can make a system resilient to external influences. It can also be used to create linear behavior out of nonlinear components, a common approach in electronics. More generally, feedback allows a system to be insensitive both to external disturbances and to variations in its individual elements.

Feedback has potential disadvantages as well. It can create dynamic instabilities in a system, causing oscillations or even runaway behavior. Another drawback, especially in engineering systems, is that feedback can introduce unwanted sensor noise into the system, requiring careful filtering of signals. It is for these reasons that a substantial portion of the study of feedback systems is devoted to developing an understanding of dynamics and a mastery of techniques in dynamical systems.

Feedback systems are ubiquitous in both natural and engineered systems. Control systems maintain the environment, lighting, and power in our buildings and factories; they regulate the operation of our cars, consumer electronics, and manufacturing processes; they enable our transportation and communications systems; and they are critical elements in our military and space systems. For the most part they are hidden from view, buried within the code of embedded microprocessors, executing their functions accurately and reliably. Feedback has also made it possible to increase dramatically the precision of instruments such as atomic force microscopes (AFMs) and telescopes.

In nature, homeostasis in biological systems maintains thermal, chemical, and biological conditions through feedback. At the other end of the size scale, global climate dynamics depend on the feedback interactions between the atmosphere, the oceans, the land, and the sun. Ecosystems are filled with examples of feedback due to the complex interactions between animal and plant life. Even the dynamics of economies are based on the feedback between individuals and corporations through markets and the exchange of goods and services.

1.2   WHAT IS FEEDFORWARD?

Feedback is reactive: there must be an error before corrective actions are taken. However, in some circumstances it is possible to measure a disturbance before it enters the system, and this information can then be used to take corrective action

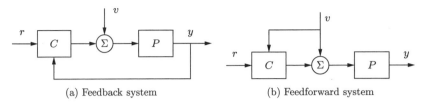

Figure 1.3: Feedback system versus feedforward system. In both systems we have a process P and a controller C. The feedback controller (a) measures the output y to determine the effect of the disturbance v, while the feedforward controller (b) measures the disturbance directly, but does not directly measure the process output.

before the disturbance has influenced the system. The effect of the disturbance is thus reduced by measuring it and generating a control signal that counteracts it. This way of controlling a system is called *feedforward*. Feedforward is particularly useful in shaping the response to command signals, which are used as external inputs to the control system, because command signals are always available. Since feedforward attempts to match two signals, it requires good process models; otherwise the corrections may have the wrong size or may be badly timed.

Figure 1.3 illustrates the difference between feedforward and feedback control. In both figures there is a reference signal r that describes the desired output of the process P and a disturbance signal v that represents an external perturbation to the process. In a feedback system, we measure the output y of the system and the controller C attempts to adjust the input to the process in a manner that causes the process output to maintain the desired the reference value. In a feedforward system, we instead measure the reference r and disturbance v and compute an input to the process that will create the desired output. Notice that the feedback controller does not directly measure the disturbance v while the feedforward controller does not measure the actual output y.

The ideas of feedback and feedforward are very general and appear in many different fields. In economics, feedback and feedforward are analogous to a market-based economy versus a planned economy. In business, a feedforward strategy corresponds to running a company based on extensive strategic planning, while a feedback strategy corresponds to a reactive approach. In biology, feedforward has been suggested as an essential element for motion control in humans that is tuned during training. Experience indicates that it is often advantageous to combine feedback and feedforward, and the correct balance requires insight and understanding of their respective properties, which are summarized in Table 1.1.

1.3 WHAT IS CONTROL?

The term *control* has many meanings and often varies between communities. In this book, we define control to be the use of algorithms and feedback in engineered systems. Thus, control includes such examples as feedback loops in electronic amplifiers, setpoint controllers in chemical and materials processing, "fly-by-wire"

Table 1.1: Properties of feedback and feedforward.

Feedback	Feedforward
Closed loop	Open loop
Acts on deviations	Acts on plans
Robust to model uncertainty	Sensitive to model uncertainty
Risk for instability	No risk for instability

systems on aircraft, and even router protocols that control traffic flow on the Internet. Emerging applications include high-confidence software systems, autonomous vehicles and robots, real-time resource management systems, and biologically engineered systems. At its core, control is an *information* science and includes the use of information in both analog and digital representations.

A modern controller senses the operation of a system, compares it against the desired behavior, computes corrective actions based on a model of the system's response to external inputs, and actuates the system to effect the desired change. This basic *feedback loop* of sensing, computation, and actuation is the central concept in control. The key issues in designing control logic are ensuring that the dynamics of the closed loop system are stable (bounded disturbances give bounded errors) and that they have additional desired behavior (good disturbance attenuation, fast responsiveness to changes in operating point, etc.). These properties are established using a variety of modeling and analysis techniques that capture the essential dynamics of the system and permit the exploration of possible behaviors in the presence of uncertainty, noise, and component failure.

A typical example of a control system is shown in Figure 1.4. The basic elements of sensing, computation, and actuation are clearly seen. In modern control systems, computation is typically implemented on a digital computer, requiring the use of analog-to-digital (A/D) and digital-to-analog (D/A) converters. Uncertainty enters the system through noise in sensing and actuation subsystems, external disturbances that affect the underlying system operation, and uncertain dynamics in the system (parameter errors, unmodeled effects, etc.). The algorithm that computes the control action as a function of the sensor values is often called a *control law*. The system can be influenced externally by an operator who introduces *command signals* to the system. These command signals can be reference values for the system output or may be more general descriptions of the task the control system is supposed to implement.

Control engineering relies on and shares tools from physics (dynamics and modeling), computer science (information and software), and operations research (optimization, probability theory, and game theory), but it is also different from these subjects in both insights and approach.

Perhaps the strongest area of overlap between control and other disciplines is in the modeling of physical systems, which is common across all areas of engineering and science. One of the fundamental differences between control-oriented modeling and modeling in other disciplines is the way in which interactions between subsystems are represented. Control relies on a type of input/output modeling that

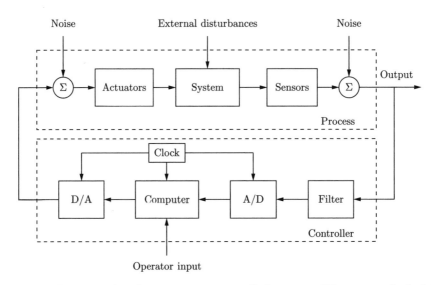

Figure 1.4: Components of a computer-controlled system. The upper dashed box represents the process dynamics, which includes the sensors and actuators in addition to the dynamical system being controlled. Noise and external disturbances can perturb the dynamics of the process. The controller is shown in the lower dashed box. It consists of a filter and analog-to-digital (A/D) and digital-to-analog (D/A) converters, as well as a computer that implements the control algorithm. A system clock controls the operation of the controller, synchronizing the A/D, D/A, and computing processes. The operator input is also fed to the computer as an external input.

allows many new insights into the behavior of systems, such as disturbance attenuation and stable interconnection. Model reduction, where a simpler (lower-fidelity) description of the dynamics is derived from a high-fidelity model, is also naturally described in an input/output framework. Perhaps most importantly, modeling in a control context allows the design of *robust* interconnections between subsystems, a feature that is crucial in the operation of all large engineered systems.

Control is also closely associated with computer science since virtually all modern control algorithms for engineering systems are implemented in software. However, control algorithms and software can be very different from traditional computer software because of the central role of the dynamics of the system and the real-time nature of the implementation.

1.4   USES OF FEEDBACK AND CONTROL

Feedback has many interesting and useful properties. It makes it possible to design precise systems from imprecise components and to make relevant quantities in a system change in a prescribed fashion. An unstable system can be stabilized using feedback, and the effects of external disturbances can be reduced. Feedback also offers new degrees of freedom to a designer by exploiting sensing, actuation, and computation. In this section we briefly survey some of the important applications and trends for feedback in the world around us. Considerably more detail is available

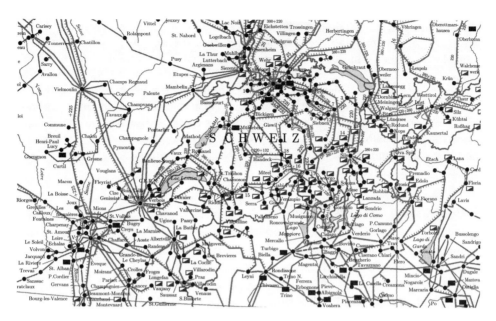

Figure 1.5: A small portion of the European power network. In 2016 European power suppliers operated a single interconnected network covering a region from the Arctic to the Mediterranean and from the Atlantic to the Urals. The installed power was more than $800\,\mathrm{GW}$ (8×10^{11} W) serving more than 500 million citizens. (Source: ENTSO-E http://www.entsoe.eu)

in several reports describing advances and directions in the field of control [158, 187, 188, 213].

Power Generation and Transmission

Access to electrical power has been one of the major drivers of technological progress in modern society. Much of the early development of control was driven by the generation and distribution of electrical power. Control is mission critical for power systems, and there are many control loops in individual power stations. Control is also important for the operation of the whole power network since it is difficult to store energy and it is thus necessary to match production to consumption. Power management is a straightforward regulation problem for a system with one generator and one power consumer, but it is more difficult in a highly distributed system with many generators and long distances between consumption and generation. Power demand can change rapidly in an unpredictable manner, and combining generators and consumers into large networks makes it possible to share loads among many suppliers and to average consumption among many customers. Large transcontinental and transnational power systems have therefore been built, such as the one shown in Figure 1.5.

Telecommunications

When telecommunications emerged in the early 20th century there was a strong need to amplify signals to enable telephone communication over long distances.

The only amplifier available at the time was based on vacuum tubes. Since the properties of vacuum tubes are nonlinear and time varying, the amplifiers created a lot of distortion. A major advance was made when Black invented the negative feedback amplifier [45, 46], which made it possible to obtain stable amplifiers with linear characteristics. Research on feedback amplifiers also generated fundamental understanding of feedback in the form of Nyquist's stability criterion [192] and Bode's methods for design of feedback amplifiers and his theorems on fundamental limits [51]. Feedback is used extensively in cellular phones and networks, and the future 5G communication networks will permit execution of real-time control systems over the networks [243].

Aerospace and Transportation

In aerospace, control has been a key technological capability tracing back to the beginning of the 20th century. Indeed, the Wright brothers are correctly famous not for demonstrating simply powered flight but *controlled* powered flight. Their early Wright Flyer incorporated moving control surfaces (vertical fins and canards) and warpable wings that allowed the pilot to regulate the aircraft's flight. In fact, the aircraft itself was not stable, so continuous pilot corrections were mandatory. This early example of controlled flight was followed by a fascinating success story of continuous improvements in flight control technology, culminating in the high-performance, highly reliable automatic flight control systems we see in modern commercial and military aircraft today.

Materials and Processing

The chemical industry is responsible for the remarkable progress in developing new materials that are key to our modern society. In addition to the continuing need to improve product quality, several other factors in the process control industry are drivers for the use of control. Environmental statutes continue to place stricter limitations on the production of pollutants, forcing the use of sophisticated pollution control devices. Environmental safety considerations have led to the design of smaller storage capacities to diminish the risk of major chemical leakage, requiring tighter control on upstream processes and, in some cases, supply chains. And large increases in energy costs have encouraged engineers to design plants that are highly integrated, coupling many processes that used to operate independently. All of these trends increase the complexity of these processes and the performance requirements for the control systems, making control system design increasingly challenging.

Instrumentation

The measurement of physical variables is of prime interest in science and engineering. Consider, for example, an accelerometer, where early instruments consisted of a mass suspended on a spring with a deflection sensor. The precision of such an instrument depends critically on accurate calibration of the spring and the sensor. There is also a design compromise because a weak spring gives high sensitivity but low bandwidth. A different way of measuring acceleration is to use *force feedback*. The spring is replaced by a voice coil that is controlled so that the mass remains at a constant position. The acceleration is proportional to the current

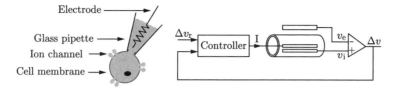

Figure 1.6: The voltage clamp method for measuring ion currents in cells using feedback. A pipette is used to place an electrode in a cell (left) and maintain the potential of the cell at a fixed level. The internal voltage in the cell is v_i, and the voltage of the external fluid is v_e. The feedback system (right) controls the current I into the cell so that the voltage drop across the cell membrane $\Delta v = v_i - v_e$ is equal to its reference value Δv_r. The current I is then equal to the ion current.

through the voice coil. In such an instrument, the precision depends entirely on the calibration of the voice coil and does not depend on the sensor, which is used only as the feedback signal. The sensitivity/bandwidth compromise is also avoided.

Another important application of feedback is in instrumentation for biological systems. Feedback is widely used to measure ion currents in cells using a device called a *voltage clamp*, which is illustrated in Figure 1.6. Hodgkin and Huxley used the voltage clamp to investigate propagation of action potentials in the giant axon of the squid. In 1963 they shared the Nobel Prize in Medicine with Eccles for "their discoveries concerning the ionic mechanisms involved in excitation and inhibition in the peripheral and central portions of the nerve cell membrane." A refinement of the voltage clamp called a *patch clamp* made it possible to measure exactly when a single ion channel is opened or closed. This was developed by Neher and Sakmann, who received the 1991 Nobel Prize in Medicine "for their discoveries concerning the function of single ion channels in cells."

Robotics and Intelligent Machines

The goal of cybernetic engineering, already articulated in the 1940s and even before, has been to implement systems capable of exhibiting highly flexible or "intelligent" responses to changing circumstances [21]. In 1948 the MIT mathematician Norbert Wiener gave a widely read account of cybernetics [253]. A more mathematical treatment of the elements of engineering cybernetics was presented by H. S. Tsien in 1954, driven by problems related to the control of missiles [242]. Together, these works and others of that time form much of the intellectual basis for modern work in robotics and control.

Two recent areas of advancement in robotics and autonomous systems are (consumer) drones and autonomous cars, some examples of which are shown in Figure 1.7. Quadrocopters such as the DJI Phantom make use of GPS receivers, accelerometers, magnetometers, and gyros to provide stable flight and also use stabilized camera platforms to provide high quality images and movies. Autonomous vehicles, such as the Google autonomous car project (now Waymo), make use of a variety of laser rangefinders, cameras, and radars to perceive their environment and then use sophisticated decision-making and control algorithms to enable them to safely drive in a variety of traffic conditions, from high-speed freeways to crowded city streets.

Figure 1.7: Autonomous vehicles. The figure on the left is a DJI Phantom 3 drone, which is able to maintain its position using GPS and inertial sensors. The figure on the right is an autonomous car that was developed by nuTonomy and is capable of driving on city streets by using sophisticated sensing and decision-making (control) software (photo courtesy Hyundai-Aptiv Autonomous Driving Joint Venture, LLC).

Networks and Computing Systems

Control of networks is a large research area spanning many topics, including congestion control, routing, data caching, and power management. Several features of these control problems make them very challenging. The dominant feature is the extremely large scale of the system: the Internet is probably the largest feedback control system humans have ever built. Another is the decentralized nature of the control problem: decisions must be made quickly and based only on local information. Stability is complicated by the presence of varying time lags, as information about the network state can be observed or relayed to controllers only after a delay, and the effect of a local control action can be felt throughout the network only after substantial delay.

Related to the control of networks is control of the servers that sit on these networks. Computers are key components of the systems of routers, web servers, and database servers used for communication, electronic commerce, advertising, and information storage. A typical example of a multilayer system for e-commerce is shown in Figure 1.8a. The system has several tiers of servers. The edge server accepts incoming requests and routes them to the HTTP server tier where they are parsed and distributed to the application servers. The processing for different requests can vary widely, and the application servers may also access external servers managed by other organizations. Control of an individual server in a layer is illustrated in Figure 1.8b. A quantity representing the quality of service or cost of operation—such as response time, throughput, service rate, or memory usage—is measured in the computer. The control variables might represent incoming messages accepted, priorities in the operating system, or memory allocation. The feedback loop then attempts to maintain quality-of-service variables within a target range of values.

Economics

The economy is a large dynamical system with many actors: governments, organizations, companies, and individuals. Governments control the economy through laws and taxes, the central banks by setting interest rates, and companies by setting

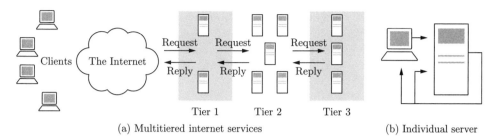

(a) Multitiered internet services   (b) Individual server

Figure 1.8: A multitier system for services on the Internet. In the complete system shown schematically in (a), users request information from a set of computers (tier 1), which in turn collect information from other computers (tiers 2 and 3). The individual server shown in (b) has a set of reference parameters set by a (human) system operator, with feedback used to maintain the operation of the system in the presence of uncertainty. (Based on Hellerstein et al. [117].)

prices and making investments. Individuals control the economy through purchases, savings, and investments. Many efforts have been made to model and control the system both at the macro level and at the micro level, but this modeling is difficult because the system is strongly influenced by the behaviors of the different actors in the system.

The financial system can be viewed as a global controller for the economy. Unfortunately this important controller does not always function as desired, as expressed in the following quote by Paul Krugman [153]:

> We have magneto trouble, said John Maynard Keynes at the start of the Great Depression: most of the economic engine was in good shape, but a crucial component, the financial system, was not working. He also said this: "We have involved ourselves in a colossal muddle, having blundered in the control of a delicate machine, the working of which we do not understand." Both statements are as true now as they were then.

One of the reasons why it is difficult to model economic systems is that conservation laws for important variables are missing. A typical example is that the value of a company as expressed by its stock can change rapidly and erratically. There are, however, some areas with conservation laws that permit accurate modeling. One example is the flow of products from a manufacturer to a retailer, as illustrated in Figure 1.9. The products are physical quantities that obey a conservation law, and the system can be modeled by accounting for the number of products in the different inventories. There are considerable economic benefits in controlling supply chains so that products are available to customers while minimizing products that are in storage. Realistic supply chain problems are more complicated than indicated in the figure because there may be many different products, there may be different factories that are geographically distributed, and the factories may require raw material or subassemblies.

Feedback in Nature

Many problems in the natural sciences involve understanding aggregate behavior in complex large-scale systems. This behavior emerges from the interaction of a

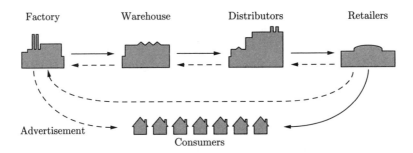

Figure 1.9: Supply chain dynamics (after Forrester [89]). Products flow from the producer to the customer through distributors and retailers as indicated by the solid lines. There are typically many factories and warehouses and even more distributors and retailers. Dashed lines represent feedback and feedforward information flowing between the various agents in the chain. Multiple feedback loops are present as each agent tries to maintain the proper inventory level.

multitude of simpler systems with intricate patterns of information flow. Representative examples can be found in fields ranging from embryology to seismology. Researchers who specialize in the study of specific complex systems often develop an intuitive emphasis on analyzing the role of feedback (or interconnection) in facilitating and stabilizing aggregate behavior. We briefly highlight three application areas here.

A major theme currently of interest to the biology community is the science of reverse (and eventually forward) engineering of biological control networks such as the one shown in Figure 1.10. There are a wide variety of biological phenomena that provide a rich source of examples of control, including gene regulation and signal transduction; hormonal, immunological, and cardiovascular feedback mechanisms; muscular control and locomotion; active sensing, vision, and proprioception; attention and consciousness; and population dynamics and epidemics. Each of these (and many more) provide opportunities to figure out what works, how it works, and what we can do to affect it.

In contrast to individual cells and organisms, emergent properties of aggregations and ecosystems inherently reflect selection mechanisms that act on multiple levels, and primarily on scales well below that of the system as a whole. Because ecosystems are complex, multiscale dynamical systems, they provide a broad range of new challenges for the modeling and analysis of feedback systems. Recent experience in applying tools from control and dynamical systems to bacterial networks suggests that much of the complexity of these networks is due to the presence of multiple layers of feedback loops that provide robust functionality to the individual cell [146, 230, 259]. Yet in other instances, events at the cell level benefit the colony at the expense of the individual. Systems level analysis can be applied to ecosystems with the goal of understanding the robustness of such systems and the extent to which decisions and events affecting individual species contribute to the robustness and/or fragility of the ecosystem as a whole.

In nature, development of organisms and their control systems have often developed in synergy. The development of birds is an interesting example, as noted by John Maynard Smith in 1952 [224]:

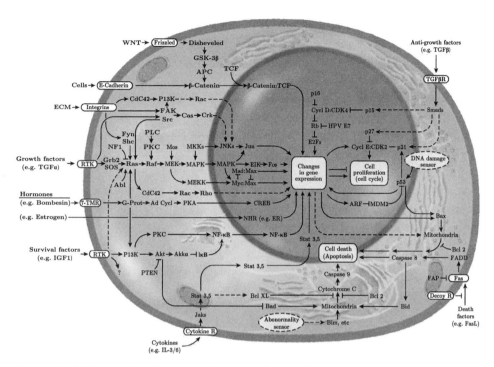

Figure 1.10: The wiring diagram of the growth-signaling circuitry of the mammalian cell [114]. The major pathways that are thought to play a role in cancer are indicated in the diagram. Lines represent interactions between genes and proteins in the cell. Lines ending in arrowheads indicate activation of the given gene or pathway; lines ending in a T-shaped head indicate repression. (Used with permission of Elsevier Ltd. and the authors.)

[T]he earliest birds, pterosaurs, and flying insects were stable. This is believed to be because in the absence of a highly evolved sensory and nervous system they would have been unable to fly if they were not. ... To a flying animal there are great advantages to be gained by instability. The greater manoeuvrability [sic] is of equal importance to an animal which catches its food in the air and to the animals upon which it preys. ... It appears that in the birds and at least in some insects [...] the evolution of the sensory and nervous systems rendered the stability found in earlier forms no longer necessary.

1.5 FEEDBACK PROPERTIES

Feedback is a powerful idea which, as we have seen, is used extensively in natural and technological systems. The principle of feedback is simple: base correcting actions on the difference between desired and actual performance. In engineering, feedback has been rediscovered and patented many times in many different contexts. The use of feedback has often resulted in vast improvements in system capability, and these improvements have sometimes been revolutionary, as discussed previously. The reason for this is that feedback has some truly remarkable properties. In this

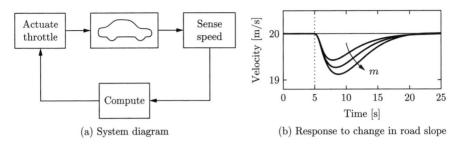

(a) System diagram   (b) Response to change in road slope

Figure 1.11: A feedback system for controlling the velocity of a vehicle. In the block diagram on the left, the velocity of the vehicle is measured and compared to the desired velocity within the "Compute" block. Based on the difference in the actual and desired velocities, the throttle (or brake) is used to modify the force applied to the vehicle by the engine, drivetrain, and wheels. The figure on the right shows how the velocity changes when the car travels on a horizontal road and the slope of the road changes to a constant uphill slope. The three different curves correspond to differing masses of the vehicle, between 1200 and 2000 kg, demonstrating that feedback can indeed compensate for the changing slope and that the closed loop system is robust to a large change in the vehicle characteristics.

section we will discuss some of the properties of feedback that can be understood intuitively. This intuition will be formalized in subsequent chapters.

Robustness to Uncertainty

One of the key uses of feedback is to provide robustness to uncertainty. For example, by measuring the difference between the sensed value of a regulated signal and its desired value, we can supply a corrective action to partially compensate for the effect of disturbances. This is precisely the effect that Watt exploited in his use of the centrifugal governor on steam engines. Another use of feedback is to provide robustness to variations in the process dynamics. If the system undergoes some change that affects the regulated signal, then we sense this change and try to force the system back to the desired operating point, even if the process parameters are not directly measured. In this way, a feedback system provides robust performance in the presence of uncertain dynamics.

As an example, consider the simple feedback system shown in Figure 1.11. In this system, the velocity of a vehicle is controlled by adjusting the amount of gas flowing to the engine. Simple *proportional-integral* (PI) feedback is used to make the amount of gas depend on both the error between the current and the desired velocity and the integral of that error. The plot on the right shows the effect of this feedback when the vehicle travels on a horizontal road and it encounters an uphill slope. When the slope changes, the car decelerates due to gravity forces and the velocity initially decreases. The velocity error is sensed by the controller, which acts to restore the velocity to the desired value by increasing the throttle. The figure also shows what happens when the same controller is used for different masses of the car, which might result from having a different number of passengers or towing a trailer. Notice that the steady-state velocity of the vehicle always approaches the desired velocity and achieves that velocity within approximately 15 s, independent of the mass (which varies by a factor of ± 25%), Thus feedback improves both performance and robustness of the system.

Another early example of the use of feedback to provide robustness is the negative feedback amplifier. When telephone communications were developed, amplifiers were used to compensate for signal attenuation in long lines. A vacuum tube was a component that could be used to build amplifiers. Distortion caused by the nonlinear characteristics of the tube amplifier together with amplifier drift were obstacles that prevented the development of line amplifiers for a long time. A major breakthrough was the invention of the feedback amplifier in 1927 by Harold S. Black, an electrical engineer at Bell Telephone Laboratories. Black used *negative feedback*, which reduces the gain but makes the amplifier insensitive to variations in tube characteristics. This invention made it possible to build stable amplifiers with linear characteristics despite the nonlinearities of the vacuum tube amplifier.

Design of Dynamics

Another use of feedback is to change the dynamics of a system. Through feedback, we can alter the behavior of a system to meet the needs of an application: systems that are unstable can be stabilized, systems that are sluggish can be made responsive, and systems that have drifting operating points can be held constant. Control theory provides a rich collection of techniques to analyze the stability and dynamic response of complex systems and to place bounds on the behavior of such systems by analyzing the gains of linear and nonlinear operators that describe their components.

An example of the use of control in the design of dynamics comes from the area of flight control. The following quote, from a lecture presented by Wilbur Wright to the Western Society of Engineers in 1901 [180], illustrates the role of control in the development of the airplane:

> Men already know how to construct wings or airplanes, which when driven through the air at sufficient speed, will not only sustain the weight of the wings themselves, but also that of the engine, and of the engineer as well. Men also know how to build engines and screws of sufficient lightness and power to drive these planes at sustaining speed ... Inability to balance and steer still confronts students of the flying problem ... When this one feature has been worked out, the age of flying will have arrived, for all other difficulties are of minor importance.

The Wright brothers thus realized that control was a key issue to enable flight. They resolved the compromise between stability and maneuverability by building an airplane, the Wright Flyer, that was unstable but maneuverable. The Flyer had a rudder in the front of the airplane, which made the plane very maneuverable. A disadvantage was the necessity for the pilot to keep adjusting the rudder to fly the plane: if the pilot let go of the stick, the plane would crash. Other early aviators tried to build stable airplanes. These would have been easier to fly, but because of their poor maneuverability they could not be brought up into the air. The Wright Brothers were well aware of the compromise between stability and maneuverability when they designed the Wright Flyer [78] and they made the first successful flight at Kitty Hawk in 1903. Modern fighter airplanes are also unstable in certain flight regimes, such as take-off and landing.

(a) Sperry autopilot   (b) 1912 Curtiss biplane

Figure 1.12: Aircraft autopilot system. The Sperry autopilot (a) contained a set of four gyros coupled to a set of air valves that controlled the wing surfaces. The 1912 Curtiss used an autopilot to stabilize the roll, pitch, and yaw of the aircraft and was able to maintain level flight as a mechanic walked on the wing (b) [125].

Since it was quite tiresome to fly an unstable aircraft, there was strong motivation to find a mechanism that would stabilize an aircraft. Such a device, invented by Sperry, was based on the concept of feedback. Sperry used a gyro-stabilized pendulum to provide an indication of the vertical. He then arranged a feedback mechanism that would pull the stick to make the plane go up if it was pointing down, and vice versa. The Sperry autopilot was the first use of feedback in aeronautical engineering, and Sperry won a prize in a competition for the safest airplane in Paris in 1914. Figure 1.12 shows the Curtiss seaplane and the Sperry autopilot. The autopilot is a good example of how feedback can be used to stabilize an unstable system and hence "design the dynamics" of the aircraft.

Creating Modularity

Feedback can be used to create modularity and shape well-defined relations between inputs and outputs in a structured hierarchical manner. A modular system is one in which individual components can be replaced without having to modify the entire system. By using feedback, it is possible to allow components to maintain their input/output properties in a manner that is robust to changes in its interconnections. A typical example is the electrical drive system shown in Figure 1.13, which has an architecture with three cascaded loops. The innermost loop is a current loop, where the current controller (CC) drives the amplifier so that the current to the motor tracks a commanded value (often called the "setpoint"). The middle feedback loop uses a velocity controller (VC) to drive the setpoint of the current controller so that velocity follows its commanded value. The outer loop

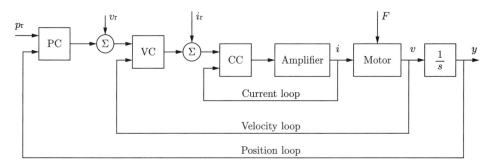

Figure 1.13: Block diagram of a system for position control. The system has three cascaded loops for control of current, velocity, and position. Each loop has an externally supplied reference value (denoted by the subscript 'r') that sets the nominal value of the input to the loop, which is added to the output from the next outermost loop to determine the commanded value for the loop (called the "setpoint").

drives the setpoint of the velocity loop to follow the setpoint of the position controller PC.

The control architecture with nested loops shown in Figure 1.13 is common. It simplifies design, commissioning, and operation. Consider for example the design of the velocity loop. With a well-designed current controller the motor current follows the setpoint of the controller CC. Since the motor torque is proportional to the current, the dynamics relating motor velocity to the input of the current controller is approximately an integrator. This simplified model can be used to design the velocity loop so that effects of friction and other disturbances are reduced. With a well-designed velocity loop, the design of the position loop is also simple. The loops can also be tuned sequentially starting with the inner loop.

This architecture illustrates how feedback can be used to simplify the overall design of the controller by breaking the problem into stages. This architecture also provides a level of modularity since each design stage depends only on the closed loop behavior of the system. If we replace the motor with a new motor, then by redesigning the current controller (CC) to give the same closed loop performance, we can leave the outer level loops unchanged. Similarly, if we need to redesign one of the outer layer controllers for an application with different specifications, we can often make use of an existing inner loop design (as long as the existing design provides enough performance to satisfy the outer loop requirements).

Challenges of Feedback

While feedback has many advantages, it also has some potential drawbacks. Chief among these is the possibility of instability if the system is not designed properly. We are all familiar with the effects of *positive feedback* when the amplification on a microphone is turned up too high in a room. This is an example of feedback instability, something that we obviously want to avoid. This is tricky because we must design the system not only to be stable under nominal conditions but also to remain stable under all possible perturbations of the dynamics.

In addition to the potential for instability, feedback inherently couples different parts of a system. One common problem is that feedback often injects measurement noise into the system. Measurements must be carefully filtered so that the actuation and process dynamics do not respond to them, while at the same time ensuring that the measurement signal from the sensor is properly coupled into the closed loop dynamics (so that the proper levels of performance are achieved).

Another potential drawback of control is the complexity of embedding a control system into a product. While the cost of sensing, computation, and actuation has decreased dramatically in the past few decades, the fact remains that control systems are often complicated, and hence one must carefully balance the costs and benefits. An early engineering example of this is the use of microprocessor-based feedback systems in automobiles.The use of microprocessors in automotive applications began in the early 1970s and was driven by increasingly strict emissions standards, which could be met only through electronic controls. Early systems were expensive and failed more often than desired, leading to frequent customer dissatisfaction. It was only through aggressive improvements in technology that the performance, reliability, and cost of these systems allowed them to be used in a transparent fashion. Even today, the complexity of these systems is such that it is difficult for an individual car owner to fix problems.

1.6   SIMPLE FORMS OF FEEDBACK

The idea of feedback to make corrective actions based on the difference between the desired and the actual values of a quantity can be implemented in many different ways. The benefits of feedback can be obtained by very simple feedback laws such as on-off control, proportional control, and proportional-integral-derivative control. In this section we provide a brief preview of some of the topics that will be studied more formally in the remainder of the text.

On-Off Control

A simple feedback mechanism can be described as follows:

$$u = \begin{cases} u_{\max} & \text{if } e > 0, \\ u_{\min} & \text{if } e < 0, \end{cases} \tag{1.1}$$

where the *control error* $e = r - y$ is the difference between the reference (or command) signal r and the output of the system y, and u is the actuation command. Figure 1.14a shows the relation between error and control. This control law implies that maximum corrective action is always used.

The feedback in equation (1.1) is called *on-off control.* One of its chief advantages is that it is simple and there are no parameters to choose. On-off control often succeeds in keeping the process variable close to the reference, such as the use of a simple thermostat to maintain the temperature of a room. It typically results in a system where the controlled variables oscillate, which is often acceptable if the oscillation is sufficiently small.

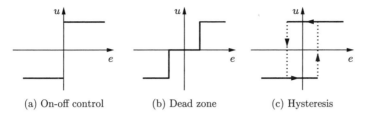

| (a) On-off control | (b) Dead zone | (c) Hysteresis |

Figure 1.14: Input/output characteristics of on-off controllers. Each plot shows the input on the horizontal axis and the corresponding output on the vertical axis. Ideal on-off control is shown in (a), with modifications for a dead zone (b) or hysteresis (c). Note that for on-off control with hysteresis, the output depends on the value of past inputs.

Notice that in equation (1.1) the control variable is not defined when the error is zero. It is common to make modifications by introducing either a dead zone or hysteresis (see Figures 1.14b and 1.14c).

PID Control

The reason why on-off control often gives rise to oscillations is that the system overreacts since a small change in the error makes the actuated variable change over the full range. This effect is avoided in *proportional control*, where the characteristic of the controller is proportional to the control error for small errors. This can be achieved with the control law

$$u = \begin{cases} u_{max} & \text{if } e \geq e_{max}, \\ k_p e & \text{if } e_{min} < e < e_{max}, \\ u_{min} & \text{if } e \leq e_{min}, \end{cases} \tag{1.2}$$

where k_p is the controller gain, $e_{min} = u_{min}/k_p$, and $e_{max} = u_{max}/k_p$. The interval (e_{min}, e_{max}) is called the *linear range* because the behavior of the controller is linear when the error is in this interval:

$$u = k_p(r - y) = k_p e \qquad \text{if } e_{min} \leq e \leq e_{max}. \tag{1.3}$$

While a vast improvement over on-off control, proportional control has the drawback that the process variable often deviates from its reference value. In particular, if some level of control signal is required for the system to maintain a desired value, then we must have $e \neq 0$ in order to generate the requisite input.

This can be avoided by making the control action proportional to the integral of the error:

$$u(t) = k_i \int_0^t e(\tau)d\tau. \tag{1.4}$$

This control form is called *integral control*, and k_i is the integral gain. It can be shown through simple arguments that a controller with integral action has zero

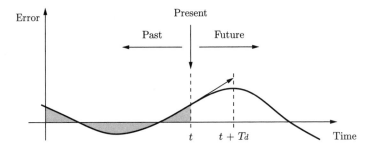

Figure 1.15: Action of a PID controller. At time t, the proportional term depends on the instantaneous value of the error. The integral portion of the feedback is based on the integral of the error up to time t (shaded portion). The derivative term provides an estimate of the growth or decay of the error over time by looking at the rate of change of the error. T_d represents the approximate amount of time in which the error is projected forward (see text).

steady-state error (Exercise 1.5). The catch is that there may not always be a steady state because the system may be oscillating. In addition, if the control action has magnitude limits, as in equation (1.2), an effect known as "integrator windup" can occur and may result in poor performance unless appropriate "anti-windup" compensation is used. Despite the potential drawbacks, which can be overcome with careful analysis and design, the benefits of integral feedback in providing zero error in the presence of constant disturbances have made it one of the most used forms of feedback.

An additional refinement is to provide the controller with an anticipative ability by using a prediction of the error. A simple prediction is given by the linear extrapolation

$$e(t + T_d) \approx e(t) + T_d \frac{de(t)}{dt},$$

which predicts the error T_d time units ahead. Combining proportional, integral, and derivative control, we obtain a controller that can be expressed mathematically as

$$u(t) = k_p e(t) + k_i \int_0^t e(\tau)\,d\tau + k_d \frac{de(t)}{dt}. \tag{1.5}$$

The control action is thus a sum of three terms: the present as represented by the proportional term, the past as represented by the integral of the error, and the future as represented by a linear extrapolation of the error (the derivative term). This form of feedback is called a *proportional-integral-derivative (PID) controller* and its action is illustrated in Figure 1.15.

A PID controller is very useful and is capable of solving a wide range of control problems. More than 95% of all industrial control problems are solved by PID control, although many of these controllers are actually *proportional-integral* (PI) *controllers* because derivative action is often not included [71].

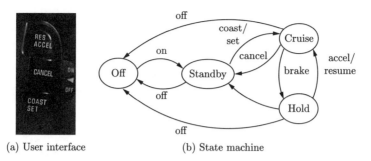

(a) User interface   (b) State machine

Figure 1.16: Finite state machine for cruise control system. The figure on the left shows some typical buttons used to control the system. The controller can be in one of four modes, corresponding to the nodes in the diagram on the right. Transition between the modes is controlled by pressing one of the four buttons on the cruise control interface: on/off, set, resume, or cancel.

1.7 COMBINING FEEDBACK WITH LOGIC

Continuous control is often combined with logic to cope with different operating conditions. Logic is typically related to changes in operating mode, equipment protection, manual interaction, and saturating actuators. One situation is when there is one variable that is of primary interest, but other variables may have to be controlled for equipment protection. For example, when controlling a compressor the outflow is the primary variable but it may be necessary to switch to a different mode to avoid compressor stall, which may damage the compressor. We illustrate some ways in which logic and feedback are combined by a few examples.

Cruise control

The basic control function in a cruise controller, such as the one shown in Figure 1.11, is to keep the velocity constant. It is typically done with a PI controller. The controller normally operates in automatic mode but it is necessary to switch it off when braking, accelerating, or changing gears. The cruise control system has a human–machine interface that allows the driver to communicate with the system. There are many different ways to implement this system; one version is illustrated in Figure 1.16a. The system has four buttons: on/off, coast/set, resume/accelerate, and cancel. The operation of the system is governed by a finite state machine that controls the modes of the PI controller and the reference generator, as shown in Figure 1.16b.

The finite state machine has four modes: off, standby, cruise, and hold. The state changes depending on actions of the driver who can brake, accelerate, and operate using the buttons. The on/off switch moves the states between off and standby. From standby the system can be moved to cruise by pushing the set/coast button. The velocity reference is set as the velocity of the car when the button is released. In the cruise state the operator can change the velocity reference; it is increased using the resume/accelerate button and decreased using the set/coast button. If the driver accelerates by pushing the gas pedal the speed increases, but it will go back to the set velocity when the gas pedal is released. If the driver

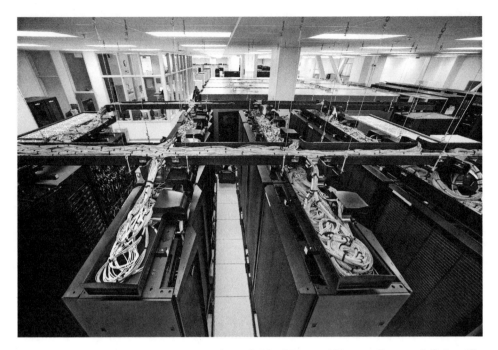

Figure 1.17: Large computer "server farm." The National Energy Research Scientific Computing Center (NERSC) at Lawrence Berkeley National Laboratory. (Figure courtesy U.S. Department of Energy.)

brakes then the car slows, and the cruise controller goes into hold but it remembers the setpoint of the controller. It can be brought to the cruise state by pushing the resume/accelerate button. The system also moves from cruise mode to standby if the cancel button is pushed. The reference for the velocity controller is remembered. The system goes into off mode by pushing on/off when the system is engaged.

The PI controller is designed to have good regulation properties and to give good transient performance when switching between resume and control modes.

Server Farms

Server farms consist of a large number of computers for providing Internet services (cloud computing). Large server farms, such as the one shown in Figure 1.17, may have thousands of processors. Power consumption for driving the servers and for cooling them is a prime concern. The cost for energy can be more than 40% of the operating cost for data centers [84]. The prime task of the server farm is to respond to a strongly varying computing demand. There are constraints given by electricity consumption and the available cooling capacity. The throughput of an individual server depends on the clock rate, which can be changed by adjusting the voltage applied to the system. Increasing the supply voltage increases the energy consumption and more cooling is required.

Control of server farms is often performed using a combination of feedback and logic. Capacity can be increased rapidly if a server is switched on simply by

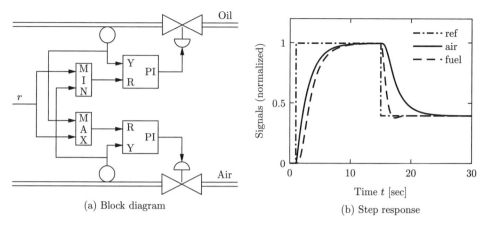

(a) Block diagram   (b) Step response

Figure 1.18: Air–fuel controller based on selectors. The left figure shows the system architecture. The letters R and Y in the PI controller denote the input ports for reference and measured signal respectively. The right figure shows a simulation where the power reference r is changed stepwise at $t = 1$ and $t = 15$. Notice that the normalized air flow is larger than the normalized fuel flow both for increasing and decreasing reference steps.

increasing the voltage to a server, but a server that is switched on consumes energy and requires cooling. To save energy it is advantageous to switch off servers that are not required, but it takes some time to switch on a new server. A control system for a server farm requires individual control of the voltage and cooling of each server and a strategy for switching servers on and off. Temperature is also important. Overheating will reduce the life time of the system and may even destroy it. The cooling system is complicated because cooling air goes through the servers in series and parallel. The measured value for the cooling system is therefore the server with the highest temperature. Temperature control is accomplished by a combination of feedforward logic to determine when servers are switched on and off and feedback using PID control.

Air–Fuel Control

Air–fuel control is an important problem for ship boilers. The control system consists of two loops for controlling air and oil (fuel) flow and a *supervisory controller* that adjusts the air–fuel ratio. The ratio should be adjusted for optimal efficiency when the ships are on open sea but it is necessary to run the system with air excess when the ships are in the harbor, since generating black smoke will result in heavy penalties.

An elegant solution to the problem can be obtained by combining PI controllers with maximum and minimum selectors, as shown in Figure 1.18a. A *selector* is a static system with several inputs and one output: a maximum selector gives an output that is the largest of the inputs, a minimum selector gives an output that is the smallest of the inputs. Consider the situation when the power demand is increased: the reference r to the air controller is selected as the commanded power level by the maximum selector, and the reference to the oil flow controller is selected as the measured airflow. The oil flow will lag the air flow and there will be

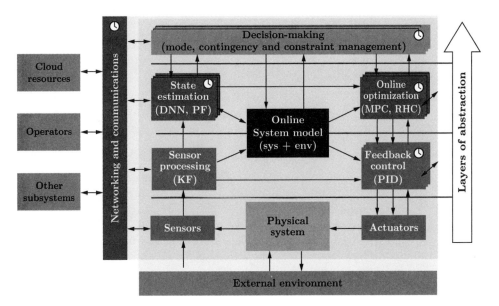

Figure 1.19: Layered decomposition of a control system.

air excess. When the commanded power level is decreased, the reference of the oil flow controller is selected as the power demand by the minimum selector and the reference for the air flow controller is selected as the oil flow by the the maximum selector. The system then operates with air excess when power is decreased.

The resulting response of the system for step changes in the desired power level is shown in Figure 1.18b, verifying that the system maintains air excess for both power increases and decreases.

Selectors are commonly used to implement logic in engines and power systems. They are also used for systems that require very high reliability: by introducing three sensors and only accepting values where two sensors agree it is possible to guard for the failure of a single sensor.

1.8   CONTROL SYSTEM ARCHITECTURES

Most of the control systems we are investigating in this book will be relatively simple feedback loops. In this section we will try to give a glimpse of the fact that in reality the simple loops combine to form a complex system that often has a hierarchical structure with controllers, logic, and optimization in different combinations. Figure 1.19 shows one representation of such a hierarchy, exposing different "layers" of the control system. The details of this class of systems is beyond the scope of this text, but we present a few representative examples to illustrate some basic points.

Freight Train Trip Optimizer

An example of two of the layers represented in Figure 1.19 can be see in the control of modern locomotives developed by General Electric (GE). Typical requirements

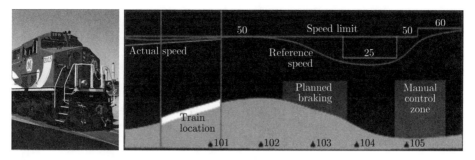

Figure 1.20: Freight train trip optimizer. GE's Trip Optimizer™ takes data about the train, the terrain, and the propulsion system and computes the best speed for the train in order to reach the destination on time while burning the least amount of diesel fuel. (Figure courtesy GE.)

for operating a freight train are to arrive on time and to use as little fuel as possible. The key issue is to avoid unnecessary braking. Figure 1.20 illustrates a system developed by GE. At the low layer the train has a speed regulator and simple logic to avoid entering a zone where there is another train. The key disturbance for the speed control is the slope of the ground. The speed controller has a model of the track, a GPS sensor, and an estimator. The setpoint for the speed controller is obtained from a trip optimizer, which computes a driving plan that minimizes the fuel consumption while arriving at the desired arrival time. The arrival time is provided by a dispatch center, which in turn may use its own optimization. These optimizations represent the second layer in Figure 1.19, with the top layer (decision-making) provided by the human operator.

Diesel-electric freight locomotives pull massive loads of freight cars, weighing more than 20,000 tons (US), and may be more than a mile in length. A typical locomotive burns about 35,000 gallons per year and can save an average 10% using the Trip Optimizer autopilot, representing a substantial savings in cost, natural resources, and pollution.

Process Control Systems

Process control systems are used to monitor and regulate the manufacturing of a wide range of chemicals and materials. One example is a paper factory, such as the one depicted in Figure 1.21. The factory produces paper for a variety of purposes from logs of wood. There are multiple fiber lines and paper machines, with a few dozen mechanical and chemical production processes that convert the logs to a slurry of fibers in different steps, and then paper machines that convert the fiber slurry to paper. Each production unit has PI(D) controllers that control flow, temperature, and tank levels. The loops typically operate on time scales from fractions of seconds to minutes. There is logic to make sure that the process is safe and there is sequencing for start, stop, and production changes. The setpoints of the low level control loops are determined from production rates and recipes, sometimes using optimization. The operation of the system is governed by a supervisory system that measures tank levels and sets the production rates of the different production units. This system performs optimization based on demanded production, measurement of

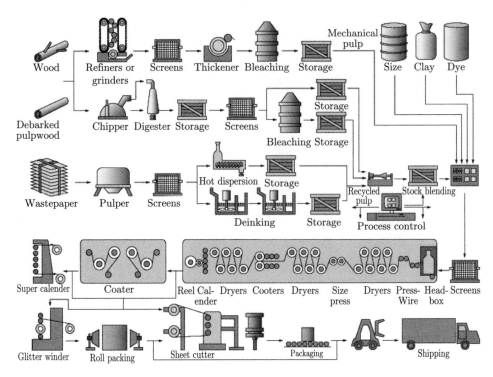

Figure 1.21: Schematic diagram for a pulp and paper manufacturing plant. The input to the plant is wood (upper left), which is then processed through a number of stages to create paper products. (Adapted from Weidenmüller [1984].)

tank levels, and flows. The optimization is performed at the time scale of minutes to hours, and it is constrained by the production rates of the different production units. Processes for continuous production in the chemical and pharmaceutical industry are similar to the paper factory but the individual production units may be very different.

One of the features of modern process control systems is that they operate across many time and spatial scales. Modern process control systems are also integrated with supply chains and product distribution chains, leading to the use of production planning systems and enterprise resource management systems. An example of an architecture for a distributed control system (DCS), typical for complex manufacturing systems, is shown in Figure 1.22.

Autonomous Driving

The cruise controller in Figure 1.11 relieves the driver of one task, to keep constant speed, but a driver still has many tasks to perform: plan the route, avoid collisions, decide the proper speed, perform lane changes, make turns, and keep proper distance to the car ahead. Car manufacturers are continuously automating more and more of these functions, going as far as automatic driving. An example of a control system for an autonomous vehicle is shown in Figure 1.23. This control system is

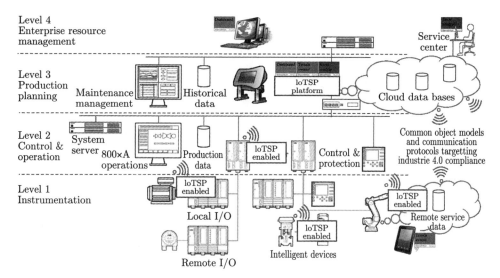

Figure 1.22: Functional architecture of process control system, implemented as a distributed control system (DCS). (Figure courtesy of ABB, Inc.)

designed for driving in urban environments. The feedback system fuses data from road and traffic sensors (cameras, laser range finders, and radar) to create a multi-layer "map" of the environment around the vehicle. This map is used to make decisions about actions that the vehicle should take (drive, stop, change lanes) and plan a specific path for the vehicle to follow. An optimization-based planner is used to compute the trajectory for the vehicle to follow, which is passed to a trajectory tracking (path following) module. A supervisory control module performs higher-level tasks such as mission planning and contingency management (if a sensor or actuator fails).

We see that this architecture has the basic features shown in Figure 1.19. The control layers are shown in the planning and control blocks, with the mission planner and traffic planner representing two levels of discrete decision-making logic, the path planner representing a trajectory optimization function, and then the lower layers of control. Similarly, there are multiple layers of sensing, with low level information, such as vehicle speed and position in the lane, being sent to the trajectory tracking controller, while higher level information about other vehicles on the road and their predicted motions is sent to the trajectory, traffic, and mission planners.

1.9 FURTHER READING

The material in the first half of this chapter draws from the report of the Panel on Future Directions on Control, Dynamics and Systems [187]. Several additional papers and reports have highlighted the successes of control [191] and new vistas in control [56, 154, 158, 213, 257]. The early development of control is described by Mayr [179] and in the books by Bennett [34, 35], which cover the period

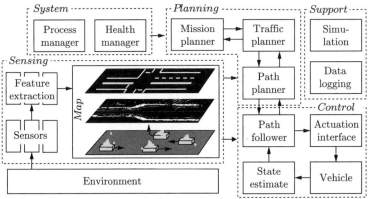

Figure 1.23: DARPA Grand Challenge. "Alice," Team Caltech's entry in the 2005 and 2007 competitions and its networked control architecture [66].

1800–1955. A fascinating examination of some of the early history of control in the United States has been written by Mindell [183]. A popular book that describes many control concepts across a wide range of disciplines is *Out of Control* by Kelly [143].

There are many textbooks available that describe control systems in the context of specific disciplines. For engineers, the textbooks by Franklin, Powell, and Emami-Naeini [93], Dorf and Bishop [73], Kuo and Golnaraghi [157], and Seborg, Edgar, and Mellichamp [219] are widely used. More mathematically oriented treatments of control theory include Sontag [225] and Lewis [163]. At the opposite end of the spectrum, the textbook *Feedback Control for Everyone* [7] provides a readable introduction with minimal mathematical background required. The books by Hellerstein et al. [117] and Janert [131] provide descriptions of the use of feedback control in computing systems. A number of books look at the role of dynamics and feedback in biological systems, including Milhorn [182] (now out of print),

J. D. Murray [186], and Ellner and Guckenheimer [83]. The book by Fradkov [91] and the tutorial article by Bechhoefer [30] cover many specific topics of interest to the physics community.

Systems that combine continuous feedback with logic and sequencing are called *hybrid systems*. The theory required to properly model and analyze such systems is outside the scope of this text, but a comprehensive description is given by Goebel, Sanfelice, and Teele [104]. It is very common that practical control systems combine feedback control with logic sequencing and selectors; many examples are given by Åström and T. Hägglund [19].

EXERCISES

1.1 Identify five feedback systems that you encounter in your everyday environment. For each system, identify the sensing mechanism, actuation mechanism, and control law. Describe the uncertainty with respect to which the feedback system provides robustness and/or the dynamics that are changed through the use of feedback.

1.2 (Balance systems) Balance yourself on one foot with your eyes closed for 15 s. Using Figure 1.4 as a guide, describe the control system responsible for keeping you from falling down. Note that the "controller" will differ from that in the diagram (unless you are an android reading this in the far future).

1.3 (Eye motion) Perform the following experiment and explain your results: Holding your head still, move one of your hands left and right in front of your face, following it with your eyes. Record how quickly you can move your hand before you begin to lose track of it. Now hold your hand still and shake your head left to right, once again recording how quickly you can move before losing track of your hand. Explain any difference in performance by comparing the control systems used to implement these behaviors.

1.4 (Cruise control) Download the MATLAB code used to produce simulations for the cruise control system in Figure 1.11 from the companion web site. Using trial and error, change the parameters of the control law so that the overshoot in speed is not more than $1\,\text{m/s}$ for a vehicle with mass $m = 1200\,\text{kg}$. Does the same controller work if we set $m = 2000\,\text{kg}$?

1.5 (Integral action) We say that a system with a constant input reaches steady state if all system variables approach constant values as time increases. Show that a controller with integral action, such as those given in equations (1.4) and (1.5), gives zero error if the closed loop system reaches steady state. Notice that there is no saturation in the controller.

1.6 (Combining feedback with logic) Consider a system for cruise control where the overall function is governed by the state machine in Figure 1.16. Assume that the system has a continuous input for vehicle velocity, discrete inputs indicating braking and gear changes, and a PI controller with inputs for the reference and measured velocities and an output for the control signal. Sketch the actions that have to be taken in the states of the finite state machine to handle the system properly. Think about if you have to store some extra variables, and if the PI controller has to be modified.

1.7 Search the web and pick an article in the popular press about a feedback and control system. Describe the feedback system using the terminology given in the article. In particular, identify the control system and describe (a) the underlying process or system being controlled, along with the (b) sensor, (c) actuator, and (d) computational element. If the some of the information is not available in the article, indicate this and take a guess at what might have been used.

Chapter Two

Feedback Principles

> Feedback – it is the fundamental principle that underlies all self-regulating systems, not only machines but also the processes of life and the tides of human affairs.
>
> —A. Tustin, "Feedback", *Scientific American*, 1952 [244].

This chapter presents examples that illustrate fundamental properties of feedback: disturbance attenuation, reference signal tracking, robustness to uncertainty, and shaping of behavior. The analysis is based on simple static and dynamical models. After reading this chapter, readers should have some insight into the power of feedback, they should know about transfer functions and block diagrams, and they should be able to design simple feedback systems. The basic concepts described in this chapter are explained in more detail in the remainder of the text, and this chapter can be skipped for readers who prefer to move directly to the more detailed analysis and design techniques.

2.1 NONLINEAR STATIC MODELS

We will start by capturing the behavior of a process and a controller using static models. Although these models are very simple, they give significant insight about the fundamental properties of feedback: negative feedback increases the range of linearity, it improves reference signal tracking, and it reduces the gain and the effects of disturbances and parameter variations. Moderate positive feedback has the opposite properties: it shrinks the range of linearity and increases the gain of the system. At a critical value the gain becomes infinite and the system behaves like a relay; larger values of the gain give hysteretic behavior. Although static models give some insight, they cannot capture dynamic phenomena like stability. Positive feedback combined with dynamics often leads to instability and oscillations, as will be discussed toward the end of the chapter.

Consider the closed loop system whose block diagram is shown in Figure 2.1. The closed loop system has a reference (or command) signal r that gives the desired system output. The controller C has an input e that is the difference between the reference r and the process output y, and the output of the controller is the control signal u. There is also a load disturbance v at the process input that perturbs the system. Although we will mostly deal with negative feedback, this simple model also permits analysis of positive feedback.

The process P is modeled as a function that is linear for inputs that are less than one in magnitude and saturates for inputs of magnitude larger than one. The controller is modeled by a constant gain k. Formally the process and the controller are described by the functions

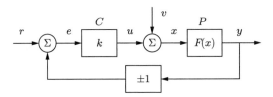

Figure 2.1: Block diagram of simple, static feedback system. The controller is a constant gain $k > 0$ and the process is modeled by a nonlinear function $F(x)$. The process output is y, the control signal is u, the external signals are the reference r, and the load disturbance v. The sign in the lower block indicates whether the feedback is positive $(+)$ or negative $(-)$.

$$y = F(x) = \text{sat}(x) = \begin{cases} -1 & \text{if } x \leq -1, \\ x & \text{if } |x| < 1, \qquad \text{and} \qquad u = ke. \\ 1 & \text{if } x \geq 1, \end{cases} \tag{2.1}$$

The process is linear for $|x| < 1$, which is called the *linear range*. In this region we have $y = x$ and the *process gain* is 1. The *controller gain* is k because the controller's output u is k times its input e.

The *open loop system* is the combination of the controller and the process when there is no feedback. Neglecting the disturbance v, it follows from equation (2.1) that the input/output relation for the open loop system is

$$y = F(kr) = \text{sat}(kr). \tag{2.2}$$

It has the gain k and linear range $|r| < 1/k$.

Response to Reference Signals

To explore how well the system output y can follow the reference signal r we assume that the load disturbance v in Figure 2.1 is zero. We will first consider *negative feedback* by setting the gain in the lower block of Figure 2.1 to -1. It follows from Figure 2.1 and equation (2.1) that the closed loop system is described by

$$y = \text{sat}(u), \qquad u = k(r - y). \tag{2.3}$$

Eliminating u in these equations we obtain

$$y = \text{sat}(k(r - y)). \tag{2.4}$$

To find the relation between the reference r and the output y we have to solve an algebraic equation. In the linear range $|k(r - y)| < 1$ we have $y = \frac{k}{k+1} r$. When $|k(r - y)| \geq 1$ the output saturates and we obtain $y = \pm 1$ (depending on the sign of $k(r - y)$). It can be shown that the overall input/output relationship satisfies

$$y = \text{sat}\left(\frac{k}{k+1} r\right) = \begin{cases} -1 & \text{if } r \leq -\frac{k+1}{k}, \\ \frac{k}{k+1} r & \text{if } |r| < \frac{k+1}{k}, \\ 1 & \text{if } r \geq \frac{k+1}{k}. \end{cases} \tag{2.5}$$

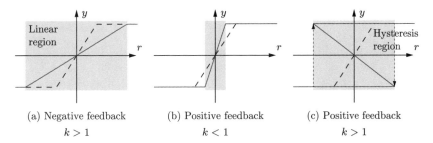

(a) Negative feedback
$k > 1$

(b) Positive feedback
$k < 1$

(c) Positive feedback
$k > 1$

Figure 2.2: Input/output behavior of the system: (a) for large negative feedback (b) positive feedback $k < 1$ and (c) large positive feedback. The solid line is the response of the closed loop system and the dashed line is the response of the open loop system. Redrawn from [221, Figure 20.5].

The linear range for the closed loop system is $|r| < \frac{k+1}{k}$. Comparing with equation (2.2) we find that negative feedback widens the linear range of the system by a factor of $k + 1$ compared to the open loop system. This is illustrated in Figure 2.2a, which shows the input/output relations of the open loop system (dashed) and the closed loop system (solid).

Robustness to Parameter Uncertainty

Next we will investigate the sensitivity of the closed loop system to gain variations. The *sensitivity* of a system describes how changes in the system parameters affect the performance of the system. For the open loop system in the linear range we have $y = kr$ and it thus follows that

$$\frac{dy}{dk} = r = \frac{y}{k} \quad \Longrightarrow \quad \frac{dy}{y} = \frac{dk}{k}. \tag{2.6}$$

The relative change of the output is thus equal to the relative change of the parameter and we say that the sensitivity is 1. Thus, for the open loop system, a change in k of 10% will lead to a change in the output of 10%.

For the closed loop system with an input in the linear range, it follows from equation (2.5) that

$$\frac{dy}{dk} = \frac{r}{k+1} - \frac{kr}{(k+1)^2} = \frac{r}{(k+1)^2} = \frac{y}{k(k+1)},$$

and hence

$$\frac{dy}{y} = \frac{1}{k+1}\frac{dk}{k}. \tag{2.7}$$

A comparison with equation (2.6) shows that negative feedback with gain k reduces the sensitivity to gain variations by a factor of $k + 1$. If k is 100, for example, a 10% change in k would lead to less than a 0.1% change in y, so the closed loop system is *much* less sensitive to parameter variation.

This type of analysis can also be used to investigate the effect of positive feedback. If the -1 in the feedback loop in Figure 2.1 is replaced by $+1$, equation (2.5) becomes

$$y = \operatorname{sat}\left(\frac{k}{-k+1}\, r\right). \tag{2.8}$$

Notice that the gain of the closed loop system is positive and larger than the open loop gain for $k < 1$, as shown in Figure 2.2b. The linear range is $|r| < (1-k)/k$. A comparison with the open loop system in equation (2.2) shows that positive feedback with $k < 1$ shrinks the linear range by a factor of $1 - k$. As k approaches 1 the closed loop gain approaches infinity, the range shrinks to zero, and the system behaves like a relay.

For positive feedback with $k > 1$ it follows from equation (2.8) that the closed loop gain is negative, as shown in Figure 2.2c, and that it approaches -1 as k approaches infinity. Positive feedback with large gains creates an input/output characteristic with multiple output values possible for inputs in the range $|r| < k/(k+1)$ and the closed loop system behaves like a switch with hysteresis. This concept is explored in more detail in Section 2.6, and it is shown that if the process has dynamics then all points where the input/output characteristics have negative slope are unstable.

We will mostly deal with negative feedback but there are systems that employ positive feedback, as illustrated in the following example.

Example 2.1 The superregenerative amplifier

Edwin Armstrong constructed a "superregenerative" radio receiver with only one vacuum tube in 1914, when he was still an undergraduate at Columbia University. The superregenerative amplifier can be modeled as an amplifier with open loop gain k and a saturated output, combined with a positive feedback loop, as shown in Figure 2.1. Using equation (2.8), we can compute the gain of the closed loop system to be $k_{cl} = k/(1-k)$. A very large closed loop gain can be obtained by selecting a feedback gain k that is just below 1. Choosing $k = 0.999$ gives $k_{cl} = 999$, which is a gain increase of almost three orders of magnitude.

The drawback of using positive feedback is that the system is highly sensitive and the gain has to be adjusted carefully to avoid oscillations. For example, if the gain k is 0.99 instead of 0.999 (a difference of less than 1%), then the closed loop gain becomes $k_{cl} = 99$, a difference of 10X (or 1000%). The oscillatory nature of this circuit requires the use of a more advanced (dynamic) model for analysis of the amplifier.

Despite its limitations, this type of amplifier is still used in simple walkie-talkies, garage door openers, and toys. ▽

Load Disturbance Attenuation

Another use of feedback is to reduce the effects of external disturbances, represented by the signal v in Figure 2.1. For the open loop system, the output when $v \neq 0$ is given by

$$y = \operatorname{sat}(kr + v).$$

In the linear region we thus have a gain of 1 between v and y, so that disturbances are passed through with no attenuation.

To investigate the effect of feedback on load disturbances we consider the system in Figure 2.1 with negative feedback and, for simplicity, we set the reference signal r to be zero. The relationship between the load disturbance v and the the output y is given by $y = \operatorname{sat}(v - ky)$, which is again an algebraic equation. In the linear range we get $y = v/(k+1)$ and more generally it can be shown that

$$y = \text{sat}\left(\frac{v}{k+1}\right). \tag{2.9}$$

In the linear region, negative feedback thus reduces the effect of load disturbances by the factor $k+1$.

Combining these three sets of analyses, we see that negative feedback increases the range of linearity of the system, decreases the sensitivity of the system to parameter uncertainty, and attenuates load disturbances. The trade-off is that the closed loop gain is decreased. Positive feedback has the opposite effect: it can increase the closed loop gain, but at the cost of increased sensitivity and amplification of disturbances.

2.2   LINEAR DYNAMICAL MODELS

The analysis in the previous section was based on static models and the dynamics of the process were neglected. We will now introduce a set of concepts and tools to analyze the effects of dynamics. To do this we will introduce block diagrams, linear differential equations, and transfer functions. The block diagram is an abstraction that describes a system as an interconnection of blocks, whose input/output behavior is described by differential equations. The transfer function, which is a function of complex variables, is a convenient representation of the differential equations describing the dynamics of the system. Transfer functions make it possible for us to find the relations between the signals of a complex system represented by block diagrams using simple algebra. The values of the transfer function on the imaginary axis gives the steady-state response to sinusoidal signals, which means that the transfer function can be determined experimentally from the steady-state response to sinusoidal signals.

Linear Differential Equations and Transfer Functions

In many practical situations, the input/output behavior of a system can be modeled by a linear differential equation of the form

$$\frac{d^n y}{dt^n} + a_1 \frac{d^{n-1}y}{dt^{n-1}} + \cdots + a_n y = b_0 \frac{d^m u}{dt^m} + b_1 \frac{d^{m-1}u}{dt^{m-1}} + \cdots + b_m u, \tag{2.10}$$

where u is the input, y is the output, and the coefficients a_k and b_k are real numbers. The differential equation (2.10) is characterized by two polynomials

$$a(s) = s^n + a_1 s^{n-1} + \cdots + a_n, \qquad b(s) = b_0 s^m + b_1 s^{m-1} + \cdots + b_m, \tag{2.11}$$

where $a(s)$ is the *characteristic polynomial* of the differential equation (2.10). We assume that the polynomials $a(s)$ and $b(s)$ do not have common roots. (The consequences of having common roots is discussed in Section 8.3.)

Equation (2.10) represents a *time-invariant system* because if the pair $u(t)$, $y(t)$ satisfies the equation so does $u(t+\tau)$, $y(t+\tau)$. The equation is also *linear* because if $u_1(t)$, $y_1(t)$, and $u_2(t)$, $y_2(t)$ satisfy the equation so does $\alpha u_1(t) + \beta u_2(t)$, $\alpha y_1(t) + \beta y_2(t)$, where α and β are real numbers. Systems that are linear and time-invariant are often called *LTI systems*. We can visualize these systems as being characterized

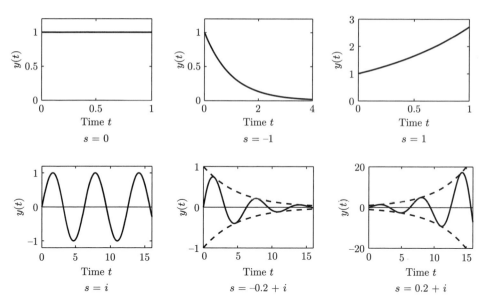

Figure 2.3: Examples of exponential signals. The top row corresponds to exponential signals with a real exponent, and the bottom row corresponds to those with complex exponents. The dashed line in the last two cases denotes the bounding envelope for the oscillatory signals. In each case, if the real part of the exponent is negative then the signal decays, while if the real part is positive then it grows.

by a huge table of corresponding input/output signal pairs. An interesting property of an LTI system is that it can be characterized by a single carefully chosen pair, for example the response of the system to a step input.

The solution to equation (2.10) is the sum of two terms: the general solution to the *homogeneous equation*, which does not depend on the input, and a *particular solution*, which depends on the input. The homogeneous equation associated with equation (2.10) is

$$\frac{d^n y}{dt^n} + a_1 \frac{d^{n-1} y}{dt^{n-1}} + \cdots + a_n y = 0. \tag{2.12}$$

Letting s_k represent the roots of the *characteristic equation* $a(s) = 0$, the solution to equation (2.12) is of the form

$$y(t) = \sum_{k=1}^{n} C_k e^{s_k t} \tag{2.13}$$

if the characteristic polynomial does not have repeated roots. The coefficients $C_1, \ldots, C_n$ can be determined from the initial conditions at $t = 0$.

Since the coefficients a_k are real, the roots of the characteristic equation are either real-valued or occur in complex conjugate pairs. A real root s_k of the characteristic polynomial corresponds to the exponential function $e^{s_k t}$. This function decreases over time if s_k is negative, is constant if $s_k = 0$, and increases if s_k is positive, as shown in the top row of Figure 2.3. For real roots s_k the parameter $T = 1/s_k$ is called the *time constant*, because it describes how quickly the signal decays.

A complex root $s_k = \sigma \pm i\omega$ corresponds to the time functions

$$e^{\sigma t} \sin(\omega t), \qquad e^{\sigma t} \cos(\omega t),$$

which have oscillatory behavior, as illustrated in the bottom row of Figure 2.3. The sine terms are shown as solid lines; they have zero crossings with the spacing π/ω. The dashed lines show the envelopes, which correspond to the exponential function $\pm e^{\sigma t}$.

When the characteristic equation has repeated roots, the solutions to the homogeneous equation (2.12) take the form

$$y(t) = \sum_{k=1}^{m} C_k(t) e^{s_k t}, \tag{2.14}$$

where $C_k(t)$ is a polynomial with degree less than the multiplicity of the root s_k. The solution (2.14) has $\sum_{k=1}^{m} (\deg C_k + 1) = n$ free parameters. This case is considered in more detail in Section 6.2.

Having explored the solution to the homogeneous equation, we now turn to the input-dependent part of the solution. The solution to equation (2.10) for an exponential input is of particular interest, as will be shown in the following. We set $u(t) = e^{st}$, where $s \neq s_k$ is a complex number, and investigate if there is a unique particular solution of the form $y(t) = G(s)e^{st}$. Assuming this to be the case, we find

$$\frac{du}{dt} = se^{st}, \qquad \frac{d^2 u}{dt^2} = s^2 e^{st}, \qquad \cdots \qquad \frac{d^m u}{dt^m} = s^m e^{st}$$

$$\frac{dy}{dt} = sG(s)e^{st}, \qquad \frac{d^2 y}{dt^2} = s^2 G(s)e^{st}, \qquad \cdots \qquad \frac{d^n y}{dt^n} = s^n G(s)e^{st}. \tag{2.15}$$

Inserting these expressions into the differential equation (2.10) gives

$$(s^n + a_1 s^{n-1} + \cdots + a_n) G(s) e^{st} = (b_0 s^m + b_1 s^{m-1} + \cdots + b_m) e^{st}$$

and hence

$$G(s) = \frac{b_0 s^m + b_1 s^{m-1} + \cdots + b_m}{s^n + a_1 s^{n-1} + \cdots + a_n} = \frac{b(s)}{a(s)}. \tag{2.16}$$

This function is called the *transfer function* of the system. It describes a particular solution to the differential equation for the input e^{st}. Combining this with the solution to the homogeneous equation, we find that a solution to the differential equation (2.10) for the exponential input $u(t) = e^{st}$ is

$$y(t) = \sum_{k=1}^{m} C_k(t) e^{s_k t} + G(s) e^{st}. \tag{2.17}$$

The relation between the transfer function (2.16) and the differential equation (2.10) is clear: the transfer function (2.16) can be obtained by inspection from the differential equation (2.10), and conversely the differential equation can be obtained from the transfer function if the polynomials $a(s)$ and $b(s)$ do not have common factors. The transfer function $G(s)$ can thus be regarded as a shorthand notation for the differential equation (2.10). It is a complete characterization of

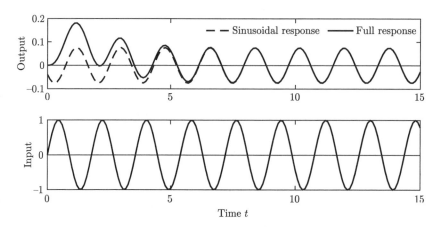

Figure 2.4: Two responses of a linear time-invariant system to a sinusoidal input. The dashed line shows the output when the initial conditions are chosen so that the output is purely sinusoidal. The solid line shows the response for the initial conditions $y(0) = 0$ and $y'(0) = 0$. The transfer function is $G(s) = 1/(s+1)^2$.

the differential equation even if it was derived as the response to a specific input $u(t) = e^{st}$. We note that the input and the initial conditions must *both* be given to obtain the full solution of the differential equation, given by equation (2.17), also referred to as the *response* of the system.

To deal with oscillatory signals, like those shown in the bottom row of Figure 2.3, we allow s to be a complex number. The transfer function G is then a function that maps complex numbers to complex numbers. We let arg represent the argument (phase, angle) of a complex number and $|\cdot|$ the magnitude, and note that the complex response to an input $u = e^{i\omega t} = \cos \omega t + i \sin \omega t$ is given by $G(i\omega)e^{i\omega t}$. Using just the imaginary parts of the signals, it follows that the particular solution for the input $u = \sin(\omega t) = \operatorname{Im} e^{i\omega t}$ is

$$y(t) = \operatorname{Im}\left(G(i\omega)\, e^{i\omega t}\right) = \operatorname{Im}\left(|G(i\omega)|\, e^{i\arg G(i\omega)}\, e^{i\omega t}\right)$$
$$= |G(i\omega)|\operatorname{Im} e^{i(\arg G(i\omega) + \omega t)} = |G(i\omega)|\sin(\omega t + \arg G(i\omega)).$$

The input is thus amplified by $|G(i\omega)|$ and the phase shift between input and output is $\arg G(i\omega)$. The functions $G(i\omega)$, $|G(i\omega)|$, and $\arg G(i\omega)$ are called the *frequency response*, *gain*, and *phase*. Gain and phase are also called *magnitude* and *angle*.

When the input and the output are constant, $u(t) = u_0$ and $y(t) = y_0$, the differential equation (2.10) has the particular solution $y(t) = (b_m/a_n)u_0 = G(0)u_0$, obtained by setting $s = 0$. The input is thus amplified by the factor $G(0)$, which is therefore called the *zero frequency gain* (or sometimes the *static gain*). If the differential equation is stable then the solution will converge to $G(0)u_0$ as t goes to infinity.

The full response to an exponential input is the sum of a particular solution and a solution to the homogeneous equation that is determined by the initial conditions, as given in equation (2.17). An illustration is given in Figure 2.4 for the transfer function $G(s) = 1/(s+1)^2$. The dashed line, which is a pure sine wave, is the solution obtained when all C_k in equation (2.17) are zero. The solid line shows the response obtained when the C_k are chosen so that $y(0)$ and its derivatives $y^{(k)}(0)$,

$k = 1, \ldots, n-1$ are all zero. Since all roots of the characteristic polynomial have negative real parts, the solution to the homogeneous equation (2.14) goes to zero as $t \to \infty$ and the general solution converges to the particular solution.

The transfer function has many interpretations that can be exploited for insight, analysis, and design. The roots s_k of the characteristic equation $a(s) = 0$ are called *poles* of the transfer function: the transfer function is infinite for $s = s_k$. The poles s_k appear as exponents in the general solution to the homogeneous equation, as seen in equations (2.13) and (2.14). Systems with poles that are "lightly damped" ($\text{Re}(s_k)$ is negative but close to zero) can exhibit resonances when a sinusoidal input is applied whose frequency is near the imaginary part of s_k.

The roots s_j of the polynomial $b(s)$ are called *zeros* of the transfer function. The reason is that if $b(s_j) = 0$ it follows that $G(s_j) = 0$, and the particular solution for the input $e^{s_k t}$ is then zero. A system theoretic interpretation is that the transmission of the exponential signal $e^{s_j t}$ is blocked by the zero $s = s_j$, which is therefore also called a *transmission zero*.

The transfer function can also convey a great deal of intuition: $G(0)$ is the zero frequency gain for constant inputs and the frequency response $G(i\omega)$ captures the steady-state response to sinusoidal functions. The frequency response of a stable system can be determined experimentally by exploring the steady-state response of a system to sinusoidal signals. This is an alternative or a complement to physical modeling. A more elaborate treatment of transfer functions and the frequency response will be given in Chapter 9.

Stability: The Routh–Hurwitz Criterion

When using feedback there is always the danger that the system may become unstable, and it is therefore important to have a stability criterion. The differential equation (2.10) is called *stable* if all solutions of the homogeneous equation (2.12) go to zero for any initial condition. It follows from equation (2.14) that this requires that all the roots of the characteristic equation

$$a(s) = s^n + a_1 s^{n-1} + \cdots + a_n = 0$$

have negative real parts.

It can often be difficult to analytically compute the roots of a high-order polynomial. The *Routh–Hurwitz criterion* is a stability criterion that does not require explicit calculation of the roots, because it gives conditions in terms of the coefficients of the characteristic polynomial.

We illustrate the Routh–Hurwitz criterion by describing it for low-order differential equations. A first-order differential equation is stable when the coefficient a_1 of the characteristic polynomial is positive, since the root of the characteristic polynomial will be $s = -a_1 < 0$. A second-order polynomial has the roots

$$s = \frac{1}{2}\left(-a_1 \pm \sqrt{a_1^2 - 4a_2}\right),$$

and it is easy to verify that the real parts of the roots are both negative if and only if $a_1 > 0$ and $a_2 > 0$. A third order differential equation is more complicated, but the roots can be shown to have negative real parts if and only if

$$a_1, a_2, a_3 > 0, \quad \text{and} \quad a_1 a_2 > a_3. \tag{2.18}$$

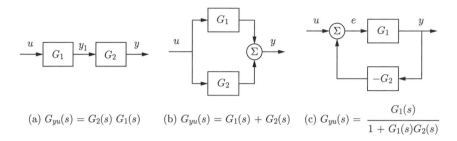

(a) $G_{yu}(s) = G_2(s)\,G_1(s)$ (b) $G_{yu}(s) = G_1(s) + G_2(s)$ (c) $G_{yu}(s) = \dfrac{G_1(s)}{1 + G_1(s)G_2(s)}$

Figure 2.5: Interconnections of linear systems. Series (a), parallel (b) and feedback (c) connections are shown. The transfer functions for the composite systems can be derived by algebraic manipulations assuming exponential functions for all signals.

The corresponding conditions for a fourth order differential equation are

$$a_1,\ a_2,\ a_3,\ a_4 > 0, \quad a_1 a_2 > a_3, \quad \text{and} \quad a_1 a_2 a_3 > a_1^2\, a_4 + a_3^2. \tag{2.19}$$

The Routh–Hurwitz criterion [97] gives similar conditions for arbitrarily high order polynomials. Stability of a linear differential equation can thus be investigated just by analyzing the signs of various combinations of the coefficients of the characteristic polynomial.

Block Diagrams and Transfer Functions

As we saw already in Chapter 1, control systems are often described using block diagrams, such as the ones shown in Figures 1.1 and 1.4. If the behavior of the blocks are represented by transfer functions, the transfer function of a system can be obtained simply by algebraic manipulations. It follows from equation (2.17) that the transfer function can be derived from the particular solution for the input e^{st}. To derive the transfer function for a system composed of several blocks, we assume that the input signal is an exponential $u(t) = e^{st}$ and compute the corresponding particular solutions for all blocks.

Consider for example the system in Figure 2.5a, which is a series connection of two systems with the transfer functions $G_1(s)$ and $G_2(s)$. Let the input of the system be $u(t) = e^{st}$ and assume the system is stable so that we focus just on the exponential response. The output of the first block is then $y_1(t) = G_1(s)e^{st}$, which is also an exponential, and the output of the second system is $y(t) = G_2(s)y_1(s) = G_2(s)G_1(s)e^{st} = G_2(s)G_1(s)u(t)$. The transfer function of the system is thus $G_{yu}(s) = G_2(s)G_1(s)$, where we use the convention that the right subscript is the input and the left subscript is the output, so that $y = G_{yu}u$.

Next we will consider parallel connections of systems as shown in Figure 2.5b. Assuming that the input is $u(t) = e^{st}$, the exponential outputs of the blocks are $y_1(t) = G_1(s)e^{st}$ and $y_2(t) = G_s(s)e^{st}$. The output of the system is then

$$y(t) = G_1(s)e^{st} + G_2(s)e^{st} = \big(G_1(s) + G_2(s)\big)\,e^{st},$$

and the transfer function of a parallel connection of systems with the transfer functions $G(s)$ and $G_2(s)$ is thus $G_{yu}(s) = G_1(s) + G_2(s)$.

Finally we will consider the feedback connection shown in Figure 2.5c. If the input $u(t) = e^{st}$ is an exponential we find

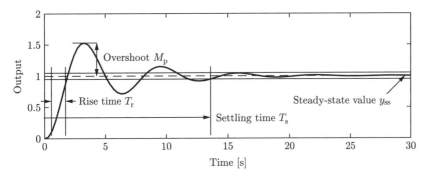

Figure 2.6: Sample step response. The rise time T_r, overshoot M_p, settling time T_s, and steady-state value y_{ss} describe important performance properties of the signal.

$$y(t) = G_1(s)e(t) = G_1(s)\big(u(t) - G_2(s)y(t)\big) = G_1(s)\big(e^{st} - G_2(s)y(t)\big).$$

Solving for $y(t)$ gives

$$y(t) = \frac{G_1(s)}{1 + G_1(s)G_2(s)} e^{st}.$$

The transfer function of a feedback connection of systems with the transfer functions $G_1(s)$ and $G_2(s)$ is thus

$$G_{yu}(s) = \frac{G_1(s)}{1 + G_1(s)G_2(s)}. \tag{2.20}$$

By using polynomials and transfer functions the relations between signals in a feedback system can thus be obtained by algebra. With some practice the transfer functions can often be obtained by inspection, as we explore in more detail in Chapter 9.

Computations Using Transfer Functions

Many software packages for control system analysis and design permit direct manipulation of transfer functions. In MATLAB the transfer function

$$G(s) = \frac{s+1}{(s^2 + 5s + 6)}$$

can be created by the commands `s = tf('s')` and `G = (s + 1)/(s^2 + 5*s + 6)`. Given two transfer functions `G1` and `G2`, we can form series, parallel, and feedback interconnections using the commands `Gs = series(G1, G2)`, `Gp = parallel(G1, G2)`, and `Gf = feedback(G1, G2)` (by default, MATLAB's `feedback()` command uses negative feedback).

Software packages can also be used to compute the response of a linear input/output system, represented by its transfer function, to different types of inputs. A common input that is used for performance characterization is a signal that is 0 for $t \leq 0$ and then 1 for $t > 0$. This type of input is called a "step input" and the response of the system to a step input is called the *step response* of the system. A typical step response for a linear system is shown in Figure 2.6. Some standard

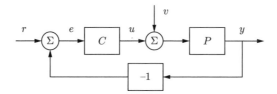

Figure 2.7: Block diagram of a simple feedback system. The controller transfer function is $C(s)$ and the process transfer function is $P(s)$. The process output is y, the external signals are the reference r and the load disturbance v.

features of a step response are the rise time T_r, settling time T_s, overshoot M_p, and steady-state value y_{ss}, as illustrated in the figure. The step response for a transfer function G is generated by the MATLAB command `y = step(G)`. If we want to specify the simulation time interval explicitly, we can instead use the command `y = step(G, T)`. The response to a specific input signal can be generated by `y = lsim(G, u, t)`, where u and t are the input and time vectors. Having a transfer function, it is thus very easy to generate time responses.

A detailed presentation of transfer functions will be given in Chapter 9, where we will see that transfer functions can also be used to represent systems with time delays and systems described by partial differential equations.

2.3   USING FEEDBACK TO ATTENUATE DISTURBANCES

Reducing the effects of disturbances is a primary use of feedback. It was used by James Watt to make steam engines run at constant speed in spite of varying load and by electrical engineers to make generators driven by water turbines deliver electricity with constant frequency and voltage. Feedback is commonly used to alleviate effects of disturbances in the process industry, for machine tools, and for engine and cruise control in cars. The human body exploits feedback to keep body temperature, blood pressure, and other important variables constant. For example the pupillary reflex guarantees that the light intensity of the retina is reasonably constant in spite of large variations in the ambient light intensity. Keeping variables close to a desired, constant reference value in spite of disturbances is called a *regulation problem*.

To discuss disturbance attenuation we consider the system shown in Figure 2.7. Since we will focus on the effects of a load disturbance v we will assume for now that the reference r is zero. To derive the transfer functions from the disturbance input v to the process output y, which we write as G_{yv}, we assume that the disturbance is an exponential function $v = e^{st}$. Applying block diagram algebra to Figure 2.7 gives

$$y(t) = P(s)e^{st} - P(s)C(s)y(t) \quad \Longrightarrow \quad y(t) = \frac{P(s)}{1 + P(s)C(s)} e^{st}.$$

The transfer function relating the output y to the load disturbance v is thus

$$G_{yv}(s) = \frac{P(s)}{1 + P(s)C(s)}. \tag{2.21}$$

To explore the use of feedback to improve disturbance attenuation, we will focus on a simple process modeled by the first-order differential equation

$$\frac{dy}{dt} + ay = bu, \qquad a > 0, \quad b > 0.$$

The corresponding transfer function is

$$P(s) = \frac{b}{s+a}. \tag{2.22}$$

This model is a reasonable approximation for a physical process if the storage of mass, momentum, or energy can be captured by a single state variable. Typical examples are the velocity of a car on a road, the angular velocity of a rotating system, and the fluid level of a tank.

Proportional Control

We will first investigate the case of proportional control, when the control signal is proportional to the output error: $u = k_{\mathrm{p}}e$, as introduced already in Section 1.6. The controller transfer function is then $C(s) = k_{\mathrm{p}}$. The process transfer function is given by equation (2.22) and the effect of the disturbance on the output is then described by the transfer function (2.21):

$$G_{yv}(s) = \frac{P(s)}{1 + P(s)C(s)} = \frac{b/(s+a)}{1 + bk_{\mathrm{p}}/(s+a)} = \frac{b}{s + (a + bk_{\mathrm{p}})}.$$

The relation between the disturbance v and the output y is thus given by the differential equation

$$\frac{dy}{dt} + (a + bk_{\mathrm{p}})y = bv.$$

The closed loop system is stable if $a + bk_{\mathrm{p}} > 0$. A constant disturbance $v = v_0$ then gives an output that exponentially approaches the value

$$y_0 = G_{yv}(0)v_0 = \frac{b}{a + bk_{\mathrm{p}}}\, v_0$$

with the time constant $T = 1/(a + bk_{\mathrm{p}})$. Without feedback, $k_{\mathrm{p}} = 0$ and for a constant disturbance v_0, the output will instead approach bv_0/a. The effect of the disturbance is thus reduced if $k_{\mathrm{p}} > 0$.

We have thus shown that a constant disturbance gives an error that can be reduced by feedback using a proportional controller. The error decreases with increasing controller gain. Figure 2.8a shows the responses for a few values of the controller gain k_{p}.

Proportional-Integral (PI) Control

The PI controller, introduced in Section 1.6, is described by

$$u(t) = k_{\mathrm{p}}e(t) + k_{\mathrm{i}} \int_0^t e(\tau)\, d\tau. \tag{2.23}$$

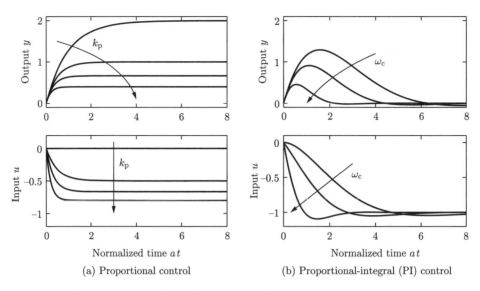

Figure 2.8: Step responses for a first-order, closed loop system with proportional control (a) and PI control (b). The process transfer function is $P = 2/(s+1)$. The controller gains for proportional control are $k_p = 0$, 0.5, 1, and 2. The PI controller is designed using equation (2.28) with $\zeta_c = 0.707$ and $\omega_c = 0.707$, 1, and 2, which gives the controller parameters $k_p = 0$, 0.207, and 0.914 and $k_i = 0.25$, 0.50, and 2.

To determine the transfer function of the controller we differentiate to obtain

$$\frac{du}{dt} = k_p \frac{de}{dt} + k_i e,$$

and we find that the transfer function is $C(s) = k_p + k_i/s$. To investigate the effect of the disturbance v on the output we use the block diagram in Figure 2.7, and the transfer function from v to y is

$$G_{yv}(s) = \frac{P(s)}{1 + P(s)C(s)} = \frac{bs}{s^2 + (a + bk_p)s + bk_i}. \tag{2.24}$$

Using the relationship between transfer functions and differential equations given by equations (2.10) and (2.16), it follows that the relation between the load disturbance and the output is given by the differential equation

$$\frac{d^2 y}{dt^2} + (a + bk_p)\frac{dy}{dt} + bk_i y = b\frac{dv}{dt}. \tag{2.25}$$

Notice that since the disturbance enters as a derivative on the right hand side, a constant disturbance gives no steady-state error. The same conclusion can be drawn from the observation that $G_{yv}(0) = 0$. This is consistent with the discussion of integral action and steady-state error in Section 1.6.

To find suitable values of the controller parameters k_p and k_i, we consider the characteristic polynomial of the differential equation (2.25),

$$a_{cl}(s) = s^2 + (a + bk_p)s + bk_i. \tag{2.26}$$

We can assign arbitrary roots to the characteristic polynomial by choosing the controller gains k_p and k_i. The most common case is that we assign complex roots that give the characteristic polynomial

$$(s + \sigma_d + i\,\omega_d)(s + \sigma_d - i\,\omega_d) = s^2 + 2\sigma_d s + \sigma_d^2 + \omega_d^2. \qquad (2.27)$$

By construction, this polynomial has roots at $s = -\sigma_d \pm i\,\omega_d$. The general solution to the homogeneous equation is then a linear combination of the terms

$$e^{-\sigma_d t}\sin(\omega_d t), \qquad e^{-\sigma_d t}\cos(\omega_d t),$$

which are damped sine and cosine functions, as shown in the lower middle plot in Figure 2.3. The coefficient σ_d determines the decay rate and the parameter ω_d, called the *damped frequency*, gives the frequency of the decaying oscillation. Identifying coefficients of equal powers of s in the polynomials (2.26) and (2.27) gives

$$k_p = \frac{2\sigma_d - a}{b}, \qquad k_i = \frac{\sigma_d^2 + \omega_d^2}{b}. \qquad (2.28)$$

We can thus choose the controller gains to give a desired closed loop response.

Instead of parameterizing the closed loop system in terms of σ_d and ω_d it is common practice to use the *(undamped) natural frequency* $\omega_c = \sqrt{\sigma_d^2 + \omega_d^2}$ and the *damping ratio* $\zeta_c = \sigma_d/\omega_c$. The closed loop characteristic polynomial is then

$$a_{cl}(s) = s^2 + 2\sigma_d s + \sigma_d^2 + \omega_d^2 = s^2 + 2\zeta_c\omega_c s + \omega_c^2.$$

This parameterization has the advantage that ζ_c, which is in the range $[-1, 1]$, determines the shape of the response and ω_c gives the response speed.

Figure 2.8 shows the output y and the control signal u for $\zeta_c = 1/\sqrt{2} \approx 0.707$ and different values of the design parameter ω_c. Proportional control gives a steady-state error that decreases with increasing controller gain k_p. With PI control the steady-state error is zero. Both the decay rate and the peak error decrease when the design parameter ω_c is increased. Larger controller gains give smaller errors and control signals that react more quickly to the disturbance.

With the controller parameters (2.28), the transfer function (2.24) from disturbance v to process output y becomes

$$G_{yv}(s) = \frac{P(s)}{1 + P(s)C(s)} = \frac{bs}{s^2 + 2\zeta_c\omega_c s + \omega_c^2}.$$

For efficient attenuation of disturbances, it is desirable that $|G_{yv}(i\omega)|$ is small for all ω. For small values of ω we have $|G_{yv}(i\omega)| \approx b\omega/\omega_c^2$, while for large ω we have $|G_{yv}(i\omega)| \approx b/\omega$. The largest value of $|G_{yv}(i\omega)|$ is $b/(2\zeta_c\omega_c)$ for $\omega = \omega_c$. It thus follows that a large value of ω_c gives good load disturbance attenuation.

In summary, we find that transfer function analysis gives a simple way to find the parameters of PI controllers for processes whose dynamics can be approximated by a first-order system. The technique can be generalized to more complicated systems but the controller will be more complex. To achieve the benefits of large control gains the model must be accurate over wide frequency ranges, as will be discussed next.

Unmodeled Dynamics

The analysis we have made so far indicates that there are no limits to the performance that can be achieved. Figure 2.8b shows that arbitrarily fast response can be obtained simply by making ω_c sufficiently large. In reality there are of course limits on what is achievable. One reason is that the controller gains increase with ω_c: the proportional gain is $k_p = (2\zeta_c\omega_c - a)/b$ and the integral gain is $k_i = \omega_c^2/b$. A large value of ω_c thus gives large controller gains and the control signal may saturate. Another reason is that the model (2.22) is a simplification: it is only valid in a given frequency range. If the model is instead

$$P(s) = \frac{b}{(s+a)(1+sT)}, \tag{2.29}$$

where the term $1 + sT$ represents the dynamics of sensors, actuators, or other dynamics that were neglected when deriving equation (2.22)—so-called *unmodeled dynamics*—the closed loop characteristic polynomial for the closed loop system becomes

$$a_{cl} = s(s+a)(1+sT) + b(k_p s + k_i) = s^3 T + s^2(1+aT) + 2\zeta_c\omega_c s + \omega_c^2.$$

It follows from the Routh–Hurwitz criterion (2.18) that the closed loop system is stable if $\omega_c^2 T < 2\zeta_c\omega_c(1+aT)$ or if

$$\omega_c T < 2\zeta_c(1+aT).$$

The frequency ω_c and the achievable response time are thus limited by the unmodeled dynamics represented by T, which typically is smaller than the time constant $1/a$ of the process. When models are developed for control it is therefore important to also consider the unmodeled dynamics.

The fact that unmodeled dynamics limit the performance of a feedback system is an important property and must be considered during the system design. It is common to use simplified models when designing components of complex systems and if the unmodeled dynamics of those components (or the other subsystems they interact with) are not properly taken into account, the implementation of the system can display poor behavior (of which instability is one extreme example). As we shall see in later chapters, it is the ability to reason about the effects of uncertainty that makes control theory a particularly powerful mathematical tool for systems design.

2.4   USING FEEDBACK TO TRACK REFERENCE SIGNALS

Another major application of feedback is to make a system output follow a reference value, which is called the *servo problem*. Cruise control, steering of a car, and tracking a satellite with an antenna or a star with a telescope are some examples. Other examples are high performance audio amplifiers, machine tools, and industrial robots.

To illustrate reference signal tracking we will consider the system in Figure 2.7 where the process is a first-order system and the controller is a PI controller with proportional gain k_p and integral gain k_i. The transfer functions of the process and the controller are

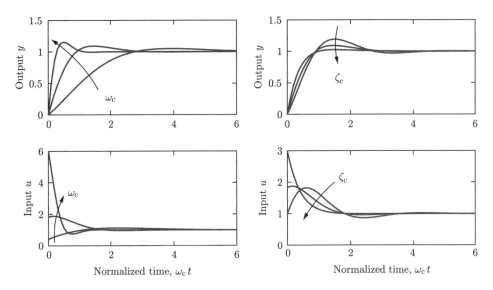

Figure 2.9: Responses to a unit step change in the reference signal for different values of the design parameters ω_c and ζ_c. The left figure shows responses for fixed $\zeta_c = 0.707$ and $\omega_c = 1$, 2, and 5. The right figure shows responses for $\omega_c = 2$ and $\zeta_c = 0.5$, 0.707, and 1. The process parameters are $a = b = 1$. The initial value of the control signal is k_p.

$$P(s) = \frac{b}{s+a}, \qquad C(s) = \frac{k_p s + k_i}{s}. \qquad (2.30)$$

Since we will focus on following the reference signal r, we will neglect the load disturbance and set $v = 0$. Applying block diagram algebra to the system in Figure 2.7, we find that the transfer function from the reference signal r to the output y is

$$G_{yr}(s) = \frac{P(s)C(s)}{1 + P(s)C(s)} = \frac{bk_p s + bk_i}{s^2 + (a + bk_p)s + bk_i}. \qquad (2.31)$$

Since $G_{yr}(0) = 1$ it follows that $r = y$ when r and y are constant, independent of the values of the parameters a and b, as long as the closed loop system is stable. The steady-state output is thus equal to the reference, a consequence of the integral action in the controller.

To determine suitable values of the controller parameters k_p and k_i, we proceed as in Section 2.3 by choosing controller parameters that make the closed loop characteristic polynomial

$$a_{cl}(s) = s^2 + (a + bk_p)s + bk_i \qquad (2.32)$$

equal to $s^2 + 2\zeta_c\omega_c s + \omega_c^2$ with $\zeta_c > 0$ and $\omega_c > 0$. Identifying coefficients of equal powers of s in these polynomials gives

$$k_p = \frac{2\zeta_c\omega_c - a}{b}, \qquad k_i = \frac{\omega_c^2}{b}, \qquad (2.33)$$

which is equivalent to equation (2.28). Notice that integral gain increases with the square of ω_c. Figure 2.9 shows the output signal y and the control signal u for different values of the design parameters ζ_c and ω_c. The response time decreases with increasing ω_c and the initial value of the control signal also increases because it takes more effort to move rapidly. The overshoot decreases with increasing ζ_c. For $\omega_c = 2$, the design choice $\zeta_c = 1$ gives a short settling time and a response without overshoot.

It is desirable that the output y will track the reference signal r for time-varying references. This means that we would like the transfer function $G_{yr}(s)$ to be close to 1 for large frequency ranges. With the controller parameters (2.33), it follows from equation (2.31) that

$$G_{yr}(s) = \frac{P(s)C(s)}{1 + P(s)C(s)} = \frac{(2\zeta_c\omega_c - a)s + \omega_c^2}{s^2 + 2\zeta_c\omega_c s + \omega_c^2}.$$

Since $G_{yr}(0) = 1$, tracking of constant inputs is perfect. In addition, if $s = i\omega$ is smaller in magnitude than ω_c, then using some approximations it can be shown that $G_{yr}(s)$ will be close to one. The frequency ω_c thus determines the upper bound of the frequency of reference signals that can be tracked with small error, and this bound is referred to as the *bandwidth* of the closed loop system. The frequency response of G_{yr} therefore provides a quantitative representation of the tracking abilities.

Controllers with Two Degrees of Freedom

The control law in Figure 2.7 has *error feedback* because the control signal u is generated from the error $e = r - y$. With proportional control, a step in the reference signal r gives an immediate step change in the control signal u. This rapid reaction can be advantageous, but it may give large overshoot, which can be avoided by a replacing the PI controller in equation (2.23) with a controller of the form

$$u(t) = k_p\big(\beta r(t) - y(t)\big) + k_i \int_0^t (r(\tau) - y(\tau))\, d\tau. \tag{2.34}$$

In this modified PI algorithm, the proportional action only acts on the fraction β of the reference signal. The signal transmissions from reference r to u and from output y to u can be represented by the (open loop) transfer functions

$$C_{ur}(s) = \beta k_p + \frac{k_i}{s}, \qquad -C_{uy}(s) = k_p + \frac{k_i}{s} = C(s). \tag{2.35}$$

The controller (2.34) is called a controller with *two degrees of freedom* since the transfer functions $C_{ur}(s)$ and $C_{uy}(s)$ are different.

A block diagram of a closed loop system with a PI controller having two degrees of freedom is shown in Figure 2.10. Let the process transfer function be $P(s) = b/(s + a)$. The transfer functions from reference r and disturbance v to output y are

$$G_{yr}(s) = \frac{b\beta k_p s + bk_i}{s^2 + (a + bk_p)s + bk_i}, \qquad G_{yv}(s) = \frac{bs}{s^2 + (a + bk_p)s + bk_i}. \tag{2.36}$$

Comparing with the corresponding transfer function for a controller with error feedback in equations (2.24) and (2.31), we find that the response to the load disturbances is the same but the response to reference signals is different.

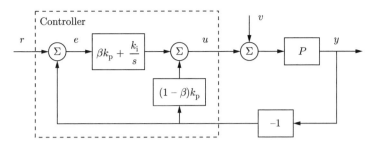

Figure 2.10: Block diagram of a closed loop system with a PI controller having an architecture with two degrees of freedom.

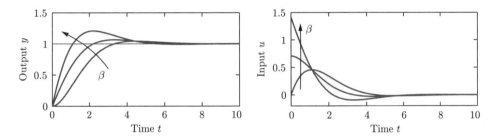

Figure 2.11: Response to a step change in the reference signal for a system with a PI controller having two degrees of freedom. The process transfer function is $P(s) = 1/s$ and the controller gains are $k_\mathrm{p} = 1.414$, $k_\mathrm{i} = 1$, and $\beta = 0$, 0.5, and 1.

A simulation of the closed loop system for $a = 0$ and $b = 1$ is shown in Figure 2.11. The figure shows that the parameter β has a significant effect on the responses. Comparing the system with error feedback ($\beta = 1$) to the system with smaller values of β we find that using a system with two degrees of freedom gives less overshoot and gentler control actions.

The example shows that reference signal response can be improved by using a controller architecture having two degrees of freedom. In Section 12.4 we will further show that the responses to reference signals and disturbances can be completely separated by using a more general system architecture. To use a system with two degrees of freedom both the reference signal r and the output signal y must be measured. There are situations where only the error signal $e = r - y$ can be measured; typical examples are DVD players, optical memories, and atomic force microscopes. In these cases, only single degree of freedom (error feedback) controllers can be used.

2.5   USING FEEDBACK TO PROVIDE ROBUSTNESS

Feedback can be used to make good systems from imprecise components. Black's invention of the feedback amplifier for the telephone network is an early example [46]. Black used negative feedback to design extremely good amplifiers with linear characteristics from components with nonlinear and time-varying properties. Since signals are transmitted over long distances they must be amplified. At the

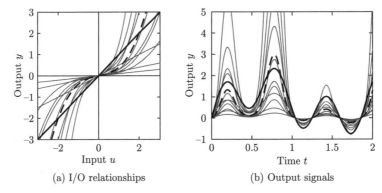

(a) I/O relationships   (b) Output signals

Figure 2.12: Responses of a static nonlinear system. The left figure shows the input/output relations of the open loop systems and the right figure shows responses to the input signal (2.38). The ideal response is shown with solid bold lines. The nominal response of the nonlinear system is shown using dashed bold lines and the responses for different parameter values are shown using thin lines. Notice the large variability in the responses.

time, the thermionic valve—a type of vacuum tube invented by Lee de Forest in 1906—was the only available technology for amplifying electric signals until the transistor was in invented in 1947. Vacuum tubes were the key to develop radio, telephony, and electronics in the first half of the 20th century. They are still used by some hi-fi aficionados in high quality audio amplifiers.

Vacuum tubes can give high gain but they have nonlinear and time varying input/output characteristics that distort the transmitted signals. Bode [52] expressed the problem as follows:

> Most of you with hi-fi systems are no doubt proud of the quality of your amplifiers, but I doubt whether many of you would care to listen to the sound after the signal had gone in succession through several dozen or several hundred even of your fine amplifiers.

The effect is illustrated in Exercise 2.9.

Black's idea to develop a good amplifier was to close a loop with negative feedback around the tube amplifier. In this way he could obtain a closed loop system with a linear input/output relation having constant gain. The general recipe is to localize the nonlinearities and the source of process variations, and to close feedback loops around them.

Reducing Effects of Parameter Variations and Nonlinearities

Consider an amplifier with a static, nonlinear input/output relation with considerable parameter variability, as illustrated in Figure 2.12a. The nominal input/output characteristic is shown as a dashed bold line and examples of variations as thin lines. The nonlinearity in the figure is given by

$$y = F(u) = \alpha(u + \beta u^3), \qquad -3 \le u \le 3. \tag{2.37}$$

The nominal values corresponding to the dashed line are $\alpha = 0.2$ and $\beta = 1$. The variations of the parameters α and β are in the ranges $0.1 \le \alpha \le 0.5$, $0 \le \beta \le 2$.

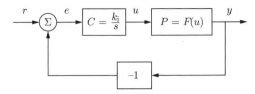

Figure 2.13: Block diagram of a nonlinear system with integral feedback.

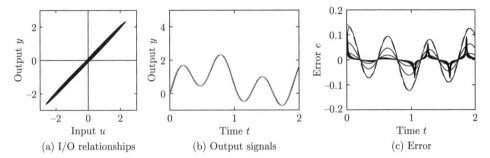

(a) I/O relationships   (b) Output signals   (c) Error

Figure 2.14: Responses of the systems with integral feedback ($k_i = 1000$). The left figure shows the input/output relationships for the closed loop systems, and the center figure shows responses to the input signal (2.38) (compare to the corresponding responses in Figure 2.12a and b). The right figure shows the individual errors (solid lines) and the approximate error given by equation (2.42) (dashed line).

The responses of the system to the input

$$u(t) = \sin(t) + \sin(\pi t) + \sin(\pi^2 t) \tag{2.38}$$

are shown in Figure 2.12b. The desired response $y = u$ is shown as a solid bold line and responses for a range of parameters are shown with thin lines. The nominal response of the nonlinear system is shown as a dashed bold line, and we see that it is distorted due to the nonlinearity. Notice in particular the heavy distortion for both small and large signal amplitudes.

The behavior of the system is clearly not satisfactory, but it can be improved significantly by introducing feedback. A block diagram of a system with a simple integral controller is shown in Figure 2.13, where the reference input is now taken as r. Figure 2.14 shows the behavior of the closed loop system with the same parameter variations as in Figure 2.12. The input/output plot in Figure 2.14a is a scatter plot of the inputs and the outputs of the feedback system. The input/output relation is practically linear and close to the desired response. There is some variability because of the dynamics introduced by the feedback. Figure 2.14b shows the responses to the reference signal; notice the dramatic improvement compared with Figure 2.12b. The tracking error is shown in Figure 2.14c.

Nonlinear Analysis and Approximations

Analysis of a closed loop system with nonlinearities is often difficult. We can, however, obtain significant insight by using approximations. We illustrate a few ideas using the nonlinear amplifier example.

We first observe that the system is linear when $\beta = 0$. In other situations we can approximate the nonlinear function by a straight line around an operating point $u = u_0$. The slope of the nonlinear function at $u = u_0$ is $F'(u_0)$ and we will approximate the process with a linear system with the gain $F'(u_0)$. The transfer functions of the process and the controller are

$$P(s) = F'(u_0) = \alpha(1 + 3\beta u_0^2) =: b, \qquad C(s) = \frac{k_i}{s}, \tag{2.39}$$

where u_0 denotes the operating condition. It follows from equation (2.21) that the transfer functions relating the output y and the error e to the reference signal r are

$$G_{yr}(s) = \frac{bk_i}{s + bk_i}, \qquad G_{er}(s) = 1 - G_{yr} = \frac{s}{s + bk_i}. \tag{2.40}$$

The closed loop system is a first-order system with the pole $s = -bk_i$. The process gain $b = \alpha(1 + 3\beta u_0^2)$ depends on the values of α, β, and u_0, and its smallest value is 0.1. If the integral gain is chosen as $k_i = 1000$, the smallest value of the closed loop pole is $100 \, \text{rad/s}$, which is fast compared to the high-frequency component $\pi^2 \, \text{rad/s}$ of the input signal. It follows from equation (2.40) that the error $e(t)$ is given by the differential equation

$$\frac{de}{dt} = -bk_i e + \frac{dr}{dt}, \qquad \frac{dr}{dt} = \cos(t) + \pi \cos(\pi t) + \pi^2 \cos(\pi^2 t). \tag{2.41}$$

Neglecting the term de/dt in equation (2.41) gives

$$e(t) \approx \frac{1}{bk_i} \frac{dr}{dt} \approx \frac{\pi^2}{bk_i} \cos(\pi^2 t). \tag{2.42}$$

An estimate of the largest error $e(t) \approx 0.1 \cos(\pi^2 t)$ is obtained for the smallest value of $b = 0.1$. It is shown as a dashed line in Figure 2.14c, and we see that it gives a good estimate of the maximum error across the uncertain parameter space.

This analysis is based on the assumption that the amplifier can be modeled by a constant gain. The closed loop system is however a dynamic system because the controller is an integrator. It follows from equation (2.40) that the closed loop dynamics have the time constant $T_{cl} = 1/(bk_i)$. If the amplifier has dynamics, its time constant must thus be small compared to T_{cl} in order to provide good tracking. It follows that the largest admissible integral gain k_i is determined by the unmodeled dynamics.

This example illustrates that feedback can be used to design an amplifier that has practically linear input/output relation even if the basic amplifier is nonlinear with strongly varying characteristics.

2.6   POSITIVE FEEDBACK

Most of this book is focused on negative feedback because of its amazingly good properties, which have been illustrated in the previous sections. In this section we will briefly discuss positive feedback, which has complementary properties. In spite of this, positive feedback has found good use in several contexts.

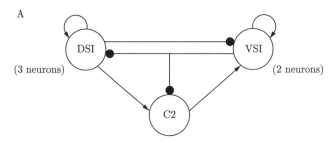

Figure 2.15: Schematic diagram of the neural network that controls swimming motions in the marine mollusk Tritonia, which has both positive and negative feedback. An excitatory connection (positive feedback) is denoted with a line ending with an arrow, an inhibitory interaction (negative feedback) is denoted with an arrow ending with a circle. (Figure adapted from [256].)

Systems with negative feedback can be well understood by linear analysis. To understand systems with positive feedback it is necessary to consider nonlinear effects, because without the nonlinearities the instability caused by positive feedback will grow without bound. The nonlinear elements can create interesting and useful effects by limiting the signals.

Positive feedback is common in many settings. Encouraging a student or a coworker when they have performed well encourages them do to even better. In biology, it is standard to distinguish inhibitory connections (negative feedback) from excitatory feedback (positive feedback) as illustrated in Figure 2.15. Neurons use a combination of positive and negative feedback to generate spikes.

Positive feedback may cause instabilities. Exponential growth, where the rate of change of a quantity x is proportional to x,

$$\frac{dx}{dt} = \alpha x,$$

is a typical example, which results in an unbounded solution $x(t) = e^{\alpha t}$. In nature, exponential growth of a species is limited by the finite amount of food. Another common example is when a microphone is placed close to a speaker in public address systems, resulting in a howling noise. Positive feedback can create stampedes in cattle herds, runs on banks, and boom-bust behavior. In all these cases there is exponential growth that is finally limited by finite resources.

The notions of positive and negative feedback are clear if the feedback is static, as we saw for example in Section 2.1. If the feedback is dynamic its action can change from positive to negative depending on the frequency of the signals and hence more care is required. Use of positive feedback will be illustrated by a few examples.

Hewlett's Oscillator

William Hewlett used positive and negative feedback very cleverly to design a stable oscillator in his master's thesis from Stanford University in 1939. The oscillator was the first product made by Hewlett-Packard, the company that Hewlett founded with David Packard in 1939 [200].

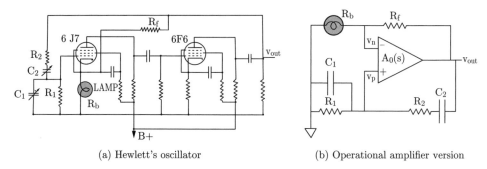

(a) Hewlett's oscillator   (b) Operational amplifier version

Figure 2.16: Circuit diagrams of William Hewlett's oscillator. (a) Original system with vacuum tubes. (b) Equivalent realization with an operational amplifier.

Electronic circuits in the 1930s and 1940s were based on vacuum tube technology. The simplest vacuum tube amplifier has three electrodes: a cathode, grid, and anode enclosed in a glass tube with vacuum. The cathode, which is heated with a filament, emits free electrons. A current is created by applying a high positive voltage between the anode and the cathode. The current can be regulated by changing the voltage on a grid positioned between the anode and the cathode. The current depends on the voltage difference between the grid and the cathode, $V_g - V_c$. Increasing this voltage difference increases the current. The vacuum tube amplifier can be regarded as a valve for controlling a current by applying a voltage to the grid.

A schematic diagram of Hewlett's oscillator is shown in Figure 2.16a. Signals are amplified by two vacuum tubes and there are two feedback loops. One loop provides positive feedback from the anode of the second tube to the grid of the first tube via the network R_1, C_1, R_2, C_2. The second feedback loop provides negative feedback from the output of the second tube to the cathode of the first tube via the resistor R_f and the lamp, which has resistance R_b. With a proper gain the positive feedback loop generates an oscillation with the frequency $\omega = 1/\sqrt{R_1 R_2 C_1 C_2}$. The gain is given by the negative feedback loop from the anode of the second loop to the cathode of the first loop, through the resistor R_f and the lamp R_b. This loop has nonlinear gain because the resistance R_b of the lamp increases with increasing temperature. An increase of the amplitude of V_{out} increases the current through the lamp, which reduces the gain. The result is that an oscillation with stable amplitude and frequency is obtained.

The feedback loops are more clearly visible in the implementation of the oscillator based on an operational amplifier, shown in Figure 2.16b.

Implementation of Integral Action by Positive Feedback

Early feedback controllers made use of integral action that was implemented by using positive feedback around a system with first order dynamics, as shown in the block diagram of Figure 2.17. Intuitively the system can be explained as follows. Proportional feedback typically gives a steady-state error. This can be overcome by adding a bias signal that cancels the steady-state error. In Figure 2.17 the bias is estimated by low-pass filtering the control signal and adding it back into the signal path. This serves to compensate for any error that is present.

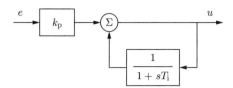

Figure 2.17: Implementation of integral action by *positive feedback*.

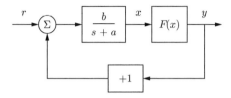

Figure 2.18: Block diagram of system with positive feedback and saturation. The parameters are $a = 1$ and $b = 4$.

The circuit can be understood better by a little analysis. Using block diagram algebra we find that the transfer function of the system is

$$G_{ue} = \frac{k_{\mathrm{p}}}{1 - 1/(1 + sT_i)} = k_{\mathrm{p}} + \frac{k_{\mathrm{p}}}{sT_i},$$

which is a transfer function of a PI controller. This way of implementing integral action is still used in many industrial regulators.

Positive Feedback Combined with Saturation

Systems with interesting and useful properties can be obtained by combining linear and nonlinear components with positive feedback. In this section we consider an example of a simple form of memory implemented using a feedback circuit.

Consider the system in Figure 2.18, which consists of a linear block with first-order dynamics and a nonlinear block with positive feedback. Assume that the nonlinearity is

$$y = F(x) = \frac{x}{1 + |x|}, \quad \text{which gives} \quad x = F^{-1}(y) = \frac{y}{1 - |y|}.$$

The system is described by the differential equation

$$\frac{dx}{dt} = -ax + b(r + y) = b(r - G(y)), \qquad G(y) := \frac{aF^{-1}(y)}{b} - y = \frac{ay}{b(1 - |y|)} - y.$$

Rewriting the dynamics in terms of the variable $y = F(x)$, we get the following relation between the input r and the output y:

$$\frac{dy}{dt} = \frac{dF(x)}{dt} = \left.\frac{dF(x)}{dx}\right|_{F^{-1}(y)} \cdot \frac{dx}{dt} = F'\big(F^{-1}(y)\big) \cdot b(r - G(y)). \tag{2.43}$$

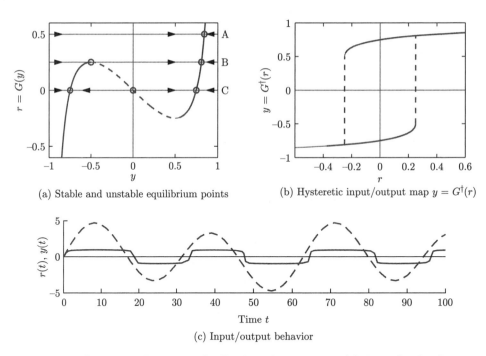

(a) Stable and unstable equilibrium points

(b) Hysteretic input/output map $y = G^\dagger(r)$

(c) Input/output behavior

Figure 2.19: System with positive feedback and saturation. (a) For a fixed reference value r, the intersections with the curve $r = G(y)$ corresponds to equilibrium points for the system. Equilibrium points at selected values of r are shown by circles (note that for some reference values there are multiple equilibrium points). Arrows indicate the sign of the derivative of y away from the equilibrium points, with the solid portions of $r = G(y)$ representing stable equilibrium points and dashed portions representing unstable equilibrium points. (b) The hysteretic input/output map given by $y = G^\dagger(r)$, showing that some values of r have single equilibrium points while others have two possible (stable) steady-state output values. (c) Simulation of the system dynamics showing the reference r (dashed curve) and the output y (solid curve).

The function F is monotone with $F'(x) > 0$ for all x and so the equilibrium points for a constant input r are given by the solutions of $r = G(y)$. The graph of the function G is shown in Figure 2.19a for $a = 1$ and $b = 4$. The function $G(y)$ has a local maximum $r_{\max} = (1 - \sqrt{a/b})^2 = 0.25$ at $y = -1 + \sqrt{a/b} = -0.5$ and a local minimum $r_{\min} = -0.25$ at $y = 0.5$. The set of possible equilibrium points for the system can be determined from Figure 2.19a by fixing r and identifying all values of y that satisfy $r = G(y)$. There is one unique equilibrium if $|r| > 0.25$, two equilibrium points if $|r| = 0.25$, and three equilibrium points if $|r| < 0.25$.

The differential equation (2.43) is of first order and the equilibrium point y_e is stable if $G'(y_e)$ is positive and unstable if $G'(y_e)$ is negative. Stable equilibrium points are shown in solid lines and unstable equilibrium points by dashed lines in Figure 2.19a. The differential equation thus has two stable equilibrium points when $r_{\min} < r < r_{\max}$ and one stable equilibrium point when $|r| \geq r_{\max}$.

To understand the behavior of the system, we will explore what happens when the reference is changed. If the reference r is zero there are two stable equilibrium points, as can be seen in Figure 2.19a by looking at the horizontal line at $r = 0$

(labeled C). We assume that the system is at the stable left equilibrium point, where y is negative. If the reference is increased, the equilibrium point moves slightly to the right. When the reference reaches the value 0.25, which corresponds an unstable equilibrium, the solution moves towards the right stable equilibrium point, where y is positive, as indicated by the line marked B in Figure 2.19a. If the value of r is increased further, the output y also increases. The static input/output relation is thus given by the "inverse function" $y = G^\dagger(r)$, which gives the value(s) of the stable output values as a function of r. The system has hysteretic behavior as shown in Figure 2.19b, where the dashed line indicates the switches between the branches of the solution curves, and they occur at $r = \pm r_{\max} = \pm 0.25$.

The temporal behavior of the system is illustrated by the simulations in Figure 2.19c, where the input r is dashed and the output y is solid. The shapes of the signals depend on the parameters; the values $a = 5$, $b = 50$ were used in the figure to give more distinct switches. The hysteresis width is $2r_{\max}$ and the parameter a gives the sharpness of the corners of the output. The circuit shown in the Figure 2.18 is commonly used as a trigger to detect changes in a signal (known as a Schmitt trigger). It is also used as a memory element in solid state memories, illustrating that feedback can be used to obtain discrete behavior.

2.7   FURTHER READING

The books by Bennett [34, 35] and Mindel [183, 184] give interesting perspective on the development of control. Much of the material touched upon in this chapter is referred to as "classical control"; see [62], [130], and [241] for early texts on this material. A more thorough introduction to the principles of feedback with minimal mathematical prerequisites is available in the textbook *Feedback Control for Everyone* [7]. The notion of controllers with two degrees of freedom was introduced by Horowitz [121].

The analysis introduced here will be elaborated in the rest of the book. Transfer functions and other descriptions of dynamics are discussed in Chapters 3 and 9, methods to investigate stability in Chapters 5 and 10. The simple method to find parameters of controllers based on matching of coefficients of the closed loop characteristic polynomial is developed further in Chapters 7, 8, and 13. Feedforward control is discussed in Sections 8.5 and 12.4.

EXERCISES

2.1 (Transfer functions and differential equations) Let $y \in \mathbb{R}$ and $u \in \mathbb{R}$. Solve the differential equations

$$\frac{dy}{dt} + ay = bu, \qquad \frac{d^2y}{dt^2} + 2\frac{dy}{dt} + y = 2\frac{du}{dt} + u,$$

for $t > 0$. Determine the responses to a unit step $u(t) = 1$ and the exponential signal $u(t) = e^{st}$ when the initial condition is zero. Derive the transfer functions of the systems.

2.2 (Effect of zeros on time responses) Let $y_0(t)$ be the response of a system with the transfer function $G_0(s)$ to a given input. The transfer function $G(s) = (1 + sT)G_0(s)$

has the same zero frequency gain but it has an additional zero at $z = -1/T$. Let $y(t)$ be the response of the system with the transfer function $G(s)$ and show that

$$y(t) = y_0(t) + T\frac{dy_0}{dt}. \tag{2.44}$$

Next consider the system with the transfer function

$$G(s) = \frac{s+a}{a(s^2 + 2s + 1)},$$

which has unit zero-frequency-gain ($G(0) = 1$). Use the result in equation (2.44) to explore the effect of a zero at $s = -1/T$ on the step response of a system.

2.3 (PI control) Consider a closed loop system with process dynamics and a PI controller modeled by

$$\frac{dy}{dt} + ay = bu, \qquad u = k_{\rm p}(r - y) + k_{\rm i}\int_0^t \big(r(\tau) - y(\tau)\big)\,d\tau,$$

where r is the reference, u is the control variable, and y is the process output.

(a) Derive a differential equation relating the output y to the reference r by direct manipulation of the equations and compute the transfer function $H_{yr}(s)$. Make the derivations both by direct manipulation of the differential equations and by polynomial algebra.

(b) Draw a block diagram of the system and derive the transfer functions of the process $P(s)$ and the controller $C(s)$.

(c) Use block diagram algebra to compute the transfer function from reference r to output y of the closed loop system and verify that your answer matches your answer in part (a).

2.4 (Zero frequency gain) Consider the system described by the differential equation (2.10) and the transfer function (2.16). Determine the zero frequency gain of the system by computing the particular solution of equation (2.10) for a constant input $u(t) = u_0$. Compare with the value of $G(0)$.

2.5 (Pupil response) The dynamics of the pupillary reflex can be approximated by a linear system with the transfer function

$$P(s) = \frac{0.2(1 - 0.1s)}{(1 + 0.1s)^3}.$$

Assume that the nervous system that controls the pupil opening is modeled as a proportional controller with the gain k. Use the Routh–Hurwitz criterion to determine the largest gain that gives a stable closed loop system.

2.6 (Parameter sensitivity) Consider the feedback system in Figure 2.7. Let the disturbance $v = 0$, $P(s) = 1$ and $C(s) = k_{\rm i}/s$. Determine the transfer function G_{yr} from reference r to output y. Also determine how much G_{yr} is changed when the process gain changes by 10%.

2.7 (PID control design) The calculations in Section 2.3 can be interpreted as a design method for a PI controller for a first-order system. A similar calculation can be made for PID control of a second-order system. Let the transfer functions of the process and the controller be

$$P(s) = \frac{b}{s^2 + a_1 s + a_2}, \qquad C(s) = k_p + \frac{k_i}{s} + k_d s.$$

Show that the controller parameters

$$k_p = \frac{(1 + 2\alpha\zeta_c)\omega_c^2 - a_2}{b}, \qquad k_i = \frac{\alpha\omega_c^3}{b}, \qquad k_d = \frac{(\alpha + 2\zeta_c)\omega_c - a_1}{b}$$

give a closed loop system with the characteristic polynomial

$$(s^2 + 2\zeta_c\omega_c s + \omega_c^2)(s + \alpha\omega_c).$$

2.8 (Linear behavior via feedback) Consider an open loop system with the non-linear input/output relation $y = F(u)$. Assume that the system is closed with the proportional controller $u = k(r - y)$. Show that the input/output relation of the closed loop system is

$$y + \frac{1}{k} F^{-1}(y) = r.$$

Estimate the largest deviation from ideal linear response $y = r$. Illustrate by plotting the input output responses for a) $F(u) = \sqrt{u}$ and b) $F(u) = u^2$ with $0 \le u \le 1$ and $k = 5$, 10, and 100.

2.9 (Nonlinear distortion) The following MATLAB commands will load and play Handel's Messiah:

```
load handel          % Load Handel's Messiah
sound(y, Fs); pause  % Play the original music through speaker
```

Write a MATLAB function that implements a nonlinear amplifier with static gain

$$y = 2(z + az(1 - z) - 0.5), \qquad z = (x + 1)/2,$$

where x is the original signal (assumed to take values between -1 and 1) and a is the amplifier gain. Compare the sound that is obtained when the music is then sent through two amplifiers with the given nonlinearity and gain $a = 1$ versus when the music is sent through the same two amplifiers with feedback $k = 10$.

2.10 (Queing systems) Consider a queuing system modeled by

$$\frac{dx}{dt} = \lambda - \mu_{max}\frac{x}{x + 1},$$

where λ is the acceptance rate of jobs and x is the length of the queue. The model is nonlinear and the dynamics of the system changes significantly with the queuing length (see Example 3.15 for a more detailed discussion). Investigate the situation when a PI controller is used for admission control. Let r be the rate of arrival of job requests and model the (average) arrival intensity λ as

$$\lambda = k_{\mathrm{p}}(r - x) + k_{\mathrm{i}} \int^{t} (r(t) - x(t)) dt.$$

The controller parameters are determined from the approximate model

$$\frac{dx}{dt} = \lambda.$$

Find controller parameters that give the closed loop characteristic polynomial $s^2 + 2s + 1$ for the approximate model. Investigate the behavior of the control strategy for the full nonlinear model by simulation for the input $r = 5 + 4\sin(0.1t)$.

Chapter Three

System Modeling

> ... I asked Fermi whether he was not impressed by the agreement between our calculated numbers and his measured numbers. He replied, "How many arbitrary parameters did you use for your calculations?" I thought for a moment about our cut-off procedures and said, "Four." He said, "I remember my friend Johnny von Neumann used to say, with four parameters I can fit an elephant, and with five I can make him wiggle his trunk."
>
> —Freeman Dyson on describing the predictions of his model for meson-proton scattering to Enrico Fermi in 1953 [80].

A model is a precise representation of a system's dynamics used to answer questions via analysis and simulation. The model we choose depends on the questions we wish to answer, and so there may be multiple models for a single dynamical system, with different levels of fidelity depending on the phenomena of interest. In this chapter we provide an introduction to the concept of modeling and present some basic material on two specific methods commonly used in feedback and control systems: differential equations and difference equations.

3.1 MODELING CONCEPTS

A *model* is a mathematical representation of a physical, biological, or information system. Models allow us to reason about a system and make predictions about how a system will behave. In this text, we will mainly be interested in models of dynamical systems describing the input/output behavior of systems, and we will often work in "state space" form. As pointed out already in Chapter 1, when using models it is important to keep in mind that they are an *approximation* of the underlying system. Analysis and design using models must always be done carefully to ensure that the limits of the model are respected.

Roughly speaking, a dynamical system is one in which the effects of actions do not occur immediately. For example, the velocity of a car does not change immediately when the gas pedal is pushed nor does the temperature in a room rise instantaneously when a heater is switched on. Similarly, a headache does not vanish right after an aspirin is taken, requiring time for it to take effect. In business systems, increased funding for a development project does not increase revenues in the short term, although it may do so in the long term (if it was a good investment). All of these are examples of dynamical systems, in which the behavior of the system evolves with time.

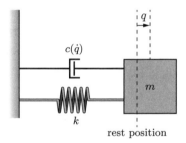

Figure 3.1: Spring–mass system with nonlinear damping. The position of the mass is denoted by q, with $q = 0$ corresponding to the rest position of the spring. The forces on the mass are generated by a linear spring with spring constant k and a damper with force dependent on the velocity $\dot{q}$.

In the remainder of this section we provide an overview of some of the key concepts in modeling. The mathematical details introduced here are explored more fully in the remainder of the chapter.

The Heritage of Mechanics

The study of dynamics originated in attempts to describe planetary motion. The basis was detailed observations of the planets by Tycho Brahe and the results of Kepler, who found empirically that the orbits of the planets could be well described by ellipses. Newton embarked on an ambitious program to try to explain why the planets move in ellipses, and he found that the motion could be explained by his law of gravitation and the formula stating that force equals mass times acceleration. In the process he also invented calculus and differential equations.

One of the triumphs of Newton's mechanics was the observation that the motion of the planets could be predicted based on the current positions and velocities of all planets. It was not necessary to know the past motion. The *state* of a dynamical system is a collection of variables that completely captures the past motion of a system for the purpose of predicting future motion. For a system of planets the state is simply the positions and the velocities of the planets. We call the set of all possible states the *state space*.

A common class of mathematical models for dynamical systems is ordinary differential equations (ODEs). In mechanics, one of the simplest such differential equations is that of a spring–mass system with damping:

$$m\ddot{q} + c(\dot{q}) + kq = 0. \tag{3.1}$$

This system is illustrated in Figure 3.1. The variable $q \in \mathbb{R}$ represents the position of the mass m with respect to its rest position. We use the notation $\dot{q}$ to denote the derivative of q with respect to time (i.e., the velocity of the mass) and $\ddot{q}$ to represent the second derivative (acceleration). The spring is assumed to satisfy Hooke's law, which says that the force is proportional to the displacement. The friction element (damper) is taken as a nonlinear function $c(\dot{q})$, which can model effects such as Coulomb friction and viscous drag. The position q and velocity $\dot{q}$ represent the instantaneous state of the system. We say that this system is a

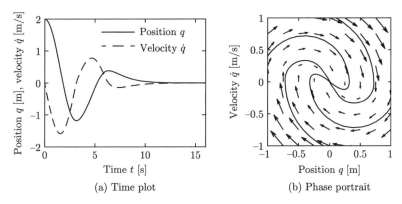

(a) Time plot   (b) Phase portrait

Figure 3.2: Illustration of a state model. A state model gives the rate of change of the state as a function of the state. The plot on the left shows the evolution of the state as a function of time. The plot on the right, called a *phase portrait*, shows the evolution of the states relative to each other, with the velocity of the state denoted by arrows.

second-order system since it has two states that we combine in the *state vector* $x = (q, \dot{q})$.

The evolution of the position and velocity can be described using either a time plot or a phase portrait, both of which are shown in Figure 3.2. The *time plot*, on the left, shows the values of the individual states as a function of time. The *phase portrait*, on the right, shows the traces of some of the states from different initial conditions: it illustrates how the states move in the state space. In the phase portrait we have also shown arrows that represent the velocity $\dot{x}$ of the state x in a few points. The phase portrait gives a strong intuitive representation of the equation as a vector field or a flow. While systems of second order (two states) can be represented in this way, unfortunately it is difficult to visualize equations of higher order using this approach.

The differential equation (3.1) is called an *autonomous* system because there are no external influences. (Note that this usage of "autonomous" is slightly different than in the phrase "autonomous vehicle.") Such a model is natural for use in celestial mechanics because it is difficult to influence the motion of the planets. In many examples it is useful to model the effects of external disturbances or controlled forces on the system. One way to capture this is to replace equation (3.1) by

$$m\ddot{q} + c(\dot{q}) + kq = u, \tag{3.2}$$

where u represents the effect of external inputs. The model (3.2) is called a *forced* or *controlled differential equation*. It implies that the rate of change of the state can be influenced by the input $u(t)$. Adding the input makes the model richer and allows new questions to be posed. For example, we can examine what influence external disturbances have on the trajectories of a system. Or, in the case where the input variable is something that can be modulated in a controlled way, we can analyze whether it is possible to "steer" the system from one point in the state space to another through proper choice of the input.

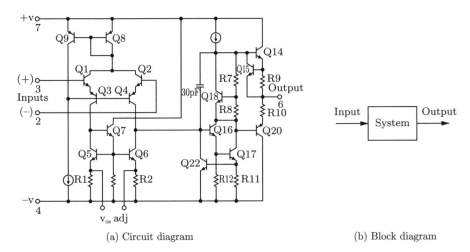

(a) Circuit diagram   (b) Block diagram

Figure 3.3: Illustration of the input/output view of a dynamical system. The figure (a) shows a detailed circuit diagram for an electronic amplifier; (b) is its representation as a block diagram.

The Heritage of Electrical Engineering

A different view of dynamics emerged from electrical engineering, where the design of electronic amplifiers led to a focus on input/output behavior. A system was considered a device that transforms inputs to outputs, as illustrated in Figure 3.3. Conceptually an input/output model can be viewed as a giant table of input and output signals. Given an input signal $u(t)$ over some interval of time, the model should produce the resulting output $y(t)$.

The input/output framework is used in many engineering disciplines since it allows us to decompose a system into individual components connected through their inputs and outputs. Thus, we can take a complicated system such as a radio or a television and break it down into manageable pieces such as the receiver, demodulator, amplifier, and speakers. Each of these pieces has a set of inputs and outputs and, through proper design, these components can be interconnected to form the entire system.

The input/output view is particularly useful for the special class of *linear time-invariant systems*. This term will be defined more carefully later in this chapter, but roughly speaking a system is linear if the superposition (addition) of two inputs yields an output that is the sum of the outputs that would correspond to individual inputs being applied separately. A system is time-invariant if the output response for a given input does not depend on when that input is applied.

Many electrical engineering systems can be modeled by linear time-invariant systems and hence a large number of tools have been developed to analyze them. One such tool is the *step response*, which describes the relationship between an input that changes from zero to a constant value abruptly (a step input) and the corresponding output. As we shall see later in the text, the step response is very useful in characterizing the performance of a dynamical system, and it is often used to specify the desired dynamics. A sample step response is shown in Figure 3.4a.

Another way to describe a linear time-invariant system is to represent it by its response to sinusoidal input signals. This is called the *frequency response*, and a

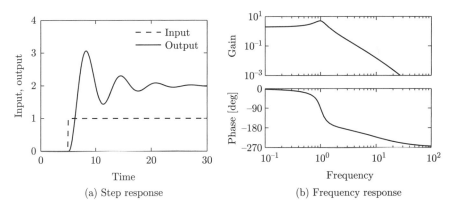

(a) Step response   (b) Frequency response

Figure 3.4: Input/output response of a linear system. The step response (a) shows the output of the system due to an input that changes from 0 to 1 at time $t = 5$ s. The frequency response (b) shows the amplitude gain and phase change due to a sinusoidal input at different frequencies.

rich, powerful theory with many concepts and strong, useful results has emerged. The results are based on the theory of complex variables and Laplace transforms. The basic idea behind the frequency response is that we can completely characterize the behavior of a system by its steady-state response to sinusoidal inputs. Roughly speaking, this is done by decomposing any arbitrary signal into a linear combination of sinusoids (e.g., by using the Fourier transform) and then using linearity to compute the output by combining the response to the individual frequencies. A sample frequency response is shown in Figure 3.4b.

The input/output view lends itself naturally to experimental determination of system dynamics, where a system is characterized by recording its response to particular inputs, e.g., a step or a set of sinusoids over a range of frequencies.

The Control View

When control theory emerged as a discipline in the 1940s, the approach to dynamics was strongly influenced by the electrical engineering (input/output) view. A second wave of developments in control, starting in the late 1950s, was inspired by mechanics, where the state space perspective was used. The emergence of space flight is a typical example, where precise control of the orbit of a spacecraft is essential. These two points of view gradually merged into what is today the state space representation of input/output systems. In the 1970s the development was influenced by advances in automation, which emphasized the need to include logic and sequencing.

The development of state space models involved modifying the models from mechanics to include external actuators and sensors and utilizing more general forms of equations. In control, the model given by equation (3.2) was replaced by

$$\frac{dx}{dt} = f(x, u), \qquad y = h(x, u), \tag{3.3}$$

where x is a vector of state variables, u is a vector of control signals, and y is a vector of measurements. The term dx/dt represents the derivative of the vector x with respect to time, and f and h are (possibly nonlinear) mappings of their arguments to vectors of the appropriate dimension. For mechanical systems, the state consists of the position and velocity of the system, so that $x = (q, \dot{q})$ in the case of a damped spring–mass system. Note that in the control formulation we model dynamics as first-order differential equations, but we will see that this can capture the dynamics of higher-order differential equations by appropriate definition of the state and the maps f and h.

Adding inputs and outputs has increased the richness of the classical problems and led to many new concepts. For example, it is natural to ask if possible states x can be reached with the proper choice of u (reachability) and if the measurement y contains enough information to reconstruct the state (observability). These topics will be addressed in greater detail in Chapters 7 and 8.

A final development in building the control point of view was the emergence of disturbances and model uncertainty as critical elements in the theory. The simple way of modeling disturbances as deterministic signals like steps and sinusoids has the drawback that such signals cannot be predicted precisely. A more realistic approach is to model disturbances as random signals. This viewpoint gives a natural connection between prediction and control. The dual views of input/output representations and state space representations are particularly useful when modeling systems with uncertainty since state models are convenient to describe a nominal model but uncertainties are easier to describe using input/output models (often via a frequency response description). Uncertainty will be a constant theme throughout the text and will be studied in particular detail in Chapter 13.

An interesting observation in the design of control systems is that feedback systems can often be analyzed and designed based on comparatively simple models. The reason for this is the inherent robustness of feedback systems. However, other uses of models may require more complexity and more accuracy. One example is feedforward control strategies, where one uses a model to precompute the inputs that cause the system to respond in a certain way. Another area is system validation, where one wishes to verify that the detailed response of the system performs as it was designed. Because of these different uses of models, it is common to use a hierarchy of models having different complexity and fidelity.

Multidomain Modeling

Modeling is an essential element of many disciplines, but traditions and methods from individual disciplines can differ from each other, as illustrated by the previous discussion of mechanical and electrical engineering. A difficulty in systems engineering is that it is frequently necessary to deal with heterogeneous systems from many different domains, including chemical, electrical, mechanical, and information systems.

To model such multidomain systems, we start by partitioning a system into smaller subsystems. Each subsystem is represented by balance equations for mass, energy, and momentum, or by appropriate descriptions of information processing in the subsystem. The behavior at the interfaces is captured by describing how the variables of the subsystem behave when the subsystems are interconnected. These interfaces act by constraining variables within the individual subsystems to be equal (such as mass, energy, or momentum fluxes). The complete model is then obtained by combining the descriptions of the subsystems and the interfaces.

Using this methodology it is possible to build up libraries of subsystems that correspond to physical, chemical, and informational components. The procedure mimics the engineering approach where systems are built from subsystems that are themselves built from smaller components. As experience is gained, the components and their interfaces can be standardized and collected in model libraries. In practice, it takes several iterations to obtain a good library that can be reused for many applications.

State models or ordinary differential equations are not suitable for component-based modeling of this form because states may disappear when components are connected. This implies that the internal description of a component may change when it is connected to other components. As an illustration we consider two capacitors in an electrical circuit. Each capacitor has a state corresponding to the voltage across the capacitors, but one of the states will disappear if the capacitors are connected in parallel. A similar situation happens with two rotating inertias, each of which is individually modeled using the angle of rotation and the angular velocity. Two states will disappear when the inertias are joined by a rigid shaft.

This difficulty can be avoided by replacing differential equations by *differential algebraic equations*, which have the form

$$F(z, \dot{z}) = 0,$$

where $z \in \mathbb{R}^n$. A simple special case is

$$\dot{x} = f(x, y), \qquad g(x, y) = 0, \tag{3.4}$$

where $z = (x, y)$ and $F = (\dot{x} - f(x, y), g(x, y))$. The key property is that the derivative $\dot{z}$ is not given explicitly and there may be pure algebraic relations between the components of the vector z. Modeling using differential algebraic equations is also called *equation-based modeling*, *acausal modeling*, or *behavioral modeling*.

The model (3.4) captures the examples of the parallel capacitors and the linked rotating inertias. For example, when two capacitors are connected, we simply add the algebraic equation expressing that the voltages across the capacitors are the same.

Modelica is a language that has been developed to support component-based modeling. Differential algebraic equations are used as the basic description, and object-oriented programming is used to structure the models. Modelica is used to model the dynamics of technical systems in domains such as mechanical, electrical, thermal, hydraulic, thermofluid, and control subsystems. Modelica is intended to serve as a standard format so that models arising in different domains can be exchanged between tools and users. A large set of free and commercial Modelica component libraries are available and are used by a growing number of people in industry, research, and academia. For further information about Modelica, see http://www.modelica.org or the books by Tiller [239] and Fritson [95].

Finite State Machines and Hybrid Systems

A final type of modeling has been developed within the computer-controlled systems community. A hybrid system (also called a *cyberphysical system*) is one that combines continuous dynamics with discrete logic. The discrete portion of the system represents logical variables that reside in a computer, such as the mode of a system (on, off, degraded, etc.).

Discrete state dynamics are often represented using a *finite state machine* that consists of a finite set of discrete states $\alpha \in \mathbb{Q}$. We can think of α as the "mode" of the system. The dynamics of a finite state machine are defined in terms of transitions between the states. One convenient representation is as a *guarded transition system*:

$$g_i(\alpha, \beta) \implies \alpha' = r_i(\alpha), \qquad i = 1, \ldots, N.$$

Here the function g is a Boolean (true/false) function that depends on the current system mode α and an input β, which might represent an environmental event (button press, component failure, etc.). If the guard g_i is true then the system transitions from the current state α to a new state α', determined by the rule (transition map) r_i. A guarded transition system can have many different rules, depending on the system state and external input.

It is also possible to combine systems that have continuous states with those having discrete states, creating a *hybrid system*. For example, if a system has a continuous state x and discrete state α, we might write the overall system dynamics as

$$\frac{dx}{dt} = f_\alpha(x, u), \quad g_i(x, \alpha, \beta) \implies \alpha' = r_i(x, \alpha), \qquad i = 1, \ldots, N.$$

In this representation, the continuous dynamics (with state x) are governed by an ordinary differential equation that may depend on the system mode α (indicated by the subscript in f_α). The discrete transition system is also influenced by the continuous state, so that the guards g_i and rules r_i now depend on the continuous state.

Many other representations are possible for hybrid systems, including models that allow a non-continuous change in the continuous variables when a change in the discrete state occurs (so-called *reset logic*). Computer modeling packages for hybrid systems include StateFlow (part of the MATLAB suite of tools), Modelica, and Ptolemy [205].

Model Uncertainty

Reducing uncertainty is one of the main reasons for using feedback, and it is therefore important to characterize uncertainty. When making measurements, there is a good tradition to assign both a nominal value and a measure of uncertainty. It is useful to apply the same principle to modeling, but unfortunately it is often difficult to express the uncertainty of a model quantitatively.

For a static system whose input/output relation can be characterized by a function, uncertainty can be expressed by an uncertainty band as illustrated in Figure 3.5a. At low signal levels there are uncertainties due to sensor resolution, friction, and quantization. For example, some models for queuing systems or cells are based on averages that exhibit significant variations for small populations. At large signal levels there are saturations or even system failures. The signal ranges where a model is reasonably accurate vary dramatically between applications, but it is rare to find models that are accurate for signal ranges larger than 10^4.

Characterization of the uncertainty of a dynamical model is much more difficult. We can try to capture uncertainties by assigning uncertainties to parameters of the model, but this is often not sufficient. There may be errors due to phenomena that have been neglected, e.g., small time delays. In control the ultimate test is how well a control system based on the model performs, and time delays can be important. There is also a frequency aspect. There are slow phenomena, such as

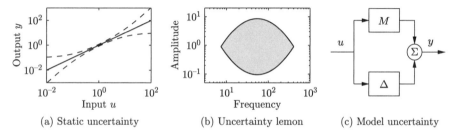

(a) Static uncertainty   (b) Uncertainty lemon   (c) Model uncertainty

Figure 3.5: Characterization of model uncertainty. Uncertainty of a static system is illustrated in (a), where the solid line indicates the nominal input/output relationship and the dashed lines indicate the range of possible uncertainty. The uncertainty lemon [101] in (b) is one way to capture uncertainty in dynamical systems emphasizing that a model is valid only in the amplitude and frequency ranges within the shaded region. In (c) a model is represented by a nominal model M and another model Δ representing the uncertainty analogous to the representation of parameter uncertainty.

aging, that can cause changes or drift in the systems. There are also high-frequency effects: a resistor will no longer be a pure resistance at very high frequencies, and a beam has stiffness and will exhibit additional dynamics when subject to high-frequency excitation. The *uncertainty lemon* [101] shown in Figure 3.5b is one way to conceptualize the uncertainty of a system. It illustrates that a model is valid only in certain amplitude and frequency ranges.

We will introduce some formal tools for representing uncertainty in Chapter 13 using figures such as Figure 3.5c. These tools make use of the concept of a transfer function, which describes the frequency response of an input/output system. For now, we simply note that one should always be careful to recognize the limits of a model and not to make use of models outside their range of applicability. For example, one can describe the uncertainty lemon and then check to make sure that signals remain in this region. In early analog computing, a system was simulated using operational amplifiers, and it was customary to give alarms when certain signal levels were exceeded. Similar features can be included in digital simulation.

3.2 STATE SPACE MODELS

In this section we describe the two primary forms of models that we use in this text: differential equations and difference equations. Both make use of the notions of state, inputs, outputs, and dynamics to describe the behavior of a system. We also briefly discuss modeling of finite state systems.

Ordinary Differential Equations

The state of a system is a collection of variables that summarize the past of a system for the purpose of predicting the future. For a physical system the state is composed of the variables required to account for storage of mass, momentum, and energy. A key issue in modeling is to decide how accurately this information has to be represented. The state variables are gathered in a vector $x \in \mathbb{R}^n$ called the *state vector*. The control variables are represented by another vector $u \in \mathbb{R}^p$, and the measured signal by the vector $y \in \mathbb{R}^q$. A system can then be represented by the differential equation

$$\frac{dx}{dt} = f(x, u), \qquad y = h(x, u), \qquad (3.5)$$

where $f : \mathbb{R}^n \times \mathbb{R}^p \to \mathbb{R}^n$ and $h : \mathbb{R}^n \times \mathbb{R}^p \to \mathbb{R}^q$ are smooth mappings. We call a model of this form a *state space model*.

The dimension of the state vector is called the *order* of the model. The model given in equation (3.5) is called *time invariant* because the functions f and h do not depend explicitly on time t; there are more general time-varying systems where the functions do depend on time. The model consists of two functions: the function f gives the rate of change of the state vector as a function of state x and control u, and the function h gives the measured values as functions of state x and control u.

A model is called a *linear* state space model (or often just a "linear system") if the functions f and h are linear in x and u. A linear state space model can thus be represented by

$$\frac{dx}{dt} = Ax + Bu, \qquad y = Cx + Du, \qquad (3.6)$$

where A, B, C, and D are constant matrices. Such a model is said to be *linear and time-invariant*, or LTI for short. (In this text we will usually omit the term time-invariant and just say the model is linear.) The matrix A is called the *dynamics matrix*, the matrix B is called the *control matrix*, the matrix C is called the *sensor matrix*, and the matrix D is called the *direct term*. Frequently models will not have a direct term, indicating that the control signal u does not influence the output directly.

A different form of linear differential equations, generalizing the second-order dynamics from mechanics, is an equation of the form

$$\frac{d^n y}{dt^n} + a_1 \frac{d^{n-1} y}{dt^{n-1}} + \cdots + a_n y = u, \qquad (3.7)$$

where t is the independent (time) variable, $y(t)$ is the dependent (output) variable and $u(t)$ is the input. The notation $d^k y/dt^k$ is used to denote the kth derivative of y with respect to t, sometimes also written as $y^{(k)}$. The controlled differential equation (3.7) is said to be an nth-order model. This model can be converted into state space form by defining

$$x = \begin{pmatrix} x_1 \\ x_2 \\ \vdots \\ x_{n-1} \\ x_n \end{pmatrix} = \begin{pmatrix} d^{n-1} y/dt^{n-1} \\ d^{n-2} y/dt^{n-2} \\ \vdots \\ dy/dt \\ y \end{pmatrix},$$

and the state space equations become

$$\frac{d}{dt} \begin{pmatrix} x_1 \\ x_2 \\ \vdots \\ x_{n-1} \\ x_n \end{pmatrix} = \begin{pmatrix} -a_1 x_1 - \cdots - a_n x_n \\ x_1 \\ \vdots \\ x_{n-2} \\ x_{n-1} \end{pmatrix} + \begin{pmatrix} u \\ 0 \\ \vdots \\ 0 \\ 0 \end{pmatrix}, \qquad y = x_n.$$

With the appropriate definitions of A, B, C, and D, this equation is in linear state space form.

An even more general model is obtained by letting the output be a linear combination of the states of the model, i.e.,

$$y = b_1 x_1 + b_2 x_2 + \cdots + b_n x_n + du.$$

This model can be represented in state space as

$$\frac{d}{dt} \begin{pmatrix} x_1 \\ x_2 \\ x_3 \\ \vdots \\ x_n \end{pmatrix} = \begin{pmatrix} -a_1 & -a_2 & \cdots & -a_{n-1} & -a_n \\ 1 & 0 & \cdots & 0 & 0 \\ 0 & 1 & & 0 & 0 \\ \vdots & & \ddots & & \vdots \\ 0 & 0 & & 1 & 0 \end{pmatrix} x + \begin{pmatrix} 1 \\ 0 \\ 0 \\ \vdots \\ 0 \end{pmatrix} u, \tag{3.8}$$

$$y = \begin{pmatrix} b_1 & b_2 & \cdots & b_n \end{pmatrix} x + du.$$

This particular form of a linear state space model is called *reachable canonical form* and will be studied in more detail in later chapters. Many other representations for a model are possible and we shall see several of these in Chapters 6–8. It is also possible to expand the form of equation (3.7) to allow derivatives of the input to appear, as we saw briefly in Chapter 2.

Example 3.1 Spring–mass system

As a simple example of converting a linear differential equation to state space form, consider the externally-driven spring mass system whose dynamics are given in equation (3.2):

$$m\ddot{q} + c(\dot{q}) + kq = u.$$

This has the same form as equation (3.7) where the output y is the position q. The state of the system can then be written as

$$x = \begin{pmatrix} x_1 \\ x_2 \end{pmatrix} = \begin{pmatrix} \dot{q} \\ q \end{pmatrix}$$

and the state space equations are

$$\frac{d}{dt} \begin{pmatrix} x_1 \\ x_2 \end{pmatrix} = \begin{pmatrix} -c/m & -k/m \\ 1 & 0 \end{pmatrix} \begin{pmatrix} x_1 \\ x_2 \end{pmatrix} + \begin{pmatrix} 1/m \\ 0 \end{pmatrix} u,$$

where we have further assumed that $c(\dot{q}) = c\dot{q}$ (corresponding to viscous friction). ∇

Example 3.2 Balance systems

A more complex example of a type of system that can be modeled using ordinary differential equations is the class of *balance systems*. A balance system is a mechanical system in which the center of mass is balanced above a pivot point. Some common examples of balance systems are shown in Figure 3.6. The Segway® Personal Transporter (Figure 3.6a) uses a motorized platform to stabilize a person standing on top of it. When the rider leans forward, the transportation device propels itself along the ground but maintains its upright position. Another example is

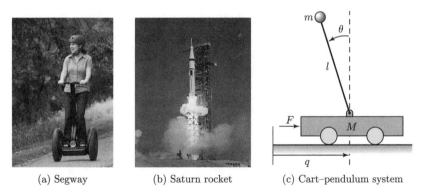

(a) Segway   (b) Saturn rocket   (c) Cart–pendulum system

Figure 3.6: Balance systems. (a) Segway® Personal Transporter, (b) Saturn rocket, and (c) inverted pendulum on a cart. Each of these examples uses forces at the bottom of the system to keep it upright.

a rocket (Figure 3.6b), in which a gimballed nozzle at the bottom of the rocket is used to stabilize the body of the rocket above it. Other examples of balance systems include humans or other animals standing upright or a person balancing a stick on their hand.

Balance systems are a generalization of the spring–mass system we saw earlier. We can write the dynamics for a mechanical system in the general form

$$M(q)\ddot{q} + C(q, \dot{q}) + K(q) = B(q)u,$$

where $M(q)$ is the inertia matrix for the system, $C(q, \dot{q})$ represents the Coriolis forces as well as the damping, $K(q)$ gives the forces due to potential energy, and $B(q)$ describes how the external applied forces couple into the dynamics. Note that q may be a vector, rather than just a scalar, and represents the *configuration variables* of the system. The specific form of the equations can be derived using Newtonian mechanics. Each of the terms depends on the configuration of the system q and these terms are often nonlinear in the configuration variables.

Figure 3.6c shows a simplified diagram for a balance system consisting of an inverted pendulum on a cart. To model this system, we choose state variables that represent the position and velocity of the base of the system, q and $\dot{q}$, and the angle and angular rate of the structure above the base, θ and $\dot{\theta}$. (Note the slight abuse of notation in using q to represent the position and (q, θ) for the full set of configuration variables.) We let F represent the force applied at the base of the system, assumed to be in the horizontal direction (aligned with q), and choose the position and angle of the system as outputs. With this set of definitions, the dynamics of the system can be computed using Newtonian mechanics and have the form

$$\begin{pmatrix} (M+m) & -ml\cos\theta \\ -ml\cos\theta & (J+ml^2) \end{pmatrix} \begin{pmatrix} \ddot{q} \\ \ddot{\theta} \end{pmatrix} + \begin{pmatrix} c\dot{q} + ml\sin\theta\,\dot{\theta}^2 \\ \gamma\dot{\theta} - mgl\sin\theta \end{pmatrix} = \begin{pmatrix} F \\ 0 \end{pmatrix}, \qquad (3.9)$$

where M is the mass of the base, m and J are the mass and moment of inertia of the system to be balanced, l is the distance from the base to the center of mass of the balanced body, c and γ are coefficients of viscous friction, and g is the acceleration due to gravity.

We can rewrite the dynamics of the system in state space form by defining the state as $x = (q, \theta, \dot{q}, \dot{\theta})$, the input as $u = F$, and the output as $y = (q, \theta)$. If we define the total mass and total inertia as

$$M_t = M + m, \qquad J_t = J + ml^2,$$

the equations of motion then become

$$
\frac{d}{dt}\begin{pmatrix} q \\ \theta \\ \dot{q} \\ \dot{\theta} \end{pmatrix} = \begin{pmatrix} \dot{q} \\ \dot{\theta} \\ \dfrac{-mls_\theta\dot{\theta}^2 + mg(ml^2/J_t)s_\theta c_\theta - c\dot{q} - (\gamma/J_t)mlc_\theta\dot{\theta} + u}{M_t - m(ml^2/J_t)c_\theta^2} \\ \dfrac{-ml^2s_\theta c_\theta\dot{\theta}^2 + M_t gls_\theta - clc_\theta\dot{q} - \gamma(M_t/m)\dot{\theta} + lc_\theta u}{J_t(M_t/m) - m(lc_\theta)^2} \end{pmatrix},
$$

$$y = \begin{pmatrix} q \\ \theta \end{pmatrix},$$

where we have used the shorthand $c_\theta = \cos\theta$ and $s_\theta = \sin\theta$.

In many cases, the angle θ will be very close to 0, and hence we can use the approximations $\sin\theta \approx \theta$ and $\cos\theta \approx 1$. Furthermore, if $\dot{\theta}$ is small, we can ignore quadratic and higher terms in $\dot{\theta}$. Substituting these approximations into our equations, we see that we are left with a *linear* state space equation

$$
\frac{d}{dt}\begin{pmatrix} q \\ \theta \\ \dot{q} \\ \dot{\theta} \end{pmatrix} = \begin{pmatrix} 0 & 0 & 1 & 0 \\ 0 & 0 & 0 & 1 \\ 0 & m^2l^2g/\mu & -cJ_t/\mu & -\gamma lm/\mu \\ 0 & M_t mgl/\mu & -clm/\mu & -\gamma M_t/\mu \end{pmatrix}\begin{pmatrix} q \\ \theta \\ \dot{q} \\ \dot{\theta} \end{pmatrix} + \begin{pmatrix} 0 \\ 0 \\ J_t/\mu \\ lm/\mu \end{pmatrix}u,
$$

$$y = \begin{pmatrix} 1 & 0 & 0 & 0 \\ 0 & 1 & 0 & 0 \end{pmatrix}x,$$

where $\mu = M_t J_t - m^2l^2$. ∇

Example 3.3 Inverted pendulum
A variation of the previous example is one in which the location of the base q does not need to be controlled. This happens, for example, if we are interested only in stabilizing a rocket's upright orientation without worrying about the location of the base of the rocket. The dynamics of this simplified system are given by

$$
\frac{d}{dt}\begin{pmatrix} \theta \\ \dot{\theta} \end{pmatrix} = \begin{pmatrix} \dot{\theta} \\ \dfrac{mgl}{J_t}\sin\theta - \dfrac{\gamma}{J_t}\dot{\theta} + \dfrac{l}{J_t}u\cos\theta \end{pmatrix}, \qquad y = \theta, \qquad (3.10)
$$

where γ is the coefficient of rotational friction, $J_t = J + ml^2$, and u is the force applied at the base. This system is referred to as an *inverted pendulum*. ∇

Difference Equations

In some circumstances, it is more natural to describe the evolution of a system at discrete instants of time rather than continuously in time. If we refer to each of these times by an integer $k = 0, 1, 2, \ldots$, then we can ask how the state of the system changes for each k. Just as in the case of differential equations, we define the state to be the set of variables that summarizes the past of the system for the purpose of predicting its future. Systems described in this manner are referred to as *discrete-time systems*.

The evolution of a discrete-time system can be written in the form

$$x[k+1] = f(x[k], u[k]), \qquad y[k] = h(x[k], u[k]), \qquad (3.11)$$

where $x[k] \in \mathbb{R}^n$ is the state of the system at time k (an integer), $u[k] \in \mathbb{R}^p$ is the input, and $y[k] \in \mathbb{R}^q$ is the output. As before, f and h are smooth mappings of the appropriate dimension. We call equation (3.11) a *difference equation* since it tells us how $x[k+1]$ differs from $x[k]$. The state $x[k]$ can be either a scalar- or a vector-valued quantity; in the case of the latter we write $x_j[k]$ for the value of the jth state at time k.

Just as in the case of differential equations, it is often the case that the equations are linear in the state and input, in which case we can describe the system by

$$x[k+1] = Ax[k] + Bu[k], \qquad y[k] = Cx[k] + Du[k].$$

As before, we refer to the matrices A, B, C, and D as the dynamics matrix, the control matrix, the sensor matrix, and the direct term. The solution of a linear difference equation with initial condition $x[0]$ and input $u[0], \ldots, u[T]$ can be computed using repeated substitution and is given by

$$x[k] = A^k x[0] + \sum_{j=0}^{k-1} A^{k-j-1} Bu[j],$$
$$\qquad\qquad\qquad\qquad\qquad\qquad k > 0. \qquad (3.12)$$
$$y[k] = CA^k x[0] + \sum_{j=0}^{k-1} CA^{k-j-1} Bu[j] + Du[k],$$

Difference equations are also useful as an approximation of differential equations, as we will show later.

Example 3.4 Predator–prey

As an example of a discrete-time system, consider a simple model for a predator–prey system. The predator–prey problem refers to an ecological system in which we have two species, one of which feeds on the other. This type of system has been studied for decades and is known to exhibit interesting dynamics. Figure 3.7 shows a historical record taken over 90 years for a population of lynxes versus a population of hares [173]. As can been seen from the graph, the annual records of the populations of each species are oscillatory in nature.

A simple model for this situation can be constructed using a discrete-time model to keep track of the rate of births and deaths of each species. Letting H represent the population of hares and L represent the population of lynxes, we can describe the state in terms of the populations at discrete periods of time. Letting k be the discrete-time index (corresponding here to each day), we can write

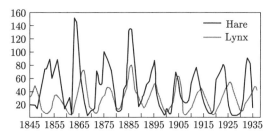

Figure 3.7: Predator versus prey. The photograph on the left shows a Canadian lynx and a snowshoe hare, the lynx's primary prey. The graph on the right shows the populations of hares and lynxes between 1845 and 1935 in a section of the Canadian Rockies [173]. The data were collected on an annual basis over a period of 90 years. (Photograph copyright Tom and Pat Leeson.)

$$H[k+1] = H[k] + b_{\mathrm{h}}(u)H[k] - aL[k]H[k],$$

$$L[k+1] = L[k] + cL[k]H[k] - d_{\mathrm{l}}L[k], \tag{3.13}$$

where $b_{\mathrm{h}}(u)$ is the hare birth rate per unit period and is a function of the food supply u, d_{l} is the lynx mortality rate, and a and c are the interaction coefficients. The interaction term $aL[k]H[k]$ models the rate of predation, which is assumed to be proportional to the rate at which predators and prey meet and is hence given by the product of the population sizes. The interaction term $cL[k]H[k]$ in the lynx dynamics has a similar form and represents the rate of growth of the lynx population. This model makes many simplifying assumptions—such as the fact that hares decrease in number only through predation by lynxes—but it often is sufficient to answer basic questions about the system.

To illustrate the use of this system, we can compute the number of lynxesand hares at each time point from some initial population. This is done by starting with $x[0] = (H_0, L_0)$ and then using equation (3.13) to compute the populations in the following period. By iterating this procedure, we can generate the population over time. The output of this process for a specific choice of parameters and initial conditions is shown in Figure 3.8. While the details of the simulation are different from the experimental data (to be expected given the simplicity of our assumptions), we see qualitatively similar trends and hence we can use the model to help explore the dynamics of the system. ∇

Example 3.5 E-mail server
The IBM Lotus (now Domino) server is a collaborative software system that administers users' e-mail, documents, and notes. Client machines interact with end users to provide access to data and applications. The server also handles other administrative tasks. In the early development of the system it was observed that the performance was poor when the central processing unit (CPU) was overloaded because of too many service requests, and mechanisms to control the load were therefore introduced.

The interaction between the client and the server is in the form of remote procedure calls (RPCs). The server maintains a log of statistics of completed requests. The total number of requests being served, called RIS (RPCs in server), is also

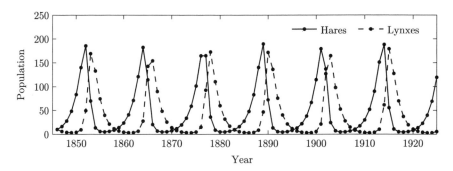

Figure 3.8: Discrete-time simulation of the predator–prey model (3.13). Using the parameters $a = c = 0.014$, $b_{\mathrm{h}}(u) = 0.6$, and $d_{\mathrm{l}} = 0.7$ in equation (3.13), the period and magnitude of the lynx and hare population cycles approximately match the data in Figure 3.7.

measured. The load on the server is controlled by a parameter called `MaxUsers`, which sets the total number of client connections to the server. This parameter is controlled by the system administrator. The server can be regarded as a dynamical system with `MaxUsers` as the input and `RIS` as the output. The relationship between input and output was first investigated by exploring the steady-state performance and was found to be linear.

In [117] a dynamical model in the form of a first-order difference equation is used to capture the dynamic behavior of this system. Using system identification techniques, they construct a model of the form

$$y[k + 1] = a y[k] + b u[k],$$

where $u = \texttt{MaxUsers} - \overline{\texttt{MaxUsers}}$ and $y = \texttt{RIS} - \overline{\texttt{RIS}}$. The parameters $a = 0.43$ and $b = 0.47$ are parameters that describe the dynamics of the system around the operating point, and $\overline{\texttt{MaxUsers}} = 165$ and $\overline{\texttt{RIS}} = 135$ represent the nominal operating point of the system. The number of requests was averaged over a sampling period of 60 s. ∇

Another application of difference equations is in the implementation of control systems on computers. Early controllers were analog physical systems, which can be modeled by differential equations. When implementing a controller described by a differential equation using a computer it is necessary to do approximations. A simple way is to approximate derivatives by finite differences, as illustrated by the following example.

Example 3.6 Difference approximation of a PI controller
Consider the proportional-integral (PI) controller

$$u(t) = k_{\mathrm{p}} e(t) + k_{\mathrm{i}} \int_0^t e(\tau)\, d\tau = k_{\mathrm{p}} e(t) + x(t), \qquad x(t) = k_{\mathrm{i}} \int_0^t e(\tau)\, d\tau,$$

where the controller state is given by the differential equation

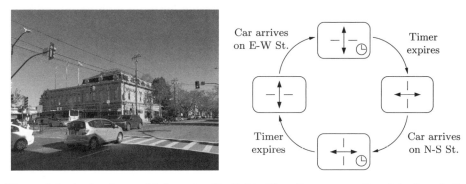

Figure 3.9: A simple model for a traffic light. The diagram on the right is a finite state machine model of the traffic light controller.

$$\frac{dx}{dt} = k_i e(t). \tag{3.14}$$

Assume that the error is measured at regular sampling intervals $t = h, 2h, 3h, \ldots$. Approximating the derivative in equation (3.14) by differences gives

$$\frac{x(jh + h) - x(jh)}{h} = k_i e(jh),$$

and the controller is then given by the difference equation

$$x[j + 1] = x[j] + h k_i e[j], \qquad u[j] = k_p e[j] + x[j],$$

where $x[j] = x(jh)$, $e[j] = e(jh)$, and $u[j] = u(jh)$ represent the discrete-time state, error, and input sampled at each time interval (and we use j as our discrete time index here to avoid confusion with the gains k_p and k_i). This controller is easy to implement on a computer since it consists of just addition and multiplication.   ∇

The approximation in the example works well provided that the sampling interval is so short that the variable $e(t)$ changes very little over a sampling interval.

Finite State Machines

In addition to systems that can be modeled by continuous variables (e.g., positions, velocities, voltages, temperatures), we often encounter systems that have discrete states (e.g., on, off, standby, fault). A finite state machine is a model in which the states of the system are chosen from a finite list of "modes." The dynamics of a finite state machine are given by transitions between these modes, possibly in response to external signals. We illustrate this concept with a simple example.

Example 3.7 Traffic light controller

Consider a finite state machine model of a traffic light control system, as shown in Figure 3.9. We represent the state of the system in terms of the set of traffic lights that are turned on (either east–west or north–south). In addition, once a light is turned on it should stay that way for a certain minimum time, and then only change when a car comes up to the intersection in the opposite direction. This gives us two states for each direction of the lights: waiting for a car to arrive and waiting for the timer to expire. Thus, we have four states for the system, as shown in Figure 3.9.

The dynamics for the light describe how the system transitions from one state to another. Starting at the leftmost state, we assume that the lights are set to allow traffic in the north–south direction. When a car arrives on the east–west street, we transition to the state at the top of the diagram, where a timer is started. Once the timer reaches the designated amount of time, we transition to the state on the right side of the diagram and turn on the lights in the east–west direction. From here we wait until a car arrives on the north–south street and continue the cycle.

Viewed as a control system, this model has a state space consisting of four discrete states: north–south waiting, north–south countdown, east–west waiting, and east–west countdown. The inputs to the controller consist of the signals that indicate whether a car is present at the roads leading up to the intersection. The outputs from the controller are the signals that change the colors of the traffic light. Finally, the dynamics of the controller are the transition diagram that controls how the states (or modes) of the system change in time.   ▽

More formally, a finite state machine can be represented as a finite set of discrete states $\alpha \in \mathbb{Q}_{\text{sys}}$, where $\mathbb{Q}_{\text{sys}}$ is a discrete set. The dynamics of the system are described by transitions between the discrete states, as in the finite state machine described in the previous example. These transitions can depend on external inputs or measurements and can generate output actions on transition into or out of a given state. If we let $\beta \in \mathbb{Q}_{\text{in}}$ represent (discrete) input events (button press, component failure, etc.) and $\gamma \in \mathbb{Q}_{\text{out}}$ represent (discrete) output actions (such as turning off a device), then the dynamics of the finite state machine can be written as a guarded command system

$$g_i(\alpha, \beta) \implies \begin{cases} \alpha' = r_i(\alpha, \beta), \\ \gamma = a_i(\alpha, \beta), \end{cases} \qquad i = 1, \ldots, N. \qquad (3.15)$$

Here the function g_i is a Boolean (true/false) function that depends on the current system mode α and an external input β. If the guard g_i is true then the system transitions from the current state α to a new state α', determined by the rule (transition map) r_i and the external input. The output action a_i is similarly dependent on the current state and external input. A guarded transition system can have many different rules, depending on the system state and external input.

The dynamics of a transition system is similar in many ways to the discrete time dynamics in equation (3.11). The major difference is that the transitions do not necessarily occur at regularly spaced intervals of time. Indeed, there is no strict notion of time in a transition system as we have described it here: it is only the sequence of events that is kept track of (through the evolution of the discrete state).

Specifications for finite transition systems are often written as logical functions describing the conditions that should be imposed on the system. For example, we might wish to say that if a specific sensor is not operating, then the system cannot transition to a mode that requires the use of that sensor. This could be written as the logical formula

$$\alpha \in \{\text{states with sensor } k \text{ not functioning}\} \implies \alpha' \notin \{\text{states requiring sensor } k\}.$$

The formula of the form $p \implies q$ where p and q are Boolean propositions can be written as the logical function $(!p) \,\|\, (p \,\&\&\, q)$, which asserts that if proposition p is

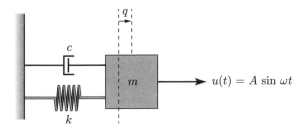

Figure 3.10: A driven spring–mass system with damping. Here we use a linear damping element with coefficient of viscous friction c. The mass is driven with a sinusoidal force of amplitude A.

true then proposition q must be true. In the sensor example, p and q are represented by whether the system mode α is in some set of states.

Finite state machines are very useful for describing logical operations and are often combined with continuous state models (differential or difference equations) to create a hybrid system model. The study of hybrid systems is beyond the scope of this text, but excellent references include Lee and Seshia [161] and Alur [8].

Simulation and Analysis

State space models can be used to answer many questions. One of the most common, as we have seen in the previous examples, involves predicting the evolution of the system state from a given initial condition. While for simple models this can be done in closed form, more often it is accomplished through computer simulation.

Consider again the damped spring–mass system from Section 3.1, but this time with an external force applied, as shown in Figure 3.10. We wish to predict the motion of the system for a periodic forcing function, with a given initial condition, and determine the amplitude, frequency, and decay rate of the resulting motion.

We choose to model the system with a linear ordinary differential equation. Using Hooke's law to model the spring and assuming that the damper exerts a force that is proportional to the velocity of the system, we have

$$m\ddot{q} + c\dot{q} + kq = u, \tag{3.16}$$

where m is the mass, q is the displacement of the mass, c is the coefficient of viscous friction, k is the spring constant, and u is the applied force. In state space form, using $x = (q, \dot{q})$ as the state and choosing $y = q$ as the output, we have

$$\frac{dx}{dt} = \begin{pmatrix} x_2 \\ -\dfrac{c}{m}x_2 - \dfrac{k}{m}x_1 + \dfrac{u}{m} \end{pmatrix}, \qquad y = x_1.$$

We see that this is a linear second-order differential equation with one input u and one output y.

We now wish to compute the response of the system to an input of the form $u = A\sin\omega t$. Although it is possible to solve for the response analytically, we instead make use of a computational approach that does not rely on the specific form of

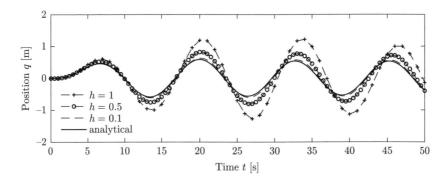

Figure 3.11: Simulation of the forced spring–mass system with different simulation time constants. The solid line represents the analytical solution. The dashed lines represent the approximate solution via the method of Euler integration, using decreasing step sizes.

this system. Consider the general state space system

$$\frac{dx}{dt} = f(x, u).$$

Given the state x at time t, we can approximate the value of the state at a short time $h > 0$ later by assuming that the rate of change $f(x, u)$ is constant over the interval t to $t + h$. This gives

$$x(t + h) = x(t) + h f(x(t), u(t)). \tag{3.17}$$

Iterating this equation, we can thus solve for x as a function of time. This approximation is known as Euler integration and is in fact a difference equation if we let h represent the time increment and write $x[k] = x(kh)$, as we saw in Example 3.6. Although modern simulation tools such as MATLAB and Mathematica use more accurate methods than Euler integration, they still have some of the same basic trade-offs.

Returning to our specific example, Figure 3.11 shows the results of computing $x(t)$ using equation (3.17), along with the analytical computation. We see that as h gets smaller, the computed solution converges to the exact solution. The form of the solution is also worth noticing: after an initial transient, the system settles into a periodic motion. The portion of the response after the transient is called the *steady-state response* to the input.

In addition to generating simulations, models can also be used to answer other types of questions. Two that are central to the methods described in this text concern the stability of an equilibrium point and the input/output frequency response. We illustrate these two computations through the examples below and return to the general computations in later chapters.

Returning to the damped spring–mass system, the equations of motion with no input forcing are given by

$$\frac{dx}{dt} = \begin{pmatrix} x_2 \\ -\dfrac{c}{m} x_2 - \dfrac{k}{m} x_1 \end{pmatrix}, \tag{3.18}$$

where x_1 is the position of the mass (relative to the rest position) and x_2 is its velocity. We wish to show that if the initial state of the system is away from the rest position, the system will return to the rest position eventually (we will later define this situation to mean that the rest position is *asymptotically stable*). While we could heuristically show this by simulating many, many initial conditions, we seek instead to prove that this is true for *any* initial condition.

To do so, we construct a function $V : \mathbb{R}^n \to \mathbb{R}$ that maps the system state to a positive real number. For mechanical systems, a convenient choice is the energy of the system,

$$V(x) = \frac{1}{2}kx_1^2 + \frac{1}{2}mx_2^2. \tag{3.19}$$

If we look at the time derivative of the energy function, we see that

$$\frac{dV}{dt} = kx_1\dot{x}_1 + mx_2\dot{x}_2 = kx_1x_2 + mx_2(-\frac{c}{m}x_2 - \frac{k}{m}x_1) = -cx_2^2,$$

which is always either negative or zero. Hence $V(x(t))$ is never increasing and, using a bit of analysis that we will see formally later, the individual states must remain bounded.

If we wish to show that the states eventually return to the origin, we must use a slightly more detailed analysis. Intuitively, we can reason as follows: suppose that for some period of time, $V(x(t))$ stops decreasing. Then it must be true that $\dot{V}(x(t)) = 0$, which in turn implies that $x_2(t) = 0$ for that same period. In that case, $\dot{x}_2(t) = 0$, and we can substitute into the second line of equation (3.18) to obtain

$$0 = \dot{x}_2 = -\frac{c}{m}x_2 - \frac{k}{m}x_1 = -\frac{k}{m}x_1.$$

Thus we must have that x_1 also equals zero, and so the only time that $V(x(t))$ can stop decreasing is if the state is at the origin (and hence this system is at its rest position). Since we know that $V(x(t))$ is never increasing (because $\dot{V} \leq 0$), we therefore conclude that the origin is stable (for *any* initial condition).

This type of analysis, called Lyapunov stability analysis, is considered in detail in Chapter 5. It shows some of the power of using models for the analysis of system

Another type of analysis that we can perform with models is to compute the output of a system to a sinusoidal input, known as the *frequency response*. We again consider the spring–mass system, but this time keeping the input and leaving the system in its original form:

$$m\ddot{q} + c\dot{q} + kq = u. \tag{3.20}$$

We wish to understand how the system responds to a sinusoidal input of the form

$$u(t) = A\sin\omega t.$$

We will see how to do this analytically in Chapter 7, but for now we make use of simulations to compute the answer.

We first begin with the observation that if $q(t)$ is the solution to equation (3.20) with input $u(t)$, then applying an input $2u(t)$ will give a solution $2q(t)$ (this is easily verified by substitution). Hence it suffices to look at an input with unit magnitude, $A = 1$. A second observation, which we will prove in Chapter 6, is that the long-term response of the system to a sinusoidal input is itself a sinusoid at the same

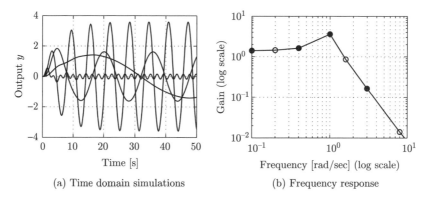

(a) Time domain simulations   (b) Frequency response

Figure 3.12: A frequency response (gain only) computed by measuring the response of individual sinusoids. The figure on the left shows the response of the system as a function of time to a number of different unit magnitude inputs (at different frequencies). The figure on the right shows this same data in a different way, with the magnitude of the response plotted as a function of the input frequency. The filled circles correspond to the particular frequencies shown in the time responses.

frequency, and so the output has the form

$$q(t) = g(\omega) \sin(\omega t + \varphi(\omega)),$$

where $g(\omega)$ is called the *gain* of the system and $\varphi(\omega)$ is called the *phase* (or phase offset).

To compute the frequency response numerically, we can simulate the system at a set of frequencies $\omega_1, \ldots, \omega_N$ and plot the gain and phase at each of these frequencies. An example of this type of computation is shown in Figure 3.12. For linear systems the frequency response does not depend on the amplitude A of the input signal. Frequency response can also be applied to nonlinear systems but the gain and phase then depend on the A.

3.3   MODELING METHODOLOGY

To deal with large, complex systems, it is useful to have different representations of the system that capture essential features and hide irrelevant details. In all branches of science and engineering it is common practice to use some graphical description of systems, called *schematic diagrams*. They can range from stylistic pictures to drastically simplified standard symbols. These pictures make it possible to get an overall view of the system and to identify the individual components. Examples of such diagrams are shown in Figure 3.13. Schematic diagrams are useful because they give an overall picture of a system, showing different subprocesses and their interconnection and indicating variables that can be manipulated and signals that can be measured.

Block Diagrams

A special graphical representation called a *block diagram* has been developed in control engineering. The purpose of a block diagram is to emphasize the information

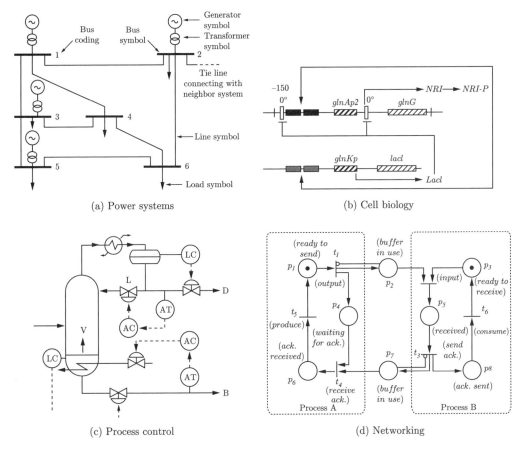

Figure 3.13: Schematic diagrams for different disciplines. Each diagram is used to illustrate the dynamics of a control system: (a) electrical schematics for a power system [156]; (b) a biological circuit diagram for a synthetic clock circuit [26]; (c) a process diagram for a distillation column [219]; and (d) a Petri net description of a communication protocol.

flow and to hide details of the system. In a block diagram, different process elements are shown as boxes, and each box has inputs denoted by lines with arrows pointing toward the box and outputs denoted by lines with arrows going out of the box. The inputs denote the variables that influence a process, and the outputs denote the signals that we are interested in or signals that influence other subsystems. Block diagrams can also be organized in hierarchies, where individual blocks may themselves contain more detailed block diagrams.

Figure 3.14 shows some of the notation that we use for block diagrams. Signals are represented as lines, with arrows to indicate inputs and outputs. The first diagram is the representation for a summation of two signals. An input/output response is represented as a rectangle with the system name (or mathematical description) in the block. Two special cases are a proportional gain, which scales the input by a multiplicative factor, and an integrator, which outputs the integral of the input signal.

Figure 3.15 illustrates the use of a block diagram, in this case for modeling the flight response of a fly. The flight dynamics of an insect are incredibly intricate,

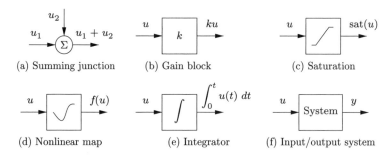

(a) Summing junction (b) Gain block (c) Saturation

(d) Nonlinear map (e) Integrator (f) Input/output system

Figure 3.14: Standard block diagram elements. The arrows indicate the inputs and outputs of each element, with the mathematical operation corresponding to the block labeled at the output. The system block (f) represents the full input/output response of a dynamical system.

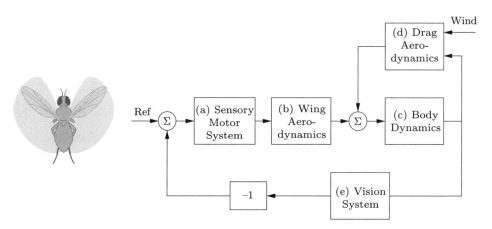

Figure 3.15: A block diagram representation of the flight control system for an insect flying against the wind. The mechanical portion of the model consists of the rigid-body dynamics of the fly, the drag due to flying through the air, and the forces generated by the wings. The motion of the body causes the visual environment of the fly to change, and this information is then used to control the motion of the wings (through the sensory motor system), closing the loop.

involving careful coordination of the muscles within the fly to maintain stable flight in response to external stimuli. One known characteristic of flies is their ability to fly upwind by making use of the optical flow in their compound eyes as a feedback mechanism. Roughly speaking, the fly controls its orientation so that the point of contraction of the visual field is centered in its visual field [207].

To understand this complex behavior, we can decompose the overall dynamics of the system into a series of interconnected subsystems (or *blocks*). Referring to Figure 3.15, we can model the insect navigation system through an interconnection of five blocks. The sensory motor system (a) takes the information from the visual system (e) and generates muscle commands that attempt to steer the fly so that the point of contraction is centered. These muscle commands are converted into forces through the flapping of the wings (b) and the resulting aerodynamic forces that are produced. The forces from the wings are combined with the drag on the fly (d) to produce a net force on the body of the fly. The wind velocity enters through the drag aerodynamics. Finally, the body dynamics (c) describe how the fly translates and rotates as a function of the net forces that are applied to it. The insect position, speed, and orientation are fed back to the drag aerodynamics and vision system blocks as inputs.

Each of the blocks in the diagram can itself be a complicated subsystem. For example, the visual system of a fruit fly consists of two complicated compound eyes (with about 700 elements per eye), and the sensory motor system has about 200,000 neurons that are used to process information. A more detailed block diagram of the insect flight control system would show the interconnections between these elements, but here we have used one block to represent how the motion of the fly affects the output of the visual system, and a second block to represent how the visual field is processed by the fly's brain to generate muscle commands. The choice of the level of detail of the blocks and what elements to separate into different blocks often depends on experience and on the questions that one wants to answer using the model. One of the powerful features of block diagrams is their ability to hide information about the details of a system that may not be needed to gain an understanding of the essential dynamics of the system.

Algebraic Loops

When analyzing or simulating a system described by a block diagram, we need to form the differential equations that describe the complete system. In many cases the equations can be obtained by combining the differential equations that describe each subsystem and substituting variables. This simple procedure cannot be used when there are closed loops of subsystems that all have a direct connection between inputs and outputs, known as an *algebraic loop*. A *direct connection* means that a change in the input u gives an instantaneous change in the output y.

To see what can happen, consider a system with two blocks, a first-order non-linear system,

$$\frac{dx}{dt} = f(x, u), \qquad y = h(x), \tag{3.21}$$

and a proportional controller described by $u = -ky$. There is no direct connection since the function h does not depend on u. In that case we can obtain the equation for the closed loop system simply by replacing u by $-ky = -kh(x)$ in equation (3.21) to give

$$\frac{dx}{dt} = f(x, -kh(x)), \qquad y = h(x),$$

which is an ordinary differential equation.

The situation is more complicated if there is a direct connection. If $y = h(x, u)$, then replacing u by $-ky$ gives

$$\frac{dx}{dt} = f(x, -ky), \qquad y = h(x, -ky).$$

To obtain a differential equation for x, the algebraic equation $y = h(x, -ky)$ must first be solved to give $y = \alpha(x)$, which in general is a complicated task.

When algebraic loops are present, it is necessary to solve algebraic equations to obtain the differential equations for the complete system. The resulting model becomes a set of differential algebraic equations, similar to equation (3.4). Resolving algebraic loops is a nontrivial problem because it requires the symbolic solution of algebraic equations. Most block diagram-oriented modeling languages cannot handle algebraic loops, and they simply give a diagnosis that such loops are present. In the era of analog computing, algebraic loops were eliminated by introducing fast dynamics between the loops. This created differential equations with fast and slow modes that are difficult to solve numerically. Advanced modeling languages like Modelica use several sophisticated methods to resolve algebraic loops.

Modeling from Experiments

Since control systems are provided with sensors and actuators, it is also possible to obtain models of system dynamics from experiments on the process. The models are restricted to input/output models since only these signals are accessible to experiments, but modeling from experiments can also be combined with modeling from physics through the use of feedback and interconnection.

A simple way to determine a system's dynamics is to observe the response to a step change in the control signal. Such an experiment begins by setting the control signal to a constant value. When the output settles to a constant value (assuming the system is stable), the control signal is changed quickly to a new level and the output is observed. The experiment gives the step response of the system, and the shape of the response gives useful information about the dynamics. It immediately gives an indication of the response time, and it tells if the system is oscillatory or if the response is monotone.

Example 3.8 Spring–mass system

The dynamics of the spring–mass system in Section 3.1 are given by

$$m\ddot{q} + c\dot{q} + kq = u. \tag{3.22}$$

We wish to determine the constants m, c, and k by measuring the response of the system to a step input of magnitude F_0.

We will show in Chapter 7 that when $c^2 < 4km$, the step response for this system from the rest configuration is given by

$$q(t) = \frac{F_0}{k}\left(1 - \frac{1}{\omega_d}\sqrt{\frac{k}{m}}\exp\left(-\frac{ct}{2m}\right)\sin(\omega_d t + \varphi)\right),$$

$$\omega_d = \frac{\sqrt{4km - c^2}}{2m}, \qquad \varphi = \tan^{-1}\left(\frac{\sqrt{4km - c^2}}{c}\right).$$

From the form of the solution, we see that the shape of the step response is determined by the parameters of the system. Hence, by measuring certain features of the step response we can determine the parameter values.

Figure 3.16 shows the response of the system to a step of magnitude $F_0 = 20$ N, along with some measurements. We start by noting that the steady-state position of the mass (after the oscillations die down) is a function of the spring constant k:

$$q(\infty) = \frac{F_0}{k}, \tag{3.23}$$

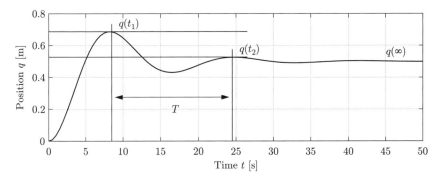

Figure 3.16: Step response for a spring–mass system. The magnitude of the step input is $F_0 = 20$ N. The period of oscillation T is determined by looking at the time between two subsequent local maxima in the response. The period combined with the steady-state value $q(\infty)$ and the relative decrease between local maxima can be used to estimate the parameters in a model of the system.

where F_0 is the magnitude of the applied force ($F_0 = 1$ for a unit step input). The parameter $1/k$ is called the *gain* of the system. The period of the oscillation can be measured between two peaks and must satisfy

$$\frac{2\pi}{T} = \frac{\sqrt{4km - c^2}}{2m}.$$

Finally, the rate of decay of the oscillations is given by the exponential factor in the solution. Measuring the amount of decay between two peaks, we have

$$\log\left(q(t_1) - \frac{F_0}{k}\right) - \log\left(q(t_2) - \frac{F_0}{k}\right) = \frac{c}{2m}(t_2 - t_1). \tag{3.24}$$

Using this set of three equations, we can solve for the parameters and determine that for the step response in Figure 3.16 we have $m \approx 250$ kg, $c \approx 60$ N s/m, and $k \approx 40$ N/m. ∇

   Modeling from experiments can also be done using many other signals. Sinusoidal signals are commonly used (particularly for systems with fast dynamics) and precise measurements can be obtained by exploiting correlation techniques. An indication of nonlinearities can be obtained by repeating experiments with input signals having different amplitudes. Modeling based on sinusoidal signals is very time consuming for systems with slow dynamics. In such situations it is advantageous to used signals that switch between two different levels. There is a whole subfield of control called *system identification* that deals with experimental determination of models. Questions like optimal inputs, experiments in open and closed loop, model accuracy, and fundamental limits are dealt with extensively.

Normalization and Scaling

When deriving a model, it is often useful to introduce dimension-free variables. Such a procedure can often simplify the equations for a system by reducing the number of parameters. It can also reveal interesting properties of the model. It is also useful to normalize variables by scaling to improve numerics and allow faster and more accurate simulations.

The procedure of scaling is straightforward in principle: choose units for each independent variable and introduce new variables by dividing the variables by the chosen normalization unit. We illustrate the procedure with two examples.

Example 3.9 Spring–mass system

Consider again the spring–mass system introduced earlier. Neglecting the damping, the system is described by

$$m\ddot{q} + kq = u.$$

The model has two parameters m and k. To normalize the model we introduce dimension-free variables $x = q/l$ and $\tau = \omega_0 t$, where $\omega_0 = \sqrt{k/m}$ and l is the chosen length scale. We scale force by $ml\omega_0^2$ and introduce $v = u/(ml\omega_0^2)$. The scaled equation then becomes

$$\frac{d^2 x}{d\tau^2} = \frac{d^2 q/l}{d(\omega_0 t)^2} = \frac{1}{ml\omega_0^2}(-kq + u) = -x + v,$$

which is the normalized undamped spring–mass system. Notice that the normalized model has no parameters, while the original model had two parameters m and k. Introducing the scaled, dimension-free state variables $z_1 = x = q/l$ and $z_2 = dx/d\tau = \dot{q}/(l\omega_0)$, the model can be written as

$$\frac{d}{d\tau} \begin{pmatrix} z_1 \\ z_2 \end{pmatrix} = \begin{pmatrix} 0 & 1 \\ -1 & 0 \end{pmatrix} \begin{pmatrix} z_1 \\ z_2 \end{pmatrix} + \begin{pmatrix} 0 \\ v \end{pmatrix}.$$

This simple linear equation describes the dynamics of any spring–mass system, independent of the particular parameters, and hence gives us insight into the fundamental dynamics of this oscillatory system. To recover the physical frequency of oscillation or its magnitude, we must invert the scaling we have applied. ∇

Example 3.10 Balance system

Consider the balance system described in Example 3.2. Neglecting damping by putting $c = 0$ and $\gamma = 0$ in equation (3.9), the model can be written as

$$(M + m)\frac{d^2 q}{dt^2} - ml\cos\theta\frac{d^2\theta}{dt^2} + ml\sin\theta\Big(\frac{d\theta}{dt}\Big)^2 = F,$$

$$-ml\cos\theta\frac{d^2 q}{dt^2} + (J + ml^2)\frac{d^2\theta}{dt^2} - mgl\sin\theta = 0.$$

Let $\omega_0 = \sqrt{mgl/(J + ml^2)}$, choose the length scale as l, let the time scale be $1/\omega_0$, choose the force scale as $(M + m)l\omega_0^2$, and introduce the scaled variables $\tau = \omega_0 t$, $x = q/l$, and $u = F/((M + m)l\omega_0^2)$. The equations then become

$$\frac{d^2 x}{d\tau^2} - \alpha\cos\theta\frac{d^2\theta}{d\tau^2} + \alpha\sin\theta\Big(\frac{d\theta}{d\tau}\Big)^2 = u, \qquad -\beta\cos\theta\frac{d^2 x}{d\tau^2} + \frac{d^2\theta}{d\tau^2} - \sin\theta = 0,$$

where $\alpha = m/(M + m)$ and $\beta = ml^2/(J + ml^2)$. Notice that the original model has five parameters m, M, J, l, and g but the normalized model has only two parameters α and β. If $M \gg m$ and $ml^2 \gg J$, we get $\alpha \approx 0$ and $\beta \approx 1$, and the model can be approximated by

$$\frac{d^2x}{d\tau^2} = u, \qquad \frac{d^2\theta}{d\tau^2} - \sin\theta = u\cos\theta.$$

The model can be interpreted as a mass combined with an inverted pendulum driven by the same input. ∇

For large systems scaling is not so easy: there are many choices and good selection of variables and normalization units require good understanding of the physics of the system and the numerical methods that will be used for analysis. Scaling of large systems is therefore still an art.

3.4   MODELING EXAMPLES

In this section we introduce additional examples that illustrate some of the different types of systems for which one can develop differential equation and difference equation models. These examples are specifically chosen from a range of different fields to highlight the broad variety of systems to which feedback and control concepts can be applied. A more detailed set of applications that serve as running examples throughout the text are given in Chapter 4.

Motion Control Systems

Motion control systems involve the use of computation and feedback to control the movement of a mechanical system. Motion control systems range from nanopositioning systems (atomic force microscopes, adaptive optics), to control systems for the read/write heads in a disk drive of a DVD player, to manufacturing systems (transfer machines and industrial robots), to automotive control systems (antilock brakes, suspension control, traction control), to air and space flight control systems (airplanes, satellites, rockets, and planetary rovers).

Example 3.11 Vehicle steering

A common problem in motion control is to control the trajectory of a vehicle through an actuator that causes a change in the orientation. A steering wheel on an automobile and the front wheel of a bicycle are two examples, but similar dynamics occur in the steering of ships or control of the pitch dynamics of an aircraft. In many cases, we can understand the basic behavior of these systems through the use of a simple model that captures the basic kinematics of the system.

Consider a conventional vehicle with a fixed rear axle and a set of front wheels that can be rotated, as shown in Figure 3.17. For the purpose of steering we are interested in a model that describes how the velocity of the vehicle depends on the steering angle δ. To be specific, let b be the wheelbase and consider the velocity v at the center of mass, a distance a from the rear wheel, as shown in Figure 3.17. Let x and y be the coordinates of the center of mass, θ the heading angle, and α the angle between the velocity vector v and the centerline of the vehicle. The point O is at the intersection of the normals to the front and rear wheels.

Assuming no slipping of the wheels, the motion of the vehicle is given by a rotation around the point O in the figure. Letting the distance from the center

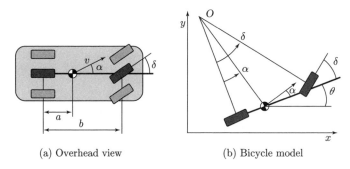

(a) Overhead view   (b) Bicycle model

Figure 3.17: Vehicle steering dynamics. The left figure shows an overhead view of a
vehicle with four wheels. The wheelbase is b and the center of mass at a distance a
forward of the rear wheels. By approximating the motion of the front and rear pairs
of wheels by a single front wheel and a single rear wheel, we obtain an abstraction
called the *bicycle model*, shown on the right. The steering angle is δ and the velocity
at the center of mass has the angle α relative to the length axis of the vehicle. The
position of the vehicle is given by (x, y) and the orientation (heading) by θ.

of rotation O to the contact point of the rear wheel be r_{r}, it the follows from
Figure 3.17 that $b = r_{\mathrm{r}} \tan \delta$ and $a = r_{\mathrm{r}} \tan \alpha$, which implies that $\tan \alpha = (a/b) \tan \delta$,
and we obtain the following relation between α and the steering angle δ:

$$\alpha = \arctan \left(\frac{a \tan \delta}{b} \right). \tag{3.25}$$

If the vehicle speed at its center of mass is v, the motion of the center of mass is
then given by

$$\frac{dx}{dt} = v \cos (\alpha + \theta),$$
$$\frac{dy}{dt} = v \sin (\alpha + \theta). \tag{3.26}$$

To see how the heading angle θ is influenced by the steering angle, we observe from
Figure 3.17 that the distance from the center of mass to the center of rotation O is
$r_{\mathrm{c}} = a / \sin \alpha$. The vehicle thus rotates around the point O with the angular velocity
$v/r_{\mathrm{c}} = (v/a) \sin \alpha$. Hence

$$\frac{d\theta}{dt} = \frac{v}{r_{\mathrm{c}}} = \frac{v \sin \alpha}{a} = \frac{v}{a} \sin \left(\arctan \left(\frac{a \tan \delta}{b} \right) \right) \approx \frac{v}{b} \delta, \tag{3.27}$$

where the approximation holds for small δ and α.

Equations (3.25)–(3.27) can be used to model an automobile under the assumptions that there is no slip between the wheels and the road and that the two front
wheels can be approximated by a single wheel at the center of the car. This model is
often called the *bicycle model*. The assumption of no slip can be relaxed by adding
an extra state variable, giving a more realistic model. Such a model also describes
the steering dynamics of ships as well as the pitch dynamics of aircraft and missiles.
It is also possible to choose coordinates so that the reference point is at the rear
wheels (corresponding to setting $a = 0$), a model often referred to as the *Dubins
car* [79].

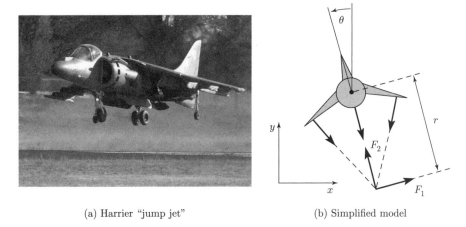

(a) Harrier "jump jet"   (b) Simplified model

Figure 3.18: Vectored thrust aircraft. The Harrier AV-8B military aircraft (a) redirects its engine thrust downward so that it can "hover" above the ground. Some air from the engine is diverted to the wing tips to be used for maneuvering. As shown in (b), the net thrust on the aircraft can be decomposed into a horizontal force F_1 and a vertical force F_2 acting at a distance r from the center of mass.

Figure 3.17 represents the situation when the vehicle moves forward and has front-wheel steering. The figure shows that the model also applies to rear wheel steering if the sign of the velocity is reversed. ∇

Example 3.12 Vectored thrust aircraft
Consider the motion of vectored thrust aircraft, such as the Harrier "jump jet" shown Figure 3.18a. The Harrier is capable of vertical takeoff by redirecting its thrust downward and through the use of smaller maneuvering thrusters located on its wings. A simplified model of the Harrier is shown in Figure 3.18b, where we focus on the motion of the vehicle in a vertical plane through the wings of the aircraft. We resolve the forces generated by the main downward thruster and the maneuvering thrusters as a pair of forces F_1 and F_2 acting at a distance r below the aircraft (determined by the geometry of the thrusters).

Let (x, y, θ) denote the position and orientation of the center of mass of the aircraft. Let m be the mass of the vehicle, J the moment of inertia, g the gravitational constant, and c the damping coefficient. Then the equations of motion for the vehicle are given by

$$m\ddot{x} = F_1 \cos\theta - F_2 \sin\theta - c\dot{x},$$
$$m\ddot{y} = F_1 \sin\theta + F_2 \cos\theta - mg - c\dot{y}, \qquad (3.28)$$
$$J\ddot{\theta} = rF_1.$$

It is convenient to redefine the inputs so that the origin is an equilibrium point of the system with zero input. Letting $u_1 = F_1$ and $u_2 = F_2 - mg$, the equations become

$$m\ddot{x} = -mg\sin\theta - c\dot{x} + u_1\cos\theta - u_2\sin\theta,$$
$$m\ddot{y} = mg(\cos\theta - 1) - c\dot{y} + u_1\sin\theta + u_2\cos\theta, \qquad (3.29)$$
$$J\ddot{\theta} = ru_1.$$

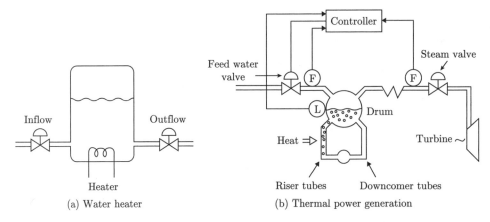

Figure 3.19: Two thermofluid systems. A schematic diagram of a simple water heater, a tank with a submerged electrical heater (a), and schematic diagram of a drum boiler (b).

These equations describe the motion of the vehicle as a set of three coupled second-order differential equations.   ▽

Thermofluid Systems

Thermofluid systems are commonly used in process control, power generation, and for heating ventilation and air conditioning in buildings and cars. The processes involve motion of fluids and transmission of energy; typical processes include heat exchangers, evaporators, chillers, and compressors. The dynamics are often complicated because of two-phase flows, and accurate modeling often requires partial differential equations and computational fluid dynamics. Two examples are given in Figure 3.19.

Example 3.13 Water heater

Consider the water heater in Figure 3.19a, which is a cylindrical tank with cross section A. The mass of the water is m and its temperature is T. The inflow and outflow rates are q_{in} and q_{out}, the temperature of the inflow is T_{in}, and the temperature of the outflow is T. The total mass is $m = \rho A h$, where ρ is its density, h is the water level, C is the specific heat capacity for water, and mCT is the total energy. The system can be modeled by a mass balance and an energy balance, and we obtain

$$\frac{dm}{dt} = q_{in} - q_{out}, \qquad \frac{d(m\,CT)}{dt} = P + q_{in}\,CT_{in} - q_{out}\,CT, \qquad (3.30)$$

where P is the power from the heater. Energy losses have been neglected and it is assumed that all water in the tank has the same temperature.

Assuming that C is constant and expanding the derivative for the energy balance we obtain

$$\frac{d(m\,CT)}{dt} = \frac{dm}{dt}\,CT + m\,C\frac{dT}{dt} = P + q_{in}\,CT_{in} - q_{out}\,CT.$$

Solving this equation for dT/dt and using the mass balance to eliminate dm/dt, we find that the mass and energy balances expressed by equation (3.30) can be written as

$$\frac{dm}{dt} = q_{\text{in}} - q_{\text{out}}, \qquad \frac{dT}{dt} = \frac{q_{\text{in}}}{m}(T_{\text{in}} - T) + \frac{1}{mC}P. \tag{3.31}$$

The state variables are the total mass m and the temperature T, the control (input) variables are the input power P and inflow rate q_{in}, and the disturbances are the temperature of the inflow T_{in} and the output flow rate q_{out}. ∇

Example 3.14 Drum Boiler

A drum boiler is a piece of equipment used to produce steam, for example as part of a power generation system where the steam drives a turbine connected to a generator. The drum in a drum boiler shares many properties with the water heater but there are two significant complications: the material constants ρ and C depend on the state, and there is a mixture of water and steam in both the riser and the drum. Modeling can still be done by mass and energy balances, but the two-phase flow leads to significant complications, which we discuss briefly (and informally) here. A diagram of a drum boiler is shown in Figure 3.19b.

Control of the drum level is a key problem: if the level is too low the tubes will burn through, and if the level is too high water may enter the turbine and cause damage to the turbine blades. We will focus on modeling of the drum level. Water entering the system is controlled by the feedwater valve; water leaves the drum as steam through the steam valve. Water circulates through the drum-downcomer-riser loop, and it is heated in the riser tubes. The differences in densities in the downcomer tubes and the riser tubes creates self-circulation. The figure shows only one riser tube and one downcomer tube, but in the boiler we discuss there are 22 downcomer tubes and 788 riser tubes, and the drum volume is $40\,\text{m}^3$. There is pure water in the downcomer tubes and at the bottom of the riser tubes. Steam is generated by heating the tubes and the amount of steam increases along the riser tubes. There is a mixture of steam and water in the drum.

Consider the situation when the system is in equilibrium and the steam valve is suddenly opened. More steam then leaves the system, and we may expect the drum level to decrease. This will not happen because the pressure in the drum will decrease when steam leaves the system. The air bubbles in the riser and the drum will then increase, and the water level will initially increase. If we continue to keep the steam valve open, the level will finally start to decrease. The dynamics relating drum level to feedwater flow has a similar characteristic. If feedwater flow is increased then the water temperature in the drum will decrease, bubbles will collapse, and the drum level will initially decrease. This effect, which is called *shrink and swell* or *inverse response*, makes it difficult to control the drum level.

The effect is illustrated in Figure 3.20, which shows simulated and experimental data for a medium sized boiler. The inverse response characteristics are clearly seen in the figure. The model used in the simulation is a fifth-order model based on mass, energy, and momentum balances; details are given in [18].

The inverse response character of the dynamics from feedwater to drum level makes it difficult to control the drum level. For this reason the system is provided with sensors of steam flow and feedwater flow as indicated in Figure 3.19b. The extra sensors make it possible to predict whether the mass of water and steam in

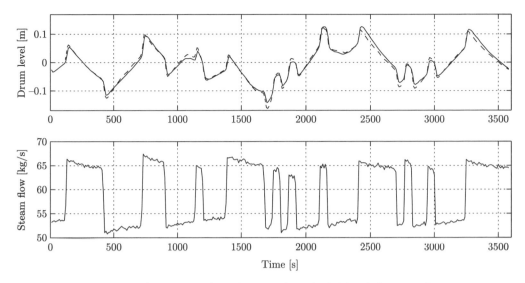

Figure 3.20: Model (dashed line) and plant data (solid line) for open loop perturbations in steam flow rate at medium load. Notice that the drum level increases initially when the steam flow is increased. The experiment was performed by removing all controllers and introducing a perturbation in the steam flow [18].

the system is decreasing or increasing. We will discuss the consequences of having dynamics with inverse response in Section 14.4.   ∇

Information Systems

Information systems range from communication systems like the Internet to software systems that manipulate data or manage enterprise-wide resources. Feedback is present in all these systems, and designing strategies for routing, flow control, and buffer management is a typical problem. Many results in queuing theory emerged from design of telecommunication systems and later from development of the Internet and computer communication systems [43, 149, 218]. Management of queues to avoid congestion is a central problem and we will therefore start by discussing the modeling of queuing systems.

Example 3.15 Queuing systems

A schematic picture of a simple queue is shown in Figure 3.21. Requests arrive and are then queued and processed. There can be large variations in arrival rates and service rates, and the queue length builds up when the arrival rate is larger than the service rate. When the queue becomes too large, service is denied using an admission control policy.

The system can be modeled in many different ways. One way is to model each incoming request, which leads to an event-based, discrete-state model where the state is an integer that represents the queue length. The queue changes when a request arrives or a request is serviced. The statistics of arrival and servicing are typically modeled as random processes. In many cases it is possible to determine statistics of quantities like queue length and service time, but the computations can be quite complicated.

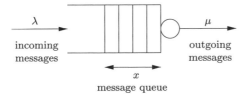

Figure 3.21: Schematic diagram of a queuing system. Messages arrive at rate λ and are stored in a queue. Messages are processed and removed from the queue at rate μ. The average length of the queue is given by $x \in \mathbb{R}$.

A significant simplification can be obtained by approximating the discrete queue length by a continuous variable. Instead of keeping track of each request we instead view service and requests as continuous flows. The model obtained is called a *flow model* because of the analogy with fluid dynamics where motion of molecules are replace by continuous flows. Hence, if the queue length x is a continuous variable and the arrivals and services are flows with rates λ and μ, the system can be modeled by the first-order differential equation

$$\frac{dx}{dt} = \lambda - \mu = \lambda - \mu_{\text{max}} f(x), \quad x \geq 0, \tag{3.32}$$

as proposed by Agnew [5]. The service rate μ depends on the queue length; if there are no capacity restrictions we have $\mu = x/T$ where T is the time it takes to serve one customer. The service rate thus increases linearly with the queue length. In reality the growth will be slower because longer queues require more resources, and the service rate has an upper limit μ_{max}. These effects are captured by modeling the service rate as $\mu_{\text{max}} f(x)$, where function $f(x)$ is monotone, approximately linear for small x, and $f(\infty) = 1$.

For a particular queue, the function $f(x)$ can be determined empirically by measuring the queue length for different arrival and service rates. A simple choice is $f(x) = x/(1+x)$, which gives the model

$$\frac{dx}{dt} = \lambda - \mu_{\text{max}} \frac{x}{x+1}. \tag{3.33}$$

It was shown by Tipper [240] that if arrival and service processes are Poisson processes, then average queue length is given by equation (3.33).

To explore the properties of the model (3.33) we will first investigate the equilibrium value of the queue length when the arrival rate λ is constant. Setting the derivative dx/dt to zero in equation (3.33) and solving for x, we find that the queue length x approaches the steady-state value

$$x_{\text{e}} = \frac{\lambda}{\mu_{\text{max}} - \lambda}. \tag{3.34}$$

Figure 3.22a shows the steady-state queue length as a function of λ/μ_{max}, the effective service rate excess. Notice that the queue length increases rapidly as λ approaches μ_{max}. To have a queue length less than 20 requires $\lambda/\mu_{\text{max}} < 0.95$. The average time to service a request can be shown to be $T_{\text{s}} = (x+1)/\mu_{\text{max}}$, and it increases dramatically as λ approaches μ_{max}.

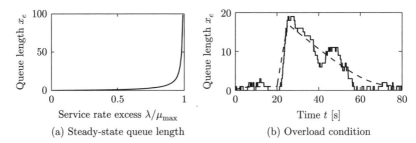

(a) Steady-state queue length   (b) Overload condition

Figure 3.22: Queuing dynamics. (a) The steady-state queue length as a function of $\lambda/\mu_{\max}$. (b) The behavior of the queue length when there is a temporary overload in the system. The solid line shows a realization of an event-based simulation, and the dashed line shows the behavior of the flow model (3.33). The maximum service rate is $\mu_{\max} = 1$, and the arrival rate starts at $\lambda = 0.5$. The arrival rate is increased to $\lambda = 4$ at time 20, and it returns to $\lambda = 0.5$ at time 25.

Figure 3.22b illustrates the behavior of the server in a typical overload situation. The figure shows that the queue builds up quickly and clears very slowly. Since the response time is proportional to queue length, it means that the quality of service is poor for a long period after an overload. This behavior is called the *rush-hour effect* and has been observed in web servers and many other queuing systems such as automobile traffic.

The dashed line in Figure 3.22b shows the behavior of the flow model, which describes the average queue length. The simple model captures behavior qualitatively, but there are variations from sample to sample when the queue length is short. ∇

Many complex systems use discrete control actions. Such systems can be modeled by characterizing the situations that correspond to each control action, as illustrated in the following example.

Example 3.16 Virtual memory paging control
An early example of the use of feedback in computer systems was applied in the operating system OS/VS for the IBM 370 [54, 67]. The system used virtual memory, which allows programs to address more memory than is physically available as fast memory. Data in current fast memory (random access memory, RAM) is accessed directly, but data that resides in slower memory (disk) is automatically loaded into fast memory. The system is implemented in such a way that it appears to the programmer as a single large section of memory. The system performed very well in many situations, but very long execution times were encountered in overload situations, as shown by the open circles in Figure 3.23a. The difficulty was resolved with a simple discrete feedback system. The load of the central processing unit (CPU) was measured together with the number of page swaps between fast memory and slow memory. The operating region was classified as being in one of three states: normal, underload, or overload. The normal state is characterized by high CPU activity, the underload state is characterized by low CPU activity and few page replacements, the overload state has moderate to low CPU load but many page replacements; see Figure 3.23b. The boundaries between the regions and the time for measuring the load were determined from simulations using typical loads.

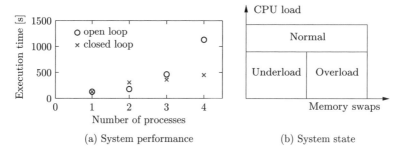

(a) System performance   (b) System state

Figure 3.23: Illustration of feedback in the virtual memory system of the IBM/370. (a) The effect of feedback on execution times in a simulation, following [54]. Results with no feedback are shown with o, and results with feedback with x. Notice the dramatic decrease in execution time for the system with feedback. (b) How the three states are obtained based on process measurements.

The control strategy was to do nothing in the normal load condition, to exclude a process from memory in the overload condition and to allow a new process or a previously excluded process in the underload condition. The crosses in Figure 3.23a show the effectiveness of the simple feedback system in simulated loads. Similar principles based on crude quantization of the state and simple heuristic algorithms are used in many other situations, e.g., in communication systems and in web server control.   ▽

Example 3.17 Consensus protocols in sensor networks
Sensor networks are used in a variety of applications where we want to collect and aggregate information over a region of space using multiple sensors that are connected together via a communications network. Examples include monitoring environmental conditions in a geographical area (or inside a building), monitoring the movement of animals or vehicles, and monitoring the resource loading across a group of computers. In many sensor networks the computational resources are distributed along with the sensors, and it can be important for the set of distributed agents to reach a consensus about a certain property, such as the average temperature in a region or the average computational load among a set of computers.

To illustrate how such a consensus might be achieved, we consider the problem of computing the average value of a set of numbers that are locally available to the individual agents. We wish to design a "protocol" (algorithm) such that all agents will agree on the average value. We consider the case in which all agents cannot necessarily communicate with each other directly, although we will assume that the communications network is connected (meaning that no two groups of agents are completely isolated from each other). Figure 3.24a shows a simple situation of this type.

We model the connectivity of the sensor network using a graph, with nodes corresponding to the sensors and edges corresponding to the existence of a direct communications link between two nodes. For any such graph, we can build an *adjacency matrix*, where each row and column of the matrix corresponds to a node and a 1 in the respective row and column indicates that the two nodes are connected.

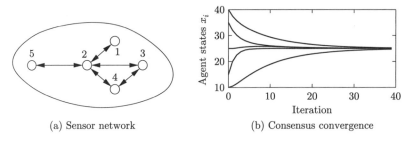

(a) Sensor network   (b) Consensus convergence

Figure 3.24: Consensus protocols for sensor networks. (a) A simple sensor network with five nodes. In this network, node 1 communicates with node 2 and node 2 communicates with nodes 1, 3, 4, 5, etc. (b) A simulation demonstrating the convergence of the consensus protocol (3.35) to the average value of the initial conditions.

For the network shown in Figure 3.24a, the corresponding adjacency matrix is

$$A = \begin{pmatrix} 0 & 1 & 0 & 0 & 0 \\ 1 & 0 & 1 & 1 & 1 \\ 0 & 1 & 0 & 1 & 0 \\ 0 & 1 & 1 & 0 & 0 \\ 0 & 1 & 0 & 0 & 0 \end{pmatrix}.$$

We use the notation $\mathcal{N}_i$ to represent the set of neighbors of a node i. For example, in the network shown in Figure 3.24a $\mathcal{N}_2 = \{1, 3, 4, 5\}$ and $\mathcal{N}_3 = \{2, 4\}$.

To solve the consensus problem, let x_i be the state of the ith sensor, corresponding to that sensor's estimate of the average value that we are trying to compute. We initialize the state to the value of the quantity measured by the individual sensor. The consensus protocol (algorithm) can now be realized as a local update law

$$x_i[k+1] = x_i[k] + \gamma \sum_{j \in \mathcal{N}_i} (x_j[k] - x_i[k]). \tag{3.35}$$

This protocol attempts to compute the average by updating the local state of each agent based on the value of its neighbors. The combined dynamics of all agents can be written in the form

$$x[k+1] = x[k] - \gamma(D - A)x[k], \tag{3.36}$$

where A is the adjacency matrix and D is a diagonal matrix with entries corresponding to the number of neighbors of each node. The constant γ describes the rate at which the estimate of the average is updated based on information from neighboring nodes. The matrix $L := D - A$ is called the *Laplacian* of the graph.

The *equilibrium points* of equation (3.36) are the set of states such that $x_e[k+1] = x_e[k]$. It can be shown that if the network is connected, $x_e = (\alpha, \alpha, \ldots, \alpha)$ is an equilibrium state for the system, corresponding to each sensor having an identical estimate α for the average. Furthermore, we can show that α is indeed the average value of the initial states. Since there can be cycles in the graph, it is possible that the state of the system could enter into an infinite loop and never converge to the desired consensus state. A formal analysis requires tools that will be introduced later in the text, but it can be shown that for any connected graph we can

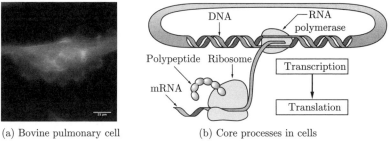

(a) Bovine pulmonary cell   (b) Core processes in cells

Figure 3.25: Biological circuitry. The cell on the left is a bovine pulmonary cell, stained so that the nucleus, actin, and chromatin are visible. The figure on the right gives an overview of the process by which proteins in the cell are made. RNA is transcribed from DNA by an RNA polymerase enzyme. The RNA is then translated into a polypeptide chain by a molecular machine called a ribosome, and then the polypeptide chain folds into a protein molecule.

always find a γ such that the states of the individual agents converge to the average. A simulation demonstrating this property is shown in Figure 3.24b. Although we have focused here on consensus to the average value of a set of measurements, other consensus states can be achieved through choice of appropriate feedback laws. Examples include finding the maximum or minimum value in a network, counting the number of nodes in a network, or computing higher-order statistical moments of a distributed quantity [64, 197]. ∇

Biological Systems

Biological systems provide perhaps the richest source of feedback and control examples. The basic problem of homeostasis, in which a quantity such as temperature or blood sugar level is regulated to a fixed value, is but one of the many types of complex feedback interactions that can occur in molecular machines, cells, organisms, and ecosystems.

Example 3.18 Transcriptional regulation

Transcription is the process by which messenger RNA (mRNA) is generated from a segment of DNA. The promoter region of a gene allows transcription to be controlled by the presence of other proteins, called *transcription factors*, which bind to the promoter region and either repress or activate RNA polymerase, the enzyme that produces an mRNA transcript from DNA. The mRNA is then translated into a protein according to its nucleotide sequence. This process is illustrated in Figure 3.25.

A simple model of the transcriptional regulation process is through the use of a Hill function [70, 186]. Consider the regulation of a protein A with a concentration given by p_a and a corresponding mRNA concentration m_a. Let B be a second protein with concentration p_b that represses the production of protein A through transcriptional regulation. The resulting dynamics of p_a and m_a can be written as

$$\frac{dm_a}{dt} = \frac{\alpha_{ab}}{1 + k_{ab}p_b^{n_{ab}}} + \alpha_{a0} - \delta_a m_a, \qquad \frac{dp_a}{dt} = \kappa_a m_a - \gamma_a p_a, \qquad (3.37)$$

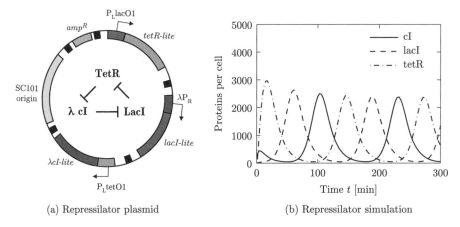

(a) Repressilator plasmid   (b) Repressilator simulation

Figure 3.26: The repressilator genetic regulatory network. (a) A schematic diagram of the repressilator, showing the layout of the genes in the plasmid that holds the circuit as well as the circuit diagram (center). (b) A simulation of a simple model for the repressilator, showing the oscillation of the individual protein concentrations. (Figure courtesy M. Elowitz.)

where $\alpha_{\mathrm{ab}} + \alpha_{\mathrm{a0}}$ is the unregulated transcription rate, δ_{a} represents the rate of degradation of mRNA, α_{ab}, k_{ab}, and n_{ab} are parameters that describe how B represses A, κ_{a} represents the rate of production of the protein from its corresponding mRNA, and γ_{a} represents the rate of degradation of the protein A. The parameter α_{a0} describes the "leakiness" of the promoter, and n_{ab} is called the Hill coefficient and relates to the cooperativity of the promoter.

A similar model can be used when a protein activates the production of another protein rather than repressing it. In this case, the equations have the form

$$\frac{dm_{\mathrm{a}}}{dt} = \frac{\alpha_{\mathrm{ab}} k_{\mathrm{ab}} p_{\mathrm{b}}^{n_{\mathrm{ab}}}}{1 + k_{\mathrm{ab}} p_{\mathrm{b}}^{n_{\mathrm{ab}}}} + \alpha_{\mathrm{a0}} - \delta_{\mathrm{a}} m_{\mathrm{a}}, \qquad \frac{dp_{\mathrm{a}}}{dt} = \kappa_{\mathrm{a}} m_{\mathrm{a}} - \gamma_{\mathrm{a}} p_{\mathrm{a}}, \qquad (3.38)$$

where the variables are the same as described previously. Note that in the case of the activator, if p_{b} is zero, then the production rate is $\alpha_{\mathrm{a0}} \ll \alpha_{\mathrm{ab}}$ (versus $\alpha_{\mathrm{ab}} + \alpha_{\mathrm{a0}}$ for the repressor). As p_{b} gets large, the first term in the expression for $\dot{m}_{\mathrm{a}}$ approaches 1 and the transcription rate becomes $\alpha_{\mathrm{ab}} + \alpha_{\mathrm{a0}}$ (versus α_{a0} for the repressor). Thus we see that the activator and repressor act in opposite fashion from each other.

As an example of how these models can be used, we consider the model of a "repressilator," originally due to Elowitz and Leibler [85]. The repressilator is a synthetic circuit in which three proteins each repress another in a cycle. This is shown schematically in Figure 3.26a, where the three proteins are TetR, λ cI, and LacI. The basic idea of the repressilator is that if TetR is present, then it represses the production of λ cI. If λ cI is absent, then LacI is produced (at the unregulated transcription rate), which in turn represses TetR. Once TetR is repressed, then λ cI is no longer repressed, and so on. If the dynamics of the circuit are designed properly, the resulting protein concentrations will oscillate.

We can model this system using three copies of equation (3.37), with A and B replaced by the appropriate combination of TetR, cI, and LacI. The state of the system is then given by $x = (m_{\mathrm{TetR}}, p_{\mathrm{TetR}}, m_{\mathrm{cI}}, p_{\mathrm{cI}}, m_{\mathrm{LacI}}, p_{\mathrm{LacI}})$. Figure 3.26b shows the traces of the three protein concentrations for parameters $n = 2$, $\alpha = 0.5$, $k =$

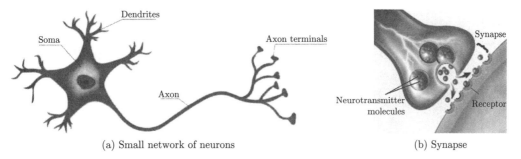

(a) Small network of neurons   (b) Synapse

Figure 3.27: Nerve cell physiology. The left figure shows a neuron and the right figure illustrates the synaptic gap between an axon terminal and a dendrite.

6.25×10^{-4}, $\alpha_0 = 5 \times 10^{-4}$, $\delta = 5.8 \times 10^{-3}$, $\kappa = 0.12$, and $\gamma = 1.2 \times 10^{-3}$ with initial conditions $x(0) = (1, 200, 0, 0, 0, 0)$ (following [85]). ∇

Example 3.19 Nerve cells
Neurons are key elements of the control systems for all humans and animals. There are different types of neurons: sensory neurons respond to stimuli; motor neurons control muscles and other organs; and interneurons act as intermediaries in passing signals between other neurons. Neurons are often connected to form networks; a human brain has close to 100 billion neurons.

A neuron has three parts: the cell body (soma), the axon, and the dendrites, as shown in Figure 3.27a. The cell body varies in size from 4 to 100 µm and axons have lengths from one millimeter to a meter. The cell has a membrane that separates it from the outside environment (extracellular space), with molecular-scale channels that let ions pass through the membrane, creating a voltage difference across the cell membrane. An electric pulse (spike) is generated when the voltage difference reaches a critical level. Pulse rates range from 1 Hz to 1 kHz and the generated pulse travels along the axon to its terminals.

Neurons receive signals from other neurons through dendrites. There are electrochemical reactions at the interface between an axon and a dendrite of another cell that allow transmission between two neurons. The axon terminal has vesicles that contain neurotransmitters, which are released in the synaptic gap when the axon is stimulated by electrical pulses, as illustrated in Figure 3.27b. The neurotransmitters stimulate ion channels in the cell membrane, causing them to open. There are many types of channels; two common ones are sodium (Na^+) channels and potassium (K^+) channels. The potassium channel has a slow excitatory action, while the sodium channel has a fast excitatory and a slow inhibitory action.

The dynamics of the neuron are a fundamental mechanism for understanding signaling in cells. The Hodgkin–Huxley equation is a model for neuron dynamics. It models the cell membrane as a capacitor,

$$C\frac{dV}{dt} = I_{Na^+} + I_{K^+} + I_{leak} + I_{input},$$

where V is the membrane potential, C is the capacitance, I_{Na^+} and I_{K^+} are the current caused by the transport of sodium and potassium ions across the cell membrane, I_{leak} is a leakage current, and I_{input} is the external stimulation of the cell. Each current obeys Ohm's law,

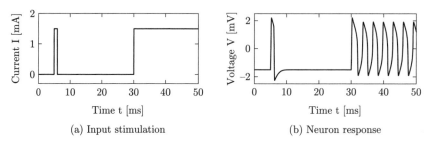

(a) Input stimulation   (b) Neuron response

Figure 3.28: Response of a neuron to a current input. The current input is shown in (a) and the neuron voltage V in (b). The simulation was done using the FitzHugh–Nagumo model (Exercise 3.11).

$$I_{\text{Na}^+} = g_{\text{Na}}(E_{\text{Na}^+} - V), \qquad I_{\text{K}^+} = g_{\text{K}}(E_{\text{K}^+} - V), \qquad I_{\text{leak}} = g_{\text{leak}}(E_{\text{leak}} - V).$$

The conductances g_{Na}, g_{K}, and g_{leak} depend on the voltage V through the variables m, n, and h, where g_{Na} is proportional to $m^3 h$, g_{K} is proportional to n^4, and g_{leak} is a constant. The variables m, n, and h are given by the differential equations

$$\frac{dm}{dt} = \frac{m_a(V) - m}{\tau_m(V)}, \qquad \frac{dh}{dt} = \frac{h_a(V) - h}{\tau_h(V)}, \qquad \frac{dn}{dt} = \frac{n_a(V) - n}{\tau_n(V)},$$

where the functions m_a, h_a, n_a, τ_m, τ_h, and τ_n are derived from experimental data. The functions m_a and n_a are monotone and increasing in V, creating excitatory behavior. The function h_a is monotone and decreasing, creating inhibitory behavior. The time constant τ_m is almost an order of magnitude smaller than the time constants τ_h and τ_n.

The equilibrium voltages E_{Na^+} and E_{K^+} are given by Nernst's law,

$$E = \frac{RT}{nF} \log \frac{c_e}{c_i},$$

where R is Boltzmann's constant, T is the absolute temperature, F is Faraday's constant, n is the charge (or valence) of the ion, and c_i and c_e are the ion concentrations inside the cell and in the external fluid. At $20\,^{\circ}\text{C}$ we have $RT/F = 20\,\text{mV}$, $E_{\text{Na}^+} = 55\,\text{mV}$, and $E_{\text{K}^+} = -92\,\text{mV}$.

The Hodgkin-Huxley equations are complicated and contain many widely different time scales, and many approximations have therefore been proposed. One approximation is the FitzHugh–Nagumo model (Exercise 3.11). A simulation of this model is shown in Figure 3.28 to illustrate the behavior of a neuron to an external current stimulation. The system is initially at rest with $I = 0$ and $V = 0$. A short current pulse enters at time $t = 5\,\text{ms}$, the neuron is excited, and responds by sending out a spike. The neuron is then excited at time $t = 30\,\text{ms}$ and the neuron then starts spiking.   ▽

The Hodgkin–Huxley model was originally developed as a means to predict the quantitative behavior of the squid giant axon [120]. Hodgkin and Huxley shared the 1963 Nobel Prize in Physiology (along with J. C. Eccles) for analysis of the electrical and chemical events in nerve cell discharges. The voltage clamp described in Section 1.4 was used to determine the functions $m_a(V)$, $n_a(V)$, and $h_a(V)$. There are many variations of models for the dynamics of neurons based on the Hodgkin–Huxley model; a recent reference is [202]. Some models combine ordinary differential equations with discrete transitions, so–called integrate-and-fire models or hybrid systems.

3.5 FURTHER READING

Modeling is ubiquitous in engineering and science and has a long history in applied mathematics. For example, the Fourier series was introduced by Fourier when he modeled heat conduction in solids [90]. A classic book on the modeling of physical systems, especially mechanical, electrical, and thermofluid systems, is Cannon [60]. The book by Aris [13] is highly original and has a detailed discussion of the use of dimension-free variables. Models of dynamics have been developed in many different fields, including mechanics [14, 105], heat conduction [61], fluids [47], vehicles [1, 48, 82], robotics [189, 226], circuits [111], power systems [156], acoustics [38], and micromechanical systems [220]. The authors' favorite books on modeling of biological systems are Keener and Sneyd [140, 141], J. D. Murray [186], and Wilson [256]. Control requires modeling from many different domains, and most control theory texts contain several chapters on modeling using ordinary differential equations and difference equations (see, for example, [93]). A good source for system identification is Ljung [166].

EXERCISES

3.1 (Chain of integrators form) Consider the linear ordinary differential equation (3.7). Show that by choosing a state space representation with $x_1 = y$, the dynamics can be written as

$$
A = \begin{pmatrix} 0 & 1 & & 0 \\ 0 & \ddots & \ddots & 0 \\ 0 & \cdots & 0 & 1 \\ -a_n & -a_{n-1} & & -a_1 \end{pmatrix}, \qquad B = \begin{pmatrix} 0 \\ 0 \\ \vdots \\ 1 \end{pmatrix}, \qquad C = \begin{pmatrix} 1 & \cdots & 0 & 0 \end{pmatrix}.
$$

This canonical form is called the *chain of integrators* form.

3.2 (Discrete-time dynamics) Consider the following discrete-time system

$$
x[k+1] = Ax[k] + Bu[k], \qquad y[k] = Cx[k],
$$

where

$$
x = \begin{pmatrix} x_1 \\ x_2 \end{pmatrix}, \qquad A = \begin{pmatrix} a_{11} & a_{12} \\ 0 & a_{22} \end{pmatrix}, \qquad B = \begin{pmatrix} 0 \\ 1 \end{pmatrix}, \qquad C = \begin{pmatrix} 1 & 0 \end{pmatrix}.
$$

In this problem, we will explore some of the properties of this discrete-time system as a function of the parameters, the initial conditions, and the inputs.

a) For the case when $a_{12} = 0$ and $u = 0$, give a closed form expression for the output of the system.

b) A discrete system is in *equilibrium* when $x[k+1] = x[k]$ for all k. Let $u = r$ be a constant input and compute the resulting equilibrium point for the system. Show that if $|a_{ii}| < 1$ for all i, all initial conditions give solutions that converge to the equilibrium point.

c) Write a computer program to plot the output of the system in response to a unit step input, $u[k] = 1$, $k \geq 0$. Plot the response of your system with $x[0] = 0$ and A given by $a_{11} = 0.5$, $a_{12} = 1$, and $a_{22} = 0.25$.

3.3 (Keynesian economics) Keynes' simple model for an economy is given by

$$Y[k] = C[k] + I[k] + G[k],$$

where Y, C, I, and G are gross national product (GNP), consumption, investment, and government expenditure for year k. Consumption and investment are modeled by difference equations of the form

$$C[k+1] = aY[k], \qquad I[k+1] = b(C[k+1] - C[k]),$$

where a and b are parameters. The first equation implies that consumption increases with GNP but that the effect is delayed. The second equation implies that investment is proportional to the rate of change of consumption.

Show that the equilibrium value of the GNP is given by

$$Y_e = \frac{1}{1-a} G_e,$$

where the parameter $1/(1-a)$ is the Keynes multiplier (the gain from G to Y). With $a = 0.75$ an increase of government expenditure will result in a fourfold increase of GNP. Also show that the model can be written as the following discrete-time state model:

$$\begin{pmatrix} C[k+1] \\ I[k+1] \end{pmatrix} = \begin{pmatrix} a & a \\ ab-b & ab \end{pmatrix} \begin{pmatrix} C[k] \\ I[k] \end{pmatrix} + \begin{pmatrix} a \\ ab \end{pmatrix} G[k],$$

$$Y[k] = C[k] + I[k] + G[k].$$

 3.4 (Least squares system identification) Consider a nonlinear differential equation that can be written in the form

$$\frac{dx}{dt} = \sum_{i=1}^{M} \alpha_i f_i(x),$$

where $f_i(x)$ are known nonlinear functions and α_i are unknown, but constant, parameters. Suppose that we have measurements (or estimates) of the full state x at time instants $t_1, t_2, \ldots, t_N$, with $N > M$. Show that the parameters α_i can be estimated by finding the least squares solution to a linear equation of the form

$$H\alpha = b,$$

where $\alpha \in \mathbb{R}^M$ is the vector of all parameters and $H \in \mathbb{R}^{N \times M}$ and $b \in \mathbb{R}^N$ are appropriately defined.

3.5 (Normalized oscillator dynamics) Consider a damped spring–mass system with dynamics

$$m\ddot{q} + c\dot{q} + kq = F.$$

Let $\omega_0 = \sqrt{k/m}$ be the natural frequency and $\zeta = c/(2\sqrt{km})$ be the damping ratio.

a) Show that by rescaling the equations, we can write the dynamics in the form

$$\ddot{q} + 2\zeta\omega_0\dot{q} + \omega_0^2 q = \omega_0^2 u, \tag{3.39}$$

where $u = F/k$. This form of the dynamics is that of a linear oscillator with natural frequency ω_0 and damping ratio ζ.

b) Show that the system can be further normalized and written in the form

$$\frac{dz_1}{d\tau} = z_2, \qquad \frac{dz_2}{d\tau} = -z_1 - 2\zeta z_2 + v. \tag{3.40}$$

The essential dynamics of the system are governed by a single damping parameter ζ. The *Q-value*, defined as $Q = 1/2\zeta$, is sometimes used instead of ζ.

3.6 (Dubins car) Show that the trajectory of a vehicle with reference point chosen as the center of the rear wheels can be modeled by dynamics of the form

$$\frac{dx}{dt} = v\cos\theta, \qquad \frac{dy}{dt} = v\sin\theta, \qquad \frac{d\theta}{dt} = \frac{v}{b}\tan\delta,$$

where the variables and constants are defined as in Example 3.11.

3.7 (Motor drive) Consider a system consisting of a motor driving two masses that are connected by a torsional spring, as shown in the diagram below.

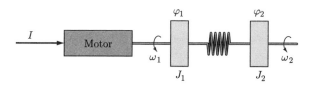

This system can represent a motor with a flexible shaft that drives a load. Assuming that the motor delivers a torque that is proportional to the current I, the dynamics of the system can be described by the equations

$$J_1\frac{d^2\varphi_1}{dt^2} + c\left(\frac{d\varphi_1}{dt} - \frac{d\varphi_2}{dt}\right) + k(\varphi_1 - \varphi_2) = k_I I,$$

$$J_2\frac{d^2\varphi_2}{dt^2} + c\left(\frac{d\varphi_2}{dt} - \frac{d\varphi_1}{dt}\right) + k(\varphi_2 - \varphi_1) = T_d, \tag{3.41}$$

where φ_1 and φ_2 are the angles of the two masses, $\omega_i = d\varphi_i/dt$ are their velocities, J_i represents moments of inertia, c is the damping coefficient, k represents the shaft stiffness, k_I is the torque constant for the motor, and T_d is the disturbance torque applied at the end of the shaft. Similar equations are obtained for a robot with flexible arms and for the arms of DVD and optical disk drives.

Derive a state space model for the system by introducing the (normalized) state variables $x_1 = \varphi_1$, $x_2 = \varphi_2$, $x_3 = \omega_1/\omega_0$, and $x_4 = \omega_2/\omega_0$, where $\omega_0 = \sqrt{k(J_1 + J_2)/(J_1 J_2)}$ is the undamped natural frequency of the system when the control signal is zero.

3.8 (Electric generator) An electric generator connected to a power grid can be modeled by a momentum balance for the rotor of the generator:

$$J\frac{d^2\varphi}{dt^2} = P_m - P_e = P_m - \frac{EV}{X}\sin\varphi,$$

where J is the effective moment of inertia of the generator, φ is the angle of rotation, P_m is the mechanical power that drives the generator, P_e is the active electrical power, E is the generator voltage, V is the grid voltage, and X is the reactance of the line. Assuming that the line dynamics are much faster than the rotor dynamics, $P_e = VI = (EV/X)\sin\varphi$, where I is the current component in phase with the voltage E and φ is the phase angle between voltages E and V. Show that the dynamics of the electric generator have a normalized form that is similar to the dynamics of a pendulum with forcing at the pivot.

3.9 (Admission control for a queue) Consider the queuing system described in Example 3.15. The long delays created by temporary overloads can be reduced by rejecting requests when the queue gets large. This allows requests that are accepted to be serviced quickly and requests that cannot be accommodated to receive a rejection quickly so that they can try another server. Consider an admission control system described by

$$\frac{dx}{dt} = \lambda u - \mu_{max}\frac{x}{x+1}, \qquad u = \text{sat}_{(0,1)}(k(r-x)), \qquad (3.42)$$

where the controller is a simple proportional control with saturation ($\text{sat}_{(a,b)}$ is defined by equation (4.10)) and r is the desired (reference) queue length. Use a simulation to show that this controller reduces the rush-hour effect and explain how the choice of r affects the system dynamics.

3.10 (Biological switch) A genetic switch can be formed by connecting two repressors together in a cycle as shown below.

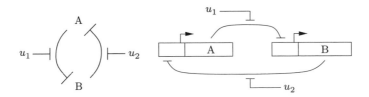

Using the models from Example 3.18—assuming that the parameters are the same for both genes and that the mRNA concentrations reach steady-state quickly—show that the dynamics can be written in normalized coordinates as

$$\frac{dz_1}{d\tau} = \frac{\mu}{1 + z_2^n} - z_1 - u_1, \qquad \frac{dz_2}{d\tau} = \frac{\mu}{1 + z_1^n} - z_2 - u_2, \qquad (3.43)$$

where z_1 and z_2 are scaled versions of the protein concentrations and the time scale has also been changed. Show that $\mu \approx 200$ using the parameters in Example 3.18, and use simulations to demonstrate the switch-like behavior of the system.

3.11 (FitzHugh–Nagumo) The second-order FitzHugh–Nagumo equations

$$\frac{dV}{dt} = 10(V - V^3/3 - R + I_{\text{in}}), \qquad \frac{dR}{dt} = 0.8(1.25V - R + 1.5)$$

are a simplified version of the Hodgkin–Huxley equations discussed in Example 3.19. The variable V is the voltage across the axon membrane and R is an auxiliary variable that approximates several ion currents that flow across the membrane. Simulate the equations and reproduce the simulation in Figure 3.28. Explore the effect of the input current I_{in}.

Chapter Four

Examples

... Don't apply any model until you understand the simplifying assumptions on which it is based, and you can test their validity. Catch phrase: Use only as directed. Don't beleive that the model is the reality. Catch phrase: You will never strike oil by drilling through the map.

—Saul Golomb, "Mathematical Models—Uses and Limitations," 1970 [106].

In this chapter we present a collection of examples spanning many different fields of science and engineering. These examples are used throughout the text and in exercises to illustrate different concepts. First-time readers may wish to focus on only a few examples with which they have had the most prior experience or insight to understand the concepts of state, input, output, and dynamics in a familiar setting.

4.1   CRUISE CONTROL

The cruise control system of a car is a common feedback system encountered in everyday life. The system attempts to maintain a constant velocity in the presence of disturbances primarily caused by changes in the slope of a road. The controller compensates for these unknowns by measuring the speed of the car and adjusting the throttle appropriately.

To model the system we start with the block diagram in Figure 4.1. Let v be the speed of the car and v_r the desired (reference) speed. The controller, which typically is of the proportional-integral (PI) type described briefly in Chapter 1, receives the signals v and v_r and generates a (normalized) control signal u that is sent to an actuator that controls the throttle position. The throttle in turn controls the torque T delivered by the engine, which is transmitted through the gears and the wheels, generating a force F that moves the car. There are disturbance forces F_d due to variations in the slope of the road, the rolling resistance, and aerodynamic forces. The cruise controller also has a human–machine interface that allows the driver to set and modify the desired speed. There are also functions that disconnect the cruise control when the brake is touched.

The system has many individual components—actuator, engine, transmission, wheels, and car body—and a detailed model can be very complicated. In spite of this, the model required to design the cruise controller can be quite simple.

To develop a mathematical model we start with a force balance for the car body. Letting m be the total mass of the car (including passengers), the equation of motion of the car is simply

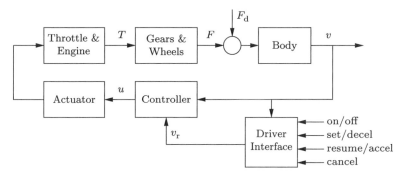

Figure 4.1: Block diagram of a cruise control system for an automobile. The throttle-controlled engine generates a torque T that is transmitted to the ground through the gearbox and wheels. Combined with the external forces from the environment, such as aerodynamic drag and gravitational forces on hills, the net force causes the car to move. The velocity of the car v is measured by a control system that adjusts the throttle through an actuation mechanism. A driver interface allows the system to be turned on and off and the reference speed v_r to be established.

$$m\frac{dv}{dt} = F - F_d. \tag{4.1}$$

Typical values for the mass of a car are in the range of 1000–2000 kg (we will use 1600 kg here).

The force F is generated by the engine, whose torque is proportional to the rate of fuel injection, which is itself proportional to a control signal $0 \le u \le 1$ that controls the throttle position. The torque also depends on engine speed ω. A simple representation of the torque at full throttle is given by the torque curve

$$T(\omega) = T_m \left(1 - \beta\left(\frac{\omega}{\omega_m} - 1\right)^2\right), \tag{4.2}$$

where the maximum torque T_m is obtained at engine speed ω_m. Typical parameters are $T_m = 190$ Nm, $\omega_m = 420$ rad/s (about 4000 RPM), and $\beta = 0.4$. Let n be the gear ratio and r the wheel radius. The engine speed is related to the velocity through the expression

$$\omega = \frac{n}{r}v =: \alpha_n v,$$

and the driving force can be written as

$$F = \frac{nu}{r}T(\omega) = \alpha_n u T(\alpha_n v).$$

Typical values of α_n for gears 1 through 5 are $\alpha_1 = 40$, $\alpha_2 = 25$, $\alpha_3 = 16$, $\alpha_4 = 12$, and $\alpha_5 = 10$. The inverse of α_n has a physical interpretation as the *effective wheel radius*. Figure 4.2 shows the torque as a function of engine speed and vehicle speed. The figure shows that the effect of the gear is to "flatten" the torque curve so that nearly full torque can be obtained over almost the whole speed range.

The disturbance force F_d has three major components: F_g, the forces due to gravity; F_r, the forces due to rolling friction; and F_a, the aerodynamic drag. Letting

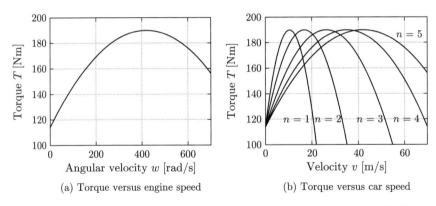

(a) Torque versus engine speed   (b) Torque versus car speed

Figure 4.2: Torque curves for typical car engine. The graph on the left shows the torque generated by the engine as a function of the angular velocity of the engine, while the curve on the right shows torque as a function of car speed for different gears.

the slope of the road be θ, gravity gives the force $F_g = mg \sin \theta$, as illustrated in Figure 4.3a, where $g = 9.8 \, \text{m/s}^2$ is the gravitational constant. A simple model of rolling friction is

$$F_r = mg C_r \, \text{sgn}(v),$$

where C_r is the coefficient of rolling friction and $\text{sgn}(v)$ is the sign of v (± 1) or zero if $v = 0$. A typical value for the coefficient of rolling friction is $C_r = 0.01$. Finally, the aerodynamic drag is proportional to the square of the speed:

$$F_a = \frac{1}{2} \rho C_d A |v| v,$$

where ρ is the density of air, C_d is the shape-dependent aerodynamic drag coefficient, and A is the frontal area of the car. Typical parameters are $\rho = 1.3 \, \text{kg/m}^3$, $C_d = 0.32$, and $A = 2.4 \, \text{m}^2$.

Summarizing, we find that the car's speed can be modeled by

$$m \frac{dv}{dt} = \alpha_n u T(\alpha_n v) - mg C_r \, \text{sgn}(v) - \frac{1}{2} \rho C_d A |v| v - mg \sin \theta, \qquad (4.3)$$

where the function T is given by equation (4.2). The model (4.3) is a dynamical system of first order. The state is the car velocity v, which is also the output. The input is the signal u that controls the throttle position, and the disturbance is the force $F_d = mg \sin \theta$, which depends on the slope of the road. The system is nonlinear because of the torque curve, the gravity term, and the nonlinear character of rolling friction and aerodynamic drag. There can also be variations in the parameters; e.g., the mass of the car depends on the number of passengers and the load being carried in the car.

We add to this model a feedback controller that attempts to regulate the speed of the car in the presence of disturbances. We use a proportional integral controller, which has the form

$$u(t) = k_p e(t) + k_i \int_0^t e(\tau) \, d\tau.$$

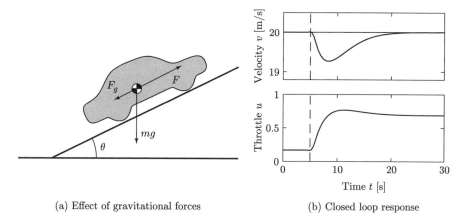

(a) Effect of gravitational forces   (b) Closed loop response

Figure 4.3: Car with cruise control encountering a sloping road. A schematic diagram is shown in (a), and (b) shows the response in speed and throttle when a slope of 4° is encountered. The hill is modeled as a net change of 4° in hill angle θ, with a linear change in the angle between $t = 5$ and $t = 6$. The PI controller has proportional gain $k_p = 0.5$ and integral gain $k_i = 0.1$.

This controller can itself be realized as an input/output dynamical system by defining a controller state z and implementing the differential equation

$$\frac{dz}{dt} = v_r - v, \qquad u = k_p(v_r - v) + k_i z, \tag{4.4}$$

where v_r is the desired (reference) speed. As discussed briefly in Section 1.6, the integrator (represented by the state z) ensures that in steady state the error will be driven to zero, even when there are disturbances or modeling errors. (The design of PI controllers is the subject of Chapter 11.) Figure 4.3b shows the response of the closed loop system, consisting of equations (4.3) and (4.4), when it encounters a hill. The figure shows that even if the hill is so steep that the throttle changes from 0.17 to almost full throttle, the largest speed error is less than 1 m/s, and the desired velocity is recovered after 20 s.

Many approximations were made when deriving the model (4.3). It may seem surprising that such a seemingly complicated system can be described by the simple model (4.3). It is important to make sure that we restrict our use of the model to the uncertainty lemon conceptualized in Figure 3.5b. The model is not valid for very rapid changes of the throttle because we have ignored the details of the engine dynamics, neither is it valid for very slow changes because the properties of the engine will change over the years. Nevertheless the model is very useful for the design of a cruise control system. As we shall see in later chapters, the reason for this is the inherent robustness of feedback systems: even if the model is not perfectly accurate, we can use it to design a controller and make use of the feedback in the controller to manage the uncertainty in the system.

The cruise control system also has a human–machine interface that allows the driver to communicate with the system. There are many different ways to implement this system; one version is illustrated in Figure 4.4. The system has four buttons: on-off, set/decelerate, resume/accelerate, and cancel. The operation of the system

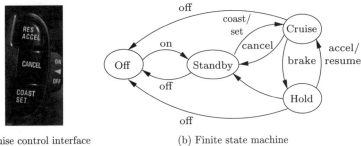

(a) Cruise control interface   (b) Finite state machine

Figure 4.4: Finite state machine for cruise control system. The figure on the left shows some typical buttons used to control the system. The controller can be in one of four modes, corresponding to the nodes in the diagram on the right. Transition between the modes is controlled by pressing one of the five buttons on the cruise control interface: on, off, set, resume, or cancel.

is governed by a finite state machine that controls the modes of the PI controller and the reference generator. Implementation of controllers and reference generators will be discussed more fully in Chapter 11.

The use of control in automotive systems goes well beyond the simple cruise control system described here. Applications include emissions control, traction control, power control (especially in hybrid vehicles), and adaptive cruise control. Many automotive applications are discussed in detail in the book by Kiencke and Nielsen [145] and in the survey papers by Powers et al. [27, 203]. New vehicles coming on the market also include many "self-driving" features, which represent even more complex feedback systems.

4.2   BICYCLE DYNAMICS

The bicycle is an interesting dynamical system with the feature that one of its key properties is due to a feedback mechanism that is created by the design of the front fork. A detailed model of a bicycle is complex because the system has many degrees of freedom and the geometry is complicated. However, a great deal of insight can be obtained from simple models.

To derive the equations of motion we assume that the bicycle rolls on the horizontal xy plane. Introduce a coordinate system that is fixed to the bicycle with the ξ-axis through the contact points of the wheels with the ground, the η-axis horizontal, and the ζ-axis vertical, as shown in Figure 4.5. Let v_0 be the velocity of the bicycle at the rear wheel, b the wheelbase, φ the tilt angle, and δ the steering angle. The coordinate system rotates around the point O with the angular velocity $\omega = v_0 \delta / b$, and an observer fixed to the bicycle experiences forces due to the motion of the coordinate system.

The tilting motion of the bicycle is similar to an inverted pendulum, as shown in the rear view in Figure 4.5b. To model the tilt, consider the rigid body obtained when the wheels, the rider, and the front fork assembly are fixed to the bicycle frame. Let m be the total mass of the system, J the moment of inertia of the body with respect to the ξ-axis, and D the product of inertia with respect to the $\xi\zeta$ axes. Furthermore, let the ξ and ζ coordinates of the center of mass with respect

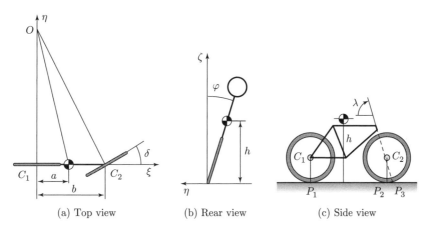

(a) Top view   (b) Rear view   (c) Side view

Figure 4.5: Schematic views of a bicycle. The steering angle is δ, and the roll angle is φ. The center of mass has height h and distance a from a vertical through the contact point P_1 of the rear wheel. The wheelbase b is the distance between P_1 and P_2, and the trail c is the distance between P_2 and P_3.

to the rear wheel contact point, P_1, be a and h, respectively. We have $J \approx mh^2$ and $D = mah$. The torques acting on the system are due to gravity and centripetal action. Assuming that the steering angle δ is small, the equation of motion becomes

$$J\frac{d^2\varphi}{dt^2} - \frac{Dv_0}{b}\frac{d\delta}{dt} = mgh \sin\varphi + \frac{mv_0^2 h}{b}\delta. \qquad (4.5)$$

The term $mgh \sin\varphi$ is the torque generated by gravity. The terms containing δ and its derivative are the torques generated by steering, with the term $(Dv_0/b)\, d\delta/dt$ due to inertial forces and the term $(mv_0^2 h/b)\, \delta$ due to centripetal forces.

The steering angle is influenced by the torque the rider applies to the handle bar. Because of the tilt of the steering axis and the shape of the front fork, the contact point of the front wheel with the road P_2 is behind the axis of rotation of the front wheel assembly, as shown in Figure 4.5c. The distance c between the contact point of the front wheel P_2 and the projection of the axis of rotation of the front fork assembly P_3 is called the *trail*. The steering properties of a bicycle depend critically on the trail. A large trail increases stability but makes the steering less agile.

A consequence of the design of the front fork is that the steering angle δ is influenced both by steering torque T and by the tilt of the frame φ. This means that a bicycle with a front fork is a *feedback system* as illustrated by the block diagram in Figure 4.6. The steering angle δ influences the tilt angle φ, and the tilt angle influences the steering angle, giving rise to the circular causality that is characteristic of reasoning about feedback. For a front fork with a positive trail, the bicycle will steer into the lean, creating a centrifugal force that attempts to diminish the lean.

Under certain conditions, the feedback can actually stabilize the bicycle. A crude empirical model is obtained by assuming that the front fork can be modeled as the static system

$$\delta = k_1 T - k_2 \varphi. \qquad (4.6)$$

Combining the model of the bicycle frame (4.5) with the model of the front fork (4.6), we get the the following system model:

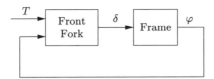

Figure 4.6: Block diagram of a bicycle with a front fork. The steering torque applied to the handlebars is T, the roll angle is φ, and the steering angle is δ. Notice that the front fork creates a feedback from the roll angle φ to the steering angle δ that under certain conditions can stabilize the system.

$$J\frac{d^2\varphi}{dt^2} + \frac{Dv_0k_2}{b}\frac{d\varphi}{dt} + \left(\frac{mv_0^2hk_2}{b} - mgh\right)\varphi = \frac{Dv_0k_1}{b}\frac{dT}{dt} + \frac{mv_0^2hk_1}{b}T, \qquad (4.7)$$

where we have approximated $\sin\varphi$ with φ. The left hand side of this equation looks like the equation for a spring mass system, where the damping term is Dv_0k_2/b and the spring term is $mv_0^2hk_2/b - mgh$. Notice that the spring term is negative if $v_0 = 0$ and that it becomes positive for $v > \sqrt{gb/k_2}$. We can thus conclude that the bicycle is unstable for small velocities but that the feedback provided by the front fork makes the bicycle stable if the velocity is sufficiently large.

The simple model given by equations (4.5) and (4.6) neglects the dynamics of the front fork, the tire–road interaction, and the fact that the parameters depend on the velocity. A more accurate model, called the *Whipple model*, is obtained using the rigid-body dynamics of the front fork and the frame. Assuming small angles, this model becomes

$$M\begin{pmatrix}\ddot{\varphi}\\\ddot{\delta}\end{pmatrix} + Cv_0\begin{pmatrix}\dot{\varphi}\\\dot{\delta}\end{pmatrix} + (K_0 + K_2v_0^2)\begin{pmatrix}\varphi\\\delta\end{pmatrix} = \begin{pmatrix}0\\T\end{pmatrix}, \qquad (4.8)$$

where the elements of the 2×2 matrices M, C, K_0, and K_2 depend on the geometry and the mass distribution of the bicycle. Note that this has a form somewhat similar to that of the spring–mass system introduced in Chapter 3 and the balance system in Example 3.2. Even this more complex model is inaccurate because the interaction between the tire and the road is neglected; taking this into account requires two additional state variables. Again, the uncertainty lemon in Figure 3.5b provides a framework for understanding the validity of the model under these assumptions.

Interesting presentations on the development of the bicycle are given in the books by D. Wilson [255] and Herlihy [118]. The model (4.8) was presented in a paper by Whipple in 1899 [249]. More details on bicycle modeling are given in the papers [20, 164], which has many additional references.

4.3 OPERATIONAL AMPLIFIER CIRCUITS

An operational amplifier (op amp) is a modern implementation of Black's feedback amplifier. It is a universal component that is widely used for instrumentation, control, and communication. It is also a key element in analog computing. Schematic diagrams of the operational amplifier are shown in Figure 4.7. The amplifier has one inverting input (v_-), one noninverting input (v_+), and one output (v_{out}). There are also connections for the supply voltages, e_- and e_+, and a zero adjustment (offset

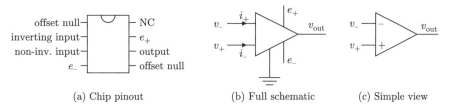

(a) Chip pinout   (b) Full schematic   (c) Simple view

Figure 4.7: An operational amplifier and two schematic diagrams. (a) The amplifier pin connections on an integrated circuit chip. (b) A schematic with all connections. (c) Only the signal connections.

null). A simple model is obtained by assuming that the input currents i_- and i_+ are zero and that the output is given by the static relation

$$v_{\text{out}} = \text{sat}_{(v_{\min}, v_{\max})}\big(k(v_+ - v_-)\big), \tag{4.9}$$

where sat denotes the saturation function

$$\text{sat}_{(a,b)}(x) = \begin{cases} a & \text{if } x < a, \\ x & \text{if } a \le x \le b, \\ b & \text{if } x > b. \end{cases} \tag{4.10}$$

We assume that the gain k is large, in the range of 10^6–10^8, and the voltages $v_{\min}$ and $v_{\max}$ satisfy

$$e_- \le v_{\min} < v_{\max} \le e_+$$

and hence are in the range of the supply voltages. More accurate models are obtained by replacing the saturation function with a smooth function as shown in Figure 4.8. For small input signals the amplifier characteristic (4.9) is linear:

$$v_{\text{out}} = k(v_+ - v_-) =: -kv. \tag{4.11}$$

Since the open loop gain k is very large, the range of input signals where the system is linear is very small.

A simple amplifier is obtained by arranging feedback around the basic operational amplifier as shown in Figure 4.9a. To model the feedback amplifier in the linear range, we assume that the current $i_0 = i_- + i_+$ is zero and that the gain of the amplifier is so large that the voltage $v = v_- - v_+$ is also zero. It follows from Ohm's law that the currents through resistors R_1 and R_2 are given by

$$\frac{v_1}{R_1} = -\frac{v_2}{R_2},$$

and hence the closed loop gain of the amplifier is

$$\frac{v_2}{v_1} = -k_{\text{cl}}, \quad \text{where} \quad k_{\text{cl}} = \frac{R_2}{R_1}. \tag{4.12}$$

A more accurate model is obtained by continuing to neglect the current i_0 but assuming that the voltage v is small but not negligible. The current balance is then

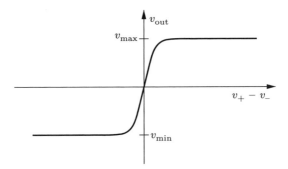

Figure 4.8: Input/output characteristics of an operational amplifier. The differential input is given by $v_+ - v_-$. The output voltage is a linear function of the input in a small range around 0, with saturation at v_{min} and v_{max}. In the linear regime the op amp has high gain.

$$\frac{v_1 - v}{R_1} = \frac{v - v_2}{R_2}. \tag{4.13}$$

Assuming that the amplifier operates in the linear range and using equation (4.11) with $v_{out} = v_2$, the gain of the closed loop system becomes

$$k_{cl} = -\frac{v_2}{v_1} = \frac{R_2}{R_1}\frac{kR_1}{R_1 + R_2 + kR_1} \approx \frac{R_2}{R_1}. \tag{4.14}$$

If the open loop gain k of the operational amplifier is large, the closed loop gain k_{cl} is the same as in the simple model given by equation (4.12). Notice that the closed loop gain depends only on the passive components and that variations in k have only a marginal effect on the closed loop gain. For example if $k = 10^6$ and $R_2/R_1 = 100$, a variation of k by 100% gives only a variation of 0.01% in the closed loop gain. The drastic reduction in sensitivity is a nice illustration of how feedback can be used to make precise systems from uncertain components. In this particular case, feedback is used to trade high gain and low robustness for low gain and high robustness. Equation (4.14) was the formula that inspired Black when he invented the feedback amplifier [45] (see the quote at the beginning of Chapter 13).

It is instructive to develop a block diagram for the feedback amplifier in Figure 4.9a. To do this we will represent the pure amplifier with input v and output v_2 as one block. To complete the block diagram, we must describe how v depends on v_1 and v_2. Solving equation (4.13) for v gives

$$v = \frac{R_2}{R_1 + R_2}v_1 + \frac{R_1}{R_1 + R_2}v_2 = \frac{R_1}{R_1 + R_2}\left(\frac{R_2}{R_1}v_1 + v_2\right),$$

and we obtain the block diagram shown in Figure 4.9b. The diagram clearly shows that the system has feedback and that the gain from v_2 to v is $R_1/(R_1 + R_2)$, which can also be read from the circuit diagram in Figure 4.9a. If the loop is stable and the gain of the amplifier is large, it follows that the error e is small, and we find that $v_2 = -(R_2/R_1)v_1$. Notice that the resistor R_1 appears in two blocks in the block diagram. This situation is typical in electrical circuits, and it is one reason why block diagrams are not always well suited for some types of physical modeling.

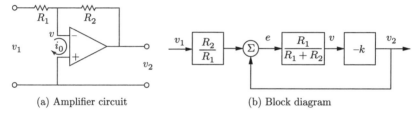

(a) Amplifier circuit   (b) Block diagram

Figure 4.9: Stable amplifier using an op amp. The circuit (a) uses negative feedback around an operational amplifier and has a corresponding block diagram (b). The resistors R_1 and R_2 determine the gain of the amplifier.

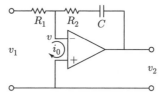

Figure 4.10: Circuit diagram of a PI controller obtained by feedback around an operational amplifier. The capacitor C is used to store charge and represents the integral of the input.

The simple model of the amplifier given by equation (4.11) provides qualitative insight, but it neglects the fact that the amplifier is a dynamical system. A more realistic model is

$$\frac{dv_{\text{out}}}{dt} = -av_{\text{out}} - bv. \tag{4.15}$$

The parameter b has dimensions of frequency and is called the *gain-bandwidth product* of the amplifier. Whether a more complicated model is used depends on the questions to be answered and the required size of the uncertainty lemon. The model (4.15) is still not valid for very high or very low frequencies since drift causes deviations at low frequencies and there are additional dynamics that appear at frequencies close to b. The model is also not valid for large signals—an upper limit is given by the voltage of the power supply, typically in the range of 5–10 V—neither is it valid for very low signals because of electrical noise. These effects can be added, if needed, but increase the complexity of the analysis.

The operational amplifier is very versatile, and many different systems can be built by combining it with resistors and capacitors. In fact, any linear system can be implemented by combining operational amplifiers with resistors and capacitors. Exercise 4.4 shows how a second-order oscillator is implemented, and Figure 4.10 shows the circuit diagram for an analog proportional-integral controller. To develop a simple model for the circuit we assume that the current i_0 is zero and that the open loop gain k is so large that the input voltage v is negligible. The current i through the capacitor is $i = Cdv_{\text{c}}/dt$, where v_{c} is the voltage across the capacitor. Since the same current goes through the resistor R_1, we get

$$i = \frac{v_1}{R_1} = C\frac{dv_{\text{c}}}{dt},$$

which implies that

$$v_c(t) = \frac{1}{C} \int i(t) \, dt = \frac{1}{R_1 C} \int_0^t v_1(\tau) d\tau.$$

The output voltage is thus given by

$$v_2(t) = -R_2 i - v_c = -\frac{R_2}{R_1} v_1(t) - \frac{1}{R_1 C} \int_0^t v_1(\tau) d\tau,$$

which is the input/output relation for a PI controller.

The development of operational amplifiers was pioneered by Philbrick [170, 201], and their usage is described in many textbooks (e.g., [65]). Good information is also available from suppliers [133, 176].

4.4 COMPUTING SYSTEMS AND NETWORKS

The application of feedback to computing systems follows the same principles as the control of physical systems, but the types of measurements and control inputs that can be used are somewhat different. Measurements (sensors) are typically related to resource utilization in the computing system or network and can include quantities such as the processor load, memory usage, or network bandwidth. Control variables (actuators) typically involve setting limits on the resources available to a process. This might be done by controlling the amount of memory, disk space, or time that a process can consume, turning processes on or off delaying availability of a resource, or rejecting incoming requests to a server process. Process modeling for networked computing systems is also challenging, and empirical models based on measurements are often used when a first-principles model is not available.

Web Server Control

Web servers respond to requests from the Internet and provide information in the form of web pages. Modern web servers start multiple processes to respond to requests, with each process assigned to a single source until no further requests are received from that source for a predefined period of time. Processes that are idle become part of a pool that can be used to respond to new requests. To provide a fast response to web requests, it is important that the web server processes do not overload the server's computational capabilities or exhaust its memory. Since other processes may be running on the server, the amount of available processing power and memory is uncertain, and feedback can be used to provide good performance in the presence of this uncertainty.

Figure 4.11 illustrates the use of feedback to modulate the operation of an Apache web server. The web server operates by placing incoming connection requests on a queue and then starting a subprocess to handle requests for each accepted connection. This subprocess responds to requests from a given connection as they come in, alternating between a Busy state and a Wait state. (Keeping the subprocess active between requests is known as the *persistence* of the connection and provides a substantial reduction in latency to requests for multiple pieces of information from a single site.) If no requests are received for a sufficiently long period of time, controlled by the KeepAlive parameter, then the connection is dropped and the

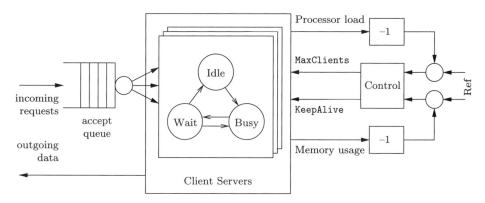

Figure 4.11: Feedback control of a web server. Connection requests arrive on an input queue, where they are sent to a server process. A finite state machine keeps track of the state of the individual server processes and responds to requests. A control algorithm can modify the server's operation by controlling parameters that affect its behavior, such as the maximum number of requests that can be serviced at a single time (`MaxClients`) or the amount of time that a connection can remain idle before it is dropped (`KeepAlive`).

subprocess enters an `Idle` state, where it can be assigned another connection. A maximum of `MaxClients` simultaneous requests will be served, with the remainder remaining on the incoming request queue.

The parameters that control the server represent a trade-off between performance (how quickly requests receive a response) and resource usage (the amount of processing power and memory used by the server). Increasing the `MaxClients` parameter allows connection requests to be pulled off of the queue more quickly but increases the amount of processing power and memory usage that is required. Increasing the `KeepAlive` timeout means that individual connections can remain idle for a longer period of time, which decreases the processing load on the machine but increases the length of the queue (and hence the amount of time required for a user to initiate a connection). Successful operation of a busy server requires a proper choice of these parameters, often based on trial and error.

To model the dynamics of this system in more detail, we create a discrete-time model with states given by the average processor load x_{cpu} and the percentage memory usage x_{mem}. The inputs to the system are taken as the maximum number of clients u_{mc} and the keep-alive time u_{ka}. If we assume a linear model around the equilibrium point, the dynamics can be written as

$$\begin{pmatrix} x_{cpu}[k+1] \\ x_{mem}[k+1] \end{pmatrix} = \begin{pmatrix} A_{11} & A_{12} \\ A_{21} & A_{22} \end{pmatrix} \begin{pmatrix} x_{cpu}[k] \\ x_{mem}[k] \end{pmatrix} + \begin{pmatrix} B_{11} & B_{12} \\ B_{21} & B_{22} \end{pmatrix} \begin{pmatrix} u_{ka}[k] \\ u_{mc}[k] \end{pmatrix}, \quad (4.16)$$

where the coefficients of the A and B matrices can be determined based on empirical measurements or detailed modeling of the web server's processing and memory usage. Using system identification, Diao et al. [72, 117] identified the linearized dynamics as

$$A = \begin{pmatrix} 0.54 & -0.11 \\ -0.026 & 0.63 \end{pmatrix}, \qquad B = \begin{pmatrix} -85 & 4.4 \\ -2.5 & 2.8 \end{pmatrix} \times 10^{-4},$$

where the system was linearized about the equilibrium point

$$x_{\text{cpu}} = 0.58, \qquad u_{\text{ka}} = 11\,\text{s}, \qquad x_{\text{mem}} = 0.55, \qquad u_{\text{mc}} = 600.$$

This model shows the basic characteristics that were described above. Looking first at the B matrix, we see that increasing the `KeepAlive` timeout (first column of the B matrix) decreases both the processor usage and the memory usage since there is more persistence in connections and hence the server spends a longer time waiting for a connection to close rather than taking on a new active connection. The `MaxClients` connection increases both the processing and memory requirements. Note that the largest effect on the processor load is the `KeepAlive` timeout. The A matrix tells us how the processor and memory usage evolve in a region of the state space near the equilibrium point. The diagonal terms describe how the individual resources return to equilibrium after a transient increase or decrease. The off-diagonal terms show that there is coupling between the two resources, so that a change in one could cause a later change in the other.

Although this model is very simple, we will see in later examples that it can be used to modify the parameters controlling the server in real time and provide robustness with respect to uncertainties in the load on the machine. Similar types of mechanisms have been used for other types of servers. It is important to remember the assumptions on the model and their role in determining when the model is valid. In particular, since we have chosen to use average quantities over a given sample time, the model will not provide an accurate representation for high-frequency phenomena.

Congestion Control

The Internet was created to provide a large, highly decentralized, efficient, and expandable communication system. The system consists of a large number of interconnected gateways. A message is split into several packets that are transmitted over different paths in the network, and the packages are rejoined to recover the message at the receiver. An acknowledgment ("ack") message is sent back to the sender when a packet is received. The operation of the system is governed by a simple but powerful decentralized control structure that has evolved over time.

The system has two control mechanisms called *protocols*: the Transmission Control Protocol (TCP) for end-to-end network communication and the Internet Protocol (IP) for routing packets and for host-to-gateway or gateway-to-gateway communication. The current protocols evolved after some spectacular congestion collapses occurred in the mid 1980s, when throughput unexpectedly could drop by a factor of 1000 [128]. The control mechanism in TCP is based on conserving the number of packets in the loop from the sender to the receiver and back to the sender. The sending rate is increased when there is no congestion, and it is dropped to a low level when there is congestion.

To derive an overall model for congestion control, we model three separate elements of the system: the rate at which packets are sent by individual sources (computers), the dynamics of the queues in the links (routers), and the admission control mechanism for the queues. Figure 4.12a is a block diagram of the system.

The current source control mechanism on the Internet is a protocol known as TCP/Reno [168]. This protocol operates by sending packets to a receiver and waiting to receive an acknowledgment from the receiver that the packet has arrived. If

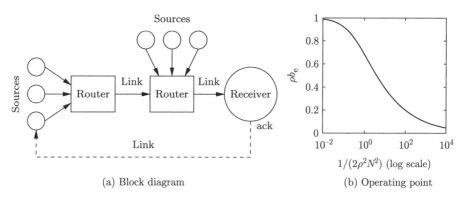

Figure 4.12: Internet congestion control. (a) Source computers send information to routers, which forward the information to other routers that eventually connect to the receiving computer. When a packet is received, an acknowledgment packet is sent back through the routers (not shown). The routers buffer information received from the sources and send the data across the outgoing link. (b) The equilibrium buffer size b_e for a set of N identical computers sending packets through a single router with drop probability ρb.

no acknowledgment is sent within a certain timeout period, the packet is retransmitted. To avoid waiting for the acknowledgment before sending the next packet, Reno transmits multiple packets up to a fixed *window* around the latest packet that has been acknowledged. If the window size is chosen properly, packets at the beginning of the window will be acknowledged before the source transmits packets at the end of the window, allowing the computer to continuously stream packets at a high rate.

To determine the size of the window to use, TCP/Reno uses a feedback mechanism in which (roughly speaking) the window size is increased at a fixed rate as long as packets are acknowledged, and the window size is cut in half when packets are lost. This mechanism allows a dynamic adjustment of the window size in which each computer acts in a greedy fashion as long as packets are being delivered but backs off quickly when congestion occurs.

A model for the behavior of the source can be developed by describing the dynamics of the window size. Suppose we have N computers (sources) and let w_i be the current window size (measured in number of packets) for the ith computer. Let q_i represent the end-to-end probability that a packet will be dropped someplace between the source and the receiver. We can model the dynamics of the window size w_i by the differential equation

$$\frac{dw_i}{dt} = (1 - q_i)\frac{r_i(t - \tau_i)}{w_i} - q_i\left(\frac{w_i}{2}r_i(t - \tau_i)\right), \qquad r_i = \frac{w_i}{\tau_i}, \qquad (4.17)$$

where τ_i is the round-trip time for a packet to reach its destination and the acknowledgment to be sent back, and r_i is the resulting rate at which packets are cleared from the list of packets that have been received. The first term in the dynamics represents the increase in window size when a packet is received, and the second term represents the decrease in window size when a packet is lost. Notice that r_i is evaluated at time $t - \tau_i$, representing the time required to receive acknowledgments that a packet has arrived.

The link dynamics are controlled by the dynamics of the router queue and the admission control mechanism for the queue. Assume that we have L links in the network and use l to index the individual links. We model the queue in terms of the current number of packets in the router's buffer b_l and assume that the router transmits packets at a rate c_l, equal to the capacity of the link. The buffer dynamics can then be written as

$$\frac{db_l}{dt} = \begin{cases} s_l - c_l & \text{if } b_l > 0, \\ 0 & \text{if } b_l = 0, \end{cases} \qquad s_l = \sum_{i=1}^{L} R_{li}\, r_i(t - \tau_{li}^{\text{f}}), \qquad (4.18)$$

where $R_{li} = 1$ if link l is used by source i and 0 otherwise, τ_{li}^{f} is the time it takes a packet from source i to reach link l, and s_l is the total rate at which packets arrive at link l. The matrix $R \in \mathbb{R}^{L \times N}$ is called the *routing matrix*.

The admission control mechanism determines whether a given packet is accepted by a router. Since our model is based on the average quantities in the network and not the individual packets, one simple model is to assume that the probability that a packet is dropped depends on how full the buffer is. If we let $b_{l,\max}$ be the maximum number of packets that the router l can buffer, we write the drop probability as $p_l = \beta_l(b_l, b_{l,\max})$, where β_l is a function with $\beta_l(0, b_{l,\max}) = 0$ and $\beta_l(b_{l,\max}, b_{l,\max}) = 1$. For simplicity, we will assume for now that $p_l = \rho_l b_l$ (see Exercise 4.5 for a more detailed model). The probability that a packet is dropped at a given link can be used to determine the end-to-end probability that a packet is lost in transmission:

$$q_i = 1 - \prod_{l=1}^{L} R_{li}(1 - p_l) \approx \sum_{l=0}^{L} R_{li}\, p_l(t - \tau_{il}^{\text{b}}), \qquad (4.19)$$

where τ_{il}^{b} is the backward delay from link l to source i and the approximation is valid as long as the individual drop probabilities are small. We use the backward delay since this represents the time required for the acknowledgment packet to be received by the source.

Together, equations (4.17), (4.18), and (4.19) represent a model of congestion control dynamics. We can obtain substantial insight by considering a special case in which we have N identical sources and one link. In addition, we assume for the moment that the forward and backward time delays can be ignored and that none of the routers are saturated or empty, in which case the dynamics can be reduced to the form

$$\frac{dw_i}{dt} = \frac{1}{\tau^{\text{p}}} - \frac{\rho c(2 + w_i^2)}{2}, \qquad \frac{db}{dt} = \sum_{i=1}^{N} \frac{w_i}{\tau^{\text{p}}} - c, \qquad \tau^{\text{p}} := \frac{b}{c}, \qquad (4.20)$$

where $w_i \in \mathbb{R}$, $i = 1, \ldots, N$, is a vector of window sizes for the sources of data, $b \in \mathbb{R}$ is the current buffer size of the router, ρ controls the rate at which packets are dropped, and c is the capacity of the link connecting the router to the computers. The variable τ^{p} represents the amount of time required for a packet to be processed by the router, based on the size of the buffer and the capacity of the link. Substituting τ^{p} into the equations, we write the state space dynamics as

$$\frac{dw_i}{dt} = \frac{c}{b} - \rho c\left(1 + \frac{w_i^2}{2}\right), \qquad \frac{db}{dt} = \sum_{i=1}^{N} \frac{cw_i}{b} - c. \qquad (4.21)$$

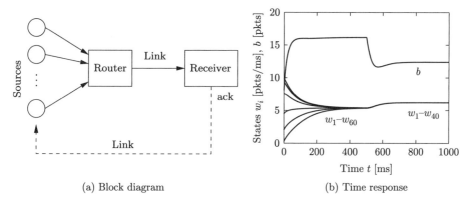

(a) Block diagram   (b) Time response

Figure 4.13: Internet congestion control for N identical sources across a single link. As shown on the left, multiple sources attempt to communicate through a router across a single link. An "ack" packet sent by the receiver acknowledges that the message was received; otherwise the message packet is resent and the sending rate is slowed down at the source. The simulation on the right is for 60 sources starting at random rates (window sizes), with 20 sources dropping out at $t = 500$ ms. The buffer size is shown at the top, and the individual source rates for 6 of the sources are shown at the bottom.

More sophisticated models can be found in [167, 168] and subsequent exercises and examples.

The nominal operating point for the system can be found by setting $\dot{w}_i = \dot{b} = 0$:

$$0 = \frac{c}{b} - \rho c \left(1 + \frac{w_i^2}{2}\right), \qquad 0 = \sum_{i=1}^{N} \frac{cw_i}{b} - c.$$

Exploiting the fact that all of the source dynamics are identical, it follows that all of the w_i should be the same, and it can be shown that there is a unique equilibrium point satisfying the equations

$$w_{i,e} = \frac{b_e}{N} = \frac{c\tau_e^{\mathrm{P}}}{N}, \qquad \frac{1}{2\rho^2 N^2}(\rho b_e)^3 + (\rho b_e) - 1 = 0. \tag{4.22}$$

The solution for the second equation is a bit messy but can easily be determined numerically. A plot of its solution as a function of $1/(2\rho^2 N^2)$ is shown in Figure 4.12b. We also note that at equilibrium we have the following additional equalities:

$$\tau_e^{\mathrm{P}} = \frac{b_e}{c} = \frac{N w_e}{c}, \qquad q_e = N p_e = N\rho b_e, \qquad r_e = \frac{w_e}{\tau_e^{\mathrm{P}}}. \tag{4.23}$$

Figure 4.13 shows a simulation of 60 sources communicating across a single link, with 20 sources dropping out at $t = 500$ ms and the remaining sources increasing their rates (window sizes) to compensate. Note that the buffer size and window sizes automatically adjust to match the capacity of the link.

A comprehensive treatment of computer networks is given in the textbook by Tannenbaum [236]. A good presentation of the ideas behind the control principles for the Internet is given by one of its designers, Van Jacobson, in [128]. F. Kelly [142] presents an early effort on the analysis of the system. The books by Hellerstein et al.

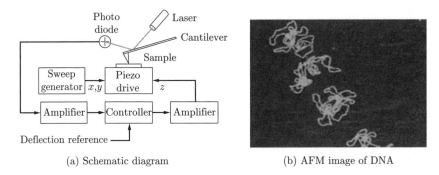

<table>
<tr><td>(a) Schematic diagram</td><td>(b) AFM image of DNA</td></tr>
</table>

Figure 4.14: Atomic force microscope. (a) A schematic diagram of an atomic force microscope, consisting of a piezo drive that scans the sample under the AFM tip. A laser reflects off of the cantilever and is used to measure the detection of the tip through a feedback controller. (b) An AFM image of strands of DNA. (Image courtesy of Bruker Corporation.)

[117] and Janert [131] give many examples of the use of feedback in computer systems.

4.5   ATOMIC FORCE MICROSCOPY

The 1986 Nobel Prize in Physics was shared by Gerd Binnig and Heinrich Rohrer for their design of the *scanning tunneling microscope*. The idea of the instrument is to bring an atomically sharp tip so close to a conducting surface that tunneling occurs. An image is obtained by traversing the tip across the sample and measuring the tunneling current as a function of tip position. This invention has stimulated the development of a family of instruments that permit visualization of surface structure at the nanometer scale, including the *atomic force microscope* (AFM), where a sample is probed by a tip on a cantilever. An AFM can operate in two modes. In *tapping mode* the cantilever is vibrated, and the amplitude of vibration is controlled by feedback. In *contact mode* the cantilever is in contact with the sample, and its bending is controlled by feedback. In both cases control is actuated by a piezo element that controls the vertical position of the cantilever base (or the sample). Control design has a direct influence on picture quality and scanning rate.

A schematic picture of an atomic force microscope is shown in Figure 4.14a. A microcantilever with a tip having a radius of the order of 10 nm is placed close to the sample. The tip can be moved vertically and horizontally using a piezoelectric scanner. It is clamped to the sample surface by attractive van der Waals forces and repulsive Pauli forces. The cantilever tilt depends on the topography of the surface and the position of the cantilever base, which is controlled by the piezo element. The tilt is measured by sensing the deflection of the laser beam using a photodiode. The signal from the photodiode is amplified and sent to a controller that drives the amplifier for the vertical position of the cantilever (z). By controlling the piezo element so that the deflection of the cantilever is constant, the signal driving the vertical deflection of the piezo element is a measure of the atomic forces between the cantilever tip and the atoms of the sample. An image of the surface is obtained by scanning the cantilever along the sample. The resolution makes it possible to

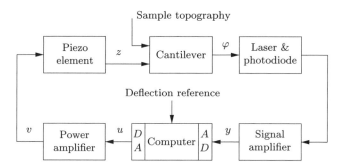

Figure 4.15: Block diagram of the system for vertical positioning of the cantilever for an atomic force microscope in contact mode. The control system attempts to keep the cantilever deflection equal to its reference value. Cantilever deflection is measured, amplified, and converted to a digital signal, then compared with its reference value. A correcting signal is generated by the computer, converted to analog form, amplified, and sent to the piezo element.

see the structure of the sample on the atomic scale, as illustrated in Figure 4.14b, which shows an AFM image of DNA.

The horizontal motion of an AFM is typically modeled as a spring–mass system with low damping. The vertical motion is more complicated. To model the system, we start with the block diagram shown in Figure 4.15. Signals that are easily accessible are the input voltage u to the power amplifier that drives the piezo element, the voltage v applied to the piezo element, and the output voltage y of the signal amplifier for the photodiode. The controller is a PI controller implemented by a computer, which is connected to the system by analog-to-digital (A/D) and digital-to-analog (D/A) converters. The deflection of the cantilever φ is also shown in the figure. The desired reference value for the deflection is an input to the computer.

There are several different configurations that have different dynamics. Here we will discuss a high-performance system from [217] where the cantilever base is positioned vertically using a piezo stack. We begin the modeling with a simple experiment on the system. Figure 4.16a shows a step response of a scanner from the power amplifier input voltage u to the output voltage y of the signal amplifier for the photodiode. This experiment captures the dynamics of the chain of blocks from u to y in the block diagram in Figure 4.15. Figure 4.16a shows that the system responds quickly but that there is a poorly damped oscillatory mode with a period of about 35 μs. A primary task of the modeling is to understand the origin of the oscillatory behavior. To do so we will explore the system in more detail.

The natural frequency of the clamped cantilever is typically several hundred kilohertz, which is much higher than the observed oscillation of about 30 kHz. As a first approximation we will model it as a static system. Since the deflections are small, we can assume that the bending φ of the cantilever is proportional to the difference in height between the cantilever tip at the probe and the piezo scanner. A more accurate model can be obtained by modeling the cantilever as a spring–mass system of the type discussed in Chapter 3.

Figure 4.16a also shows that the response of the power amplifier is fast. The photodiode and the signal amplifier also have fast responses and can thus be modeled as static systems. The remaining block is a piezo system with suspension. A

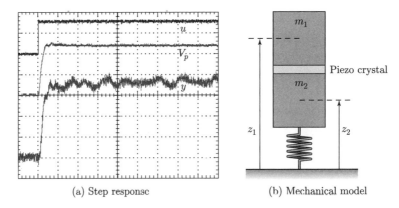

(a) Step response   (b) Mechanical model

Figure 4.16: Modeling of an atomic force microscope. (a) A measured step response. The top curve shows the voltage u applied to the drive amplifier (50 mV/div), the middle curve is the output V_p of the power amplifier (500 mV/div), and the bottom curve is the output y of the signal amplifier (500 mV/div). The time scale is 25 μs/div. Data have been supplied by Georg Schitter. (b) A simple mechanical model for the vertical positioner and the piezo crystal.

schematic mechanical representation of the vertical motion of the scanner is shown in Figure 4.16b. We will model the system as two masses separated by an ideal piezo element. The mass m_1 is half of the piezo system, and the mass m_2 is the other half of the piezo system plus the mass of the support.

A simple model is obtained by assuming that the piezo crystal generates a force F between the masses and that there is a damping c_2 in the spring. Let the positions of the center of the masses be z_1 and z_2. A momentum balance gives the following model for the system:

$$m_1 \frac{d^2 z_1}{dt^2} = F, \qquad m_2 \frac{d^2 z_2}{dt^2} = -c_2 \frac{dz_2}{dt} - k_2 z_2 - F.$$

Let the elongation of the piezo element $l = z_1 - z_2$ be the control variable and the height z_1 of the cantilever base be the output. Eliminating the variable F in the equations above and substituting $z_1 - l$ for z_2 gives the model

$$(m_1 + m_2) \frac{d^2 z_1}{dt^2} + c_2 \frac{dz_1}{dt} + k_2 z_1 = m_2 \frac{d^2 l}{dt^2} + c_2 \frac{dl}{dt} + k_2 l. \qquad (4.24)$$

Summarizing, we find that a simple model of the system is obtained by modeling the piezo by equation (4.24) and all the other blocks by static models. Introducing the linear equations $l = k_3 u$ and $y = k_4 z_1$, we now have a complete model relating the output y to the control signal u. A more accurate model can be obtained by introducing the dynamics of the cantilever and the power amplifier. As in the previous examples, the concept of the uncertainty lemon in Figure 3.5b provides a framework for describing the uncertainty: the model will be accurate up to the frequencies of the fastest modeled modes and over a range of motion in which linearized stiffness models can be used.

The experimental results in Figure 4.16a can be explained qualitatively as follows. When a voltage is applied to the piezo, it expands by l_0, the mass m_1 moves up, and the mass m_2 moves down instantaneously. The system settles after a poorly damped oscillation.

It is highly desirable to design a control system for the vertical motion so that it responds quickly with little oscillation. The instrument designer has several choices: to accept the oscillation and have a slow response time, to design a control system that can damp the oscillations, or to redesign the mechanics to give resonances of higher frequency. The last two alternatives give a faster response and faster imaging.

Since the dynamic behavior of the system changes with the properties of the sample, it is necessary to tune the feedback loop. In simple systems this is currently done manually by adjusting parameters of a PI controller. There are interesting possibilities for making AFM systems easier to use by introducing automatic tuning and adaptation.

The book by Sarid [214] gives a broad coverage of atomic force microscopes. The interaction of atoms close to surfaces is fundamental to solid state physics; see Kittel [147]. The model discussed in this section is based on Schitter [216].

4.6   DRUG ADMINISTRATION

The phrase "take two pills three times a day" is a recommendation with which we are all familiar. Behind this recommendation is a solution of an open loop control problem. The key issue is to make sure that the concentration of a medicine in a part of the body is sufficiently high to be effective but not so high that it will cause undesirable side effects. The control action is quantized, *take two pills*, and sampled, *every 8 hours*. The prescriptions are based on simple models captured in empirical tables, and the dose is based on the age and weight of the patient.

Drug administration is a control problem. To solve it we must understand how a drug spreads in the body after it is administered. This topic, called *pharmacokinetics*, is now a discipline of its own, and the models used are called *compartment models*. They go back to the 1920s when Widmark modeled the propagation of alcohol in the body [252]. Compartment models are now important for the screening of all drugs used by humans. The schematic diagram in Figure 4.17 illustrates the idea of a compartment model. The body is viewed as a number of compartments like blood plasma, kidney, liver, and tissues that are separated by membranes. It is assumed that there is perfect mixing so that the drug concentration is constant in each compartment. The complex transport processes are approximated by assuming that the flow rates between the compartments are proportional to the concentration differences in the compartments.

To describe the effect of a drug it is necessary to know both its concentration and how it influences the body. The relation between concentration c and its effect e is typically nonlinear. A simple model is

$$e = \frac{c}{\mathrm{EC}_{50} + c} e_{\max}. \tag{4.25}$$

The effect is linear for low concentrations, and it saturates at high concentrations. The parameter EC_{50} represents the concentration of the drug that gives half (50%) maximal response. The relation can also be dynamic, and it is then called *pharmacodynamics*.

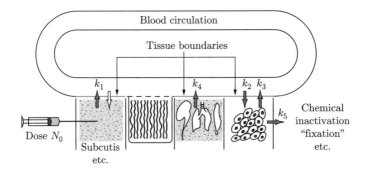

Figure 4.17: Abstraction used to compartmentalize the body for the purpose of describing drug distribution (based on Teorell [237]). The body is abstracted by a number of compartments with perfect mixing, and the complex transport processes are approximated by assuming that the flow is proportional to the concentration differences in the compartments. The constants k_i parameterize the rates of flow between different compartments.

Compartment Models

The simplest dynamical model for drug administration is obtained by assuming that the drug is evenly distributed in a single compartment after it has been administered and that the drug is removed at a rate proportional to the concentration. The compartments behave like stirred tanks with perfect mixing. Let c be the concentration, V the volume, and q the outflow rate. Converting the description of the system into differential equations gives the model

$$V\frac{dc}{dt} = -qc, \qquad c \geq 0. \tag{4.26}$$

This equation has the solution $c(t) = c_0 e^{-qt/V} = c_0 e^{-kt}$, which shows that the concentration decays exponentially with the time constant $T = V/q$ after an injection. The input is introduced implicitly as an initial condition in the model (4.26). More generally, the way the input enters the model depends on how the drug is administered. For example, the input can be represented as a mass flow into the compartment where the drug is injected. A pill that is dissolved can also be interpreted as an input in terms of a mass flow rate.

The model (4.26) is called a *one-compartment model* or a *single-pool model*. The parameter $k = q/V$ is called the elimination rate constant. This simple model is often used to model the concentration in the blood plasma. By measuring the concentration at a few times, the initial concentration can be obtained by extrapolation. If the total amount of injected substance m is known, the volume V can then be determined as $V = m/c_0$.

The simple one-compartment model captures the gross behavior of drug distribution, but it is based on many simplifications. Improved models can be obtained by considering the body as composed of several compartments. Examples of such systems are shown in Figure 4.18, where the compartments are represented as circles and the flows by arrows.

Modeling will be illustrated using the two-compartment model in Figure 4.18a. We assume that there is perfect mixing in each compartment and that the transport between the compartments is driven by concentration differences. We further assume

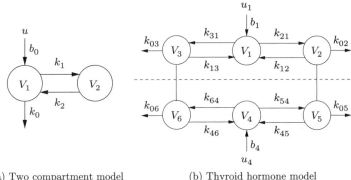

(a) Two compartment model   (b) Thyroid hormone model

Figure 4.18: Schematic diagrams of compartment models. (a) A simple two-compartment model. Each compartment is labeled by its volume, and arrows indicate the flow of chemical into, out of, and between compartments. (b) A system with six compartments used to study the metabolism of thyroid hormone [103]. The notation k_{ij} denotes the transport from compartment j to compartment i.

that a drug with concentration c_0 is injected in compartment 1 at a volume flow rate of u and that the concentration in compartment 2 is the output. Let c_1 and c_2 be the concentrations of the drug in the compartments, and let V_1 and V_2 be the volumes of the compartments. The mass balances for the compartments are

$$V_1 \frac{dc_1}{dt} = q(c_2 - c_1) - q_0 c_1 + c_0 u, \qquad c_1 \geq 0,$$

$$V_2 \frac{dc_2}{dt} = q(c_1 - c_2), \qquad c_2 \geq 0, \qquad (4.27)$$

$$y = c_2,$$

where q represents flow rate between the compartments and q_0 represents the flow rate out of compartment 1 that is not going to compartment 2. Introducing the variables $k_0 = q_0/V_1$, $k_1 = q/V_1$, $k_2 = q/V_2$, and $b_0 = c_0/V_1$ and using matrix notation, the model can be written as

$$\frac{dc}{dt} = \begin{pmatrix} -k_0 - k_1 & k_1 \\ k_2 & -k_2 \end{pmatrix} c + \begin{pmatrix} b_0 \\ 0 \end{pmatrix} u, \qquad y = \begin{pmatrix} 0 & 1 \end{pmatrix} c. \qquad (4.28)$$

Comparing this model with its graphical representation in Figure 4.18a, we find that the mathematical representation (4.28) can be written by inspection.

It should also be emphasized that simple compartment models such as the one in equation (4.28) have a limited range of validity. Low-frequency limits exist because the human body changes with time, and since the compartment model uses average concentrations, they will not accurately represent rapid changes. There are also nonlinear effects that influence transportation between the compartments.

Compartment models are widely used in medicine, engineering, and environmental science. An interesting property of these systems is that variables like concentration and mass are always positive. An essential difficulty in compartment modeling is deciding how to divide a complex system into compartments. Compartment models can also be nonlinear, as illustrated in the next section.

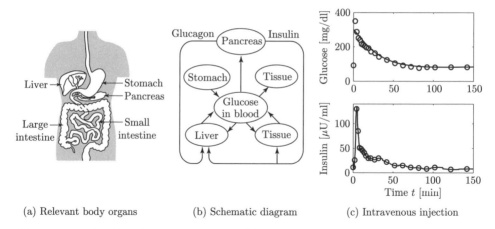

(a) Relevant body organs   (b) Schematic diagram   (c) Intravenous injection

Figure 4.19: Insulin–glucose dynamics. (a) Sketch of body parts involved in the control of glucose. (b) Schematic diagram of the system. (c) Responses of insulin and glucose when glucose in injected intravenously. From Pacini and Bergman [199].

The papers by Widmark and Tandberg [252] and Teorell [237] are classics in pharmacokinetics, which is now an established discipline with many textbooks [74, 102, 129]. Because of its medical importance, pharmacokinetics is now an essential component of drug development. The book by Riggs [208] is a good source for the modeling of physiological systems, and a more mathematical treatment is given in Keener and Sneyd [140, 141]. Compartment models are discussed in Godfrey [103]. The problem of determining rate coefficients from experimental data is discussed in Bellman and Åström [32] and Godfrey [103].

Insulin–Glucose Dynamics

Glucose provides energy to all cells in the body. It is influenced by many factors: body constitution, food intake, digestion, stress, and exercise. Healthy individuals have sophisticated mechanisms that regulate glucose concentration in the blood. A schematic picture of the relevant parts of the body involved are shown in Figures 4.19a and 4.19b. The pancreas secretes the hormones insulin and glucagon. Glucagon is released into the bloodstream when the glucose level is low. It acts on cells in the liver that release glucose. Insulin is secreted when the glucose level is high, and the glucose level is lowered by causing the liver and other cells to take up more glucose. There are also other hormones that influence glucose concentration. It is important that the blood glucose concentration is regulated to be in the range 70–110 mg/L.

Diabetes is a disease where the body's ability to produce or respond to insulin is impaired, resulting in blood sugar levels that are too high. There are several varieties of diabetes: production of insulin can be impaired (type 1) or the ability of the body to absorb insulin can be reduced (type 2). Long exposure to high blood sugar concentration is serious and may result in cardiovascular diseases, stroke, chronic kidney disease, foot ulcers, and blindness. Low blood sugar is also serious and can give headaches, fatigue, dizziness, lethargy, and blurred vision. Very low blood sugar levels can result in a coma.

The mechanisms that regulate glucose and insulin are complicated. Models of different complexity have been developed. The models are typically tested with data from experiments where glucose is injected intravenously and insulin and glucose concentrations are measured at regular time intervals, as shown in Figure 4.19c.

A simple *minimal model* was developed by Bergman and coworkers [39, 40]. It is a compartment model with two state variables: concentration of glucose in the blood plasma G and the variable X representing the effect of insulin on glucose removal, which is proportional to the concentration of insulin I in the interstitial fluid. The minimal model is given by the equations

$$\frac{dG}{dt} = -p_1(G - G_e) - XG + u_G, \qquad \frac{dX}{dt} = -p_2 X + p_3(I - I_e). \qquad (4.29)$$

The first equation is a compartment model for glucose. The right-hand side has three terms: a linear clearance term that models glucose removal at a rate proportional to $G - G_e$, the nonlinear term XG, and the external input u_G that represents injection of glucose. The nonlinear term XG captures the fact that removal rate of glucose is enhanced by insulin. The second equation represents how the variable X depends on the insulin concentration I in the interstitial fluid. If the external input u_G is zero and $I = I_e$ there is an equilibrium with $G = G_e$ and $X = 0$.

A model that is slightly more complicated than the minimal model is given in Exercise 4.8 and includes a model for insulin dynamics. Figure 4.19c shows a fit of the model to a test on a normal person where glucose was injected intravenously at time $t = 0$ and samples of concentrations of insulin and glucose are taken at different times. The glucose concentration rises rapidly, and the pancreas responds with a rapid spike-like injection of insulin. The glucose and insulin levels then gradually approach the equilibrium values.

There are many more complicated models that capture dynamics of food intake and measurement dynamics [63, 69, 86, 96, 174]. The models are used in many different ways for insight, analysis, and treatment of diabetes. A model for type 1 diabetes developed at the University of Virginia [160] has been approved by the U.S. Food and Drug Administration (FDA) as a replacement for animal testing of closed loop control strategies for regulation of blood sugar (*in silico* testing).

A simple way to measure blood sugar is to analyze glucose concentration in a drop of blood obtained by a fingerstick. Diabetic patients can also be provided with a continuous glucose monitor (GCM), which is a tiny sensor wire under the skin with an adhesive patch and a wireless transmitter. The sensor measures glucose concentration in the interstitial fluid near the sensor wire; calibration is required to obtain the glucose concentration in the bloodstream. The sensor is often placed in the upper arm where it can be connected wirelessly to a smartphone. An application on the phone can then generate advice on how much insulin has to be injected, for example long-lasting insulin for maintenance of a base level and rapid-acting insulin taken at meal times. The advice is based on a model of the glucose-insulin system that is matched to the patient. Devices of this type are increasingly available and widely used by patients with diabetes.

Patients with type 1 diabetes can also be provided with an *artificial pancreas*, a fully automatic system that regulates the blood sugar [63, 150]. An artificial pancreas consists of a glucose monitor that measures blood sugar, an insulin infusion pump, and a control algorithm that computes the amount of insulin to be injected based on the measured blood sugar value. The Medtronic MiniMed 670G was approved by FDA for use by adults in 2016 and for children over seven years old in 2018. The system has a sampling period of 5 minutes and a PID algorithm to control the injection rate [228]. Similar devices with model predictive control have also been tested [37]. The glucose monitor requires frequent observation, the wire has to be replaced regularly, and the sensor must be calibrated frequently using a fingerstick. There are extreme safety requirements on an artificial pancreas [36, 150], and it is absolutely essential to ensure that the glucose level does not get too low (hypoglycemia). All these additions make the system more complicated.

4.7 POPULATION DYNAMICS

Population growth is a complex dynamic process that involves the interaction of one or more species with their environment and the larger ecosystem. The dynamics of population groups are interesting and important in many different areas of social and environmental policy. There are examples where new species have been introduced into new habitats, sometimes with disastrous results. There have also been attempts to control population growth both through incentives and through legislation. In this section we describe some of the models that can be used to understand how populations evolve with time and as a function of their environments.

Logistic Growth Model

Let x be the population of a species at time t. A simple model is to assume that the birth rates and mortality rates are proportional to the total population. This gives the linear model

$$\frac{dx}{dt} = bx - dx = (b - d)x = rx, \qquad x \geq 0, \tag{4.30}$$

where birth rate b and mortality rate d are parameters. The model gives an exponential increase if $b > d$ or an exponential decrease if $b < d$. A more realistic model is to assume that the birth rate decreases when the population is large. The following modification of the model (4.30) has this property:

$$\frac{dx}{dt} = rx\left(1 - \frac{x}{k}\right), \qquad x \geq 0, \tag{4.31}$$

where k is the *carrying capacity* of the environment. The model (4.31) is called the *logistic growth model*.

Predator–Prey Models

A more sophisticated model of population dynamics includes the effects of competing populations, where one species may feed on another. This situation, referred to as the *predator–prey problem*, was introduced in Example 3.4, where we developed a discrete-time model that captured some of the features of historical records of lynx and hare populations.

In this section, we replace the difference equation model used there with a more sophisticated differential equation model. Let $H(t)$ represent the number of hares (prey) and let $L(t)$ represent the number of lynxes (predator). The dynamics of the system are modeled as

$$\frac{dH}{dt} = rH\left(1 - \frac{H}{k}\right) - \frac{aHL}{c+H}, \qquad H \geq 0,$$

$$\frac{dL}{dt} = b\frac{aHL}{c+H} - dL, \qquad\qquad L \geq 0. \tag{4.32}$$

In the first equation, r represents the growth rate of the hares, k represents the maximum population of the hares (in the absence of lynxes), a represents the inter-action term that describes how the hares are diminished as a function of the lynx population, and c controls the prey consumption rate for low hare population. In the second equation, b represents the growth coefficient of the lynxes and d represents the mortality rate of the lynxes. Note that the hare dynamics include a term that resembles the logistic growth model (4.31).

Of particular interest are the values at which the population values remain constant, called *equilibrium points*. The equilibrium points for this system can be determined by setting the right-hand side of the above equations to zero. Letting H_e and L_e represent the equilibrium state, from the second equation we have

$$L_e = 0 \quad \text{or} \quad H_e^* = \frac{cd}{ab-d}. \tag{4.33}$$

Substituting this into the first equation, we have that for $L_e = 0$ either $H_e = 0$ or $H_e = k$. For $L_e \neq 0$, we obtain

$$L_e^* = \frac{rH_e(c+H_e)}{aH_e}\left(1 - \frac{H_e}{k}\right) = \frac{bcr(abk - cd - dk)}{(ab-d)^2 k}. \tag{4.34}$$

Thus, we have three possible equilibrium points $x_e = (L_e, H_e)$:

$$x_e = \begin{pmatrix} 0 \\ 0 \end{pmatrix}, \qquad x_e = \begin{pmatrix} k \\ 0 \end{pmatrix}, \qquad x_e = \begin{pmatrix} H_e^* \\ L_e^* \end{pmatrix},$$

where H_e^* and L_e^* are given in equations (4.33) and (4.34). Note that the equilibrium populations may be negative for some parameter values, corresponding to a unachievable equilibrium point.

Figure 4.20 shows a simulation of the dynamics starting from a set of population values near the nonzero equilibrium values. We see that for this choice of parameters, the simulation predicts an oscillatory population count for each species, reminiscent of the data shown in Figure 3.7.

Volume I of the two-volume set by J. D. Murray [186] give a broad coverage of population dynamics.

EXERCISES

4.1 (Cruise control) Consider the cruise control example described in Section 4.1. Build a simulation that re-creates the response to a hill shown in Figure 4.3b and show the effects of increasing and decreasing the mass of the car by 25%. Redesign the controller (using trial and error is fine) so that it returns to within 1% of the desired speed within 3 s of encountering the beginning of the hill.

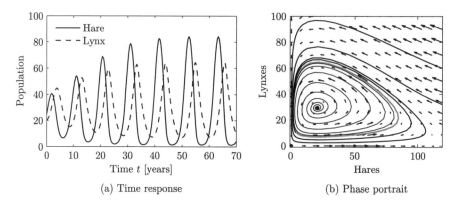

(a) Time response   (b) Phase portrait

Figure 4.20: Simulation of the predator–prey system. The figure on the left shows a simulation of the two populations as a function of time. The figure on the right shows the populations plotted against each other, starting from different values of the population. The oscillation seen in both figures is an example of a *limit cycle*. The parameter values used for the simulations are $a = 3.2$, $b = 0.6$, $c = 50$, $d = 0.56$, $k = 125$, and $r = 1.6$.

4.2 (Bicycle dynamics) Show that the dynamics of a bicycle frame given by equation (4.5) can be approximated in state space form as

$$\frac{d}{dt}\begin{pmatrix} x_1 \\ x_2 \end{pmatrix} = \begin{pmatrix} 0 & 1 \\ mgh/J & 0 \end{pmatrix}\begin{pmatrix} x_1 \\ x_2 \end{pmatrix} + \begin{pmatrix} Dv_0/(bJ) \\ mv_0^2 h/(bJ) \end{pmatrix} u,$$

$$y = \begin{pmatrix} 1 & 0 \end{pmatrix} x,$$

where the input u is the steering angle δ and the output y is the tilt angle φ. What do the states x_1 and x_2 represent?

4.3 (Operational amplifier circuit) Consider the op amp circuit shown below.

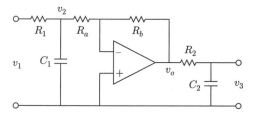

Show that the dynamics can be written in state space form as

$$\frac{dx}{dt} = \begin{pmatrix} -\dfrac{1}{R_1 C_1} - \dfrac{1}{R_a C_1} & 0 \\ -\dfrac{R_b}{R_a} \dfrac{1}{R_2 C_2} & -\dfrac{1}{R_2 C_2} \end{pmatrix} x + \begin{pmatrix} \dfrac{1}{R_1 C_1} \\ 0 \end{pmatrix} u, \quad y = \begin{pmatrix} 0 & 1 \end{pmatrix} x,$$

where $u = v_1$ and $y = v_3$. (Hint: Use v_2 and v_3 as your state variables.)

4.4 (Operational amplifier oscillator) The op amp circuit shown below is an implementation of an oscillator.

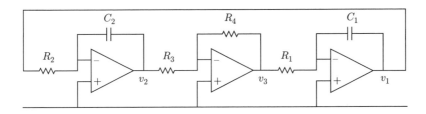

Show that the dynamics can be written in state space form as

$$\frac{dx}{dt} = \begin{pmatrix} 0 & \dfrac{R_4}{R_1 R_3 C_1} \\ -\dfrac{1}{R_2 C_2} & 0 \end{pmatrix} x,$$

where the state variables represent the voltages across the capacitors $x_1 = v_1$ and $x_2 = v_2$.

4.5 (Congestion control using RED [169]) A number of improvements can be made to the congestion control model presented in Section 4.4. To ensure that the router's buffer size remains positive, we can modify the buffer dynamics to satisfy

$$\frac{db_l}{dt} = \begin{cases} s_l - c_l & \text{if } 0 < b_l < b_{l,\text{max}}, \\ 0 & \text{otherwise.} \end{cases}$$

In addition, we can model the drop probability of a packet based on how close a filtered estimate of the buffer size is to the buffer limits, a mechanism known as random early detection (RED):

$$p_l = \beta_l(a_l) = \begin{cases} 0 & \text{if } a_l \leq b_l^{\text{low}}, \\ \rho_l(a_i - b_l^{\text{low}}) & \text{if } b_l^{\text{low}} < a_l < b_l^{\text{mid}}, \\ \eta_l(a_i - b_l^{\text{mid}}) + \rho_l(b_l^{\text{mid}} - b_l^{\text{low}}) & \text{if } b_l^{\text{mid}} \leq a_l < b_l^{\text{max}}, \\ 1 & \text{if } a_l \geq b_l^{\text{max}}, \end{cases}$$

$$\frac{da_l}{dt} = -\alpha_l c_l(a_l - b_l),$$

where α_l, ρ_l, η_l, b_l^{low}, b_l^{mid}, and b_l^{max} are parameters for the RED protocol. The variable a_l is a smoothed version of the buffer size b_l. Using the model above, write a simulation for the system and find a set of parameter values for which there is a stable equilibrium point and a set for which the system exhibits oscillatory solutions. The following sets of parameters should be explored:

$N = 20, 30, \ldots, 60,$	$b_l^{\text{low}} = 40 \text{ pkts},$	$\alpha_l = 10^{-4},$
$c = 8, 9, \ldots, 15 \text{ pkts/ms},$	$b_l^{\text{mid}} = 540 \text{ pkts},$	$\rho_l = 0.0002,$
$\tau^{\text{P}} = 55, 60, \ldots, 100 \text{ ms}$	$b_l^{\text{max}} = 1080 \text{ pkts},$	$\eta_l = 0.00167.$

4.6 (Atomic force microscope with piezo tube) A schematic diagram of an AFM where the vertical scanner is a piezo tube with preloading is shown below.

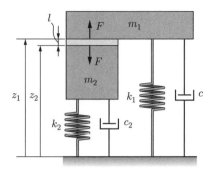

Show that the dynamics can be written as

$$(m_1 + m_2)\frac{d^2 z_1}{dt^2} + (c_1 + c_2)\frac{dz_1}{dt} + (k_1 + k_2)z_1 = m_2\frac{d^2 l}{dt^2} + c_2\frac{dl}{dt} + k_2 l,$$

where z_1 is the displacement of the first mass and $l = z_1 - z_2$ is the difference in displacement between the first and second masses. Are there parameter values that make the dynamics particularly simple?

4.7 (Drug administration) The metabolism of alcohol in the body can be modeled by the nonlinear compartment model

$$V_b\frac{dc_b}{dt} = q(c_l - c_b) + q_{iv}, \qquad V_l\frac{dc_l}{dt} = q(c_b - c_l) - q_{max}\frac{c_l}{c_0 + c_l} + q_{gi},$$

where $V_b = 48\,\text{L}$ and $V_l = 0.6\,\text{L}$ are the apparent volumes of distribution of body water and liver water, c_b and c_l are the concentrations of alcohol in the compartments, q_{iv} and q_{gi} are the injection rates for intravenous and gastrointestinal intake, $q = 1.5\,\text{L/min}$ is the total hepatic blood flow, $q_{max} = 2.75\,\text{mmol/min}$, and $c_0 = 0.1\,\text{mmol/L}$. Simulate the system and compute the concentration in the blood for oral and intravenous doses of 12 g and 40 g of alcohol.

4.8 (Insulin-glucose dynamics) The following model for insulin glucose dynamics by Gaetano and colleagues [96] has three states: glucose concentration in the blood plasma $G\,[\text{mg/dL}]$, insulin concentration in the interstitial fluid $I\,[\mu UI/\text{ml}]$, and $X\,[\text{min}^{-1}]$ that represents the increased removal rate of glucose due to insulin. The state X is proportional to the concentration of interstitial insulin. The dynamics are:

$$\frac{dG}{dt} = -(p_1 + X)G + p_1 G_b + u_G$$

$$\frac{dX}{dt} = -p_2 X + p_3(I - I_b)$$

$$\frac{dI}{dt} = p_4 \max(G - p_5, 0) - p_6(I - I_b) + u_I.$$

Use the parameters

$$G_b = 87, \quad I_b = 37.9, \quad p_1 = 0.05, \quad p_2 = 0.5, \quad p_3 = 10^{-4},$$
$$p_4 = 10^{-5}, \quad p_5 = 150, \quad p_6 = 0.05, \quad p_7 = 199.$$

Simulate the system with the initial conditions $G(0) = 400$, $I(0) = 200$ and $X(0) = 0$. This corresponds to a person having taken a large initial dose of glucose.

4.9 (Fisheries management) Some features of the dynamics of a commercial fishery can be described by the following simple model:

$$\frac{dx}{dt} = f(x) - h(x, u), \quad y = bh(x, u) - cu,$$

where x is the total biomass, $f(x) = rx(1 - x/k)$ is the growth rate, and r and k are constant parameters. The harvesting rate is $h(x, u) = axu$, where a is a constant parameter and u is the fishing effort. The output y is the rate of revenue, where b and c are constants representing the price of fish and the cost of fishing.

a) Find a sustainable equilibrium point where the revenue is as large as possible. Determine the equilibrium value of the biomass and the fishing effort at the equilibrium.

b) With the parameters $a = 0.1$, $b = 1$, $c = 1$, $k = 100$, and $r = 0.2$ the sustainable equilibrium point corresponds to $x_e = 55$ and $u_e = 0.9$. For an individual fisherman it is profitable to fish as long as the rate of revenue $y = (abx - c)u$ is positive. Explore by simulation what happens if the fishing intensity is much higher than the sustainable fishing rate u_e, say $u = 3$. Use the results to discuss the role of having a fishing quota.

Chapter Five

Dynamic Behavior

> It Don't Mean a Thing (If It Ain't Got That Swing).
>
> —Duke Ellington (1899–1974).

In this chapter we present a broad discussion of the behavior of dynamical systems focused on systems modeled by nonlinear differential equations. This allows us to consider equilibrium points, stability, limit cycles, and other key concepts in understanding dynamic behavior. We also introduce some methods for analyzing the global behavior of solutions.

5.1 SOLVING DIFFERENTIAL EQUATIONS

In the previous two chapters we saw that one of the methods of modeling dynamical systems is through the use of ordinary differential equations (ODEs). A state space, input/output system has the form

$$\frac{dx}{dt} = f(x, u), \qquad y = h(x, u), \tag{5.1}$$

where $x = (x_1, \ldots, x_n) \in \mathbb{R}^n$ is the state, $u \in \mathbb{R}^p$ is the input, and $y \in \mathbb{R}^q$ is the output. The smooth maps $f : \mathbb{R}^n \times \mathbb{R}^p \to \mathbb{R}^n$ and $h : \mathbb{R}^n \times \mathbb{R}^p \to \mathbb{R}^q$ represent the dynamics and measurements for the system. In general, they can be nonlinear functions of their arguments. Systems with many inputs and many outputs are called multi-input, multi-output systems (MIMO) systems. We will usually focus on single-input, single-output (SISO) systems, for which $p = q = 1$.

We begin by investigating systems in which the input has been set to a function of the state, $u = \alpha(x)$. This is one of the simplest types of feedback, in which the system regulates its own behavior. The differential equations in this case become

$$\frac{dx}{dt} = f(x, \alpha(x)) =: F(x). \tag{5.2}$$

To understand the dynamic behavior of this system, we need to analyze the features of the solutions of equation (5.2). While in some simple situations we can write down the solutions in analytical form, often we must rely on computational approaches. We begin by describing the class of solutions for this problem.

We say that $x(t)$ is a *solution* of the differential equation (5.2) on the time interval $t_0 \in \mathbb{R}$ to $t_f \in \mathbb{R}$ if

$$\frac{dx(t)}{dt} = F(x(t)) \quad \text{for all } t_0 < t < t_f.$$

A given differential equation may have many solutions. We will most often be interested in the *initial value problem*, where $x(t)$ is prescribed at a given time $t_0 \in \mathbb{R}$
and we wish to find a solution valid for all *future* time $t > t_0$.

We say that $x(t)$ is a solution of the differential equation (5.2) with initial value
$x_0 \in \mathbb{R}^n$ at $t_0 \in \mathbb{R}$ if

$$x(t_0) = x_0 \quad \text{and} \quad \frac{dx(t)}{dt} = F(x(t)) \quad \text{for all } t_0 < t < t_f.$$

For most differential equations we will encounter, there is a *unique* solution that is
defined for $t_0 \leq t < t_f$. The solution may be defined for all time $t > t_0$, in which case
we take $t_f = \infty$. Because we will primarily be interested in solutions of the initial
value problem for differential equations, we will usually refer to this simply as the
solution of a differential equation.

We will typically assume that t_0 is equal to 0. In the case when F is independent
of time (as in equation (5.2)), we can do so without loss of generality by choosing
a new independent (time) variable, $\tau = t - t_0$ (Exercise 5.1).

Example 5.1 Damped oscillator
Consider a damped linear oscillator with dynamics of the form

$$\ddot{q} + 2\zeta\omega_0\dot{q} + \omega_0^2 q = 0,$$

where q is the displacement of the oscillator from its rest position. These dynamics
are equivalent to those of a spring–mass system, as shown in Exercise 3.5. We
assume that $\zeta < 1$, corresponding to a lightly damped system (the reason for this
particular choice will become clear later). We can rewrite this in state space form
by setting $x_1 = q$ and $x_2 = \dot{q}/\omega_0$, giving

$$\frac{dx_1}{dt} = \omega_0 x_2, \qquad \frac{dx_2}{dt} = -\omega_0 x_1 - 2\zeta\omega_0 x_2.$$

In vector form, the right-hand side can be written as

$$F(x) = \begin{pmatrix} \omega_0 x_2 \\ -\omega_0 x_1 - 2\zeta\omega_0 x_2 \end{pmatrix}.$$

The solution to the initial value problem can be written in a number of different
ways and will be explored in more detail in Chapter 6. Here we simply assert that
the solution can be written as

$$x_1(t) = e^{-\zeta\omega_0 t}\left(x_{10}\cos\omega_d t + \frac{1}{\omega_d}(\omega_0\zeta x_{10} + x_{20})\sin\omega_d t\right),$$

$$x_2(t) = e^{-\zeta\omega_0 t}\left(x_{20}\cos\omega_d t - \frac{1}{\omega_d}(\omega_0^2 x_{10} + \omega_0\zeta x_{20})\sin\omega_d t\right),$$

where $x_0 = (x_{10}, x_{20})$ is the initial condition and $\omega_d = \omega_0\sqrt{1 - \zeta^2}$. This solution
can be verified by substituting it into the differential equation. We see that the
solution is explicitly dependent on the initial condition, and it can be shown that
this solution is unique. A plot of the initial condition response is shown in Figure 5.1.
We note that this form of the solution holds only for $0 < \zeta < 1$, corresponding to an
"underdamped" oscillator. ∇

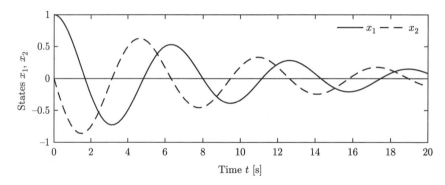

Figure 5.1: Response of the damped oscillator to the initial condition $x_0 = (1, 0)$. The solution is unique for the given initial conditions and consists of an oscillatory solution for each state, with an exponentially decaying magnitude.

Without imposing some mathematical conditions on the function F, the differential equation (5.2) may not have a solution for all t, and there is no guarantee that the solution is unique. We illustrate these possibilities with two examples.

Example 5.2 Finite escape time
Let $x \in \mathbb{R}$ and consider the differential equation

$$\frac{dx}{dt} = x^2 \tag{5.3}$$

with the initial condition $x(0) = 1$. By differentiation we can verify that the function

$$x(t) = \frac{1}{1-t}$$

satisfies the differential equation and that it also satisfies the initial condition. A graph of the solution is given in Figure 5.2a; notice that the solution goes to infinity as t goes to 1. We say that this system has *finite escape time*. Thus the solution exists only in the time interval $0 \leq t < 1$. ▽

Example 5.3 Nonunique solution
Let $x \in \mathbb{R}$ and consider the differential equation

$$\frac{dx}{dt} = 2\sqrt{x} \tag{5.4}$$

with initial condition $x(0) = 0$. We can show that the function

$$x(t) = \begin{cases} 0 & \text{if } 0 \leq t \leq a, \\ (t-a)^2 & \text{if } t > a \end{cases}$$

satisfies the differential equation for all values of the parameter $a \geq 0$. To see this, we differentiate $x(t)$ to obtain

$$\frac{dx}{dt} = \begin{cases} 0 & \text{if } 0 \leq t \leq a, \\ 2(t-a) & \text{if } t > a, \end{cases}$$

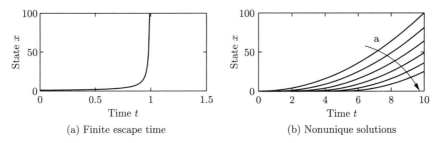

(a) Finite escape time (b) Nonunique solutions

Figure 5.2: Existence and uniqueness of solutions. Equation (5.3) has a solution only for time $t < 1$, at which point the solution goes to infinity, as shown in (a). Equation (5.4) is an example of a system with many solutions, as shown in (b). For each value of a, we get a different solution starting from the same initial condition.

and hence $\dot{x} = 2\sqrt{x}$ for all $t \geq 0$ with $x(0) = 0$. A graph of some of the possible solutions is given in Figure 5.2b. Notice that in this case there are many solutions to the differential equation. ∇

These simple examples show that there may be difficulties even with simple differential equations. Existence and uniqueness can be guaranteed by requiring that the function F have the property that for some fixed $c \in \mathbb{R}$,

$$\|F(x) - F(y)\| < c\|x - y\| \quad \text{for all } x, y,$$

which is called *Lipschitz continuity*. A sufficient condition for a function to be Lipschitz is that the Jacobian $\partial F / \partial x$ is uniformly bounded for all x. The difficulty in Example 5.2 is that the derivative $\partial F / \partial x$ becomes large for large x, and the difficulty in Example 5.3 is that the derivative $\partial F / \partial x$ is infinite at the origin.

5.2 QUALITATIVE ANALYSIS

The qualitative behavior of nonlinear systems is important in understanding some of the key concepts of stability in nonlinear dynamics. We will focus on an important class of systems known as planar dynamical systems. These systems have two state variables $x \in \mathbb{R}^2$, allowing their solutions to be plotted in the (x_1, x_2) plane. The basic concepts that we describe hold more generally and can be used to understand dynamical behavior in higher dimensions.

Phase Portraits

A convenient way to understand the behavior of dynamical systems with state $x \in \mathbb{R}^2$ is to plot the phase portrait of the system, briefly introduced in Chapter 3. We start by introducing the concept of a *vector field*. For a system of ordinary differential equations

$$\frac{dx}{dt} = F(x),$$

the right-hand side of the differential equation defines at every $x \in \mathbb{R}^n$ a velocity $F(x) \in \mathbb{R}^n$. This velocity tells us how x changes and can be represented as a vector $F(x) \in \mathbb{R}^n$.

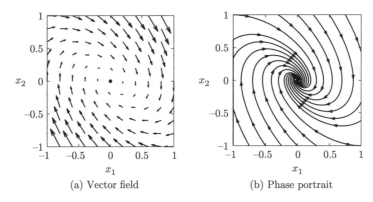

(a) Vector field   (b) Phase portrait

Figure 5.3: Phase portraits. (a) This plot shows the vector field for a planar dynamical system. Each arrow shows the velocity at that point in the state space. (b) This plot includes the solutions (sometimes called streamlines) from different initial conditions, with the vector field superimposed.

For planar dynamical systems, each state corresponds to a point in the plane and $F(x)$ is a vector representing the velocity of that state. We can plot these vectors on a grid of points in the plane and obtain a visual image of the dynamics of the system, as shown in Figure 5.3a. The points where the velocities are zero are of particular interest since they define stationary points of the flow: if we start at such a state, we stay at that state.

A *phase portrait* is constructed by plotting the flow of the vector field corresponding to the planar dynamical system. That is, for a set of initial conditions, we plot the solution of the differential equation in the plane $\mathbb{R}^2$. This corresponds to following the arrows at each point in the phase plane and drawing the resulting trajectory. By plotting the solutions for several different initial conditions, we obtain a phase portrait, as shown in Figure 5.3b. Phase portraits are also sometimes called *phase plane diagrams*.

Phase portraits give insight into the dynamics of the system by showing the solutions plotted in the (two-dimensional) state space of the system. For example, we can see whether all trajectories tend to a single point as time increases or whether there are more complicated behaviors. In the example in Figure 5.3, corresponding to a damped oscillator, the solutions approach the origin for all initial conditions. This is consistent with our simulation in Figure 5.1, but it allows us to infer the behavior for all initial conditions rather than a single initial condition. However, the phase portrait does not readily tell us the rate of change of the states (although this can be inferred from the lengths of the arrows in the vector field plot).

Equilibrium Points and Limit Cycles

An *equilibrium point* of a dynamical system represents a stationary condition for the dynamics. We say that a state x_e is an equilibrium point for a dynamical system

$$\frac{dx}{dt} = F(x)$$

if $F(x_e) = 0$. If a dynamical system has an initial condition $x(0) = x_e$, then it will stay at the equilibrium point: $x(t) = x_e$ for all $t \geq 0$, where we have taken $t_0 = 0$.

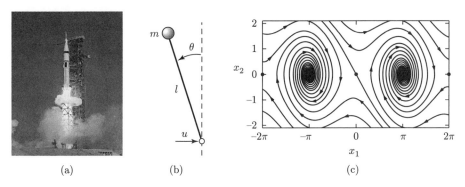

(a)   (b)   (c)

Figure 5.4: Equilibrium points for an inverted pendulum. An inverted pendulum is a model for a class of balance systems in which we wish to keep a system upright, such as a rocket (a). Using a simplified model of an inverted pendulum (b), we can develop a phase portrait that shows the dynamics of the system (c). The system has multiple equilibrium points, marked by the solid dots along the $x_2 = 0$ line.

Equilibrium points are one of the most important features of a dynamical system since they define the states corresponding to constant operating conditions. A dynamical system can have zero, one, or more equilibrium points.

Example 5.4 Inverted pendulum

Consider the inverted pendulum in Figure 5.4, which is a part of the balance system we considered in Chapter 3. The inverted pendulum is a simplified version of the problem of stabilizing a rocket: by applying forces at the base of the rocket, we seek to keep the rocket stabilized in the upright position. The state variables are the angle $\theta = x_1$ and the angular velocity $d\theta/dt = x_2$, the control variable is the acceleration u of the pivot, and the output is the angle θ.

For simplicity we assume that $mgl/J_t = 1$, $l/J_t = 1$ and set $c = \gamma/J_t$, so that the dynamics (equation (3.10)) become

$$\frac{dx}{dt} = \begin{pmatrix} x_2 \\ \sin x_1 - cx_2 + u\cos x_1 \end{pmatrix}. \tag{5.5}$$

This is a nonlinear time-invariant system of second order. This same set of equations can also be obtained by appropriate normalization of the system dynamics, as illustrated in Example 3.10.

We consider the open loop dynamics by setting $u = 0$. The equilibrium points for the system are given by

$$x_e = \begin{pmatrix} \pm n\pi \\ 0 \end{pmatrix},$$

where $n = 0, 1, 2, \ldots$. The equilibrium points for n even correspond to the pendulum pointing up and those for n odd correspond to the pendulum hanging down. A phase portrait for this system (without corrective inputs) is shown in Figure 5.4c. The phase portrait shows $-2\pi \leq x_1 \leq 2\pi$, so five of the equilibrium points are shown. ∇

Nonlinear systems can exhibit rich behavior. Apart from equilibrium points they can also exhibit stationary periodic solutions. This is of great practical value in generating sinusoidally varying voltages in power systems or in generating periodic signals for animal locomotion. A simple example is given in Exercise 5.11, which

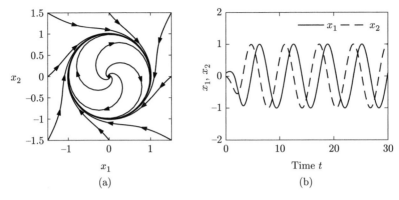

(a)   (b)

Figure 5.5: Phase portrait and time domain simulation for a system with a limit cycle. The phase portrait (a) shows the states of the solution plotted for different initial conditions. The limit cycle corresponds to a closed loop trajectory. The simulation (b) shows a single solution plotted as a function of time, with the limit cycle corresponding to a steady oscillation of fixed amplitude.

shows the circuit diagram for an electronic oscillator. A normalized model of the oscillator is given by the equation

$$\frac{dx_1}{dt} = x_2 + x_1(1 - x_1^2 - x_2^2), \qquad \frac{dx_2}{dt} = -x_1 + x_2(1 - x_1^2 - x_2^2). \qquad (5.6)$$

The phase portrait and time domain solutions are given in Figure 5.5. The figure shows that the solutions in the phase plane converge to a circular trajectory. In the time domain this corresponds to an oscillatory solution. Mathematically the circle is called a *limit cycle*. More formally, we call a nonconstant solution $x_p(t)$ a limit cycle of period $T > 0$ if $x_p(t + T) = x_p(t)$ for all $t \in \mathbb{R}$ and nearby trajectories converge to $x_p(\cdot)$ as $t \to \infty$ (stable limit cycle) or $t \to -\infty$ (unstable limit cycle).

There are methods for determining limit cycles for second-order systems, but for general higher-order systems we have to resort to computational analysis. Computer algorithms find limit cycles by searching for periodic trajectories in state space that satisfy the dynamics of the system. In many situations, stable limit cycles can be found by simulating the system with different initial conditions.

5.3   STABILITY

The stability of a solution determines whether or not solutions nearby the solution remain close, get closer, or move further away. We now give a formal definition of stability and describe tests for determining whether a solution is stable.

Definitions

Let $x(t; a)$ be a solution to the differential equation with initial condition a. A solution is *stable* if other solutions that start near a stay close to $x(t; a)$. Formally, we say that the solution $x(t; a)$ is stable if for all $\epsilon > 0$, there exists a $\delta > 0$ such that

$$\|b - a\| < \delta \quad \Longrightarrow \quad \|x(t; b) - x(t; a)\| < \epsilon \quad \text{for all } t > 0.$$

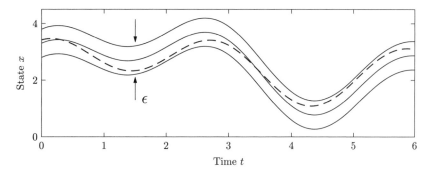

Figure 5.6: Illustration of Lyapunov's concept of a stable solution. The solution represented by the solid line is stable if we can guarantee that all solutions remain within a tube of diameter ϵ by choosing initial conditions sufficiently close the solution.

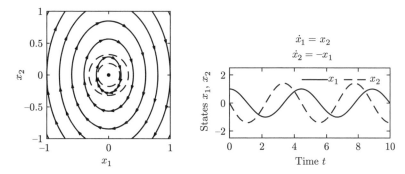

Figure 5.7: Phase portrait and time domain simulation for a system with a single stable equilibrium point. The equilibrium point x_e at the origin is stable since all trajectories that start near x_e stay near x_e.

Note that this definition does not imply that $x(t; b)$ approaches $x(t; a)$ as time increases but just that it stays nearby. Furthermore, the value of δ may depend on ϵ, so that if we wish to stay very close to the solution, we may have to start very, very close ($\delta \ll \epsilon$). This type of stability, which is illustrated in Figure 5.6, is also called *stability in the sense of Lyapunov*. If a solution is stable in this sense and the trajectories do not converge, we say that the solution is *neutrally stable*.

An important special case is when the solution $x(t; a) = x_e$ is an equilibrium solution. In this case the condition for stability becomes

$$\|x(0) - x_e\| < \delta \quad \implies \quad \|x(t) - x_e\| < \epsilon \quad \text{for all } t > 0. \tag{5.7}$$

Instead of saying that the solution is stable, we simply say that the equilibrium point is stable. An example of a neutrally stable equilibrium point is shown in Figure 5.7. From the phase portrait, we see that if we start near the equilibrium point, then we stay near the equilibrium point. Furthermore, if we choose an initial condition from within the inner dashed circle (of radius δ) then all trajectories will remain inside the region defined by the outer dashed circle (of radius ϵ). Note, however, that trajectories may not remain confined inside the individual circles (and hence we must choose $\delta < \epsilon$).

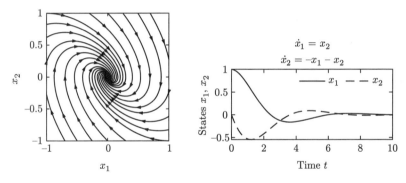

Figure 5.8: Phase portrait and time domain simulation for a system with a single asymptotically stable equilibrium point. The equilibrium point x_e at the origin is asymptotically stable since the trajectories converge to this point as $t \to \infty$.

A solution $x(t; a)$ is *asymptotically stable* if it is stable in the sense of Lyapunov and, in addition, $x(t; b)$ approaches $x(t; a)$ as t approaches infinity for b sufficiently close to a. Hence, the solution $x(t; a)$ is asymptotically stable if for every $\epsilon > 0$ there exists a $\delta > 0$ such that

$$\|b - a\| < \delta \quad \implies \quad \|x(t; b) - x(t; a)\| < \epsilon \quad \text{and} \quad \lim_{t \to \infty} \|x(t; b) - x(t; a)\| = 0.$$

This corresponds to the case where all nearby trajectories converge to the stable solution for large time. In the case of an equilibrium solution x_e, we can write this condition as

$$\|x(0) - x_e\| < \delta \quad \implies \quad \|x(t) - x_e\| < \epsilon \quad \text{and} \quad \lim_{t \to \infty} x(t) = x_e. \qquad (5.8)$$

Figure 5.8 shows an example of an asymptotically stable equilibrium point. Indeed, as seen in the phase portrait, not only do all trajectories stay near the equilibrium point at the origin, but they also all approach the origin as t gets large (the directions of the arrows on the phase portrait show the direction in which the trajectories move).

A solution $x(t; a)$ is *unstable* if it is not stable. More specifically, we say that a solution $x(t; a)$ is unstable if given some $\epsilon > 0$, there does *not* exist a $\delta > 0$ such that if $\|b - a\| < \delta$, then $\|x(t; b) - x(t; a)\| < \epsilon$ for all t. An example of an unstable equilibrium point x_e is shown in Figure 5.9. Note that no matter how small we make δ, there is always an initial condition with $\|x(0) - x_e\| < \delta$ that flows away from x_e.

The definitions above are given without careful description of their domain of applicability. More formally, we define a solution to be *locally stable* (or *locally asymptotically stable*) if it is stable for all initial conditions $x \in B_r(a)$, where

$$B_r(a) = \{x : \|x - a\| < r\}$$

is a ball of radius r around a and $r > 0$. A solution is *globally asymptotically stable* if it is asymptotically stable for *all* $r > 0$.

For planar dynamical systems, equilibrium points have been assigned names based on their stability type. An asymptotically stable equilibrium point is called a *sink* or sometimes an *attractor*. An unstable equilibrium point can be either a *source*, if all trajectories lead away from the equilibrium point, or a *saddle*, if

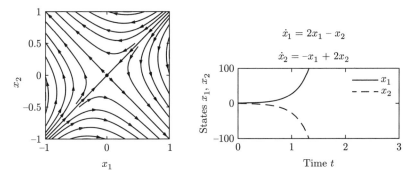

Figure 5.9: Phase portrait and time domain simulation for a system with a single unstable equilibrium point. The equilibrium point x_e at the origin is unstable since not all trajectories that start near x_e stay near x_e. The sample trajectory on the right shows that the trajectories very quickly depart from zero.

some trajectories lead to the equilibrium point and others move away (this is the situation pictured in Figure 5.9). Finally, an equilibrium point that is stable but not asymptotically stable (i.e., neutrally stable, such as the one in Figure 5.7) is called a *center*.

Example 5.5 Congestion control

The TCP protocol is used to adjust the rate of packet transmission on the Internet. Stability of this system is important to insure smooth and efficient flow of information across the network.

The model for congestion control in a network consisting of N identical computers connected to a single router, described in more detail in Section 4.4, is given by

$$\frac{dw}{dt} = \frac{c}{b} - \rho c \left(1 + \frac{w^2}{2} \right), \qquad \frac{db}{dt} = N \frac{wc}{b} - c,$$

where w is the window size and b is the buffer size of the router. The equilibrium points are given by

$$b_e = N w_e, \quad \text{where} \quad w_e \left(1 + \frac{w_e^2}{2} \right) = \frac{1}{N\rho}.$$

Since $w(1 + w^2/2)$ is monotone, there is only one equilibrium point. Phase portraits are shown in Figure 5.10 for two different sets of parameter values. In each case we see that the system converges to an equilibrium point in which the buffer is below its full capacity of 500 packets. The equilibrium size of the buffer represents a balance between the transmission rates for the sources and the capacity of the link. We see from the phase portraits that the equilibrium points are asymptotically stable since all initial conditions result in trajectories that converge to these points. ∇

Stability of Linear Systems

A linear dynamical system has the form

$$\frac{dx}{dt} = Ax, \quad x(0) = x_0, \tag{5.9}$$

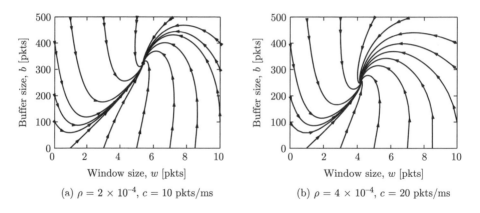

(a) $\rho = 2 \times 10^{-4}$, $c = 10$ pkts/ms   (b) $\rho = 4 \times 10^{-4}$, $c = 20$ pkts/ms

Figure 5.10: Phase portraits for a congestion control protocol running with $N = 60$ identical source computers. The equilibrium values correspond to a fixed window at the source, which results in a steady-state buffer size and corresponding transmission rate. A faster link (b) uses a smaller buffer size since it can handle packets at a higher rate.

where $A \in \mathbb{R}^{n \times n}$ is a square matrix, corresponding to the dynamics matrix of a linear control system (3.6). For a linear system, the stability of the equilibrium point at the origin can be determined from the eigenvalues of the matrix A:

$$\lambda(A) := \{s \in \mathbb{C} : \det(sI - A) = 0\}.$$

The polynomial $\det(sI - A)$ is the *characteristic polynomial* and the eigenvalues are its roots. We use the notation λ_j for the jth eigenvalue of A, so that $\lambda_j \in \lambda(A)$. In general λ can be complex-valued, although if A is real-valued, then for any eigenvalue λ, its complex conjugate λ^* will also be an eigenvalue. The origin is always an equilibrium point for a linear system. Since the stability of a linear system depends only on the matrix A, we find that stability is a property of the system. For a linear system we can therefore talk about the stability of the system rather than the stability of a particular solution or equilibrium point.

The easiest class of linear systems to analyze are those whose system matrices are in diagonal form. In this case, the dynamics have the form

$$\frac{dx}{dt} = \begin{bmatrix} \lambda_1 & & & 0 \\ & \lambda_2 & & \\ & & \ddots & \\ 0 & & & \lambda_n \end{bmatrix} x. \tag{5.10}$$

It is easy to see that the state trajectories for this system are independent of each other, so that we can write the solution in terms of n individual systems $\dot{x}_j = \lambda_j x_j$. Each of these scalar solutions is of the form

$$x_j(t) = e^{\lambda_j t} x_j(0).$$

We see that the equilibrium point $x_e = 0$ is stable if $\lambda_j \leq 0$ and asymptotically stable if $\lambda_j < 0$.

Another simple case is when the dynamics are in the block diagonal form

$$\frac{dx}{dt} = \begin{pmatrix} \sigma_1 & \omega_1 & & & & \\ -\omega_1 & \sigma_1 & & & \large 0 & \\ & & \ddots & & & \\ & & & & \sigma_m & \omega_m \\ & \large 0 & & & -\omega_m & \sigma_m \end{pmatrix} x.$$

In this case, the eigenvalues can be shown to be $\lambda_j = \sigma_j \pm i\omega_j$. We once again can separate the state trajectories into independent solutions for each pair of states, and the solutions are of the form

$$x_{2j-1}(t) = e^{\sigma_j t}\big(x_{2j-1}(0)\cos\omega_j t + x_{2j}(0)\sin\omega_j t\big),$$

$$x_{2j}(t) = e^{\sigma_j t}\big(-x_{2j-1}(0)\sin\omega_j t + x_{2j}(0)\cos\omega_j t\big),$$

where $j = 1, 2, \ldots, m$. We see that this system is asymptotically stable if and only if $\sigma_j = \operatorname{Re}\lambda_j < 0$. It is also possible to combine real and complex eigenvalues in (block) diagonal form, resulting in a mixture of solutions of the two types.

Very few systems are in one of the diagonal forms above, but many systems can be transformed into these forms via coordinate transformations. One such class of systems is those for which the dynamics matrix has distinct (nonrepeating) eigenvalues. In this case there is a matrix $T \in \mathbb{R}^{n \times n}$ such that the matrix TAT^{-1} is in (block) diagonal form, with the block diagonal elements corresponding to the eigenvalues of the original matrix A (see Exercise 5.14). If we choose new coordinates $z = Tx$, then

$$\frac{dz}{dt} = T\dot{x} = TAx = TAT^{-1}z$$

and the linear system has a (block) diagonal dynamics matrix. Furthermore, the eigenvalues of the transformed system are the same as those of the original system since if v is an eigenvector of A, then $w = Tv$ can be shown to be an eigenvector of TAT^{-1}. We can reason about the stability of the original system by noting that $x(t) = T^{-1}z(t)$, and so if the transformed system is stable (or asymptotically stable), then the original system has the same type of stability.

This analysis shows that for linear systems with distinct eigenvalues, the stability of the system can be completely determined by examining the real part of the eigenvalues of the dynamics matrix. For more general systems, we make use of the following theorem, proved in the next chapter:

Theorem 5.1 (Stability of a linear system). *The system*

$$\frac{dx}{dt} = Ax$$

is asymptotically stable if and only if all eigenvalues of A have a strictly negative real part and is unstable if any eigenvalue of A has a strictly positive real part.

Note that it is not enough to have eigenvalues with $\operatorname{Re}(\lambda) \le 0$. As a simple example, consider the system $\ddot{q} = 0$, which can be written in state space form as

$$\frac{d}{dt}\begin{pmatrix} x_1 \\ x_2 \end{pmatrix} = \begin{pmatrix} 0 & 1 \\ 0 & 0 \end{pmatrix}\begin{pmatrix} x_1 \\ x_2 \end{pmatrix}.$$

The system has eigenvalues $\lambda = 0$ but the solutions are not bounded since we have

$$x_1(t) = x_{1,0} + x_{2,0}t, \quad x_2(t) = x_{2,0}.$$

Example 5.6 Compartment model

Consider the two-compartment module for drug delivery described in Section 4.6. Using concentrations as state variables and denoting the state vector by x, the system dynamics are given by

$$\frac{dx}{dt} = \begin{pmatrix} -k_0 - k_1 & k_1 \\ k_2 & -k_2 \end{pmatrix} x + \begin{pmatrix} b_0 \\ 0 \end{pmatrix} u, \qquad y = \begin{pmatrix} 0 & 1 \end{pmatrix} x,$$

where the input u is the rate of injection of a drug into compartment 1 and the concentration of the drug in compartment 2 is the measured output y. We wish to design a feedback control law that maintains a constant output given by $y = y_d$.

We choose an output feedback control law of the form

$$u = -k(y - y_d) + u_d,$$

where u_d is the rate of injection required to maintain the desired concentration $y = y_d$, and k is a feedback gain that should be chosen such that the closed loop system is stable. Substituting the control law into the system, we obtain

$$\frac{dx}{dt} = \begin{pmatrix} -k_0 - k_1 & k_1 - b_0 k \\ k_2 & -k_2 \end{pmatrix} x + \begin{pmatrix} b_0 \\ 0 \end{pmatrix} (u_d + k y_d) =: Ax + B u_e,$$

$$y = \begin{pmatrix} 0 & 1 \end{pmatrix} x =: Cx.$$

The equilibrium concentration $x_e \in \mathbb{R}^2$ can be obtained by solving the equation $Ax_e + Bu_e = 0$ and some simple algebra yields

$$x_{1,e} = x_{2,e} = y_d, \qquad u_e = u_d = \frac{k_0}{b_0} y_d.$$

To analyze the system around the equilibrium point, we choose new coordinates $z = x - x_e$. In these coordinates the equilibrium point is at the origin and the dynamics become

$$\frac{dz}{dt} = \begin{pmatrix} -k_0 - k_1 & k_1 - b_0 k \\ k_2 & -k_2 \end{pmatrix} z.$$

We can now apply the results of Theorem 5.1 to determine the stability of the system. The eigenvalues of the system are given by the roots of the characteristic polynomial

$$\lambda(s) = s^2 + (k_0 + k_1 + k_2)s + (k_0 k_2 + b_0 k_2 k).$$

While the specific form of the roots is messy, it can be shown using the Routh–Hurwitz criterion that the roots have negative real part as long as the linear term and the constant term are both positive (see Section 2.2). Hence the system is stable for any $k > 0$.   ▽

Stability Analysis via Linear Approximation

An important feature of differential equations is that it is often possible to determine the local stability of an equilibrium point by approximating the system by a linear system. The following example illustrates the basic idea.

Example 5.7 Inverted pendulum

Consider again an inverted pendulum whose open loop dynamics are given by

$$\frac{dx}{dt} = \begin{pmatrix} x_2 \\ \sin x_1 - cx_2 \end{pmatrix},$$

where we have defined the state as $x = (\theta, \dot{\theta})$. We first consider the equilibrium point at $x = (0,0)$, corresponding to the straight-up position. If we assume that the angle $\theta = x_1$ remains small, then we can replace $\sin x_1$ with x_1 and $\cos x_1$ with 1, which gives the approximate system

$$\frac{dx}{dt} = \begin{pmatrix} x_2 \\ x_1 - cx_2 \end{pmatrix} = \begin{pmatrix} 0 & 1 \\ 1 & -c \end{pmatrix} x. \tag{5.11}$$

Intuitively, this system should behave similarly to the more complicated model as long as x_1 is small. In particular, it can be verified that the equilibrium point $(0,0)$ is unstable by plotting the phase portrait or computing the eigenvalues of the dynamics matrix in equation (5.11).

We can also approximate the system around the stable equilibrium point at $x = (\pi, 0)$. In this case we have to expand $\sin x_1$ and $\cos x_1$ around $x_1 = \pi$, according to the expansions

$$\sin(\pi + \theta) = -\sin\theta \approx -\theta, \qquad \cos(\pi + \theta) = -\cos(\theta) \approx -1.$$

If we define $z_1 = x_1 - \pi$ and $z_2 = x_2$, the resulting approximate dynamics are given by

$$\frac{dz}{dt} = \begin{pmatrix} z_2 \\ -z_1 - c\,z_2 \end{pmatrix} = \begin{pmatrix} 0 & 1 \\ -1 & -c \end{pmatrix} z. \tag{5.12}$$

It can be shown that the eigenvalues of the dynamics matrix have negative real parts, confirming that the downward equilibrium point is asymptotically stable.

Figure 5.11 shows the phase portraits for the original system and the approximate system around the stable equilibrium point. Note that $z = (0,0)$ is the equilibrium point for this system and that it has the same basic form as the dynamics shown in Figure 5.8. The solutions for the original system and the approximate are very similar, although not exactly the same. It can be shown that if a linear approximation has either asymptotically stable or unstable equilibrium points, then the local stability of the original system must be the same (see Theorem 5.3 for the case of asymptotic stability). $\qquad\qquad\qquad\qquad\qquad\qquad\qquad\qquad\qquad\qquad\nabla$

More generally, suppose that we have a nonlinear system

$$\frac{dx}{dt} = F(x)$$

that has an equilibrium point at x_e. Computing the Taylor series expansion of the vector field, we can write

$$\frac{dx}{dt} = F(x_e) + \left.\frac{\partial F}{\partial x}\right|_{x_e} (x - x_e) + \text{higher-order terms in } (x - x_e).$$

Since $F(x_e) = 0$, we can approximate the system by choosing a new state variable $z = x - x_e$ and writing

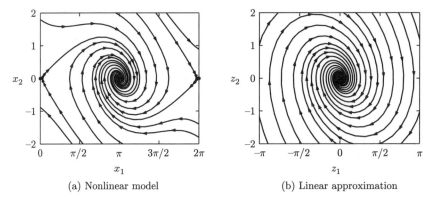

(a) Nonlinear model   (b) Linear approximation

Figure 5.11: Comparison between the phase portraits for the full nonlinear system (a) and its linear approximation around the origin (b). Notice that near the equilibrium point at the center of the plots, the phase portraits (and hence the dynamics) are almost identical.

$$\frac{dz}{dt} = Az, \quad \text{where} \quad A = \left.\frac{\partial F}{\partial x}\right|_{x_e}. \tag{5.13}$$

We call the system (5.13) the *linear approximation* of the original nonlinear system or the *linearization* at x_e. The following example illustrates the idea.

Example 5.8 Stability of a tanker

The normalized steering dynamics of a large ship can be modeled by the following equations:

$$\frac{dv}{dt} = a_1 v + a_2 r + \alpha v |v| + b_1 \delta, \qquad \frac{dr}{dt} = a_3 v + a_4 r + b_2 \delta,$$

where v is the component of the velocity vector that is orthogonal to the ship direction, r is the turning rate, and δ is the rudder angle. The variables are normalized by using the ship length l as the length unit and the time to travel one ship length as the time unit. The mass is normalized by $\rho l^3/2$, where ρ is the density of water. The normalized parameters are $a_1 = -0.6$, $a_2 = -0.3$, $a_3 = -5$, $a_4 = -2$, $\alpha = -2$, $b_1 = 0.1$, and $b_2 = -0.8$.

Setting the rudder angle $\delta = 0$, we find that the equilibrium points are given by the equations

$$a_1 v + a_2 r + \alpha v |v| = 0, \qquad a_3 v + a_4 r = 0.$$

Elimination of the variable r in these equations gives

$$(a_1 a_4 - a_2 a_3)v + \alpha a_4 v |v| = 0$$

There are three equilibrium solutions: $v_e = 0$ and $v_e = \pm 0.075$. Linearizing the equation gives a second-order system with dynamics matrices

$$A_0 = \begin{pmatrix} -0.6 & -0.3 \\ -5 & -2 \end{pmatrix}, \qquad A_1 = \begin{pmatrix} -0.9 & -0.3 \\ -5 & -2 \end{pmatrix}.$$

The linearized matrix A_0, for the equilibrium point $v_e = 0$, has the characteristic polynomial $s^2 + 2.6s - 0.3$, which has one root in the right half-plane. The

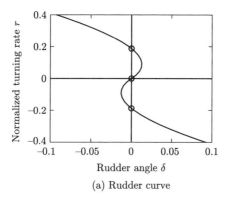 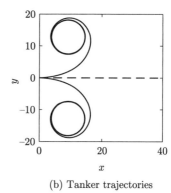

(a) Rudder curve   (b) Tanker trajectories

Figure 5.12: Stability analysis for a tanker. The rudder characteristics are shown in (a), where the equilibrium points are marked by circles, and the tanker trajectories are shown in (b).

equilibrium point is thus unstable. The matrix A_1, for the equilibrium points $v_e = \pm 0.075$, has the characteristic polynomial $s^2 + 2.9s + 0.3$, which has all roots in the left half-plane. These equilibrium points are stable.

Summarizing, we find that the equilibrium point $v_e = r_e = 0$, which corresponds to the ship moving forward at constant speed, is unstable. The other equilibrium points, $v_e = -0.075$, $r_e = 0.1875$ and $v_e = 0.075$, $r_e = -0.1875$, are stable (see Figure 5.12a). These equilibrium points correspond to the tanker moving in a circle to the left or to the right. Hence if the rudder is set to $\delta = 0$ and the ship is moving forward it will either turn to the right or to the left and approach one of the stable equilibrium points. Which way it goes depends on the exact value of the initial condition. The trajectories are shown in Figure 5.12b.   ∇

The fact that a linear model can be used to study the behavior of a nonlinear system near an equilibrium point is a powerful one. Indeed, we can take this even further and use a local linear approximation of a nonlinear system to design a feedback law that keeps the system near its equilibrium point (design of dynamics). Thus, feedback can be used to make sure that solutions remain close to the equilibrium point, which in turn ensures that the linear approximation used to stabilize it is valid.

Stability of Limit Cycles

Stability of nonequilibrium solutions can also be investigated, as illustrated by the following example.

Example 5.9 Stability of an oscillation

Consider the system given by equation (5.6),

$$\frac{dx_1}{dt} = x_2 + x_1(1 - x_1^2 - x_2^2), \qquad \frac{dx_2}{dt} = -x_1 + x_2(1 - x_1^2 - x_2^2),$$

whose phase portrait is shown in Figure 5.5. The differential equation has a periodic solution

$$x_{\mathrm{p}} = \begin{pmatrix} x_1(0)\cos t + x_2(0)\sin t \\ x_2(0)\cos t - x_1(0)\sin t \end{pmatrix}, \tag{5.14}$$

with $x_1^2(0) + x_2^2(0) = 1$. Notice that the nonlinear terms disappear on the periodic solution.

To explore the stability of this solution, we introduce polar coordinates $r \geq 0$ and φ, which are related to the state variables x_1 and x_2 by

$$x_1 = r\cos\varphi, \qquad x_2 = r\sin\varphi.$$

Differentiation gives the following linear equations for $\dot{r}$ and $\dot{\varphi}$:

$$\dot{x}_1 = \dot{r}\cos\varphi - r\dot{\varphi}\sin\varphi, \qquad \dot{x}_2 = \dot{r}\sin\varphi + r\dot{\varphi}\cos\varphi.$$

Solving this linear system for $\dot{r}$ and $\dot{\varphi}$ gives, after some calculation,

$$\frac{dr}{dt} = r(1 - r^2), \qquad \frac{d\varphi}{dt} = -1. \tag{5.15}$$

Notice that the equations are decoupled; hence we can analyze the stability of each state separately.

The equation for r has two equilibrium points: $r = 0$ and $r = 1$ (notice that r is assumed to be nonnegative). The derivative dr/dt is positive for $0 < r < 1$ and negative for $r > 1$. The variable r will therefore increase if $0 < r < 1$ and decrease if $r > 1$, and we find that the equilibrium point $r = 0$ is unstable and the equilibrium point $r = 1$ is stable. Solutions with initial conditions different from 0 will thus all converge to the stable equilibrium point $r = 1$ as time increases.

To study the stability of the full system, we must also investigate the behavior of angle φ. The equation for $\dot{\varphi}$ can be integrated analytically to give $\varphi(t) = -t + \varphi(0)$, which shows that solutions starting at different initial angles $\varphi(0)$ will grow linearly with time, remaining separated by a constant amount. The solution $r = 1$, $\varphi = -t$ is thus stable but not asymptotically stable. The unit circle in the phase plane is *attracting*, in the sense that all solutions with $r(0) > 0$ converge to the unit circle, as illustrated in the simulation in Figure 5.13. Notice that the solutions approach the circle rapidly, but that there is a constant phase shift between the solutions. $\triangledown$

5.4   LYAPUNOV STABILITY ANALYSIS

We now return to the study of the full nonlinear system

$$\frac{dx}{dt} = F(x), \quad x \in \mathbb{R}^n. \tag{5.16}$$

Having defined when a solution for a nonlinear dynamical system is stable, we can now ask how to *prove* that a given solution is stable, asymptotically stable, or unstable. For physical systems, one can often argue about stability based on dissipation of energy. The generalization of that technique to arbitrary dynamical systems is based on the use of Lyapunov functions in place of energy.

In this section we will describe techniques for determining the stability of solutions for a nonlinear system (5.16). We will generally be interested in stability of

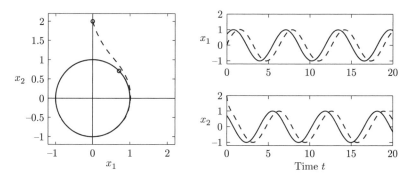

Figure 5.13: Solution curves for a stable limit cycle. The phase portrait on the left shows that the trajectory for the system rapidly converges to the stable limit cycle. The starting points for the trajectories are marked by circles in the phase portrait. The time domain plots on the right show that the states do not converge to the solution but instead maintain a constant phase error.

equilibrium points, and it will be convenient to assume that $x_e = 0$ is the equilibrium point of interest. (If not, rewrite the equations in a new set of coordinates $z = x - x_e$.)

Lyapunov Functions

A *Lyapunov function* $V : \mathbb{R}^n \to \mathbb{R}$ is an energy-like function that can be used to determine the stability of a system. Roughly speaking, if we can find a nonnegative function that always decreases along trajectories of the system, we can conclude that the minimum of the function is a stable equilibrium point (locally).

To describe this more formally, we start with a few definitions. Let $B_r = B_r(0)$ be a ball of radius r around the origin. We say that a continuous function V is *positive definite* on B_r if $V(x) > 0$ for all $x \in B_r$, $x \neq 0$ and $V(0) = 0$. Similarly, a function is *negative definite* on B_r if $V(x) < 0$ for all $x \in B_r$, $x \neq 0$ and $V(0) = 0$. We say that a function V is *positive semidefinite* if $V(x) \geq 0$ for all $x \in B_r$, but $V(x)$ can be zero at points other than just $x = 0$.

To illustrate the difference between a positive definite function and a positive semidefinite function, suppose that $x \in \mathbb{R}^2$ and let

$$V_1(x) = x_1^2, \qquad V_2(x) = x_1^2 + x_2^2.$$

Both V_1 and V_2 are always nonnegative. However, it is possible for V_1 to be zero even if $x \neq 0$. Specifically, if we set $x = (0, c)$, where $c \in \mathbb{R}$ is any nonzero number, then $V_1(x) = 0$. On the other hand, $V_2(x) = 0$ if and only if $x = (0, 0)$. Thus V_1 is positive semidefinite and V_2 is positive definite.

We can now characterize the stability of an equilibrium point $x_e = 0$ for the system (5.16).

Theorem 5.2 (Lyapunov stability theorem)**.** *Let V be a function on $\mathbb{R}^n$ and let $\dot{V}$ represent the time derivative of V along trajectories of the system dynamics (5.16):*

$$\dot{V} = \frac{\partial V}{\partial x}\frac{dx}{dt} = \frac{\partial V}{\partial x}F(x).$$

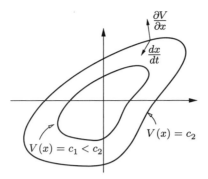

Figure 5.14: Geometric illustration of Lyapunov's stability theorem. The closed contours represent the level sets of the Lyapunov function $V(x) = c$. If dx/dt points inward to these sets at all points along the contour, then the trajectories of the system will always cause $V(x)$ to decrease along the trajectory.

If there exists $r > 0$ such that V is positive definite and $\dot{V}$ is negative semidefinite on B_r, then $x = 0$ is (locally) stable in the sense of Lyapunov. If V is positive definite and $\dot{V}$ is negative definite in B_r, then $x = 0$ is (locally) asymptotically stable.

If V satisfies one of the conditions above, we say that V is a (local) *Lyapunov function* for the system. These results have a nice geometric interpretation. The level curves for a positive definite function are the curves defined by $V(x) = c$, $c > 0$, and for each c this gives a closed contour, as shown in Figure 5.14. The condition that $\dot{V}(x)$ is negative simply means that the vector field points toward lower-level contours. This means that the trajectories move to smaller and smaller values of V and if $\dot{V}$ is negative definite then x must approach 0.

Finding Lyapunov functions is not always easy. For example, consider the linear system

$$\frac{dx_1}{dt} = x_2, \qquad \frac{dx_2}{dt} = -x_1 - \alpha x_2, \qquad \alpha > 0.$$

Since the system is linear, it can be easily verified that the eigenvalues of the corresponding dynamics matrix are given by

$$\lambda = \frac{-\alpha \pm \sqrt{\alpha^2 - 4}}{2}.$$

These eigenvalues always have negative real part for $\alpha > 0$ and hence the system is asymptotically stable. It follows that $x(t) \to 0$ and $t \to \infty$ and so a natural Lyapunov function candidate would be the squared magnitude of the state:

$$V(x) = \frac{1}{2}x_1^2 + \frac{1}{2}x_2^2.$$

Taking the time derivative of this function and evaluating along the trajectories of the system we find that

$$\dot{V}(x) = -\alpha x_2^2.$$

But this function is not positive definite, as can be seen by evaluating $\dot{V}$ at the point $x = (1, 0)$, which gives $\dot{V}(x) = 0$. Hence even though the system is asymptotically stable, a Lyapunov function that proves stability is not as simple as the squared magnitude of the state.

We now consider some additional examples.

Example 5.10 Scalar nonlinear system
Consider the scalar nonlinear system

$$\frac{dx}{dt} = \frac{2}{1+x} - x.$$

This system has equilibrium points at $x = 1$ and $x = -2$. We consider the equilibrium point at $x = 1$ and rewrite the dynamics using $z = x - 1$:

$$\frac{dz}{dt} = \frac{2}{2+z} - z - 1,$$

which has an equilibrium point at $z = 0$. Now consider the candidate Lyapunov function

$$V(z) = \frac{1}{2}z^2,$$

which is globally positive definite. The derivative of V along trajectories of the system is given by

$$\dot{V}(z) = z\dot{z} = \frac{2z}{2+z} - z^2 - z.$$

If we restrict our analysis to an interval B_r, where $r < 2$, then $2 + z > 0$ and we can multiply through by $2 + z$ to obtain

$$2z - (z^2 + z)(2 + z) = -z^3 - 3z^2 = -z^2(z + 3) < 0, \qquad z \in B_r, \ r < 2.$$

It follows that $\dot{V}(z) < 0$ for all $z \in B_r$, $z \neq 0$, and hence the equilibrium point $x = 1$ is locally asymptotically stable. ∇

A slightly more complicated situation occurs if $\dot{V}$ is negative semidefinite. In this case it is possible that $\dot{V}(x) = 0$ when $x \neq 0$, and hence x could stop decreasing in value. The following example illustrates this case.

Example 5.11 Hanging pendulum
A normalized model for a hanging pendulum is

$$\frac{dx_1}{dt} = x_2, \qquad \frac{dx_2}{dt} = -\sin x_1,$$

where x_1 is the angle between the pendulum and the vertical, with positive x_1 corresponding to counterclockwise rotation. The equation has an equilibrium point $x_1 = x_2 = 0$, which corresponds to the pendulum hanging straight down. To explore the stability of this equilibrium point we choose the total energy as a Lyapunov function:

$$V(x) = 1 - \cos x_1 + \frac{1}{2}x_2^2 \approx \frac{1}{2}x_1^2 + \frac{1}{2}x_2^2.$$

The Taylor series approximation shows that the function is positive definite for small x. The time derivative of $V(x)$ is

$$\dot{V} = \dot{x}_1 \sin x_1 + \dot{x}_2 x_2 = x_2 \sin x_1 - x_2 \sin x_1 = 0.$$

Since this function is negative semidefinite, it follows from Lyapunov's theorem that the equilibrium point is stable but not necessarily asymptotically stable. When perturbed, the pendulum actually moves in a trajectory that corresponds to constant energy.   ▽

As demonstrated already, Lyapunov functions are not always easy to find, and they are also not unique. In many cases energy functions can be used as a starting point, as was done in Example 5.11. It turns out that Lyapunov functions can always be found for any stable system (under certain conditions), and hence one knows that if a system is stable, a Lyapunov function exists (and vice versa). Recent results using sum-of-squares methods have provided systematic approaches for finding Lyapunov systems [204]. Sum-of-squares techniques can be applied to a broad variety of systems, including systems whose dynamics are described by polynomial equations, as well as hybrid systems, which can have different models for different regions of state space.

For a linear dynamical system of the form

$$\frac{dx}{dt} = Ax,$$

it is possible to construct Lyapunov functions in a systematic manner. To do so, we consider quadratic functions of the form

$$V(x) = x^T P x,$$

where $P \in \mathbb{R}^{n \times n}$ is a symmetric matrix $(P = P^T)$. The condition that V be positive definite on B_r for some $r > 0$ is equivalent to the condition that P be a *positive definite matrix*:

$$x^T P x > 0, \quad \text{for all } x \neq 0,$$

which we write as $P \succ 0$. It can be shown that if P is symmetric, then P is positive definite if and only if all of its eigenvalues are real and positive.

Given a candidate Lyapunov function $V(x) = x^T P x$, we can now compute its derivative along flows of the system:

$$\dot{V} = \frac{\partial V}{\partial x} \frac{dx}{dt} = x^T (A^T P + PA) x =: -x^T Q x.$$

The requirement that $\dot{V}$ be negative definite on B_r (for asymptotic stability) becomes a condition that the matrix Q be positive definite. Thus, to find a Lyapunov function for a linear system it is sufficient to choose a $Q \succ 0$ and solve the *Lyapunov equation*:

$$A^T P + PA = -Q. \tag{5.17}$$

This is a linear equation in the entries of P, and hence it can be solved using linear algebra. It can be shown that the equation always has a solution if all of the eigenvalues of the matrix A are in the left half-plane. Moreover, the solution P is positive definite if Q is positive definite. It is thus always possible to find a quadratic Lyapunov function for a stable linear system. We will defer a proof of this until Chapter 6, where more tools for analysis of linear systems will be developed.

Example 5.12 Spring–mass system

Consider a simple spring–mass system, whose state space dynamics are given by

$$\frac{dx_1}{dt} = x_2, \qquad \frac{dx_2}{dt} = -\frac{k}{m}x_1 - \frac{b}{m}x_2, \qquad m, b, k > 0.$$

Note that this is equivalent to the example we used after Theorem 5.2 if $k = m$ and $b/m = \alpha$.

To find a Lyapunov function for the system, we choose $Q = I$ and equation (5.17) becomes

$$\begin{pmatrix} 0 & -k/m \\ 1 & -b/m \end{pmatrix} \begin{pmatrix} p_{11} & p_{12} \\ p_{12} & p_{22} \end{pmatrix} + \begin{pmatrix} p_{11} & p_{12} \\ p_{12} & p_{22} \end{pmatrix} \begin{pmatrix} 0 & 1 \\ -k/m & -b/m \end{pmatrix} = \begin{pmatrix} -1 & 0 \\ 0 & -1 \end{pmatrix}.$$

By evaluating each element of this matrix equation, we can obtain a set of linear equations for p_{ij}:

$$-\frac{2k}{m}p_{12} = -1, \qquad p_{11} - \frac{b}{m}p_{12} - \frac{k}{m}p_{22} = 0, \qquad 2p_{12} - \frac{2b}{m}p_{22} = -1.$$

These equations can be solved for p_{11}, p_{12}, and p_{22} to obtain

$$P = \begin{pmatrix} \dfrac{b^2 + k(k+m)}{2bk} & \dfrac{m}{2k} \\[3ex] \dfrac{m}{2k} & \dfrac{m(k+m)}{2bk} \end{pmatrix}.$$

Finally, it follows that

$$V(x) = \frac{b^2 + k(k+m)}{2bk}x_1^2 + \frac{m}{k}x_1 x_2 + \frac{m(k+m)}{2bk}x_2^2.$$

Notice that while it can be verified that this function is positive definite, its level sets are rotated ellipses. ∇

Knowing that we have a direct method to find Lyapunov functions for linear systems, we can now investigate the stability of nonlinear systems. Consider the system

$$\frac{dx}{dt} = F(x) =: Ax + \tilde{F}(x), \tag{5.18}$$

where $F(0) = 0$ and $\tilde{F}(x)$ contains terms that are second order and higher in the elements of x. The function Ax is an approximation of $F(x)$ near the origin, and we can determine the Lyapunov function for the linear approximation and investigate if it is also a Lyapunov function for the full nonlinear system. The following example illustrates the approach.

Example 5.13 Genetic switch

Consider the dynamics of a set of repressors connected together in a cycle, as shown in Figure 5.15a. The normalized dynamics for this system were given in Exercise 3.10:

$$\frac{dz_1}{d\tau} = \frac{\mu}{1 + z_2^n} - z_1, \qquad \frac{dz_2}{d\tau} = \frac{\mu}{1 + z_1^n} - z_2, \tag{5.19}$$

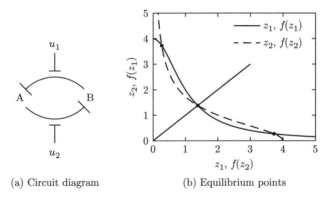

(a) Circuit diagram (b) Equilibrium points

Figure 5.15: Stability of a genetic switch. The circuit diagram in (a) represents two proteins that are each repressing the production of the other. The inputs u_1 and u_2 interfere with this repression, allowing the circuit dynamics to be modified. The equilibrium points for this circuit can be determined by the intersection of the two curves shown in (b).

where z_1 and z_2 are scaled versions of the protein concentrations, $n > 0$ and $\mu > 0$ are parameters that describe the interconnection between the genes, and we have set the external inputs u_1 and u_2 to zero.

The equilibrium points for the system are found by equating the time derivatives to zero. We define

$$f(u) = \frac{\mu}{1 + u^n}, \qquad f'(u) = \frac{df}{du} = \frac{-\mu n u^{n-1}}{(1 + u^n)^2},$$

so that our dynamics become

$$\frac{dz_1}{d\tau} = f(z_2) - z_1, \qquad \frac{dz_2}{d\tau} = f(z_1) - z_2,$$

and the equilibrium points are defined as the solutions of the equations

$$z_1 = f(z_2), \qquad z_2 = f(z_1).$$

If we plot the curves $(z_1, f(z_1))$ and $(f(z_2), z_2)$ on a graph, then these equations will have a solution when the curves intersect, as shown in Figure 5.15b. Because of the shape of the curves, it can be shown that there will always be three solutions: one at $z_{1e} = z_{2e}$, one with $z_{1e} < z_{2e}$, and one with $z_{1e} > z_{2e}$. If $\mu \gg 1$, then we can show that the solutions are given approximately by

$$z_{1e} \approx \mu, \quad z_{2e} \approx \frac{1}{\mu^{n-1}}; \qquad z_{1e} = z_{2e}; \qquad z_{1e} \approx \frac{1}{\mu^{n-1}}, \quad z_{2e} \approx \mu. \qquad (5.20)$$

To check the stability of the system, we write $f(u)$ in terms of its Taylor series expansion about u_e:

$$f(u) = f(u_e) + f'(u_e) \cdot (u - u_e) + \frac{1}{2} f''(u_e) \cdot (u - u_e)^2 + \text{higher-order terms},$$

where f' represents the first derivative of the function, and f'' the second. Using these approximations, the dynamics can then be written as

$$\frac{dw}{dt} = \begin{pmatrix} -1 & f'(z_{2e}) \\ f'(z_{1e}) & -1 \end{pmatrix} w + \tilde{F}(w),$$

where $w = z - z_e$ is the shifted state and $\tilde{F}(w)$ represents quadratic and higher-order terms.

We now use equation (5.17) to search for a Lyapunov function. Choosing $Q = I$ and letting $P \in \mathbb{R}^{2\times 2}$ have elements p_{ij}, we search for a solution of the equation

$$\begin{pmatrix} -1 & f'_1 \\ f'_2 & -1 \end{pmatrix} \begin{pmatrix} p_{11} & p_{12} \\ p_{12} & p_{22} \end{pmatrix} + \begin{pmatrix} p_{11} & p_{12} \\ p_{12} & p_{22} \end{pmatrix} \begin{pmatrix} -1 & f'_2 \\ f'_1 & -1 \end{pmatrix} = \begin{pmatrix} -1 & 0 \\ 0 & -1 \end{pmatrix},$$

where $f'_1 = f'(z_{1e})$ and $f'_2 = f'(z_{2e})$. Note that we have set $p_{21} = p_{12}$ to force P to be symmetric. Multiplying out the matrices, we obtain

$$\begin{pmatrix} -2p_{11} + 2f'_1 p_{12} & p_{11}f'_2 - 2p_{12} + p_{22}f'_1 \\ p_{11}f'_2 - 2p_{12} + p_{22}f'_1 & -2p_{22} + 2f'_2 p_{12} \end{pmatrix} = \begin{pmatrix} -1 & 0 \\ 0 & -1 \end{pmatrix},$$

which is a set of *linear* equations for the unknowns p_{ij}. We can solve these linear equations to obtain

$$p_{11} = -\frac{f'^2_1 - f'_2 f'_1 + 2}{4(f'_1 f'_2 - 1)}, \qquad p_{12} = -\frac{f'_1 + f'_2}{4(f'_1 f'_2 - 1)}, \qquad p_{22} = -\frac{f'^2_2 - f'_1 f'_2 + 2}{4(f'_1 f'_2 - 1)}.$$

To check that $V(w) = w^T P w$ is a Lyapunov function, we must verify that $V(w)$ is a positive definite function or equivalently that $P \succ 0$. Since P is a 2×2 symmetric matrix, it has two real eigenvalues λ_1 and λ_2 that satisfy

$$\lambda_1 + \lambda_2 = \mathrm{trace}(P), \qquad \lambda_1 \cdot \lambda_2 = \det(P).$$

In order for P to be positive definite λ_1 and λ_2 must be positive, and we thus require that

$$\mathrm{trace}(P) = \frac{f'^2_1 - 2f'_2 f'_1 + f'^2_2 + 4}{4 - 4f'_1 f'_2} > 0, \quad \det(P) = \frac{f'^2_1 - 2f'_2 f'_1 + f'^2_2 + 4}{16 - 16f'_1 f'_2} > 0.$$

We see that $\mathrm{trace}(P) = 4\det(P)$ and the numerator of the expressions is just $(f_1 - f_2)^2 + 4 > 0$, so it suffices to check the sign of $1 - f'_1 f'_2$. In particular, for P to be positive definite, we require that

$$f'(z_{1e}) f'(z_{2e}) < 1.$$

We can now make use of the expressions for f' defined earlier and evaluate at the approximate locations of the equilibrium points derived in equation (5.20). For the equilibrium points where $z_{1e} \neq z_{2e}$, we can show that

$$f'(z_{1e}) f'(z_{2e}) \approx f'(\mu) f'(\frac{1}{\mu^{n-1}}) = \frac{-\mu n \mu^{n-1}}{(1 + \mu^n)^2} \cdot \frac{-\mu n \mu^{-(n-1)^2}}{(1 + \mu^{-n(n-1)})^2} \approx n^2 \mu^{-n^2 + n}.$$

Using $n = 2$ and $\mu \approx 200$ from Exercise 3.10, we see that $f'(z_{1e}) f'(z_{2e}) \ll 1$ and hence P is positive definite. This implies that V is a positive definite function and hence a potential Lyapunov function for the system.

To determine if the equilibrium points $z_{1e} \neq z_{2e}$ are stable for the system (5.19), we now compute $\dot{V}$ at the equilibrium point. By construction,

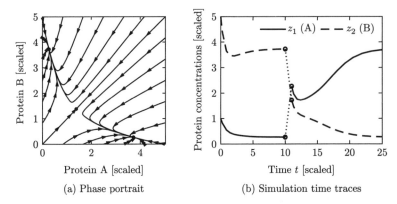

(a) Phase portrait

(b) Simulation time traces

Figure 5.16: Dynamics of a genetic switch. The phase portrait on the left shows that the switch has three equilibrium points, corresponding to protein A having a concentration greater than, equal to, or less than protein B. The equilibrium point with equal protein concentrations is unstable, but the other equilibrium points are stable. The simulation on the right shows the time response of the system starting from two different initial conditions. The initial portion of the curve corresponds to initial concentrations $z(0) = (1, 5)$ and converges to the equilibrium point where $z_{1e} < z_{2e}$. At time $t = 10$, the concentrations are perturbed by $+2$ in z_1 and -2 in z_2, moving the state into the region of the state space whose solutions converge to the equilibrium point where $z_{2e} < z_{1e}$.

$$\dot{V} = w^T(PA + A^TP)w + \tilde{F}^T(w)Pw + w^TP\tilde{F}(w)$$
$$= -w^Tw + \tilde{F}^T(w)Pw + w^TP\tilde{F}(w).$$

Since all terms in $\tilde{F}$ are quadratic or higher order in w, it follows that $\tilde{F}^T(w)Pw$ and $w^TP\tilde{F}(w)$ consist of terms that are at least third order in w. Therefore if w is sufficiently close to zero, then the cubic and higher-order terms will be smaller than the quadratic terms. Hence, sufficiently close to $w = 0$, $\dot{V}$ is negative definite, allowing us to conclude that these equilibrium points are both stable.

Figure 5.16 shows the phase portrait and time traces for a system with $\mu = 4$, illustrating the bistable nature of the system. When the initial condition starts with a concentration of protein B greater than that of A, the solution converges to the equilibrium point at (approximately) $(1/\mu^{n-1}, \mu)$. If A is greater than B, then it goes to $(\mu, 1/\mu^{n-1})$. The equilibrium point with $z_{1e} = z_{2e}$ is unstable. ∇

More generally, we can investigate what the linear approximation tells about the stability of a solution to a nonlinear equation. The following theorem gives a partial answer for the case of stability of an equilibrium point.

Theorem 5.3. *Consider the dynamical system (5.18) with $F(0) = 0$ and $\tilde{F}$ such that $\lim \|\tilde{F}(x)\|/\|x\| \to 0$ as $\|x\| \to 0$. If the real parts of all eigenvalues of A are strictly less than zero, then $x_e = 0$ is a locally asymptotically stable equilibrium point of equation (5.18).*

This theorem implies that asymptotic stability of the linear approximation implies *local* asymptotic stability of the original nonlinear system. The theorem is very important for control because it implies that stabilization of a linear approximation of a nonlinear system results in a stable equilibrium point for the nonlinear system. The proof of this theorem follows the technique used in Example 5.13. A formal proof can be found in [144].

It can also be shown that if A has one or more eigenvalues with strictly positive real part, then $x_e = 0$ is an unstable equilibrium point for the nonlinear system.

Krasovski–Lasalle Invariance Principle

For general nonlinear systems, especially those in symbolic form, it can be difficult to find a positive definite function V whose derivative is strictly negative definite. The Krasovski–Lasalle theorem enables us to conclude the asymptotic stability of an equilibrium point under less restrictive conditions, namely, in the case where $\dot{V}$ is negative semidefinite, which is often easier to construct. It only applies to time-invariant or periodic systems, which are the cases we consider here. This section makes use of some additional concepts from dynamical systems; see Hahn [113] or Khalil [144] for a more detailed description.

We will deal with the time-invariant case and begin by introducing a few more definitions. We denote the solution trajectories of the time-invariant system

$$\frac{dx}{dt} = F(x) \tag{5.21}$$

as $x(t; a)$, which is the solution of equation (5.21) at time t starting from a at $t_0 = 0$. The ω *limit set* of a trajectory $x(t; a)$ is the set of all points $z \in \mathbb{R}^n$ such that there exists a strictly increasing sequence of times t_n such that $x(t_n; a) \to z$ as $n \to \infty$. A set $M \subset \mathbb{R}^n$ is said to be an *invariant set* if for all $b \in M$, we have $x(t; b) \in M$ for all $t \geq 0$. It can be proved that the ω limit set of every trajectory is closed and invariant. We may now state the Krasovski–Lasalle principle.

Theorem 5.4 (Krasovski–Lasalle principle). *Let $V : \mathbb{R}^n \to \mathbb{R}$ be a locally positive definite function such that on the compact set $\Omega_r = \{x \in \mathbb{R}^n : V(x) \leq r\}$ we have $\dot{V}(x) \leq 0$. Define*

$$S = \{x \in \Omega_r : \dot{V}(x) = 0\}.$$

As $t \to \infty$, the trajectory tends to the largest invariant set inside S; i.e., its ω limit set is contained inside the largest invariant set in S. In particular, if S contains no invariant sets other than $x = 0$, then 0 is asymptotically stable.

Proofs are given in [151] and [159].

Lyapunov functions can often be used to design stabilizing controllers, as is illustrated by the following example, which also illustrates how the Krasovski–Lasalle principle can be applied.

Example 5.14 Inverted pendulum

Following the analysis in Example 3.10, an inverted pendulum can be described by the following normalized model:

$$\frac{dx_1}{dt} = x_2, \qquad \frac{dx_2}{dt} = \sin x_1 + u \cos x_1, \tag{5.22}$$

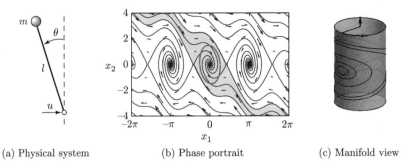

(a) Physical system   (b) Phase portrait   (c) Manifold view

Figure 5.17: Stabilized inverted pendulum. A control law applies a force u at the bottom of the pendulum to stabilize the inverted position (a). The phase portrait (b) shows that the equilibrium point corresponding to the vertical position is stabilized. The shaded region indicates the set of initial conditions that converge to the origin. The ellipse corresponds to a level set of a Lyapunov function $V(x)$ for which $V(x) > 0$ and $\dot{V}(x) < 0$ for all points inside the ellipse. This can be used as an estimate of the region of attraction of the equilibrium point. The actual dynamics of the system evolve on a manifold (c).

where x_1 is the angular deviation from the upright position and u is the (scaled) acceleration of the pivot, as shown in Figure 5.17a. The system has an equilibrium point at $x_1 = x_2 = 0$, which corresponds to the pendulum standing upright. This equilibrium point is unstable.

To find a stabilizing controller we consider the following candidate for a Lyapunov function:

$$V(x) = (\cos x_1 - 1) + a(1 - \cos^2 x_1) + \frac{1}{2}x_2^2 \approx \left(a - \frac{1}{2}\right)x_1^2 + \frac{1}{2}x_2^2.$$

The Taylor series expansion shows that the function is positive definite near the origin if $a > 0.5$. The time derivative of $V(x)$ is

$$\dot{V} = -\dot{x}_1 \sin x_1 + 2a\dot{x}_1 \sin x_1 \cos x_1 + \dot{x}_2 x_2 = x_2(u + 2a \sin x_1) \cos x_1.$$

Choosing the feedback law

$$u = -2a \sin x_1 - x_2 \cos x_1$$

gives

$$\dot{V} = -x_2^2 \cos^2 x_1.$$

It follows from Lyapunov's theorem that the equilibrium point is (locally) stable. However, since the function is only negative semidefinite, we cannot conclude asymptotic stability using Theorem 5.2. However, note that $\dot{V} = 0$ implies that $x_2 = 0$ or $x_1 = \pi/2 \pm n\pi$.

If we restrict our analysis to a small neighborhood of the origin Ω_r, $r \ll \pi/2$, then we can define

$$S = \{(x_1, x_2) \in \Omega_r : x_2 = 0\}$$

and we can compute the largest invariant set inside S. For a trajectory to remain in this set we must have $x_2 = 0$ for all t and hence $\dot{x}_2(t) = 0$ as well. Using the dynamics of the system (5.22), we see that $x_2(t) = 0$ and $\dot{x}_2(t) = 0$ implies $x_1(t) = 0$ as well. Hence the largest invariant set inside S is $(x_1, x_2) = 0$, and we can use the Krasovski–Lasalle principle to conclude that the origin is locally asymptotically stable. A phase portrait of the closed loop system is shown in Figure 5.17b.

In the analysis and the phase portrait, we have treated the angle of the pendulum $\theta = x_1$ as a real number. In fact, θ is an angle with $\theta = 2\pi$ equivalent to $\theta = 0$. Hence the dynamics of the system actually evolve on a *manifold* (smooth surface) as shown in Figure 5.17c. Analysis of nonlinear dynamical systems on manifolds is more complicated, but uses many of the same basic ideas presented here. ∇

5.5 PARAMETRIC AND NONLOCAL BEHAVIOR

Most of the tools that we have explored are focused on the local behavior of a fixed system near an equilibrium point. In this section we briefly introduce some concepts regarding the global behavior of nonlinear systems and the dependence of a system's behavior on parameters in the system model.

Regions of Attraction

To get some insight into the behavior of a nonlinear system we can start by finding the equilibrium points. We can then proceed to analyze the local behavior around the equilibrium points. The behavior of a system near an equilibrium point is called the *local* behavior of the system.

The solutions of the system can be very different far away from an equilibrium point. This is seen, for example, in the stabilized pendulum in Example 5.14. The inverted equilibrium point is stable, with small oscillations that eventually converge to the origin. But far away from this equilibrium point there are trajectories that converge to other equilibrium points or even cases in which the pendulum swings around the top multiple times, giving very long oscillations that are topologically different from those near the origin.

To better understand the dynamics of the system, we can examine the set of all initial conditions that converge to a given asymptotically stable equilibrium point. This set is called the *region of attraction* for the equilibrium point. An example is shown by the shaded region of the phase portrait in Figure 5.17b. In general, computing regions of attraction is difficult. However, even if we cannot determine the region of attraction, we can often obtain patches around the stable equilibrium points that are attracting. This gives partial information about the behavior of the system.

One method for approximating the region of attraction is through the use of Lyapunov functions. Suppose that V is a local Lyapunov function for a system around an equilibrium point x_0. Let Ω_r be a set on which $V(x)$ has a value less than r,

$$\Omega_r = \{x \in \mathbb{R}^n : V(x) \leq r\},$$

and suppose that $\dot{V}(x) \leq 0$ for all $x \in \Omega_r$, with equality only at the equilibrium point x_0. Then Ω_r is inside the region of attraction of the equilibrium point. Since this approximation depends on the Lyapunov function and the choice of Lyapunov function is not unique, it can sometimes be a very conservative estimate.

It is sometimes the case that we can find a Lyapunov function V such that V is positive definite and $\dot{V}$ is negative (semi-) definite for all $x \in \mathbb{R}^n$. In many instances it can then be shown that the region of attraction for the equilibrium point is the entire state space, and the equilibrium point is *globally* asymptotically stable. More detailed conditions for global stability can be found in [144] and other textbooks.

Example 5.15 Stabilized inverted pendulum

Consider again the stabilized inverted pendulum from Example 5.14. The Lyapunov function for the system was

$$V(x) = (\cos x_1 - 1) + a(1 - \cos^2 x_1) + \frac{1}{2}x_2^2.$$

With $a > 0.5$, $\dot{V}$ was negative semidefinite for all x and nonzero when $x_1 \neq \pm\pi/2$. Hence any x such that $|x_1| < \pi/2$ and $V(x) > 0$ will be inside the invariant set defined by the level curves of $V(x)$. One of these level sets is shown in Figure 5.17b. ∇

Bifurcations

Another important property of nonlinear systems is how their behavior changes as the parameters governing the dynamics change. We can study this in the context of models by exploring how the location of equilibrium points, their stability, their regions of attraction, and other dynamic phenomena, such as limit cycles, vary based on the values of the parameters in the model.

Consider a differential equation of the form

$$\frac{dx}{dt} = F(x, \mu), \quad x \in \mathbb{R}^n, \ \mu \in \mathbb{R}^k, \tag{5.23}$$

where x is the state and μ is a set of parameters that describe the family of equations. The equilibrium solutions satisfy

$$F(x, \mu) = 0,$$

and as μ is varied, the corresponding solutions $x_e(\mu)$ can also vary. We say that the system (5.23) has a *bifurcation* at $\mu = \mu^*$ if the behavior of the system changes qualitatively at μ^*. This can occur either because of a change in stability type or a change in the number of solutions at a given value of μ.

Example 5.16 Predator–prey

Consider the predator–prey system described in Example 3.4 and modeled as a continuous time system as described in Section 4.7. The dynamics of the system are given by

$$\frac{dH}{dt} = rH\left(1 - \frac{H}{k}\right) - \frac{aHL}{c + H}, \qquad \frac{dL}{dt} = b\frac{aHL}{c + H} - dL, \tag{5.24}$$

where H and L are the numbers of hares (prey) and lynxes (predators) and a, b, c, d, k, and r are parameters that model a given predator–prey system (described in more detail in Section 4.7). The system has an equilibrium point at $H_e > 0$ and $L_e > 0$ that can be found numerically.

To explore how the parameters of the model affect the behavior of the system, we choose to focus on two specific parameters of interest: a, the interaction coefficient

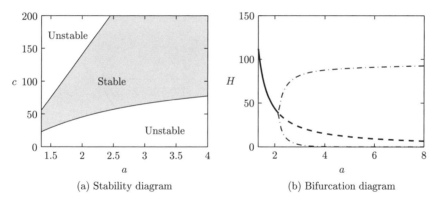

(a) Stability diagram   (b) Bifurcation diagram

Figure 5.18: Bifurcation analysis of the predator–prey system. (a) Parametric stability diagram showing the regions in parameter space for which the system is
stable. (b) Bifurcation diagram showing the location and stability of the equilibrium point as a function of a. The solid line represents a stable equilibrium point,
and the dashed line represents an unstable equilibrium point. The dash-dotted lines
indicate the upper and lower bounds for the limit cycle at that parameter value
(computed via simulation). The nominal values of the parameters in the model are
$a = 3.2$, $b = 0.6$, $c = 50$, $d = 0.56$, $k = 125$, and $r = 1.6$.

between the populations and c, a parameter affecting the prey consumption rate.
Figure 5.18a is a numerically computed *parametric stability diagram* showing the
regions in the chosen parameter space for which the equilibrium point is stable
(leaving the other parameters at their nominal values). We see from this figure that
for certain combinations of a and c we get a stable equilibrium point, while at other
values this equilibrium point is unstable.

Figure 5.18a is a numerically computed *bifurcation diagram* for the system. In
this plot, we choose one parameter to vary (a) and then plot the equilibrium value
of one of the states (H) on the vertical axis. The remaining parameters are set to
their nominal values. A solid line indicates that the equilibrium point is stable; a
dashed line indicates that the equilibrium point is unstable. Note that the stability
in the bifurcation diagram matches that in the parametric stability diagram for
$c = 50$ (the nominal value) and a varying from 1.35 to 4. For the predator–prey
system, when the equilibrium point is unstable, the solution converges to a stable
limit cycle. The amplitude of this limit cycle is shown by the dash-dotted line in
Figure 5.18b.   ∇

A particular form of bifurcation that is very common when controlling linear
systems is that the equilibrium point remains fixed but the stability of the equilibrium point changes as the parameters are varied. In such a case it is revealing to plot
the eigenvalues of the system as a function of the parameters. Such plots are called
root locus diagrams because they give the locus of the eigenvalues when parameters change. Bifurcations occur when parameter values are such that there are
eigenvalues with zero real part. Computing environments such LABVIEW, MAT-
LAB, Mathematica, and Python have tools for plotting root loci. A more detailed
discussion of the root locus is given in Section 12.5.

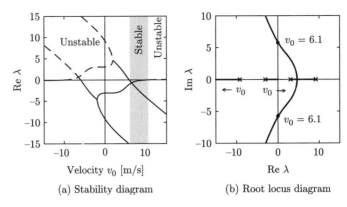

(a) Stability diagram　　　　　　　　(b) Root locus diagram

Figure 5.19: Stability plots for a bicycle moving at constant velocity. The plot in (a) shows the real part of the system eigenvalues as a function of the bicycle velocity v_0. The system is stable when all eigenvalues have negative real part (shaded region). The plot in (b) shows the locus of eigenvalues on the complex plane as the velocity v is varied and gives a different view of the stability of the system. This type of plot is called a *root locus diagram*.

Example 5.17 Root locus diagram for a bicycle model

Consider the linear bicycle model given by equation (4.8) in Section 4.2. Introducing the state variables $x_1 = \varphi$, $x_2 = \delta$, $x_3 = \dot{\varphi}$, and $x_4 = \dot{\delta}$ and setting the steering torque $T = 0$, the equations can be written as

$$\frac{dx}{dt} = \begin{pmatrix} 0 & I \\ -M^{-1}(K_0 + K_2 v_0^2) & -M^{-1} C v_0 \end{pmatrix} x =: Ax,$$

where I is a 2×2 identity matrix and v_0 is the velocity of the bicycle. Figure 5.19a shows the real parts of the eigenvalues as a function of velocity. Figure 5.19b shows the dependence of the eigenvalues of A on the velocity v_0. The figures show that the bicycle is unstable for low velocities because two eigenvalues are in the right half-plane. As the velocity increases, these eigenvalues move into the left half-plane, indicating that the bicycle becomes self-stabilizing. As the velocity is increased further, there is an eigenvalue close to the origin that moves into the right half-plane, making the bicycle unstable again. However, this eigenvalue is small and so it can easily be stabilized by a rider. Figure 5.19a shows that the bicycle is self-stabilizing for velocities between 6 and 10 m/s.　　　　　　　　　　∇

Parametric stability diagrams and bifurcation diagrams can provide valuable insights into the dynamics of a nonlinear system. It is usually necessary to carefully choose the parameters that one plots, including combining the natural parameters of the system to eliminate extra parameters when possible. Computer programs such as AUTO, LOCBIF, and XPPAUT provide numerical algorithms for producing stability and bifurcation diagrams.

Design of Nonlinear Dynamics Using Feedback

In most of the text we will rely on linear approximations to design feedback laws that stabilize an equilibrium point and provide a desired level of performance.

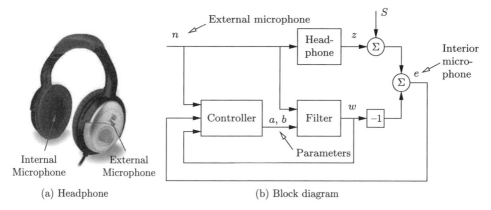

(a) Headphone   (b) Block diagram

Figure 5.20: Headphones with noise cancellation. Noise is sensed by the exterior microphone (a) and sent to a filter in such a way that it cancels the noise that penetrates the headphone (b). The filter parameters a and b are adjusted by the controller. S represents the input signal to the headphones.

However, for some classes of problems the feedback controller must be nonlinear to accomplish its function. By making use of Lyapunov functions we can often design a nonlinear control law that provides stable behavior, as we saw in Example 5.14.

One way to systematically design a nonlinear controller is to begin with a candidate Lyapunov function $V(x)$ and a control system $\dot{x} = f(x, u)$. We say that $V(x)$ is a *control Lyapunov function* if for every x there exists a u such that $\dot{V}(x) = \frac{\partial V}{\partial x} f(x, u) < 0$. In this case, it may be possible to find a function $\alpha(x)$ such that $u = \alpha(x)$ stabilizes the system. The following example illustrates the approach.

Example 5.18 Noise cancellation
Noise cancellation is used in consumer electronics and in industrial systems to reduce the effects of noise and vibrations. The idea is to locally reduce the effect of noise by generating opposing signals. A pair of headphones with noise cancellation such as those shown in Figure 5.20a is a typical example. A schematic diagram of the system is shown in Figure 5.20b. The system has two microphones, one outside the headphones that picks up exterior noise n and another inside the headphones that picks up the signal e, which is a combination of the desired signal S and the external noise that penetrates the headphone. The signal from the exterior microphone is filtered and sent to the headphones in such a way that it cancels the external noise that penetrates into the headphones. The parameters of the filter are adjusted by a feedback mechanism to make the noise signal in the internal microphone as small as possible. The feedback is inherently nonlinear because it acts by changing the parameters of the filter.

To analyze the system we assume for simplicity that the propagation of external noise into the headphones is modeled by the first-order dynamical system

$$\frac{dz}{dt} = a_0 z + b_0 n, \qquad (5.25)$$

where n is the external noise signal, z is the sound level inside the headphones, and the parameters $a_0 < 0$ and b_0 are not known. Assume that the filter is a dynamical system of the same type:

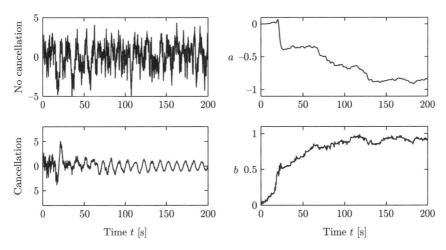

Figure 5.21: Simulation of noise cancellation. The upper left figure shows the headphone signal without noise cancellation, and the lower left figure shows the signal with noise cancellation. The right figures show the parameters a and b of the filter.

$$\frac{dw}{dt} = aw + bn,$$

where the parameters a and b are adjustable. We wish to find a controller that updates a and b so that they converge to the (unknown) parameters a_0 and b_0. If $a = a_0$ and $b = b_0$ we have $e = S$ and the noise effect of the noise is eliminated. Assuming for simplicity that $S = 0$, introduce $x_1 = e = z - w$, $x_2 = a - a_0$, and $x_3 = b - b_0$. Then

$$\frac{dx_1}{dt} = a_0(z - w) + (a - a_0)w + (b - b_0)n = a_0x_1 + x_2w + x_3n. \qquad (5.26)$$

We will achieve noise cancellation if we can find a feedback law for changing the parameters a and b so that the error e goes to zero. To do this we choose

$$V(x_1, x_2, x_3) = \frac{1}{2}\big(\alpha x_1^2 + x_2^2 + x_3^2\big)$$

as a candidate Lyapunov function for equation (5.26). The derivative of V is

$$\dot{V} = \alpha x_1 \dot{x}_1 + x_2 \dot{x}_2 + x_3 \dot{x}_3 = \alpha a_0 x_1^2 + x_2(\dot{x}_2 + \alpha w x_1) + x_3(\dot{x}_3 + \alpha n x_1).$$

Choosing

$$\dot{a} = \dot{x}_2 = -\alpha w x_1 = -\alpha w e, \qquad \dot{b} = \dot{x}_3 = -\alpha n x_1 = -\alpha n e, \qquad (5.27)$$

we find that $\dot{V} = \alpha a_0 x_1^2 < 0$, and it follows that the quadratic function will decrease as long as $e = x_1 = w - z \neq 0$. The nonlinear feedback (5.27) thus attempts to change the parameters so that the error between the signal and the noise is small. Notice that feedback law (5.27) does not use the model (5.25) explicitly.

A simulation of the system is shown in Figure 5.21. In the simulation we have represented the signal as a pure sinusoid and the noise as broad band noise. The figure shows the dramatic improvement with noise cancellation. The sinusoidal signal is not visible without noise cancellation. The filter parameters change quickly from their initial values $a = b = 0$. Filters of higher order with more coefficients are used in practice. $\triangledown$

5.6 FURTHER READING

The field of dynamical systems has a rich literature that characterizes the possible features of dynamical systems and describes how parametric changes in the dynamics can lead to topological changes in behavior. Readable introductions to dynamical systems are given by Strogatz [234] and the highly illustrated text by Abraham and Shaw [2]. More technical treatments include Andronov, Vitt, and Khaikin [10], Guckenheimer and Holmes [110], and Wiggins [254]. For students with a strong interest in mechanics, the texts by Arnold [15] and Marsden and Ratiu [178] provide an elegant approach using tools from differential geometry. Finally, good treatments of dynamical systems methods in biology are given by Wilson [256] and Ellner and Guckenheimer [83]. There is a large literature on Lyapunov stability theory, including the classic texts by Malkin [175], Hahn [113], and Krasovski [151]. We highly recommend the comprehensive treatment by Khalil [144].

EXERCISES

5.1 (Time-invariant systems) Show that if we have a solution of the differential equation (5.2) given by $x(t)$ with initial condition $x(t_0) = x_0$, then $\tilde{x}(\tau) = x(t - t_0)$ is a solution of the differential equation

$$\frac{d\tilde{x}}{d\tau} = F(\tilde{x})$$

with initial condition $\tilde{x}(0) = x_0$, where $\tau = t - t_0$.

5.2 (Flow in a tank) Consider a cylindrical tank with cross sectional area $A\,\mathrm{m}^2$, effective outlet area $a\,\mathrm{m}^2$, and inflow $q_{\mathrm{in}}\,\mathrm{m}^3/\mathrm{s}$. An energy balance shows that the outlet velocity is $v = \sqrt{2gh}\,\mathrm{m/s}$, where $g\,\mathrm{m/s}^2$ is the acceleration of gravity and h is the distance between the outlet and the water level in the tank (in meters). Show that the system can be modeled by

$$\frac{dh}{dt} = -\frac{a}{A}\sqrt{2gh} + \frac{1}{A}q_{\mathrm{in}}, \qquad q_{\mathrm{out}} = a\sqrt{2gh}.$$

Use the parameters $A = 0.2$, $a = 0.01$. Simulate the system when the inflow is zero and the initial level is $h = 0.2$. Do you expect any difficulties in the simulation?

5.3 (Lyapunov functions) Consider the second-order system

$$\frac{dx_1}{dt} = -ax_1, \qquad \frac{dx_2}{dt} = -bx_1 - cx_2,$$

where $a, b, c > 0$. Investigate whether the functions

$$V_1(x) = \frac{1}{2}x_1^2 + \frac{1}{2}x_2^2, \qquad V_2(x) = \frac{1}{2}x_1^2 + \frac{1}{2}\left(x_2 + \frac{b}{c-a}x_1\right)^2$$

are Lyapunov functions for the system and give any conditions that must hold.

5.4 (Damped spring–mass system) Consider a damped spring–mass system with dynamics

$$m\ddot{q} + c\dot{q} + kq = 0.$$

A natural candidate for a Lyapunov function is the total energy of the system, given by

$$V = \frac{1}{2}m\dot{q}^2 + \frac{1}{2}kq^2.$$

Use the Krasovski–Lasalle theorem to show that the system is asymptotically stable.

5.5 (Electric generator) The following simple model for an electric generator connected to a strong power grid was given in Exercise 3.8:

$$J\frac{d^2\varphi}{dt^2} = P_{\mathrm{m}} - P_{\mathrm{e}} = P_{\mathrm{m}} - \frac{EV}{X}\sin\varphi.$$

The parameter

$$a = \frac{P_{\mathrm{max}}}{P_{\mathrm{m}}} = \frac{EV}{XP_{\mathrm{m}}}$$

is the ratio between the maximum deliverable power $P_{\mathrm{max}} = EV/X$ and the mechanical power P_{m}.

a) Consider a as a bifurcation parameter and discuss how the equilibrium points depend on a.

b) For $a > 1$, show that there is a center at $\varphi_0 = \arcsin(1/a)$ and a saddle at $\varphi = \pi - \varphi_0$.

c) Assume $a > 1$ and show that there is a solution through the saddle that satisfies

$$\frac{J}{2}\left(\frac{d\varphi}{dt}\right)^2 - P_{\mathrm{m}}(\varphi - \varphi_0) - \frac{EV}{X}(\cos\varphi - \cos\varphi_0) = 0. \tag{5.28}$$

Set $J/P_{\mathrm{m}} = 1$ and use simulation to show that the stability region is the interior of the area enclosed by this solution. Investigate what happens if the system is in equilibrium with a value of a that is slightly larger than 1 and a suddenly decreases, corresponding to the reactance of the line suddenly increasing.

5.6 (Lyapunov equation) Show that Lyapunov equation (5.17) always has a solution if all of the eigenvalues of A are in the left half-plane. (Hint: Use the fact that the Lyapunov equation is linear in P and start with the case where A has distinct eigenvalues.)

5.7 (Shaping behavior by feedback) An inverted pendulum can be modeled by the differential equation

$$\frac{dx_1}{dt} = x_2, \qquad \frac{dx_2}{dt} = \sin x_1 + u\cos x_1,$$

where x_1 is the angle of the pendulum clockwise), and x_2 is its angular velocity (see Example 5.14). Qualitatively discuss the behavior of the open loop system and how the behavior changes when the feedback $u = -2\sin(x)$ is introduced. (Hint: use phase portraits.)

5.8 (Swinging up a pendulum) Consider the inverted pendulum, discussed in Example 5.4, that is described by

$$\ddot{\theta} = \sin\theta + u\cos\theta,$$

where θ is the angle between the pendulum and the vertical and the control signal u is the acceleration of the pivot. Using the energy function

$$V(\theta, \dot{\theta}) = \cos\theta - 1 + \frac{1}{2}\dot{\theta}^2,$$

show that the state feedback $u = k(V_0 - V)\dot{\theta}\cos\theta$ causes the pendulum to "swing up" to the upright position.

5.9 (Root locus diagram) Consider the linear system

$$\frac{dx}{dt} = \begin{pmatrix} 0 & 1 \\ 0 & -3 \end{pmatrix} x + \begin{pmatrix} -1 \\ 4 \end{pmatrix} u, \qquad y = \begin{pmatrix} 1 & 0 \end{pmatrix} x,$$

with the feedback $u = -ky$. Plot the location of the eigenvalues as a function the parameter k.

5.10 (Discrete-time Lyapunov function) Consider a nonlinear discrete-time system with dynamics $x[k+1] = f(x[k])$ and equilibrium point $x_e = 0$. Suppose there exists a smooth, positive definite function $V : \mathbb{R}^n \to \mathbb{R}$ such that $V(f(x)) - V(x) < 0$ for $x \neq 0$ and $V(0) = 0$. Show that $x_e = 0$ is (locally) asymptotically stable.

5.11 (Operational amplifier oscillator) An op amp circuit for an oscillator was shown in Exercise 4.4. The oscillatory solution for that linear circuit was stable but not asymptotically stable. A schematic of a modified circuit that has nonlinear elements is shown in the figure below.

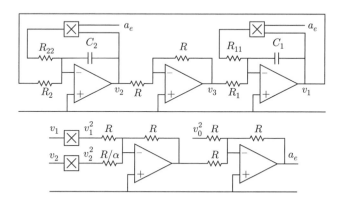

The modification is obtained by making a feedback around each of the operational amplifiers that has capacitors and making use of multipliers. The signal $a_e = v_1^2 + \alpha v_2^2 - v_0^2$ is the amplitude error. Show that the system is modeled by

$$\frac{dv_1}{dt} = \frac{1}{R_1 C_1} v_2 + \frac{1}{R_{11} C_1} v_1 (v_0^2 - v_1^2 - \alpha v_2^2),$$

$$\frac{dv_2}{dt} = -\frac{1}{R_2 C_2} v_1 + \frac{1}{R_{22} C_2} v_2 (v_0^2 - v_1^2 - \alpha v_2^2).$$

Determine α so that the circuit gives an oscillation with a stable limit cycle with amplitude v_0. (Hint: Use the results of Example 5.9.)

5.12 (Congestion control) Consider the congestion control problem described in Section 4.4. Confirm that the equilibrium point for the system is given by equation (4.22) and compute the stability of this equilibrium point using a linear approximation.

5.13 (Self-activating genetic circuit) Consider the dynamics of a genetic circuit that implements *self-activation*: the protein produced by the gene is an activator for the protein, thus stimulating its own production through positive feedback. Using the models presented in Example 3.18, the dynamics for the system can be written as

$$\frac{dm}{dt} = \frac{\alpha p^2}{1 + kp^2} + \alpha_0 - \delta m, \qquad \frac{dp}{dt} = \kappa m - \gamma p, \tag{5.29}$$

for $p, m \geq 0$. Find the equilibrium points for the system and analyze the local stability of each using Lyapunov analysis.

5.14 (Diagonal systems) Let $A \in \mathbb{R}^{n \times n}$ be a square matrix with real eigenvalues $\lambda_1, \ldots, \lambda_n$ and corresponding eigenvectors $v_1, \ldots, v_n$. Assume that the eigenvalues are distinct ($\lambda_i \neq \lambda_j$ for $i \neq j$).

a) Show that $v_i \neq v_j$ for $i \neq j$.

b) Show that the eigenvectors form a basis for $\mathbb{R}^n$ so that any vector x can be written as $x = \sum \alpha_i v_i$ for $\alpha_i \in \mathbb{R}$.

c) Let $T = \begin{pmatrix} v_1 & v_2 & \cdots & v_n \end{pmatrix}$ and show that $T^{-1}AT$ is a diagonal matrix of the form (5.10).

d) Show that if some of the λ_i are complex numbers, then A can be written as

$$A = \begin{pmatrix} \Lambda_1 & & 0 \\ & \ddots & \\ 0 & & \Lambda_k \end{pmatrix}, \quad \text{where} \quad \Lambda_i = \lambda \in \mathbb{R} \quad \text{or} \quad \Lambda_i = \begin{pmatrix} \sigma & \omega \\ -\omega & \sigma \end{pmatrix}$$

in an appropriate set of coordinates.

This form of the dynamics of a linear system is often referred to as *block diagonal form*.

Chapter Six

Linear Systems

> Few physical elements display truly linear characteristics. For example the
> relation between force on a spring and displacement of the spring is
> always nonlinear to some degree. The relation between current through a
> resistor and voltage drop across it also deviates from a straight-line
> relation. However, if in each case the relation is *reasonably* linear, then it
> will be found that the system behavior will be very close to that obtained
> by assuming an ideal, linear physical element, and the analytical
> simplification is so enormous that we make linear assumptions wherever
> we can possibly do so in good conscience.
>
> —Robert H. Cannon, *Dynamics of Physical Systems*, 1967 [60].

In Chapters 3–5 we considered the construction and analysis of differential equation models for dynamical systems. In this chapter we specialize our results to the case of linear, time-invariant input/output systems. Two central concepts are the matrix exponential and the convolution equation, through which we can completely characterize the behavior of a linear system. We also describe some properties of the input/output response and show how to approximate a nonlinear system by a linear one.

6.1 BASIC DEFINITIONS

We have seen several instances of linear differential equations in the examples in the previous chapters, including the spring–mass system (damped oscillator) and the operational amplifier in the presence of small (nonsaturating) input signals. More generally, many dynamical systems can be modeled accurately by linear differential equations. Electrical circuits are one example of a broad class of systems for which linear models can be used effectively. Linear models are also broadly applicable in mechanical engineering, for example, as models of small deviations from equilibrium points in solid and fluid mechanics. Signal-processing systems, including digital filters of the sort used in MP3 players and streaming audio, are another source of good examples, although these are often best modeled in discrete time (as described in more detail in the exercises).

In many cases, we *create* systems with a linear input/output response through the use of feedback. Indeed, it was the desire for linear behavior that led Harold S. Black to the invention of the negative feedback amplifier. Almost all modern signal processing systems, whether analog or digital, use feedback to produce linear or near-linear input/output characteristics. For these systems, it is often useful to represent the input/output characteristics as linear, ignoring the internal details required to get that linear response.

For other systems, nonlinearities cannot be ignored, especially if one cares about the global behavior of the system. The predator–prey problem is one example of this: to capture the oscillatory behavior of the interdependent populations we must include the nonlinear coupling terms. Other examples include switching behavior and generating periodic motion for locomotion. However, if we care about what happens near an equilibrium point, it often suffices to approximate the nonlinear dynamics by their local linearization, as we already explored briefly in Section 5.3. The linearization is essentially an approximation of the nonlinear dynamics around the desired operating point.

Linearity

We now proceed to define linearity of input/output systems more formally. Consider a state space system of the form

$$\frac{dx}{dt} = f(x, u), \qquad y = h(x, u), \tag{6.1}$$

where $x \in \mathbb{R}^n$, $u \in \mathbb{R}^p$, and $y \in \mathbb{R}^q$. As in the previous chapters, we will usually restrict ourselves to the single-input, single-output case by taking $p = q = 1$. We also assume that all functions are smooth and that for a reasonable class of inputs (e.g., piecewise continuous functions of time) the solutions of equation (6.1) exist for all time.

It will be convenient to assume that the origin $x = 0$, $u = 0$ is an equilibrium point for this system ($\dot{x} = 0$) and that $h(0, 0) = 0$. Indeed, we can do so without loss of generality. To see this, suppose that $(x_e, u_e) \neq (0, 0)$ is an equilibrium point of the system with output $y_e = h(x_e, u_e)$. Then we can define a new set of states, inputs, and outputs,

$$\tilde{x} = x - x_e, \qquad \tilde{u} = u - u_e, \qquad \tilde{y} = y - y_e,$$

and rewrite the equations of motion in terms of these variables:

$$\frac{d}{dt}\tilde{x} = f(\tilde{x} + x_e, \tilde{u} + u_e) =: \tilde{f}(\tilde{x}, \tilde{u}),$$

$$\tilde{y} = h(\tilde{x} + x_e, \tilde{u} + u_e) - y_e =: \tilde{h}(\tilde{x}, \tilde{u}).$$

In the new set of variables, the origin is an equilibrium point with output 0, and hence we can carry out our analysis in this set of variables. Once we have obtained our answers in this new set of variables, we simply "translate" them back to the original coordinates using $x = \tilde{x} + x_e$, $u = \tilde{u} + u_e$, and $y = \tilde{y} + y_e$.

Returning to the original equations (6.1), now assuming without loss of generality that the origin is the equilibrium point of interest, we write the output $y(t)$ corresponding to the initial condition $x(0) = x_0$ and input $u(t)$ as $y(t; x_0, u)$. Using this notation, a system is said to be a *linear input/output system* if the following conditions are satisfied:

$$\text{(i)} \quad y(t; \alpha x_1 + \beta x_2, 0) = \alpha y(t; x_1, 0) + \beta y(t; x_2, 0),$$

$$\text{(ii)} \quad y(t; \alpha x_0, \delta u) = \alpha y(t; x_0, 0) + \delta y(t; 0, u), \text{ and} \tag{6.2}$$

$$\text{(iii)} \quad y(t; 0, \delta u_1 + \gamma u_2) = \delta y(t; 0, u_1) + \gamma y(t; 0, u_2).$$

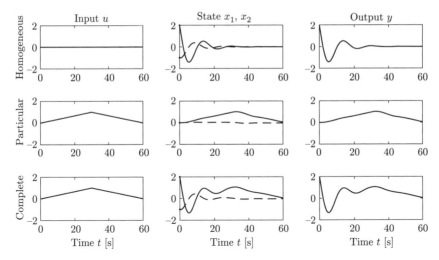

Figure 6.1: Superposition of homogeneous and particular solutions. The first row shows the input, state, and output corresponding to the initial condition response. The second row shows the same variables corresponding to zero initial condition but nonzero input. The third row is the complete solution, which is the sum of the two individual solutions.

Thus, we define a system to be linear if the outputs are jointly linear in the initial condition response $(u=0)$ and the forced response $(x(0)=0)$. Property (iii) is a statement of the *principle of superposition*: the response of a linear system to the sum of two inputs u_1 and u_2 is the sum of the outputs y_1 and y_2 corresponding to the individual inputs.

The general form of a linear state space system is

$$\frac{dx}{dt} = Ax + Bu, \qquad y = Cx + Du, \tag{6.3}$$

where $A \in \mathbb{R}^{n \times n}$, $B \in \mathbb{R}^{n \times p}$, $C \in \mathbb{R}^{q \times n}$ and $D \in \mathbb{R}^{q \times p}$. In the special case of a single-input, single-output system, B is a column vector, C is a row vector, and D is scalar. Equation (6.3) is a system of linear first-order differential equations with input u, state x, and output y. It is easy to show that given solutions $x_1(t)$ and $x_2(t)$ for this set of equations, the corresponding outputs satisfy the linearity conditions (6.2).

We define $x_h(t)$ to be the solution with zero input (the general solution to the *homogeneous system*),

$$\frac{dx_h}{dt} = Ax_h, \qquad x_h(0) = x_0,$$

and the solution $x_p(t)$ to be the input dependent solution with zero initial condition (the *particular solution* or *forced solution*),

$$\frac{dx_p}{dt} = Ax_p + Bu, \qquad x_p(0) = 0.$$

Figure 6.1 illustrates how these two individual solutions can be superimposed to form the complete solution.

It is also possible to show that if a dynamical system with a finite number of states is input/output linear in the sense we have described, it can always be represented by a state space equation of the form (6.3) through an appropriate choice of state variables. In Section 6.2 we will give an explicit solution of equation (6.3), but we illustrate the basic form through a simple example.

Example 6.1 Linearity of solutions for a scalar system
Consider the first-order differential equation

$$\frac{dx}{dt} = ax + u, \qquad y = x,$$

with $x(0) = x_0$. Let $u_1 = A \sin \omega_1 t$ and $u_2 = B \cos \omega_2 t$. The solution to the homogeneous system is $x_h(t) = e^{at} x_0$, and two particular solutions with $x(0) = 0$ are

$$x_{p1}(t) = -A \frac{-\omega_1 e^{at} + \omega_1 \cos \omega_1 t + a \sin \omega_1 t}{a^2 + \omega_1^2},$$

$$x_{p2}(t) = B \frac{a e^{at} - a \cos \omega_2 t + \omega_2 \sin \omega_2 t}{a^2 + \omega_2^2}.$$

Suppose that we now choose $x(0) = \alpha x_0$ and $u = u_1 + u_2$. Then the resulting solution is the weighted sum of the individual solutions:

$$
\begin{aligned}
x(t) = e^{at} &\left(\alpha x_0 + \frac{A \omega_1}{a^2 + \omega_1^2} + \frac{Ba}{a^2 + \omega_2^2} \right) \\
&- A \frac{\omega_1 \cos \omega_1 t + a \sin \omega_1 t}{a^2 + \omega_1^2} + B \frac{-a \cos \omega_2 t + \omega_2 \sin \omega_2 t}{a^2 + \omega_2^2}.
\end{aligned}
\tag{6.4}
$$

To see this, substitute equation (6.4) into the differential equation. Thus, the properties of a linear system are satisfied.   ▽

Time Invariance

Time invariance is an important concept that is used to describe a system whose properties do not change with time. More precisely, for a time-invariant system if the input $u(t)$ gives output $y(t)$, then if we shift the time at which the input is applied by a constant amount a, $u(t + a)$ gives the output $y(t + a)$. Systems that are linear and time-invariant, often called *LTI systems*, have the interesting property that their response to an arbitrary input is completely characterized by their response to step inputs or their response to short "impulses."

To explore the consequences of time invariance, we first compute the response to a piecewise constant input. Assume that the system has zero initial condition and consider the piecewise constant input shown in Figure 6.2a. The input has jumps at times t_k, and its values after the jumps are $u(t_k)$. The input can be viewed as a combination of steps: the first step at time t_0 has amplitude $u(t_0)$, the second step at time t_1 has amplitude $u(t_1) - u(t_0)$, etc.

Assuming that the system is initially at an equilibrium point (so that the initial condition response is zero), the response to the input can be obtained by superimposing the responses to a combination of step inputs. Let $H(t)$ be the response to a unit step applied at time 0, and assume that $H(0) = 0$. The response to the first step is then $H(t - t_0)u(t_0)$, the response to the second step is $H(t - t_1)\big(u(t_1) - u(t_0)\big)$, and we find that the complete response is given by

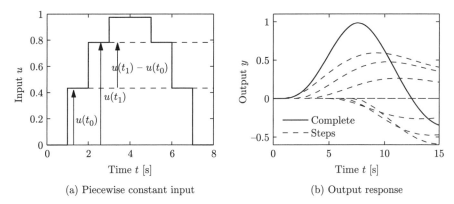

(a) Piecewise constant input   (b) Output response

Figure 6.2: Response to piecewise constant inputs. A piecewise constant signal can be represented as a sum of step signals (a), and the resulting output is the sum of the individual outputs (b).

$$y(t) = H(t - t_0)u(t_0) + H(t - t_1)\big(u(t_1) - u(t_0)\big) + \cdots$$

$$= \big(H(t - t_0) - H(t - t_1)\big)u(t_0) + \big(H(t - t_1) - H(t - t_2)\big)u(t_1) + \cdots$$

$$= \sum_{k=1}^{n} \big(H(t - t_{k-1}) - H(t - t_k)\big)u(t_{k-1}) + H(t - t_n)u(t_n)$$

$$= \sum_{k=1}^{n} \frac{H(t - t_{k-1}) - H(t - t_k)}{t_k - t_{k-1}} u(t_{k-1})\big(t_k - t_{k-1}\big) + H(t - t_n)u(t_n),$$

where n is such that $t_n \leq t$. An example of this computation is shown in Figure 6.2b.

The response to a continuous input signal is obtained by taking the limit $n \to \infty$ in such a way that $t_k - t_{k-1} \to 0$ and $t_n \to t$, which gives

$$y(t) = \int_0^t H'(t - \tau)u(\tau)d\tau, \qquad (6.5)$$

where H' is the derivative of the step response, also called the *impulse response*. The response of a linear time-invariant system to any input can thus be computed from the step response. Notice that the output depends only on the input since we assumed the system was initially at rest, $x(0) = 0$. We will derive equation (6.5) in a slightly different way in Section 6.3.

6.2   THE MATRIX EXPONENTIAL

Equation (6.5) shows that the output of a linear system with zero initial state can be written as an integral over the inputs $u(t)$. In this section and the next we derive a more general version of this formula, which includes nonzero initial conditions. We begin by exploring the initial condition response using the matrix exponential.

Initial Condition Response

We will now explicitly show that the output of a linear system depends linearly on the input and the initial conditions. We begin by considering the general solution to the homogeneous system corresponding to the dynamics

$$\frac{dx}{dt} = Ax. \tag{6.6}$$

For the *scalar* differential equation

$$\frac{dx}{dt} = ax, \qquad x \in \mathbb{R},\ a \in \mathbb{R},$$

the solution is given by the exponential

$$x(t) = e^{at}x(0).$$

We wish to generalize this to the vector case, where A becomes a matrix. We define the *matrix exponential* as the infinite series

$$e^X = I + X + \frac{1}{2}X^2 + \frac{1}{3!}X^3 + \cdots = \sum_{k=0}^{\infty}\frac{1}{k!}X^k, \tag{6.7}$$

where $X \in \mathbb{R}^{n \times n}$ is a square matrix and I is the $n \times n$ identity matrix. We make use of the notation

$$X^0 = I, \qquad X^2 = XX, \qquad X^n = X^{n-1}X,$$

which defines what we mean by the "power" of a matrix. Equation (6.7) is easy to remember since it is just the Taylor series for the scalar exponential, applied to the matrix X. It can be shown that the series in equation (6.7) converges for any matrix $X \in \mathbb{R}^{n \times n}$ in the same way that the normal exponential is defined for any scalar $a \in \mathbb{R}$.

Replacing X in equation (6.7) by At, where $t \in \mathbb{R}$, we find that

$$e^{At} = I + At + \frac{1}{2}A^2t^2 + \frac{1}{3!}A^3t^3 + \cdots = \sum_{k=0}^{\infty}\frac{1}{k!}A^kt^k,$$

and differentiating this expression with respect to t gives

$$\frac{d}{dt}e^{At} = A + A^2t + \frac{1}{2}A^3t^2 + \cdots = A\sum_{k=0}^{\infty}\frac{1}{k!}A^kt^k = Ae^{At}. \tag{6.8}$$

Multiplying by $x(0)$ from the right, we find that $x(t) = e^{At}x(0)$ is the solution to the differential equation (6.6) with initial condition $x(0)$. We summarize this important result as a proposition.

Proposition 6.1. *The solution to the homogeneous system of differential equations (6.6) is given by*

$$x(t) = e^{At}x(0).$$

Notice that the form of the solution is exactly the same as for scalar equations, but we must be sure to put the vector $x(0)$ on the right of the matrix e^{At}.

The form of the solution immediately allows us to see that the solution is linear in the initial condition. In particular, if $x_{h1}(t)$ is the solution to equation (6.6) with initial condition $x(0) = x_{01}$ and $x_{h2}(t)$ with initial condition $x(0) = x_{02}$, then the solution with initial condition $x(0) = \alpha x_{01} + \beta x_{02}$ is given by

$$x(t) = e^{At}(\alpha x_{01} + \beta x_{02}) = (\alpha e^{At} x_{01} + \beta e^{At} x_{02}) = \alpha x_{h1}(t) + \beta x_{h2}(t).$$

Similarly, we see that the corresponding output is given by

$$y(t) = Cx(t) = \alpha y_{h1}(t) + \beta y_{h2}(t),$$

where $y_{h1}(t)$ and $y_{h2}(t)$ are the outputs corresponding to $x_{h1}(t)$ and $x_{h2}(t)$.

We illustrate computation of the matrix exponential by two examples.

Example 6.2 Double integrator

A very simple linear system that is useful in understanding basic concepts is the second-order system given by

$$\ddot{q} = u, \qquad y = q.$$

This system is called a *double integrator* because the input u is integrated twice to determine the output y.

In state space form, we write $x = (q, \dot{q})$ and

$$\frac{dx}{dt} = \begin{pmatrix} 0 & 1 \\ 0 & 0 \end{pmatrix} x + \begin{pmatrix} 0 \\ 1 \end{pmatrix} u.$$

The dynamics matrix of a double integrator is

$$A = \begin{pmatrix} 0 & 1 \\ 0 & 0 \end{pmatrix},$$

and we find by direct calculation that $A^2 = 0$ and hence

$$e^{At} = I + At = \begin{pmatrix} 1 & t \\ 0 & 1 \end{pmatrix}.$$

Thus the solution of the homogeneous system $(u = 0)$ for the double integrator is given by

$$x(t) = \begin{pmatrix} 1 & t \\ 0 & 1 \end{pmatrix} \begin{pmatrix} x_1(0) \\ x_2(0) \end{pmatrix} = \begin{pmatrix} x_1(0) + t x_2(0) \\ x_2(0) \end{pmatrix},$$

$$y(t) = x_1(0) + t x_2(0).$$

∇

Example 6.3 Undamped oscillator

A model for an oscillator, such as the spring–mass system with zero damping, is

$$\ddot{q} + \omega_0^2 q = u.$$

Putting the system into state space form using $x_1 = q$, $x_2 = \dot{q}/\omega_0$, the dynamics matrix for this system can be written as

$$A = \begin{pmatrix} 0 & \omega_0 \\ -\omega_0 & 0 \end{pmatrix} \quad \text{and} \quad e^{At} = \begin{pmatrix} \cos\omega_0 t & \sin\omega_0 t \\ -\sin\omega_0 t & \cos\omega_0 t \end{pmatrix}.$$

This expression for e^{At} can be verified by differentiation:

$$\frac{d}{dt} e^{At} = \begin{pmatrix} -\omega_0 \sin\omega_0 t & \omega_0 \cos\omega_0 t \\ -\omega_0 \cos\omega_0 t & -\omega_0 \sin\omega_0 t \end{pmatrix}$$

$$= \begin{pmatrix} 0 & \omega_0 \\ -\omega_0 & 0 \end{pmatrix} \begin{pmatrix} \cos\omega_0 t & \sin\omega_0 t \\ -\sin\omega_0 t & \cos\omega_0 t \end{pmatrix} = Ae^{At}.$$

The solution to the initial value problem is then given by

$$x(t) = e^{At}x(0) = \begin{pmatrix} \cos\omega_0 t & \sin\omega_0 t \\ -\sin\omega_0 t & \cos\omega_0 t \end{pmatrix} \begin{pmatrix} x_1(0) \\ x_2(0) \end{pmatrix}.$$

The solution is more complicated if the system has damping:

$$\ddot{q} + 2\zeta\omega_0\dot{q} + \omega_0^2 q = u.$$

If $\zeta < 1$ we have

$$\exp\begin{pmatrix} -\zeta\omega_0 & \omega_d \\ -\omega_d & -\zeta\omega_0 \end{pmatrix} t = e^{-\zeta\omega_0 t} \begin{pmatrix} \cos\omega_d t & \sin\omega_d t \\ -\sin\omega_d t & \cos\omega_d t \end{pmatrix},$$

where $\omega_d = \omega_0\sqrt{1-\zeta^2}$. The result can be proven by differentiating the exponential matrix. The corresponding results for $\zeta \geq 1$ are given in Exercise 6.4. ∇

An important class of linear systems are those that can be converted into diagonal form by a linear change of coordinates. Suppose that we are given a system

$$\frac{dx}{dt} = Ax$$

such that all the eigenvalues of A are distinct. It can be shown (Exercise 5.14) that there exists an invertible matrix T such that TAT^{-1} is diagonal. If we choose a set of coordinates $z = Tx$, then in the new coordinates the dynamics become

$$\frac{dz}{dt} = T\frac{dx}{dt} = TAx = TAT^{-1}z.$$

By definition of T, this system will be diagonal.

Now consider a diagonal matrix A and the corresponding kth power of At, which is also diagonal:

$$A = \begin{pmatrix} \lambda_1 & & & \\ & \lambda_2 & & 0 \\ & & \ddots & \\ 0 & & & \lambda_n \end{pmatrix}, \quad (At)^k = \begin{pmatrix} \lambda_1^k t^k & & & \\ & \lambda_2^k t^k & & 0 \\ & & \ddots & \\ 0 & & & \lambda_n^k t^k \end{pmatrix}.$$

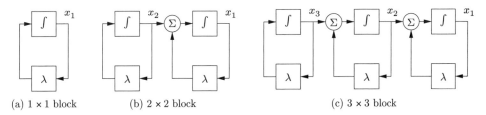

(a) 1 × 1 block   (b) 2 × 2 block   (c) 3 × 3 block

Figure 6.3: Representations of linear systems where the dynamics matrices are Jordan blocks. A 1×1 Jordan block corresponds to an integrator with feedback λ, as shown on the left. 2×2 and 3×3 Jordan blocks correspond to cascade connections of integrators with identical feedback, as shown in the middle and right diagrams.

It follows from the series expansion that the matrix exponential is given by

$$
e^{At} = \begin{pmatrix} e^{\lambda_1 t} & & & \\ & e^{\lambda_2 t} & & 0 \\ & & \ddots & \\ 0 & & & e^{\lambda_n t} \end{pmatrix}.
$$

A similar expansion can be done in the case where the eigenvalues are complex, using a block diagonal matrix, similar to what was done in Section 5.3.

Given the solution to the dynamics in the z coordinates, the solution in the original x coordinates can be obtained using the expression $x = T^{-1}z$. We can thus obtain an explicit solution for a linear system whose dynamics matrix is diagonalizable.

Jordan Form

Some matrices with repeated eigenvalues cannot be transformed to diagonal form. They can, however, be transformed to a closely related form, called the *Jordan form*, in which the dynamics matrix has the eigenvalues along the diagonal. When there are equal eigenvalues, there may be 1's appearing in the superdiagonal indicating that there is coupling between the states.

Specifically, we define a matrix to be in Jordan form if it can be written as

$$
J = \begin{pmatrix} J_1 & & & \\ & J_2 & & 0 \\ & & \ddots & \\ 0 & & & J_k \end{pmatrix}, \quad \text{where} \quad J_i = \begin{pmatrix} \lambda_i & 1 & & 0 \\ & \ddots & \ddots & \\ & & \ddots & 1 \\ 0 & & & \lambda_i \end{pmatrix}, \tag{6.9}
$$

and λ_i is an eigenvalue of J_i. Each matrix J_i is called a *Jordan block*. A first-order Jordan block can be represented as a system consisting of an integrator with feedback λ. A Jordan block of higher order can be represented as series connections of such systems, as illustrated in Figure 6.3.

Theorem 6.2 (Jordan decomposition). *Any matrix $A \in \mathbb{R}^{n \times n}$ can be transformed into Jordan form with the eigenvalues of A determining λ_i in the Jordan form.*

Proof. See any standard text on linear algebra, such as Strang [233]. The special case where the eigenvalues are distinct is examined in Exercise 5.14. □

Converting a matrix into Jordan form can be complicated, although MATLAB can do this conversion for numerical matrices using the `jordan` function. There is no requirement that the individual λ_i's be distinct, and hence for a given eigenvalue we can have one or more Jordan blocks of different sizes.

Once a matrix is in Jordan form, the exponential of the matrix can be computed in terms of the Jordan blocks:

$$
e^{Jt} = \begin{pmatrix} e^{J_1 t} & & & \\ & e^{J_2 t} & & \text{\huge 0} \\ & & \ddots & \\ \text{\huge 0} & & & e^{J_k t} \end{pmatrix}. \tag{6.10}
$$

This follows from the block diagonal form of J. The exponentials of the Jordan blocks can in turn be written as

$$
e^{J_i t} = \begin{pmatrix} 1 & t & \frac{t^2}{2!} & \cdots & \frac{t^{n-1}}{(n-1)!} \\ & 1 & t & \cdots & \frac{t^{n-2}}{(n-2)!} \\ & & \ddots & \ddots & \vdots \\ & \text{\huge 0} & & \ddots & t \\ & & & & 1 \end{pmatrix} e^{\lambda_i t}. \tag{6.11}
$$

As before, we can express the solution to a linear system that can be converted into this form by making use of the transformations $z = Tx$ and $x = T^{-1}z$.

When there are multiple eigenvalues, the invariant subspaces associated with each eigenvalue correspond to the Jordan blocks of the matrix A. Note that some eigenvalues of A may be complex, in which case the transformation T that converts a matrix into Jordan form will also be complex. When λ has a nonzero imaginary component, the solutions will have oscillatory components since

$$
e^{(\sigma + i\omega)t} = e^{\sigma t}(\cos \omega t + i \sin \omega t).
$$

We can now use these results to prove Theorem 5.1, which states that the equilibrium point $x_e = 0$ of a linear system is asymptotically stable if and only if $\text{Re } \lambda_i < 0$ for all i.

Proof of Theorem 5.1. Let $T \in \mathbb{C}^{n \times n}$ be an invertible matrix that transforms A into Jordan form, $J = TAT^{-1}$. Using coordinates $z = Tx$, we can write the solution $z(t)$ as

$$
z(t) = e^{Jt} z(0),
$$

where $z(0) = Tx(0)$, so that $x(t) = T^{-1} e^{Jt} z(0)$.

The solution $z(t)$ can be written in terms of the elements of the matrix exponential. From equation (6.11) these elements all decay to zero for arbitrary $z(0)$ if and only if $\text{Re } \lambda_i < 0$ for all i. Furthermore, if any λ_i has positive real part, then there exists an initial condition $z(0)$ such that the corresponding solution increases without bound. Since we can scale this initial condition to be arbitrarily small, it follows that the equilibrium point is unstable if any eigenvalue has positive real part. □

The existence of a canonical form allows us to prove many properties of linear systems by changing to a set of coordinates in which the A matrix is in Jordan form. We illustrate this in the following proposition, which follows along the same lines as the proof of Theorem 5.1.

Proposition 6.3. *Suppose that the system*

$$\frac{dx}{dt} = Ax$$

has no eigenvalues with strictly positive real part and one or more eigenvalues with zero real part. Then the system is stable (in the sense of Lyapunov) if and only if the Jordan blocks corresponding to each eigenvalue with zero real part are scalar (1×1) blocks.

Proof. See Exercise 6.6b. □

The following example illustrates the use of the Jordan form.

Example 6.4 Linear model of a vectored thrust aircraft
Consider the dynamics of a vectored thrust aircraft such as that described in Example 3.12. Suppose that we choose $u_1 = u_2 = 0$ so that the dynamics of the system become

$$\frac{dz}{dt} = \begin{pmatrix} z_4 \\ z_5 \\ z_6 \\ -g \sin z_3 - \frac{c}{m} z_4 \\ g(\cos z_3 - 1) - \frac{c}{m} z_5 \\ 0 \end{pmatrix}, \qquad (6.12)$$

where $z = (x, y, \theta, \dot{x}, \dot{y}, \dot{\theta})$. The equilibrium points for the system are given by setting the velocities $\dot{x}$, $\dot{y}$, and $\dot{\theta}$ to zero and choosing the remaining variables to satisfy

$$\begin{aligned} -g \sin z_{3,e} &= 0 \\ g(\cos z_{3,e} - 1) &= 0 \end{aligned} \qquad \Longrightarrow \qquad z_{3,e} = \theta_e = 0.$$

This corresponds to the upright orientation for the aircraft. Note that x_e and y_e are not specified. This is because we can translate the system to a new (upright) position and still obtain an equilibrium point.

To compute the stability of the equilibrium point, we compute the linearization using equation (5.13):

$$A = \frac{\partial F}{\partial z}\bigg|_{z_e} = \begin{pmatrix} 0 & 0 & 0 & 1 & 0 & 0 \\ 0 & 0 & 0 & 0 & 1 & 0 \\ 0 & 0 & 0 & 0 & 0 & 1 \\ 0 & 0 & -g & -c/m & 0 & 0 \\ 0 & 0 & 0 & 0 & -c/m & 0 \\ 0 & 0 & 0 & 0 & 0 & 0 \end{pmatrix}.$$

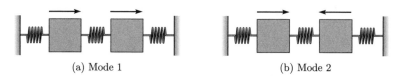

(a) Mode 1   (b) Mode 2

Figure 6.4: Modes of vibration for a system consisting of two masses connected by springs. In (a) the masses move left and right in synchronization in (b) they move toward or against each other.

The eigenvalues of the system can be computed as

$$\lambda(A) = \{0, 0, 0, 0, -c/m, -c/m\}.$$

We see that the linearized system is not asymptotically stable since not all of the eigenvalues have strictly negative real part.

To determine whether the system is stable in the sense of Lyapunov, we must make use of the Jordan form. It can be shown that the Jordan form of A is given by

$$
J = \left(
\begin{array}{cccc|c|c}
0 & 0 & 0 & 0 & 0 & 0 \\
0 & 0 & 1 & 0 & 0 & 0 \\
0 & 0 & 0 & 1 & 0 & 0 \\
0 & 0 & 0 & 0 & 0 & 0 \\
\hline
0 & 0 & 0 & 0 & -c/m & 0 \\
\hline
0 & 0 & 0 & 0 & 0 & -c/m
\end{array}
\right).
$$

Since the second Jordan block has eigenvalue 0 and is not a simple eigenvalue, the linearization is unstable (Exercise 6.6).   ∇

Eigenvalues and Modes

The eigenvalues and eigenvectors of a system provide a description of the types of behavior the system can exhibit. For oscillatory systems, the term *mode* is often used to describe the vibration patterns that can occur. Figure 6.4 illustrates the modes for a system consisting of two masses connected by springs. One pattern is when both masses oscillate left and right in unison, and another is when the masses move toward and away from each other.

The initial condition response of a linear system can be written in terms of a matrix exponential involving the dynamics matrix A. The properties of the matrix A therefore determine the resulting behavior of the system. Given a matrix $A \in \mathbb{R}^{n \times n}$, recall that v is an eigenvector of A with eigenvalue λ if

$$Av = \lambda v.$$

In general λ and v may be complex-valued, although if A is real-valued, then for any eigenvalue λ its complex conjugate λ^* will also be an eigenvalue (with v^* as the corresponding eigenvector).

Suppose first that λ and v are a real-valued eigenvalue/eigenvector pair for A. If we look at the solution of the differential equation for $x(0) = v$, it follows from the definition of the matrix exponential that

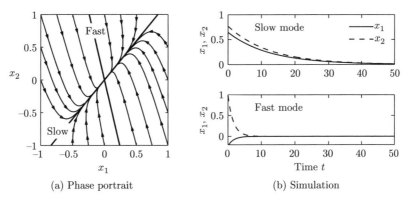

(a) Phase portrait   (b) Simulation

Figure 6.5: The notion of modes for a second-order system with real eigenvalues. The left figure shows the phase portrait and the modes corresponding to solutions that start on the eigenvectors (bold lines). The corresponding time functions are shown on the right.

$$e^{At}v = \left(I + At + \frac{1}{2}A^2t^2 + \cdots\right)v = v + \lambda tv + \frac{\lambda^2 t^2}{2}v + \cdots = e^{\lambda t}v.$$

The solution thus lies in the subspace spanned by the eigenvector. The eigenvalue λ describes how the solution varies in time, and this solution is often called a *mode* of the system. (In the literature, the term "mode" is also often used to refer to the eigenvalue rather than the solution.)

If we look at the individual elements of the vectors x and v, it follows that

$$\frac{x_i(t)}{x_j(t)} = \frac{e^{\lambda t}v_i}{e^{\lambda t}v_j} = \frac{v_i}{v_j},$$

and hence the ratios of the components of the state x are constants for a (real) mode. The eigenvector thus gives the "shape" of the solution and is also called a *mode shape* of the system. Figure 6.5 illustrates the modes for a second-order system consisting of a fast mode and a slow mode. Notice that the state variables have the same sign for the slow mode and different signs for the fast mode.

The situation is more complicated when the eigenvalues of A are complex. Since A has real elements, the eigenvalues and the eigenvectors are complex conjugates $\lambda = \sigma \pm i\omega$ and $v = u \pm iw$, which implies that

$$u = \frac{v + v^*}{2}, \qquad w = \frac{v - v^*}{2i}.$$

Making use of the matrix exponential, we have

$$e^{At}v = e^{\lambda t}(u + iw) = e^{\sigma t}\big((u\cos\omega t - w\sin\omega t) + i(u\sin\omega t + w\cos\omega t)\big),$$

from which it follows that

$$e^{At}u = \frac{1}{2}\left(e^{At}v + e^{At}v^*\right) = ue^{\sigma t}\cos\omega t - we^{\sigma t}\sin\omega t,$$

$$e^{At}w = \frac{1}{2i}\left(e^{At}v - e^{At}v^*\right) = ue^{\sigma t}\sin\omega t + we^{\sigma t}\cos\omega t.$$

A solution with initial conditions in the subspace spanned by the real part u and imaginary part w of the eigenvector will thus remain in that subspace. The solution will be a logarithmic spiral characterized by σ and ω. We again call the solution corresponding to λ a mode of the system and v the mode shape.

If a matrix A has n distinct eigenvalues $\lambda_1, \ldots, \lambda_n$, then the initial condition response can be written as a linear combination of the modes. To see this, suppose for simplicity that we have all real eigenvalues with corresponding unit eigenvectors $v_1, \ldots, v_n$. From linear algebra, these eigenvectors are linearly independent, and we can write the initial condition $x(0)$ as

$$x(0) = \alpha_1 v_1 + \alpha_2 v_2 + \cdots + \alpha_n v_n.$$

Using linearity, the initial condition response can be written as

$$x(t) = \alpha_1 e^{\lambda_1 t} v_1 + \alpha_2 e^{\lambda_2 t} v_2 + \cdots + \alpha_n e^{\lambda_n t} v_n.$$

Thus, the response is a linear combination of the modes of the system, with the amplitude of the individual modes growing or decaying as $e^{\lambda_i t}$. The case for distinct complex eigenvalues follows similarly (the case for nondistinct eigenvalues is more subtle and requires making use of the Jordan form discussed in the previous section).

Example 6.5 Coupled spring–mass system

Consider the spring–mass system shown in Figure 6.4, but with the addition of dampers on each mass. The equations of motion of the system are

$$m\ddot{q}_1 = -2kq_1 - c\dot{q}_1 + kq_2, \qquad m\ddot{q}_2 = kq_1 - 2kq_2 - c\dot{q}_2.$$

In state space form, we define the state to be $x = (q_1, q_2, \dot{q}_1, \dot{q}_2)$, and we can rewrite the equations as

$$\frac{dx}{dt} = \begin{pmatrix} 0 & 0 & 1 & 0 \\ 0 & 0 & 0 & 1 \\ -\dfrac{2k}{m} & \dfrac{k}{m} & -\dfrac{c}{m} & 0 \\ \dfrac{k}{m} & -\dfrac{2k}{m} & 0 & -\dfrac{c}{m} \end{pmatrix} x.$$

We now define a transformation $z = Tx$ that puts this system into a simpler form. Let $z_1 = \frac{1}{2}(q_1 + q_2)$, $z_2 = \dot{z}_1$, $z_3 = \frac{1}{2}(q_1 - q_2)$ and $z_4 = \dot{z}_3$, so that

$$z = Tx = \frac{1}{2} \begin{pmatrix} 1 & 1 & 0 & 0 \\ 0 & 0 & 1 & 1 \\ 1 & -1 & 0 & 0 \\ 0 & 0 & 1 & -1 \end{pmatrix} x.$$

In the new coordinates, the dynamics become

$$\frac{dz}{dt} = \begin{pmatrix} 0 & 1 & 0 & 0 \\ -\dfrac{k}{m} & -\dfrac{c}{m} & 0 & 0 \\ 0 & 0 & 0 & 1 \\ 0 & 0 & -\dfrac{3k}{m} & -\dfrac{c}{m} \end{pmatrix} z,$$

and we see that the model is now in block diagonal form.

In the z coordinates, the states z_1 and z_2 parameterize one mode with eigenvalues $\lambda \approx -c/(2m) \pm i\sqrt{k/m}$, and the states z_3 and z_4 another mode with $\lambda \approx -c/(2m) \pm i\sqrt{3k/m}$. From the form of the transformation T we see that these modes correspond exactly to the modes in Figure 6.4, in which q_1 and q_2 move either toward or against each other. The real and imaginary parts of the eigenvalues give the decay rates σ and frequencies ω for each mode. ∇

6.3 INPUT/OUTPUT RESPONSE

In the previous section we saw how to compute the initial condition response using the matrix exponential. In this section we derive the convolution equation, which includes the inputs and outputs as well.

The Convolution Equation

We return to the general input/output case in equation (6.3), repeated here:

$$\frac{dx}{dt} = Ax + Bu, \qquad y = Cx + Du. \tag{6.13}$$

Using the matrix exponential, the solution to equation (6.13) can be written as follows.

Theorem 6.4. *The solution to the linear differential equation* (6.13) *is given by*

$$x(t) = e^{At}x(0) + \int_0^t e^{A(t-\tau)}Bu(\tau)d\tau. \tag{6.14}$$

Proof. To prove this, we differentiate both sides and use the property (6.8) of the matrix exponential. This gives

$$\frac{dx}{dt} = Ae^{At}x(0) + \int_0^t Ae^{A(t-\tau)}Bu(\tau)d\tau + Bu(t) = Ax + Bu,$$

which proves the result since the initial conditions are also met. Notice that the calculation is essentially the same as for proving the result for a first-order equation. $\square$

It follows from equations (6.13) and (6.14) that the input/output relation for a linear system is given by

$$y(t) = Ce^{At}x(0) + \int_0^t Ce^{A(t-\tau)}Bu(\tau)d\tau + Du(t). \tag{6.15}$$

It is easy to see from this equation that the output is jointly linear in both the initial conditions and the input, which follows from the linearity of matrix/vector multiplication and integration.

Equation (6.15) is called the *convolution equation*, and it represents the general form of the solution of a system of coupled linear differential equations. We see immediately that the dynamics of the system, as characterized by the matrix A,

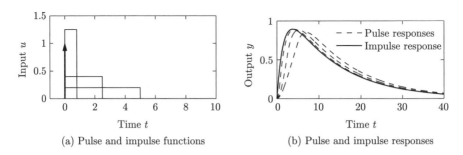

(a) Pulse and impulse functions   (b) Pulse and impulse responses

Figure 6.6: Pulse response and impulse response. (a) The rectangles show pulses of width 5, 2.5, and 0.8, each with total area equal to 1. The arrow denotes an impulse $\delta(t)$ defined by equation (6.17). The corresponding pulse responses for a linear system with eigenvalues $\lambda = \{-0.08, -0.62\}$ are shown in (b) as dashed lines. The solid line is the true impulse response, which is well approximated by a pulse of duration 0.8.

play a critical role in both the stability and performance of the system. Indeed, the matrix exponential describes *both* what happens when we perturb the initial condition and how the system responds to inputs.

 Another interpretation of the convolution equation can be given using the concept of the *impulse response* of a system. Consider the application of an input signal $u(t)$ given by the following equation:

$$u(t) = p_\epsilon(t) = \begin{cases} 0 & \text{if } t < 0, \\ 1/\epsilon & \text{if } 0 \leq t < \epsilon, \\ 0 & \text{if } t \geq \epsilon. \end{cases} \tag{6.16}$$

This signal is a *pulse* of duration ϵ and amplitude $1/\epsilon$, as illustrated in Figure 6.6a. We define an *impulse* $\delta(t)$ to be the limit of this signal as $\epsilon \to 0$:

$$\delta(t) = \lim_{\epsilon \to 0} p_\epsilon(t). \tag{6.17}$$

This signal, sometimes called a *delta function,* is not physically achievable but provides a convenient abstraction in understanding the response of a system. Note that the integral of an impulse is 1:

$$\int_0^t \delta(\tau)\, d\tau = \int_0^t \lim_{\epsilon \to 0} p_\epsilon(t)\, d\tau = \lim_{\epsilon \to 0} \int_0^t p_\epsilon(t)\, d\tau$$

$$= \lim_{\epsilon \to 0} \int_0^\epsilon 1/\epsilon\, d\tau = 1, \qquad t > 0.$$

In particular, the integral of an impulse over an arbitrarily short period of time that includes the origin is identically 1.

We define the *impulse response* $h(t)$ for a system as the output of the system with zero initial condition and having an impulse as its input:

$$h(t) = \int_0^t Ce^{A(t-\tau)}B\delta(\tau)\, d\tau + D\delta(t) = Ce^{At}B + D\delta(t), \tag{6.18}$$

where the second equality follows from the fact that $\delta(t)$ is zero everywhere except the origin and its integral is identically 1. We can now write the convolution equation in terms of the initial condition response and the convolution of the impulse response and the input signal:

$$y(t) = Ce^{At}x(0) + \int_0^t h(t-\tau)u(\tau)\,d\tau. \tag{6.19}$$

One interpretation of this equation, explored in Exercise 6.2, is that the response of the linear system is the superposition of the response to an infinite set of shifted impulses whose magnitudes are given by the input $u(t)$. This is essentially the argument used in analyzing Figure 6.2 and deriving equation (6.5). Note that the second term in equation (6.19) is identical to equation (6.5), and it can be shown that the impulse response is the derivative of the step response.

The use of pulses $p_\epsilon(t)$ as approximations of the impulse function $\delta(t)$ also provides a mechanism for identifying the dynamics of a system from experiments. Figure 6.6b shows the pulse responses of a system for different pulse widths. Notice that the pulse responses approach the impulse response as the pulse width goes to zero. As a general rule, if the fastest eigenvalue of a stable system has real part $-\sigma_{\max}$, then a pulse of length ϵ will provide a good estimate of the impulse response if $\epsilon\sigma_{\max} \ll 1$. Note that for Figure 6.6, a pulse width of $\epsilon = 1$ s gives $\epsilon\sigma_{\max} = 0.62$ and the pulse response is already close to the impulse response.

Coordinate Invariance

The components of the input vector u and the output vector y are determined by the chosen inputs and outputs of a model, but the state variables depend on the coordinate frame chosen to represent the state. This choice of coordinates affects the values of the matrices A, B, and C that are used in the model. (The direct term D is not affected since it maps inputs to outputs.) We now investigate some of the consequences of changing coordinate systems.

Introduce new coordinates z by the transformation $z = Tx$, where T is an invertible matrix. It follows from equation (6.3) that

$$\frac{dz}{dt} = T(Ax + Bu) = TAT^{-1}z + TBu =: \tilde{A}z + \tilde{B}u,$$

$$y = Cx + Du = CT^{-1}z + Du =: \tilde{C}z + Du.$$

The transformed system has the same form as equation (6.3), but the matrices A, B, and C are different:

$$\tilde{A} = TAT^{-1}, \qquad \tilde{B} = TB, \qquad \tilde{C} = CT^{-1}. \tag{6.20}$$

There are often special choices of coordinate systems that allow us to see a particular property of the system, hence coordinate transformations can be used to gain new insight into the dynamics. The eigenvalues of $\tilde{A}$ are the same as those of A, so stability is not affected.

We can also compare the solution of the system in transformed coordinates to that in the original state coordinates. We make use of an important property of the exponential map,

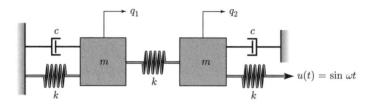

Figure 6.7: Coupled spring mass system. Each mass is connected to two springs with stiffness k and a viscous damper with damping coefficient c. The mass on the right is driven through a spring connected to a sinusoidally varying attachment.

$$e^{TST^{-1}} = Te^S T^{-1},$$

which can be verified by substitution in the definition of the matrix exponential. Using this property, it is easy to show that

$$x(t) = T^{-1} z(t) = T^{-1} e^{\tilde{A}t} T x(0) + T^{-1} \int_0^t e^{\tilde{A}(t-\tau)} \tilde{B} u(\tau) \, d\tau.$$

From this form of the equation, we see that if it is possible to transform A into a form $\tilde{A}$ for which the matrix exponential is easy to compute, we can use that computation to solve the general convolution equation for the untransformed state x by simple matrix multiplications. This technique is illustrated in the following example.

Example 6.6 Coupled spring–mass system

Consider the coupled spring–mass system shown in Figure 6.7. The input to this system is the sinusoidal motion of the position of the rightmost spring, and the output is the position of each mass, q_1 and q_2. The equations of motion are given by

$$m\ddot{q}_1 = -2kq_1 - c\dot{q}_1 + kq_2, \qquad m\ddot{q}_2 = kq_1 - 2kq_2 - c\dot{q}_2 + ku.$$

In state space form, we define the state to be $x = (q_1, q_2, \dot{q}_1, \dot{q}_2)$, and we can rewrite the equations as

$$\frac{dx}{dt} = \begin{pmatrix} 0 & 0 & 1 & 0 \\ 0 & 0 & 0 & 1 \\ -\dfrac{2k}{m} & \dfrac{k}{m} & -\dfrac{c}{m} & 0 \\ \dfrac{k}{m} & -\dfrac{2k}{m} & 0 & -\dfrac{c}{m} \end{pmatrix} x + \begin{pmatrix} 0 \\ 0 \\ 0 \\ \dfrac{k}{m} \end{pmatrix} u.$$

This is a coupled set of four differential equations and is quite complicated to solve in analytical form.

The dynamics matrix is the same as in Example 6.5, and we can use the coordinate transformation defined there to put the system in block diagonal form:

$$\frac{dz}{dt} = \begin{pmatrix} 0 & 1 & 0 & 0 \\ -\dfrac{k}{m} & -\dfrac{c}{m} & 0 & 0 \\ 0 & 0 & 0 & 1 \\ 0 & 0 & -\dfrac{3k}{m} & -\dfrac{c}{m} \end{pmatrix} z + \begin{pmatrix} 0 \\ \dfrac{k}{2m} \\ 0 \\ -\dfrac{k}{2m} \end{pmatrix} u.$$

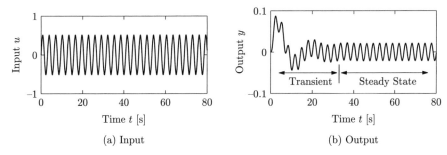

Figure 6.8: Transient versus steady-state response. The input to a linear system is shown in (a), and the corresponding output with $x(0) = 0$ is shown in (b). The output signal initially undergoes a transient before settling into its steady-state behavior.

Note that the resulting matrix equations are decoupled, and we can solve for the solutions by computing the solutions of two sets of second-order systems represented by the states (z_1, z_2) and (z_3, z_4). Indeed, the functional form of each set of equations is identical to that of a single spring–mass system. (The explicit solution is derived in Section 7.3.)

Once we have solved the two sets of independent second-order equations, we can recover the dynamics in the original coordinates by inverting the state transformation and writing $x = T^{-1}z$. We can also determine the stability of the system by looking at the stability of the independent second-order systems. ∇

Steady-State Response

A common practice in evaluating the response of a linear system is to separate out the short-term response from the long-term response. Given a linear input/output system

$$\frac{dx}{dt} = Ax + Bu, \qquad y = Cx + Du, \tag{6.21}$$

the general form of the solution to equation (6.21) is given by the convolution equation:

$$y(t) = Ce^{At}x(0) + \int_0^t Ce^{A(t-\tau)}Bu(\tau)d\tau + Du(t).$$

We see from the form of this equation that the solution consists of an initial condition response and an input response.

The input response, corresponding to the last two terms in the equation above, itself consists of two components—the *transient response* and the *steady-state response*. The transient response occurs in the first period of time after the input is applied and reflects the mismatch between the initial condition and the steady-state solution. The steady-state response is the portion of the output response that reflects the long-term behavior of the system under the given inputs. For inputs that are periodic the steady-state response will often be periodic, and for constant inputs the response will often be constant. An example of the transient and the steady-state response for a periodic input is shown in Figure 6.8.

A particularly common form of input is a *step input*, which represents an abrupt change in input from one value to another. A *unit step* (sometimes called the Heaviside step function) is defined as

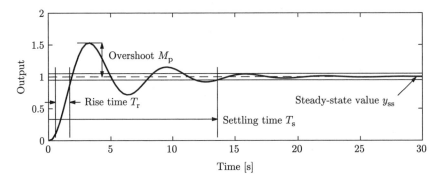

Figure 6.9: Sample step response. The rise time, overshoot, settling time, and steady-state value give the key performance properties of the signal.

$$u(t) = S(t) = \begin{cases} 0 & \text{if } t = 0, \\ 1 & \text{if } t > 0. \end{cases}$$

The *step response* of the system (6.21) is defined as the output $y(t)$ starting from zero initial condition (or the appropriate equilibrium point) and given a step input. We note that the step input is discontinuous and hence is not practically implementable. However, it is a convenient abstraction that is widely used in studying input/output systems.

We can compute the step response to a linear system using the convolution equation. Setting $x(0) = 0$ and using the definition of the step input above, we have

$$y(t) = \int_0^t C e^{A(t-\tau)} B u(\tau) d\tau + D u(t) = C \int_0^t e^{A(t-\tau)} B d\tau + D$$

$$= C \int_0^t e^{A\sigma} B d\sigma + D = C \left(A^{-1} e^{A\sigma} B \right) \big|_{\sigma=0}^{\sigma=t} + D$$

$$= C A^{-1} e^{At} B - C A^{-1} B + D.$$

We can rewrite the solution as

$$y(t) = \underbrace{C A^{-1} e^{At} B}_{\text{transient}} + \underbrace{D - C A^{-1} B}_{\text{steady-state}}, \qquad t > 0. \tag{6.22}$$

The first term is the transient response and it decays to zero as $t \to \infty$ if all eigenvalues of A have negative real parts (implying that the origin is a stable equilibrium point in the absence of any input). The second term, computed under the assumption that the matrix A is invertible, is the steady-state step response and represents the value of the output for large time.

A sample step response is shown in Figure 6.9. Several key properties are used when describing a step response. The *steady-state value* y_{ss} of a step response is the final level of the output, assuming it converges. The *rise time* T_r is the amount of time required for the signal to first go from 10% of its final value to 90% of its final value. (It is possible to define other limits as well, but in this book we shall use these percentages unless otherwise indicated.) The *overshoot* M_p is the percentage of the final value by which the signal initially rises above the final value. This

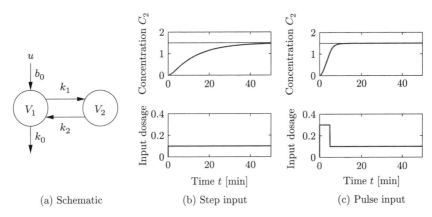

(a) Schematic   (b) Step input   (c) Pulse input

Figure 6.10: Response of a compartment model to a constant drug infusion. A simple diagram of the system is shown in (a). The step response (b) shows the rate of concentration buildup in compartment 2. In (c) a pulse of initial concentration is used to speed up the response.

usually assumes that future values of the signal do not overshoot the final value by more than this initial transient, otherwise the term can be ambiguous. Finally, the *settling time* T_s is the amount of time required for the signal to stay within 2% of its final value for all future times. The settling time is also sometimes defined as reaching 1% or 5% of the final value (see Exercise 6.7). In general these performance measures can depend on the amplitude of the input step, but for linear systems the last three quantities defined above are independent of the size of the step.

Example 6.7 Compartment model
Consider the compartment model illustrated in Figure 6.10 and described in more detail in Section 4.6. Assume that a drug is administered by constant infusion in compartment V_1 and that the drug has its effect in compartment V_2. To assess how quickly the concentration in the compartment reaches steady state we compute the step response, which is shown in Figure 6.10b. The step response is quite slow, with a settling time of 39 min. It is possible to obtain the steady-state concentration much faster by having a faster injection rate initially, as shown in Figure 6.10c. The response of the system in this case can be computed by combining two step responses (Exercise 6.3). ∇

Frequency Response

Another common input signal to a linear system is a sinusoid (or a combination of sinusoids). The *frequency response* of an input/output system measures the way in which the system responds to a sinusoidal excitation on one of its inputs. As we have already seen for scalar systems, the particular solution associated with a sinusoidal excitation is itself a sinusoid at the same frequency. Hence we can compare the magnitude and phase of the output sinusoid to the input.

To see this in more detail, we must evaluate the convolution equation (6.15) for $u = \cos \omega t$. This turns out to be a very messy calculation, but we can make use of the fact that the system is linear to simplify the derivation. It follows from Euler's formula that

$$\cos \omega t = \frac{1}{2} \left(e^{i\omega t} + e^{-i\omega t} \right).$$

Since the system is linear, it suffices to compute the response of the system to the complex input $u(t) = e^{st}$ and we can then reconstruct the input to a sinusoid by averaging the responses corresponding to $s = i\omega$ and $s = -i\omega$.

Applying the convolution equation to the input $u = e^{st}$ we have

$$y(t) = Ce^{At} x(0) + \int_0^t Ce^{A(t-\tau)} Be^{s\tau} d\tau + De^{st}$$

$$= Ce^{At} x(0) + Ce^{At} \int_0^t e^{(sI-A)\tau} B d\tau + De^{st}.$$

If we assume that none of the eigenvalues of A are equal to $\pm i\omega$, then the matrix $sI - A$ is invertible, and we can write

$$y(t) = Ce^{At} x(0) + Ce^{At} \left((sI - A)^{-1} e^{(sI-A)\tau} B \right) \Big|_0^t + De^{st}$$

$$= Ce^{At} x(0) + Ce^{At} (sI - A)^{-1} \left(e^{(sI-A)t} - I \right) B + De^{st}$$

$$= Ce^{At} x(0) + C(sI - A)^{-1} e^{st} B - Ce^{At} (sI - A)^{-1} B + De^{st},$$

and we obtain

$$y(t) = \underbrace{Ce^{At} \left(x(0) - (sI - A)^{-1} B \right)}_{\text{transient}} + \underbrace{\left(C(sI - A)^{-1} B + D \right) e^{st}}_{\text{steady-state}}. \tag{6.23}$$

Notice that once again the solution consists of both a transient component and a steady-state component. The transient component decays to zero if the system is asymptotically stable and the steady-state component is proportional to the (complex) input $u = e^{st}$.

We can simplify the form of the solution slightly further by rewriting the steady-state response as

$$y_{\text{ss}}(t) = Me^{i\theta} e^{st} = Me^{(st+i\theta)},$$

where

$$Me^{i\theta} = G(s) = C(sI - A)^{-1} B + D, \tag{6.24}$$

and M and θ represent the magnitude and phase of the complex number $G(s)$. When $s = i\omega$, we say that $M = |G(i\omega)|$ is the *gain* and $\theta = \arg G(i\omega)$ is the *phase* of the system at a given forcing frequency ω. Using linearity and combining the solutions for $s = +i\omega$ and $s = -i\omega$, we can show that if we have an input $u = A_u \sin(\omega t + \psi)$ and an output $y = A_y \sin(\omega t + \varphi)$, then

$$\text{gain}(\omega) = \frac{A_y}{A_u} = M, \qquad \text{phase}(\omega) = \varphi - \psi = \theta.$$

The steady-state solution for a sinusoid $u = \cos \omega t = \sin(\omega t + \pi/2)$ is now given by

$$y_{\text{ss}}(t) = \text{Re}\left(G(i\omega) e^{i\omega t} \right) = M \cos(\omega t + \theta). \tag{6.25}$$

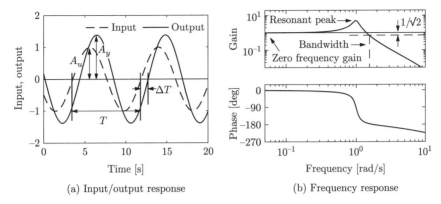

(a) Input/output response   (b) Frequency response

Figure 6.11: Steady-state response of an asymptotically stable linear system to a sinusoid. (a) A sinusoidal input of magnitude A_u (dashed) gives a sinusoidal output of magnitude A_y (solid), delayed by ΔT seconds. (b) Frequency response, showing gain and phase. The gain is given by the ratio of the output amplitude to the input amplitude, $M = A_y/A_u$. The phase lag is given by $\theta = -2\pi\Delta T/T$; it is negative for the case shown because the output lags the input.

If the phase θ is positive, we say that the output *leads* the input, otherwise we say it *lags* the input.

A sample steady-state sinusoidal response is illustrated in Figure 6.11a. The dashed line shows the input sinusoid, which has amplitude 1. The output sinusoid is shown as a solid line and has a different amplitude plus a shifted phase. The gain is the ratio of the amplitudes of the sinusoids, which can be determined by measuring the height of the peaks. The phase is determined by comparing the ratio of the time between zero crossings of the input and output to the overall period of the sinusoid:

$$\theta = -2\pi \cdot \frac{\Delta T}{T}.$$

A convenient way to view the frequency response is to plot how the gain and phase in equation (6.24) depend on ω (through $s = i\omega$). Figure 6.11b shows an example of this type of representation (called a Bode plot and discussed in more detail in Section 9.6).

Example 6.8 Active band-pass filter

Consider the op amp circuit shown in Figure 6.12a. We can derive the dynamics of the system by writing the *nodal equations*, which state that the sum of the currents at any node must be zero. Assuming that $v_- = v_+ = 0$, as we did in Section 4.3, we have

$$0 = \frac{v_1 - v_2}{R_1} - C_1\frac{dv_2}{dt}, \qquad 0 = C_1\frac{dv_2}{dt} + \frac{v_3}{R_2} + C_2\frac{dv_3}{dt}.$$

Choosing v_2 and v_3 as our states and using these equations, we obtain

$$\frac{dv_2}{dt} = \frac{v_1 - v_2}{R_1C_1}, \qquad \frac{dv_3}{dt} = \frac{-v_3}{R_2C_2} - \frac{v_1 - v_2}{R_1C_2}.$$

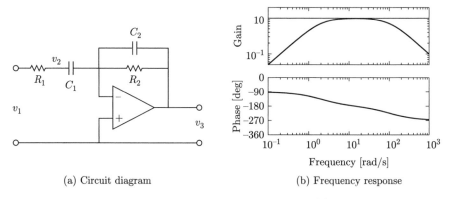

(a) Circuit diagram   (b) Frequency response

Figure 6.12: Active band-pass filter. The circuit diagram (a) shows an op amp with two RC filters arranged to provide a band-pass filter. The plot in (b) shows the gain and phase of the filter as a function of frequency. Note that the phase starts at $-90°$ due to the negative gain of the operational amplifier.

Rewriting these in linear state space form, we obtain

$$\frac{dx}{dt} = \begin{pmatrix} -\dfrac{1}{R_1 C_1} & 0 \\ \dfrac{1}{R_1 C_2} & -\dfrac{1}{R_2 C_2} \end{pmatrix} x + \begin{pmatrix} \dfrac{1}{R_1 C_1} \\ -\dfrac{1}{R_1 C_2} \end{pmatrix} u, \qquad y = \begin{pmatrix} 0 & 1 \end{pmatrix} x, \qquad (6.26)$$

where $x = (v_2, v_3)$, $u = v_1$, and $y = v_3$.

The frequency response for the system can be computed using equation (6.24):

$$M e^{i\theta} = C(sI - A)^{-1} B + D = -\frac{R_2}{R_1} \frac{R_1 C_1 s}{(1 + R_1 C_1 s)(1 + R_2 C_2 s)}, \qquad s = i\omega.$$

The magnitude and phase are plotted in Figure 6.12 for $R_1 = 100$ Ω, $R_2 = 5$ kΩ, and $C_1 = C_2 = 100$ µF. We see that signals with frequencies around 15 rad/s pass through the circuit with small attenuation but that signals below 2 rad/s or above 100 rad/s are attenuated. At 0.1 rad/s the input signal is attenuated by a factor of 20. This type of circuit is called a *band-pass filter* since it passes through signals in the band of frequencies between 5 and 50 rad/s (approximately).   ∇

As in the case of the step response, a number of standard properties are defined for frequency responses. The gain of a system at $\omega = 0$ is called the *zero frequency gain* and corresponds to the ratio between a constant input and the steady output:

$$M_0 = G(0) = -CA^{-1}B + D$$

(compare to equation (6.24)). The zero frequency gain is well defined only if A is invertible (i.e., if it does not have eigenvalues at 0). It is also important to note that the zero frequency gain is a relevant quantity only when a system is stable about the corresponding equilibrium point. So, if we apply a constant input $u = r$, then the corresponding equilibrium point $x_e = -A^{-1}Br$ must be stable in order to talk about the zero frequency gain. (In electrical engineering, the zero frequency gain is often called the *DC gain*. DC stands for direct current and reflects the common separation of signals in electrical engineering into a direct current [zero frequency] term and an alternating current [AC] term.)

The *bandwidth* ω_b of a system is the frequency range over which the gain has decreased by no more than a factor of $1/\sqrt{2}$ from its reference value. For systems with nonzero, finite zero frequency gain, the reference value is taken as the zero frequency gain. For systems that attenuate low frequencies but pass through high frequencies, the reference gain is taken as the high-frequency gain. For a system such as the band-pass filter in Example 6.8, bandwidth is defined as the range of frequencies where the gain is larger than $1/\sqrt{2}$ of the gain at the center of the band. (For Example 6.8 this would give a bandwidth of approximately 2 to 100 rad/s.)

Other important properties of the frequency response are the *resonant peak* M_r, the largest value of the frequency response, and the *peak frequency* ω_{mr}, the frequency where the maximum occurs. These two properties describe the frequency of the sinusoidal input that produce the largest possible output and the gain at the frequency.

Example 6.9 Atomic force microscope in contact mode
Consider the model for the vertical dynamics of the atomic force microscope in contact mode, discussed in Section 4.5. The basic dynamics are given by equation (4.24). The piezo stack can be modeled by a second-order system with undamped natural frequency ω_3 and damping ratio ζ_3. The dynamics are then described by the linear system

$$
\frac{dx}{dt} = \begin{pmatrix} 0 & 1 & 0 & 0 \\ -k_2/(m_1+m_2) & -c_2/(m_1+m_2) & 1/m_2 & 0 \\ 0 & 0 & 0 & \omega_3 \\ 0 & 0 & -\omega_3 & -2\zeta_3\omega_3 \end{pmatrix} x + \begin{pmatrix} 0 \\ 0 \\ 0 \\ \omega_3 \end{pmatrix} u,
$$

$$
y = \frac{m_2}{m_1+m_2} \begin{pmatrix} \dfrac{m_1 k_2}{m_1+m_2} & \dfrac{m_1 c_2}{m_1+m_2} & 1 & 0 \end{pmatrix} x,
$$

where the input is the drive signal to the amplifier and the output is the elongation of the piezo. The frequency response of the system is shown in Figure 6.13b. The zero frequency gain of the system is $M_0 = 1$. There are two resonant poles with peaks $M_{r1} = 2.12$ at $\omega_{mr1} = 238$ krad/s and $M_{r2} = 4.29$ at $\omega_{mr2} = 746$ krad/s. There is also a dip in the gain $M_d = 0.556$ for $\omega_{md} = 268$ krad/s. This dip, called an *antiresonance*, is associated with a dip in the phase and limits the performance when the system is controlled by simple controllers, as we will see in Chapter 11. The bandwidth is the frequency range over which the gain has decreased by no more than a factor of $1/\sqrt{2}$ from its reference value, which in this case is the zero frequency gain. Neglecting the slight dip at the antiresonance, the bandwidth becomes $\omega_b = 1.12$ Mrad/s. ∇

So far we have used the frequency response to compute the output for a single sinusoid. The transfer function can also be used to compute the output for any periodic signal. Consider a system with the frequency response $G(i\omega)$. Let the input signal $u(t)$ be periodic and decompose it into a sum of a set of sines and cosines,

$$
u(t) = \sum_{k=0}^{\infty} a_k \sin(k\omega_f t) + b_k \cos(k\omega_f t),
$$

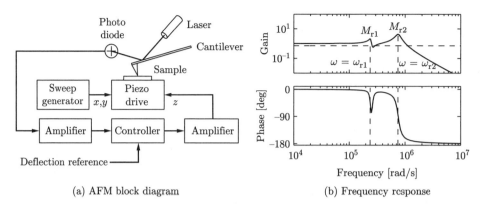

(a) AFM block diagram   (b) Frequency response

Figure 6.13: AFM frequency response. (a) A block diagram for the vertical dynamics of an atomic force microscope in contact mode. The plot in (b) shows the gain and phase for the piezo stack. The response contains two frequency peaks at resonances of the system, along with an antiresonance at $\omega = 268$ krad/s. The combination of a resonant peak followed by an antiresonance is common for systems with multiple lightly damped modes. The dashed horizontal line represents the gain equal to the zero frequency gain divided by $\sqrt{2}$.

where ω_{f} is the fundamental frequency of the periodic input. Using equation (6.25) and superposition, we find that the input $u(t)$ generates the steady-state output

$$y(t) = \sum_{k=0}^{\infty} |G(ik\omega_{\mathrm{f}})| \Big(a_k \sin\big(k\omega_{\mathrm{f}} t + \arg G(ik\omega_{\mathrm{f}})\big) + b_k \cos\big(k\omega_{\mathrm{f}} t + \arg G(ik\omega_{\mathrm{f}})\big) \Big).$$

The gain and phase at each frequency are determined by the frequency response $G(i\omega)$, as given in equation (6.24). If we know the steady-state frequency response $G(i\omega)$, we can thus compute the response to any (periodic) signal using superposition.

 We can go even further to approximate the response to a transient signal. Consider a system with the transfer function $G(s)$ and the input u. Approximate the initial part of the function $u(t)$ by the periodic signal

$$u_{\mathrm{p}}(t) = \begin{cases} u(t) & \text{if } 0 \leq t < T/2, \\ 0 & \text{if } T/2 \leq t < T, \end{cases}$$

with period T. Since u_{p} is periodic it has a Fourier transform $u_{\mathrm{F}}(i\omega)$, and it follows from equation (6.25) that the Fourier transform of y_{p} is $y_{\mathrm{F}}(i\omega) = G(i\omega)u_{\mathrm{F}}(i\omega)$, where u_{F} and y_{F} represent the Fourier transforms of u_{p} and y_{p}, respectively. Taking the inverse Fourier transform then gives the time response $y_{\mathrm{p}}(t)$. Efficient algorithms can be obtained using fast Fourier transforms (Exercise 6.12).

Sampling

It is often convenient to use both differential and difference equations in modeling and control. For linear systems it is straightforward to transform from one to the other. Consider the general linear system described by equation (6.13) and assume that the control signal is constant over a sampling interval of constant length h. It follows from equation (6.14) of Theorem 6.4 that

$$x(t+h) = e^{Ah}x(t) + \int_t^{t+h} e^{A(t+h-\tau)}Bu(\tau)\,d\tau =: \Phi x(t) + \Gamma u(t), \qquad (6.27)$$

where we have assumed that the discontinuous control signal is continuous from the right. The behavior of the system at the sampling times $t = kh$ is described by the difference equation

$$x[k+1] = \Phi x[k] + \Gamma u[k], \qquad y[k] = Cx[k] + Du[k], \qquad (6.28)$$

where

$$\Phi = e^{Ah}, \qquad \Gamma = \left(\int_0^h e^{As}\,ds\right)B.$$

Notice that the difference equation (6.28) is an exact representation of the behavior of the system at the sampling instants. Similar expressions can also be obtained if the control signal is linear over the sampling interval.

The transformation from equation (6.27) to equation (6.28) is called *sampling*. The relations between the system matrices in the continuous and sampled representations are as follows:

$$A = \frac{1}{h}\log\Phi, \qquad B = \left(\int_0^h e^{As}\,ds\right)^{-1}\Gamma. \qquad (6.29)$$

Notice that if A is invertible, we have

$$\Gamma = A^{-1}\left(e^{Ah} - I\right)B.$$

All continuous-time systems can be sampled to obtain a discrete-time version, but there are discrete-time systems that do not have a continuous-time equivalent. The issue is related to logarithms of matrices and there are several subtleties; for example, there may be many solutions. A necessary but not sufficient condition is that the matrix Φ is nonsingular [97]. A key result is that a real matrix has a real logarithm if and only if it is invertible and if each Jordan block associated with a negative eigenvalue occurs an even number of times [68]. This implies that the matrix Φ cannot have isolated eigenvalues on the negative real axis. A detailed discussion of sampling is given in [231].

Example 6.10 IBM Lotus server
In Example 3.5 we described how the dynamics of an IBM Lotus server were obtained as the discrete-time system

$$x[k+1] = ax[k] + bu[k],$$

where $a = 0.43$, $b = 0.47$, the sampling period is $h = 60\,\mathrm{s}$, and x denotes the total requests being served. A differential equation model is needed if we would like to design control systems based on continuous-time theory. Such a model is obtained by applying equation (6.29); hence

$$A = \frac{\log a}{h} = -0.0141, \qquad B = \left(\int_0^h e^{At}\,dt\right)^{-1}b = 0.0116,$$

and we find that the difference equation can be interpreted as a sampled version of the ordinary differential equation

$$\frac{dx}{dt} = -0.0141x + 0.0116u.$$

∇

6.4   LINEARIZATION

As described at the beginning of the chapter, a common source of linear system models is through the approximation of a nonlinear system by a linear one. It is common practice in control engineering to design controllers based on an approximate linear model and to verify the results by simulating the closed loop system using a nonlinear model. In this section we describe how to locally approximate a nonlinear system by a linear one, and discuss what can be inferred about the stability of the original system. We begin with an illustration that controllers can successfully be designed from approximate linear models using the cruise control example, which is described in more detail in Chapter 4.

Example 6.11 Cruise control
The dynamics for the cruise control system are derived in Section 4.1 and have the form

$$m\frac{dv}{dt} = \alpha_n uT(\alpha_n v) - mgC_r \operatorname{sgn}(v) - \frac{1}{2}\rho C_d Av|v| - mg\sin\theta, \qquad (6.30)$$

where the first term on the right-hand side of the equation is the force generated by the engine and the remaining three terms are the rolling friction, aerodynamic drag, and gravitational disturbance force. There is an equilibrium point (v_e, u_e) when the force applied by the engine balances the disturbance forces.

To explore the behavior of the system near the equilibrium point we will linearize the system. A Taylor series expansion of equation (6.30) around the equilibrium point gives

$$\frac{d(v - v_e)}{dt} = -a(v - v_e) - b_g(\theta - \theta_e) + b(u - u_e) + \text{higher-order terms}, \qquad (6.31)$$

where

$$a = \frac{\rho C_d A|v_e| - u_e\alpha_n^2 T'(\alpha_n v_e)}{m}, \qquad b_g = g\cos\theta_e, \qquad b = \frac{\alpha_n T(\alpha_n v_e)}{m}. \qquad (6.32)$$

Notice that the term corresponding to rolling friction disappears if $v > 0$. For a car in fourth gear with $v_e = 20$ m/s, $\theta_e = 0$, and the numerical values for the car from Section 4.1, the equilibrium value for the throttle is $u_e = 0.1687$ and the parameters are $a = 0.01$, $b = 1.32$, and $b_g = 9.8$. This linear model describes how small perturbations in the velocity about the nominal speed evolve in time.

We will later describe how to design a proportional-integral (PI) controller for the system. Here we will simply assume that we have obtained a good controller and we will compare the behaviors when the closed loop system is simulated using the nonlinear model and the linear approximation. The simulation scenario is that the car is running with constant speed on a horizontal road and the system has stabilized so that the vehicle speed and the controller output are constant. Figure 6.14 shows

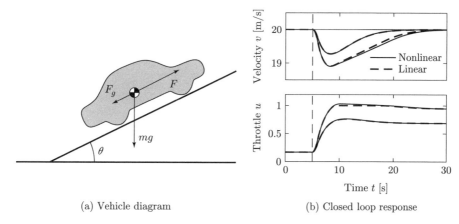

(a) Vehicle diagram   (b) Closed loop response

Figure 6.14: Simulated response of a vehicle with PI cruise control as it climbs a hill with a slope of 4° (smaller velocity deviation/throttle) and a slope of 6° (larger velocity deviation/throttle). The solid line is the simulation based on a nonlinear model, and the dashed line shows the corresponding simulation using a linear model. The controller gains are $k_{\mathrm{p}} = 0.5$ and $k_{\mathrm{i}} = 0.1$ and include anti-windup compensation (described in more detail in Example 11.6).

what happens when the car encounters a hill with a slope of 4° and a hill with a slope of 6° at time $t = 5\,\mathrm{s}$. The results for the nonlinear model are solid curves and those for the linear model are dashed curves. The differences between the curves are very small (especially for $\theta = 4°$), and control design based on the linearized model is thus validated. ∇

Jacobian Linearization Around an Equilibrium Point

To proceed more formally, consider a single-input, single-output nonlinear system

$$\begin{aligned} \frac{dx}{dt} &= f(x, u), \qquad x \in \mathbb{R}^n, u \in \mathbb{R}, \\ y &= h(x, u), \qquad y \in \mathbb{R}, \end{aligned} \tag{6.33}$$

with an equilibrium point at $x = x_{\mathrm{e}}$, $u = u_{\mathrm{e}}$. Without loss of generality we can assume that $x_{\mathrm{e}} = 0$ and $u_{\mathrm{e}} = 0$, although initially we will consider the general case to make the shift of coordinates explicit.

To study the *local* behavior of the system around the equilibrium point $(x_{\mathrm{e}}, u_{\mathrm{e}})$, we suppose that $x - x_{\mathrm{e}}$ and $u - u_{\mathrm{e}}$ are both small, so that nonlinear perturbations around this equilibrium point can be ignored compared with the (lower-order) linear terms. This is roughly the same type of argument that is used when we do small-angle approximations, replacing $\sin\theta$ with θ and $\cos\theta$ with 1 for θ near zero.

We define a new set of state variables z, as well as inputs v and outputs w:

$$z = x - x_{\mathrm{e}}, \qquad v = u - u_{\mathrm{e}}, \qquad w = y - h(x_{\mathrm{e}}, u_{\mathrm{e}}).$$

These variables are all close to zero when we are near the equilibrium point, and so in these variables the nonlinear terms can be thought of as the higher-order terms in a Taylor series expansion of the relevant vector fields (assuming for now that these exist).

Formally, the *Jacobian linearization* of the nonlinear system (6.33) is

$$\frac{dz}{dt} = Az + Bv, \qquad w = Cz + Dv, \tag{6.34}$$

where

$$A = \left.\frac{\partial f}{\partial x}\right|_{(x_e, u_e)}, \quad B = \left.\frac{\partial f}{\partial u}\right|_{(x_e, u_e)}, \quad C = \left.\frac{\partial h}{\partial x}\right|_{(x_e, u_e)}, \quad D = \left.\frac{\partial h}{\partial u}\right|_{(x_e, u_e)}. \tag{6.35}$$

The system (6.34) approximates the original system (6.33) when we are near the equilibrium point about which the system was linearized. It follows from Theorem 5.3 that if the linearization is asymptotically stable, then the equilibrium point x_e is locally asymptotically stable for the full nonlinear system.

Example 6.12 Cruise control using Jacobian linearization
Consider again the cruise control system from Example 6.11 with θ taken as a constant θ_e. We can write the dynamics as a first-order, nonlinear differential equation:

$$\frac{dx}{dt} = f(x, u) = \frac{\alpha_n}{m} u T(\alpha_n x) - g C_r \operatorname{sgn}(x) - \frac{1}{2}\frac{\rho C_d A}{m} x|x| - g \sin \theta_e,$$
$$y = h(x, u) = x,$$

where $x = v$ is the velocity of the vehicle and u is the throttle. We use the velocity as the output of the system (since this is what we are trying to control).

If we linearize the dynamics of the system about an equilibrium point $x = v_e > 0$, $u = u_e$, using equation (6.35) and the previous formula we obtain

$$A = \left.\frac{\partial f}{\partial x}\right|_{(x_e, u_e)} = \frac{u_e \alpha_n^2 T'(\alpha_n x_e) - \rho C_d A |x_e|}{m}, \quad B = \left.\frac{\partial f}{\partial u}\right|_{(x_e, u_e)} = \frac{\alpha_n T(\alpha_n x_e)}{m},$$
$$C = \left.\frac{\partial h}{\partial x}\right|_{(x_e, u_e)} = 1 \qquad\qquad\qquad D = \left.\frac{\partial h}{\partial u}\right|_{(x_e, u_e)} = 0,$$

where we have used the fact that $\operatorname{sgn}(x) = 1$ for $x > 0$. This matches the results in Example 6.11, remembering that we have used x as the system state (vehicle velocity). ∇

It is important to note that we can define the linearization of a system only near a solution of the differential equations for the system, of which an equilibrium point is a particularly common case. To see this, consider a polynomial system

$$\frac{dx}{dt} = a_0 + a_1 x + a_2 x^2 + a_3 x^3 + u,$$

where $a_0 \neq 0$. A set of equilibrium points for this system is given by $(x_e, u_e) = (x_e, -a_0 - a_1 x_e - a_2 x_e^2 - a_3 x_e^3)$, and we can linearize around any of them. Suppose that we try to linearize around the origin of the system $x = 0$, $u = 0$ (which does not correspond to a solution of the differential equation if $a_0 \neq 0$). If we drop the higher-order terms in x, then we get

$$\frac{dx}{dt} = a_0 + a_1 x + u,$$

which is *not* the Jacobian linearization if $a_0 \neq 0$. The constant term must be kept, and it is not present in equation (6.34). Furthermore, even if we kept the constant term in the approximate model, the system would quickly move away from this point (since it is "driven" by the constant term a_0), and hence the approximation could soon fail to hold.

Software for modeling and simulation frequently has facilities for performing linearization symbolically or numerically. The MATLAB command `trim` finds the equilibrium point, and `linmod` extracts linear state space models from a SIMULINK system around an equilibrium point. The more general case of linearizing around a trajectory leads to a time-varying linear system.

Example 6.13 Vehicle steering

Consider the vehicle steering system introduced in Example 3.11. The nonlinear equations of motion for the system are given by equations (3.25)–(3.27) and can be written as

$$
\frac{d}{dt}\begin{pmatrix} x \\ y \\ \theta \end{pmatrix} = \begin{pmatrix} v \cos\left(\alpha(\delta) + \theta\right) \\ v \sin\left(\alpha(\delta) + \theta\right) \\ \dfrac{v \sin \alpha(\delta)}{a} \end{pmatrix}, \qquad \alpha(\delta) = \arctan\left(\frac{a \tan \delta}{b}\right).
$$

The state of the system is the position x, y of the center of mass and the orientation θ of the vehicle. The control variable is the steering angle δ. Furthermore b is the wheelbase and a is the distance between the center of mass and the rear wheel.

We are interested in the motion of the vehicle about a straight-line path $(\theta = \theta_0)$ with constant velocity $v_0 \neq 0$. To find the relevant equilibrium point, we first set $\dot{\theta} = 0$ and we see that we must have $\delta = 0$, corresponding to the steering wheel being straight. This also yields $\alpha = 0$. Looking at the first two equations in the dynamics, we see that the motion in the xy plane is by definition *not* at equilibrium since $\dot{x}^2 + \dot{y}^2 = v^2 \neq 0$. Therefore we cannot formally linearize the full model.

Suppose instead that we are concerned with the lateral deviation of the vehicle from a straight line. For simplicity, we let $\theta_e = 0$, which corresponds to driving along the x axis. We can then focus on the equations of motion in the y and θ directions. With some abuse of notation we introduce the state $x = (y, \theta)$ and $u = \delta$. The system is then in standard form with

$$
f(x, u) = \begin{pmatrix} v_0 \sin\left(\alpha(u) + x_2\right) \\ \dfrac{v_0 \sin \alpha(u)}{a} \end{pmatrix}, \qquad \alpha(u) = \arctan\left(\frac{a \tan u}{b}\right), \qquad h(x, u) = x_1.
$$

The equilibrium point of interest is given by $x = (0, 0)$ and $u = 0$. To compute the linearized model around this equilibrium point, we make use of the formulas (6.35). A straightforward calculation yields

$$
A = \left.\frac{\partial f}{\partial x}\right|_{\substack{x=0 \\ u=0}} = \begin{pmatrix} 0 & v_0 \\ 0 & 0 \end{pmatrix}, \qquad B = \left.\frac{\partial f}{\partial u}\right|_{\substack{x=0 \\ u=0}} = \begin{pmatrix} av_0/b \\ v_0/b \end{pmatrix},
$$

$$
C = \left.\frac{\partial h}{\partial x}\right|_{\substack{x=0 \\ u=0}} = \begin{pmatrix} 1 & 0 \end{pmatrix}, \qquad D = \left.\frac{\partial h}{\partial u}\right|_{\substack{x=0 \\ u=0}} = 0,
$$

and the linearized system

$$\frac{dx}{dt} = Ax + Bu, \qquad y = Cx + Du \qquad (6.36)$$

thus provides an approximation to the original nonlinear dynamics.

The linearized model can be simplified further by introducing normalized variables, as discussed in Section 3.3. For this system, we choose the wheelbase b as the length unit and the time unit as the time required to travel a wheelbase. The normalized state is thus $z = (x_1/b, x_2)$, and the new time variable is $\tau = v_0 t/b$. The model (6.36) then becomes

$$\frac{dz}{d\tau} = \begin{pmatrix} z_2 + \gamma u \\ u \end{pmatrix} = \begin{pmatrix} 0 & 1 \\ 0 & 0 \end{pmatrix} z + \begin{pmatrix} \gamma \\ 1 \end{pmatrix} u, \qquad y = \begin{pmatrix} 1 & 0 \end{pmatrix} z, \qquad (6.37)$$

where $\gamma = a/b$. The normalized linear model for vehicle steering with nonslipping wheels is thus a linear system with only one parameter γ. ∇

Feedback Linearization

Another type of linearization is the use of feedback to convert the dynamics of a nonlinear system into those of a linear one. We illustrate the basic idea with an example.

Example 6.14 Cruise control
Consider again the cruise control system from Example 6.11, whose dynamics are given in equation (6.30):

$$m\frac{dv}{dt} = \alpha_n u T(\alpha_n v) - mgC_r \operatorname{sgn}(v) - \frac{1}{2}\rho C_d Av|v| - mg\sin\theta.$$

If we choose u as a feedback law of the form

$$u = \frac{1}{\alpha_n T(\alpha_n v)}\left(\tilde{u} + mgC_r \operatorname{sgn}(v) + \frac{1}{2}\rho C_d Av|v|\right), \qquad (6.38)$$

then the resulting dynamics become

$$m\frac{dv}{dt} = \tilde{u} + d, \qquad (6.39)$$

where $d(t) = -mg\sin\theta(t)$ is the disturbance force due the slope of the road (which may be changing as we drive). If we now define a feedback law for $\tilde{u}$ (such as a proportional-integral-derivative [PID] controller), we can use equation (6.38) to compute the final input that should be commanded.

Equation (6.39) is a linear differential equation. We have essentially "inverted" the nonlinearity through the use of the feedback law (6.38). This requires that we have an accurate measurement of the vehicle velocity v as well as an accurate model of the torque characteristics of the engine, gear ratios, drag and friction characteristics, and mass of the car. While such a model is not generally available (remembering that the parameter values can change), if we design a good feedback law for $\tilde{u}$, then we can achieve robustness to these uncertainties. ∇

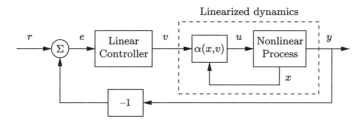

Figure 6.15: Feedback linearization. A nonlinear feedback of the form $u = \alpha(x, v)$ is used to modify the dynamics of a nonlinear process so that the response from the input v to the output y is linear. A linear controller can then be used to regulate the system's dynamics.

More generally, we say that a system of the form

$$\frac{dx}{dt} = f(x, u), \qquad y = h(x),$$

is *feedback linearizable* if there exists a control law $u = \alpha(x, v)$ such that the resulting closed loop system is input/output linear with input v and output y, as shown in Figure 6.15. To fully characterize such systems is beyond the scope of this text, but we note that in addition to changes in the input, the general theory also allows for (nonlinear) changes in the states that are used to describe the system, keeping only the input and output variables fixed. More details of this process can be found in the textbooks by Isidori [126] and Khalil [144].

One case that comes up relatively frequently, and is hence worth special mention, is the set of mechanical systems of the form

$$M(q)\ddot{q} + C(q, \dot{q}) = B(q)u.$$

Here $q \in \mathbb{R}^n$ is the configuration of the mechanical system, $M(q) \in \mathbb{R}^{n \times n}$ is the configuration-dependent inertia matrix, $C(q, \dot{q}) \in \mathbb{R}^n$ represents the Coriolis forces and additional nonlinear forces (such as stiffness and friction), and $B(q) \in \mathbb{R}^{n \times p}$ is the input matrix. If $p = n$, then we have the same number of inputs and configuration variables, and if we further have that $B(q)$ is an invertible matrix for all configurations q, then we can choose

$$u = B^{-1}(q)\big(M(q)v + C(q, \dot{q})\big). \tag{6.40}$$

The resulting dynamics become

$$M(q)\ddot{q} = M(q)v \qquad \Longrightarrow \qquad \ddot{q} = v,$$

which is a linear system. We can now use the tools of linear system theory to analyze and design control laws for the linearized system, remembering to apply equation (6.40) to obtain the actual input that will be applied to the system.

This type of control is common in robotics, where it goes by the name of *computed torque*, and in aircraft flight control, where it is called *dynamic inversion*. Some modeling tools like Modelica can generate the code for the inverse model automatically. One caution is that feedback linearization can often cancel out beneficial terms in the natural dynamics, and hence it must be used with care. Extensions that do not require complete cancellation of nonlinearities are discussed in Khalil [144] and Krstić et al. [152].

6.5 FURTHER READING

The majority of the material in this chapter is classical and can be found in most books on dynamics and control theory, including early works on control such as James, Nichols, and Phillips [130] and more recent textbooks such as Dorf and Bishop [73], Franklin, Powell, and Emami-Naeini [93], and Ogata [195]. An excellent presentation of linear systems based on the matrix exponential is given in the book by Brockett [55], a more comprehensive treatment is given by Rugh [211], and an elegant mathematical treatment is given in Sontag [225]. Material on feedback linearization can be found in books on nonlinear control theory such as Isidori [126] and Khalil [144]. The idea of characterizing dynamics by considering the responses to step inputs is due to Heaviside, who also introduced an operator calculus to analyze linear systems. The unit step is therefore also called the *Heaviside step function*. Analysis of linear systems was simplified significantly, but Heaviside's work was heavily criticized because of lack of mathematical rigor, as described in the biography by Nahin [190]. The difficulties were cleared up later by the mathematician Laurent Schwartz who developed *distribution theory* in the late 1940s. In engineering, linear systems have traditionally been analyzed using Laplace transforms as described in Gardner and Barnes [98]. Use of the matrix exponential started with developments of control theory in the 1960s, strongly stimulated by a textbook by Zadeh and Desoer [262]. Use of matrix techniques expanded rapidly when the powerful methods of numeric linear algebra were packaged in programs like LABVIEW, MATLAB, and Mathematica. The books by Gantmacher [97] are good sources for matrix theory.

EXERCISES

6.1 (Response to the derivative of a signal) Show that if $y(t)$ is the output of a linear time-invariant system corresponding to input $u(t)$, then the output corresponding to an input $\dot{u}(t)$ is given by $\dot{y}(t)$. (Hint: Use the definition of the derivative: $\dot{z}(t) = \lim_{\epsilon \to 0} \big(z(t+\epsilon) - z(t)\big)/\epsilon$.)

6.2 (Impulse response and convolution) Show that a signal $u(t)$ can be decomposed in terms of the impulse function $\delta(t)$ as

$$u(t) = \int_0^t \delta(t-\tau)u(\tau)\,d\tau$$

and use this decomposition plus the principle of superposition to show that the response of a linear, time-invariant system to an input $u(t)$ (assuming a zero initial condition) can be written as a convolution equation

$$y(t) = \int_0^t h(t-\tau)u(\tau)\,d\tau,$$

where $h(t)$ is the impulse response of the system. (Hint: Use the definition of the Riemann integral.)

6.3 (Pulse response for a compartment model) Consider the compartment model given in Example 6.7. Compute the step response for the system and compare it with Figure 6.10b. Use the principle of superposition to compute the response to the 5 s pulse input shown in Figure 6.10c. Use the parameter values $k_0 = 0.1$, $k_1 = 0.1$, $k_2 = 0.5$, and $b_0 = 1.5$.

6.4 (Matrix exponential for second-order system) Assume that $\zeta < 1$ and let $\omega_d = \omega_0\sqrt{1-\zeta^2}$. Show that

$$\exp\begin{pmatrix} -\zeta\omega_0 & \omega_d \\ -\omega_d & -\zeta\omega_0 \end{pmatrix} t = e^{-\zeta\omega_0 t}\begin{pmatrix} \cos\omega_d t & \sin\omega_d t \\ -\sin\omega_d t & \cos\omega_d t \end{pmatrix}.$$

Also show that

$$\exp\left(\begin{pmatrix} -\omega_0 & \omega_0 \\ 0 & -\omega_0 \end{pmatrix} t\right) = e^{-\omega_0 t}\begin{pmatrix} 1 & \omega_0 t \\ 0 & 1 \end{pmatrix}.$$

6.5 (Lyapunov function for a linear system) Consider a linear system $\dot{x} = Ax$ with $\operatorname{Re}\lambda_j < 0$ for all eigenvalues λ_j of the matrix A. Show that the matrix

$$P = \int_0^\infty e^{A^T\tau}Qe^{A\tau}\,d\tau$$

defines a Lyapunov function of the form $V(x) = x^T Px$ with $Q \succ 0$ (positive definite).

6.6 (Nondiagonal Jordan form) Consider a linear system with a Jordan form that is non-diagonal.

a) Prove Proposition 6.3 by showing that if the system contains a real eigenvalue $\lambda = 0$ with a nontrivial Jordan block, then there exists an initial condition with a solution that grows in time.

b) Extend this argument to the case of complex eigenvalues with $\operatorname{Re}\lambda = 0$ by using the block Jordan form

$$J_i = \begin{pmatrix} 0 & \omega & 1 & 0 \\ -\omega & 0 & 0 & 1 \\ 0 & 0 & 0 & \omega \\ 0 & 0 & -\omega & 0 \end{pmatrix}.$$

6.7 (Rise time and settling time for a first-order system) Consider a first-order system of the form

$$\tau\frac{dx}{dt} = -x + u, \qquad y = x.$$

We say that the parameter τ is the *time constant* for the system since the zero input system approaches the origin as $e^{-t/\tau}$. For a first-order system of this form, show that the rise time for a step response of the system is approximately 2τ, and that 1%, 2%, and 5% settling times approximately corresponds to 4.6τ, 4τ, and 3τ.

6.8 (Discrete-time systems) Consider a linear discrete-time system of the form

$$x[k+1] = Ax[k] + Bu[k], \qquad y[k] = Cx[k] + Du[k].$$

a) Show that the general form of the output of a discrete-time linear system is given by the discrete-time convolution equation:

$$y[k] = CA^k x[0] + \sum_{j=0}^{k-1} CA^{k-j-1} Bu[j] + Du[k].$$

b) Show that a discrete-time linear system is asymptotically stable if and only if all the eigenvalues of A have a magnitude strictly less than 1.

c) Show that a discrete-time linear system is unstable if any of the eigenvalues of A have magnitude greater than 1.

d) Derive conditions for stability of a discrete-time linear system having one or more eigenvalues with magnitude identically equal to 1. (Hint: Use Jordan form.)

6.9 (Keynesian economics) Consider the following simple Keynesian macroeconomic model in the form of a linear discrete-time system discussed in Exercise 6.8:

$$\begin{pmatrix} C[t+1] \\ I[t+1] \end{pmatrix} = \begin{pmatrix} a & a \\ ab-b & ab \end{pmatrix} \begin{pmatrix} C[t] \\ I[t] \end{pmatrix} + \begin{pmatrix} a \\ ab \end{pmatrix} G[t],$$

$$Y[t] = C[t] + I[t] + G[t].$$

Determine the eigenvalues of the dynamics matrix. When are the magnitudes of the eigenvalues less than 1? Assume that the system is in equilibrium with constant values capital spending C, investment I, and government expenditure G. Explore what happens when government expenditure increases by 10%. Use the values $a = 0.25$ and $b = 0.5$.

6.10 (Keynes model in continuous time) A continuous version of the Keynes model is given by the equations

$$Y = C + I + G, \qquad T\frac{dC}{dt} + C = ay, \qquad T\frac{dI}{dt} + I = b\frac{dc}{dt}.$$

Write the equations in state space form, and give the conditions for stability.

6.11 (State variables in compartment models) Consider the compartment model described by equation (4.28). Let x_1 and x_2 be the total mass of the drug in the compartments. Show that the system can be described by the equation

$$\frac{dx}{dt} = \begin{pmatrix} -k_0 - k_1 & k_2 \\ k_1 & -k_2 \end{pmatrix} x + \begin{pmatrix} c_0 \\ 0 \end{pmatrix} u, \qquad y = \begin{pmatrix} 0 & 1/V_2 \end{pmatrix} x. \tag{6.41}$$

Compare the this equation with equation (4.28), where the state variables were concentrations. Mass is called an *extensive variable*, and concentration is called an *intensive variable*.

6.12 (Time responses from frequency responses) Consider the following MAT-LAB program, which computes the approximate step response from the frequency response. Explain how it works and explore the effects of the parameter `tmax`.

```
P = '1./(s+1).^2';      % process dynamics
tmax = 20;              % simulation time
N = 2^(12);             % number of points for simulation
dt = tmax/N;            % time interval
dw = 2*pi/tmax;         % frequency interval

% Compute the time and frequency vectors
t = dt*(0:N-1);
omega = -pi/dt:dw:(pi/dt-dw);
s = i*omega;

% Evaluate the frequency response
pv=eval(P);

% Compute the input and output signals using the frequency response
u = [ones(1,N/2) zeros(1,N/2)]; U = fft(u);
y = ifft(fftshift(pv) .* U); y = real(y);

%  Analytic solution in the time domain
ye = 1 - exp(-t) - t .* exp(-t);

% Plot analytic and approximate step responses
subplot(211); plot(t, y, 'b-', t, ye, 'r--');

% Zoom in on the first half of the response
tp = t(1:N/2); yp = y(1:N/2); ye = 1-exp(-t) - t .* exp(-t);
subplot(212); plot(tp, yp, 'b-', t, ye, 'r--');
```

6.13 Consider a scalar system

$$\frac{dx}{dt} = 1 - x^3 + u.$$

Compute the equilibrium points for the unforced system ($u = 0$) and use a Taylor series expansion around the equilibrium point to compute the linearization. Verify that this agrees with the linearization in equation (6.34).

6.14 Consider the model for queuing dynamics in Example 3.15. Let the admission rate λ be the control variable. Linearize the system around an equilibrium point, compute the time constant of the system and determine how it depends on the queue length.

6.15 (Transcriptional regulation) Consider the dynamics of a genetic circuit that implements *self-repression*: the protein produced by a gene is a repressor for that gene, thus restricting its own production. Using the models presented in Example 3.18, the dynamics for the system can be written as

$$\frac{dm}{dt} = \frac{\alpha}{1 + kp^2} + \alpha_0 - \delta m - u, \qquad \frac{dp}{dt} = \kappa m - \gamma p, \qquad (6.42)$$

where u is a disturbance term that affects RNA transcription and $m, p \geq 0$. Find the equilibrium points for the system and use the linearized dynamics around each equilibrium point to determine the local stability of the equilibrium point and the step response of the system to a disturbance.

6.16 (Monotone step response) Consider a stable linear system with monotone step response $S(t)$. Let the input signal be bounded: $|u(t)| \leq u_{\max}$. Assuming that the initial conditions are zero, show that $|y(t)| \leq S(\infty)u_{\max}$. (Hint: Use the convolution integral.)

Chapter Seven

State Feedback

Intuitively, the state may be regarded as a kind of information storage or memory or accumulation of past causes. We must, of course, demand that the set of internal states Σ be sufficiently rich to carry all information about the past history of Σ to predict the effect of the past upon the future. We do not insist, however, that the state is the *least* such information although this is often a convenient assumption.

—R. E. Kalman, P. L. Falb, and M. A. Arbib,
Topics in Mathematical System Theory, 1969 [138].

This chapter describes how the feedback of a system's state can be used to shape the local behavior of a system. The concept of reachability is introduced and used to investigate how to design the dynamics of a system through assignment of its eigenvalues. In particular, we show that under certain conditions it is possible to assign the system eigenvalues arbitrarily by appropriate feedback of the system state.

7.1   REACHABILITY

One of the fundamental properties of a control system is what set of points in the state space can be reached through the choice of a control input. It turns out that the property of reachability is also fundamental in understanding the extent to which feedback can be used to design the dynamics of a system.

Definition of Reachability

We begin by disregarding the output measurements of the system and focusing on the evolution of the state, given by

$$\frac{dx}{dt} = Ax + Bu, \tag{7.1}$$

where $x \in \mathbb{R}^n$, $u \in \mathbb{R}$, A is an $n \times n$ matrix, and B a column vector. A fundamental question is whether it is possible to find control signals so that any point in the state space can be reached through some choice of input. To study this, we define the *reachable set* $\mathcal{R}(x_0, \leq T)$ as the set of all points x_f such that there exists an input $u(t)$, $0 \leq t \leq T$ that steers the system from $x(0) = x_0$ to $x(T) = x_f$, as illustrated in Figure 7.1a.

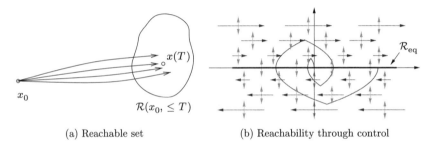

(a) Reachable set   (b) Reachability through control

Figure 7.1: The reachable set for a control system. The set $\mathcal{R}(x_0, \leq T)$ shown in (a) is the set of points reachable from x_0 in time less than T. The phase portrait in (b) shows the dynamics for a double integrator, with the natural dynamics drawn as horizontal arrows and the control inputs drawn as vertical arrows. The set of achievable equilibrium points is the x axis. By setting the control inputs as a function of the state, it is possible to steer the system to the origin, as shown on the sample path.

Definition 7.1 (Reachability). A linear system is *reachable* if for any $x_0, x_{\mathrm{f}} \in \mathbb{R}^n$ there exists a $T > 0$ and $u \colon [0, T] \to \mathbb{R}$ such that if $x(0) = x_0$ then the corresponding solution satisfies $x(T) = x_{\mathrm{f}}$.

The definition of reachability addresses whether it is possible to reach all points in the state space in a *transient* fashion. In many applications, the set of points that we are most interested in reaching is the set of equilibrium points of the system (since we can remain at those points with constant input u). The set of all possible equilibrium points for constant controls is given by

$$\mathcal{R}_{\mathrm{eq}} = \{x_{\mathrm{e}} : Ax_{\mathrm{e}} + Bu_{\mathrm{e}} = 0 \text{ for some } u_{\mathrm{e}} \in \mathbb{R}\}.$$

This means that possible equilibrium points lie in a one- (or possibly higher) dimensional subspace. If the matrix A is invertible, this subspace is one-dimensional and is spanned by $A^{-1}B$.

The following example provides some insight into the possibilities.

Example 7.1 Double integrator
Consider a linear system consisting of a double integrator whose dynamics are given by

$$\frac{dx_1}{dt} = x_2, \qquad \frac{dx_2}{dt} = u.$$

Figure 7.1b shows a phase portrait of the system. The open loop dynamics ($u = 0$) are shown as horizontal arrows pointed to the right for $x_2 > 0$ and to the left for $x_2 < 0$. The control input is represented by a double-headed arrow in the vertical direction, corresponding to our ability to set the value of $\dot{x}_2$. The set of equilibrium points $\mathcal{E}$ corresponds to the x_1 axis, with $u_{\mathrm{e}} = 0$.

Suppose first that we wish to reach the origin from an initial condition $(a, 0)$. We can directly move the state up and down in the phase plane, but we must rely on the natural dynamics to control the motion to the left and right. If $a > 0$, we can move toward the origin by first setting $u < 0$, which will cause x_2 to become negative. Once $x_2 < 0$, the value of x_1 will begin to decrease and we will move to the left. After a while, we can set u to be positive, moving x_2 back toward zero and slowing the motion in the x_1 direction. If we bring x_2 to a positive value, we can move the system state in the opposite direction.

Figure 7.1b shows a sample trajectory bringing the system to the origin. Note that if we steer the system to an equilibrium point, it is possible to remain there indefinitely (since $\dot{x}_1 = 0$ when $x_2 = 0$), but if we go to a point in the state space with $x_2 \neq 0$, we can pass through the point only in a transient fashion. ∇

To find general conditions under which a linear system is reachable, we will first give a heuristic argument based on formal calculations with impulse functions. We note that if we can reach all points in the state space through some choice of input, then we can also reach all equilibrium points.

Testing for Reachability

When the initial state is zero, the response of the system (7.1) to an input $u(t)$ is given by

$$x(t) = \int_0^t e^{A(t-\tau)} B u(\tau)\, d\tau. \tag{7.2}$$

If we choose the input to be a impulse function $\delta(t)$ as defined in Section 6.3, the state becomes

$$x_\delta(t) = \int_0^t e^{A(t-\tau)} B \delta(\tau)\, d\tau = e^{At} B.$$

(Note that the state changes instantaneously in response to the impulse.) We can find the response to the derivative of an impulse function by taking the derivative of the impulse response (Exercise 6.1):

$$x_{\dot{\delta}}(t) = \frac{dx_\delta}{dt} = A e^{At} B.$$

Continuing this process and using the linearity of the system, the input

$$u(t) = \alpha_1 \delta(t) + \alpha_2 \dot{\delta}(t) + \alpha_3 \ddot{\delta}(t) + \cdots + \alpha_n \delta^{(n-1)}(t)$$

gives the state

$$x(t) = \alpha_1 e^{At} B + \alpha_2 A e^{At} B + \alpha_3 A^2 e^{At} B + \cdots + \alpha_n A^{n-1} e^{At} B.$$

Taking the limit as t goes to zero through positive values, we get

$$\lim_{t \to 0+} x(t) = \alpha_1 B + \alpha_2 AB + \alpha_3 A^2 B + \cdots + \alpha_n A^{n-1} B.$$

On the right is a linear combination of the columns of the matrix

$$W_\mathrm{r} = \begin{pmatrix} B & AB & \cdots & A^{n-1}B \end{pmatrix}. \tag{7.3}$$

To reach an arbitrary point in the state space, we thus require that W_r has n independent columns (full rank). The matrix W_r is called the *reachability matrix* and it is full rank if and only if its determinant is nonzero.

Although we have only considered the scalar input case, it turns out that this same test works in the multi-input case, where we require that W_r be full column rank (have n linearly independent columns). In addition, it can be shown that only the terms up to $A^{n-1}B$ must be computed; additional terms add no new directions to W_r (see Exercise 7.3).

An input consisting of a sum of impulse functions and their derivatives is a very violent signal. To see that an arbitrary point can be reached with smoother signals we can make use of the convolution equation. Assuming that the initial condition is zero, the state of a linear system is given by

$$x(t) = \int_0^t e^{A(t-\tau)} Bu(\tau) d\tau = \int_0^t e^{A\tau} Bu(t-\tau) d\tau.$$

It follows from the theory of matrix functions, specifically the Cayley–Hamilton theorem (Exercise 7.3), that

$$e^{A\tau} = I\alpha_0(\tau) + A\alpha_1(\tau) + \cdots + A^{n-1}\alpha_{n-1}(\tau),$$

where $\alpha_i(\tau)$ are scalar functions, and we find that

$$x(t) = B \int_0^t \alpha_0(\tau)u(t-\tau) \, d\tau + AB \int_0^t \alpha_1(\tau)u(t-\tau) \, d\tau$$
$$+ \cdots + A^{n-1}B \int_0^t \alpha_{n-1}(\tau)u(t-\tau) \, d\tau.$$

Again we observe that the right-hand side is a linear combination of the columns of the reachability matrix W_r given by equation (7.3). This basic approach leads to the following theorem.

Theorem 7.1 (Reachability rank condition). *A linear system of the form (7.1) is reachable if and only if the reachability matrix W_r is invertible (full column rank).*

The formal proof of this theorem is beyond the scope of this text but follows along the lines of the previous sketch and can be found in most books on linear control theory, such as Callier and Desoer [59] or Lewis [163]. It is also interesting to note that Theorem 7.1 makes no mention of the time T that was in our definition of reachability. For a linear system, it turns out that we can find an input taking x_0 to x_f for *any* $T > 0$, though the size of the input required can be very large when T is very small.

We illustrate the concept of reachability with the following example.

Example 7.2 Balance system
Consider the balance system introduced in Example 3.2 and shown in Figure 7.2. Recall that this system is a model for a class of examples in which the center of mass is balanced above a pivot point. One example is the Segway® Personal Transporter shown in Figure 7.2a, about which a natural question to ask is whether we can move from one stationary point to another by appropriate application of forces through the wheels.

The nonlinear equations of motion for the system are given in equation (3.9) and repeated here:

$$(M+m)\ddot{q} - ml\cos\theta\,\ddot{\theta} = -c\dot{q} - ml\sin\theta\,\dot{\theta}^2 + F,$$
$$(J+ml^2)\ddot{\theta} - ml\cos\theta\,\ddot{q} = -\gamma\dot{\theta} + mgl\sin\theta. \tag{7.4}$$

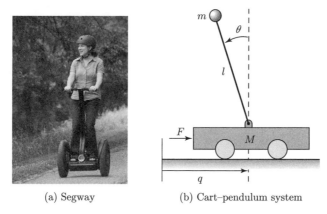

(a) Segway   (b) Cart–pendulum system

Figure 7.2: Balance system. The Segway® Personal Transporter shown in (a) is an example of a balance system that uses torque applied to the wheels to keep the rider upright. A simplified diagram for a balance system is shown in (b). The system consists of a mass m on a rod of length l connected by a pivot to a cart with mass M.

For simplicity, we take $c = \gamma = 0$. Linearizing around the equilibrium point $x_e = (0, 0, 0, 0)$, the dynamics matrix and the control matrix are

$$A = \begin{pmatrix} 0 & 0 & 1 & 0 \\ 0 & 0 & 0 & 1 \\ 0 & m^2 l^2 g/\mu & 0 & 0 \\ 0 & M_t mgl/\mu & 0 & 0 \end{pmatrix}, \qquad B = \begin{pmatrix} 0 \\ 0 \\ J_t/\mu \\ lm/\mu \end{pmatrix},$$

where $\mu = M_t J_t - m^2 l^2$, $M_t = M + m$, and $J_t = J + ml^2$. The reachability matrix is

$$W_r = \begin{pmatrix} 0 & J_t/\mu & 0 & gl^3 m^3/\mu^2 \\ 0 & lm/\mu & 0 & gl^2 m^2 M_t/\mu^2 \\ J_t/\mu & 0 & gl^3 m^3/\mu^2 & 0 \\ lm/\mu & 0 & gl^2 m^2 M_t/\mu^2 & 0 \end{pmatrix}. \tag{7.5}$$

To compute the determinant we permute the first and the last columns of the matrix W_r and use the fact that such a permutation changes the determinant by a factor of -1. This gives a block diagonal matrix with two identical blocks and the determinant becomes

$$\det(W_r) = -\left(\frac{gl^4 m^4}{\mu^3} - \frac{gl^2 m^2 J_t M_t}{\mu^3} \right)^2 = -\frac{g^2 l^4 m^4}{\mu^6}(MJ + mJ + Mml^2)^2,$$

and we can conclude that the system is reachable. This implies that we can move the system from any initial state to any final state and, in particular, that we can always find an input to bring the system from an initial state to an equilibrium point.   ∇

It is useful to have an intuitive understanding of the mechanisms that make a system unreachable. An example of such a system is given in Figure 7.3. The system consists of two identical systems with the same input. We cannot separately cause the first and the second systems to do something different since they have the same

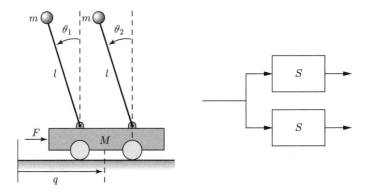

Figure 7.3: An unreachable system. The cart–pendulum system shown on the left has a single input that affects two pendula of equal length and mass. Since the forces affecting the two pendula are the same and their dynamics are identical, it is not possible to arbitrarily control the state of the system. The figure on the right is a block diagram representation of this situation.

input. Hence we cannot reach arbitrary states, and so the system is not reachable (Exercise 7.4).

More subtle mechanisms for nonreachability can also occur. For example, if there is a linear combination of states that always remains constant, then the system is not reachable. To see this, suppose that there exists a row vector H such that

$$0 = \frac{d}{dt} H x = H(Ax + Bu), \quad \text{for all } x \text{ and } u.$$

Then H is in the left null space of both A and B and it follows that

$$H W_{\mathrm{r}} = H \begin{pmatrix} B & AB & \cdots & A^{n-1}B \end{pmatrix} = 0.$$

Hence the reachability matrix is not full rank. In this case, if we have an initial condition x_0 and we wish to reach a state x_{f} for which $H x_0 \neq H x_{\mathrm{f}}$, then since $H x(t)$ is constant, no input u can move the state from x_0 to x_{f}.

Reachable Canonical Form

As we have already seen in previous chapters, it is often convenient to change coordinates and write the dynamics of the system in the transformed coordinates $z = Tx$. One application of a change of coordinates is to convert a system into a canonical form in which it is easy to perform certain types of analysis.

A linear state space system is in *reachable canonical form* if its dynamics are given by

$$\frac{dz}{dt} = \begin{pmatrix} -a_1 & -a_2 & -a_3 & \cdots & -a_n \\ 1 & 0 & & & \\ & 1 & 0 & & 0 \\ & & \ddots & \ddots & \\ 0 & & & 1 & 0 \end{pmatrix} z + \begin{pmatrix} 1 \\ 0 \\ 0 \\ \vdots \\ 0 \end{pmatrix} u, \tag{7.6}$$

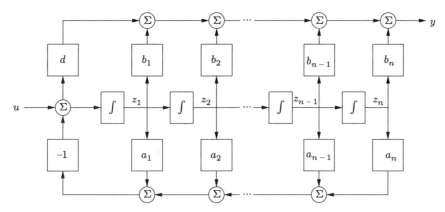

Figure 7.4: Block diagram for a system in reachable canonical form. The individual states of the system are represented by a chain of integrators whose input depends on the weighted values of the states. The output is given by an appropriate combination of the system input and other states.

$$y = \begin{pmatrix} b_1 & b_2 & b_3 & \dots & b_n \end{pmatrix} z + du.$$

A block diagram for a system in reachable canonical form is shown in Figure 7.4. We see that the coefficients that appear in the A and B matrices show up directly in the block diagram. Furthermore, the output of the system is a simple linear combination of the outputs of the integration blocks.

The characteristic polynomial for a system in reachable canonical form is given by

$$\lambda(s) = s^n + a_1 s^{n-1} + \cdots + a_{n-1} s + a_n. \tag{7.7}$$

The reachability matrix also has a relatively simple structure:

$$\tilde{W}_{\mathrm{r}} = \begin{pmatrix} \tilde{B} & \tilde{A}\tilde{B} & \dots & \tilde{A}^{n-1}\tilde{B} \end{pmatrix} = \begin{pmatrix} 1 & -a_1 & a_1^2 - a_2 & & \\ 0 & 1 & -a_1 & & * \\ & & \ddots & \ddots & \\ & 0 & & 1 & -a_1 \\ & & & & 1 \end{pmatrix},$$

where $*$ indicates a possibly nonzero term and we use a tilde to remind us that A and B are in a special form. The matrix W_{r} is full rank since no column can be written as a linear combination of the others because of the triangular structure of the matrix.

We now consider the problem of finding a change of coordinates such that the dynamics of a system can be written in reachable canonical form. Let A, B represent the dynamics of a given system and $\tilde{A}, \tilde{B}$ be the dynamics in reachable canonical form. Suppose that we wish to transform the original system into reachable canonical form using a coordinate transformation $z = Tx$. As shown in the previous chapter, the dynamics matrix and the control matrix for the transformed system are

$$\tilde{A} = TAT^{-1}, \qquad \tilde{B} = TB.$$

The reachability matrix for the transformed system then becomes

$$\tilde{W}_r = \begin{pmatrix} \tilde{B} & \tilde{A}\tilde{B} & \cdots & \tilde{A}^{n-1}\tilde{B} \end{pmatrix}.$$

Transforming each element individually, we have

$$\tilde{A}\tilde{B} = TAT^{-1}TB = TAB,$$
$$\tilde{A}^2\tilde{B} = (TAT^{-1})^2 TB = TAT^{-1}TAT^{-1}TB = TA^2 B,$$

$$\vdots$$

$$\tilde{A}^n\tilde{B} = TA^n B,$$

and hence the reachability matrix for the transformed system is

$$\tilde{W}_r = T \begin{pmatrix} B & AB & \cdots & A^{n-1}B \end{pmatrix} = TW_r. \tag{7.8}$$

If W_r is invertible, we can thus solve for the transformation T that takes the system into reachable canonical form:

$$T = \tilde{W}_r W_r^{-1}.$$

The following example illustrates the approach.

Example 7.3 Transformation to reachable form
Consider a simple two-dimensional system of the form

$$\frac{dx}{dt} = \begin{pmatrix} \alpha & \omega \\ -\omega & \alpha \end{pmatrix} x + \begin{pmatrix} 0 \\ 1 \end{pmatrix} u.$$

We wish to find the transformation that converts the system into reachable canonical form:

$$\tilde{A} = \begin{pmatrix} -a_1 & -a_2 \\ 1 & 0 \end{pmatrix}, \qquad \tilde{B} = \begin{pmatrix} 1 \\ 0 \end{pmatrix}.$$

The coefficients a_1 and a_2 can be determined from the characteristic polynomial for the original system:

$$\lambda(s) = \det(sI - A) = s^2 - 2\alpha s + (\alpha^2 + \omega^2) \quad \Longrightarrow \quad \begin{aligned} a_1 &= -2\alpha, \\ a_2 &= \alpha^2 + \omega^2. \end{aligned}$$

The reachability matrix for each system is

$$W_r = \begin{pmatrix} 0 & \omega \\ 1 & \alpha \end{pmatrix}, \qquad \tilde{W}_r = \begin{pmatrix} 1 & -a_1 \\ 0 & 1 \end{pmatrix}.$$

The transformation T becomes

$$T = \tilde{W}_r W_r^{-1} = \begin{pmatrix} -(a_1 + \alpha)/\omega & 1 \\ 1/\omega & 0 \end{pmatrix} = \begin{pmatrix} \alpha/\omega & 1 \\ 1/\omega & 0 \end{pmatrix},$$

and hence the coordinates

$$\begin{pmatrix} z_1 \\ z_2 \end{pmatrix} = Tx = \begin{pmatrix} \alpha x_1/\omega + x_2 \\ x_1/\omega \end{pmatrix}$$

put the system in reachable canonical form. ∇

We summarize the results of this section in the following theorem.

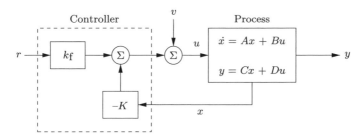

Figure 7.5: A feedback control system with state feedback. The controller uses the system state x and the reference input r to command the process through its input u. We model disturbances via the additive input v.

Theorem 7.2 (Reachable canonical form). *Let A and B be the dynamics and control matrices for a reachable system and suppose that the characteristic polynomial for A is given by*

$$\det(sI - A) = s^n + a_1 s^{n-1} + \cdots + a_{n-1}s + a_n.$$

Then there exists a transformation $z = Tx$ such that in the transformed coordinates the dynamics and control matrices are in reachable canonical form (7.6).

One important implication of this theorem is that for any reachable system, we can assume without loss of generality that the coordinates are chosen such that the system is in reachable canonical form. This is particularly useful for proofs, as we shall see later in this chapter. However, for high-order systems, small changes in the coefficients a_i can give large changes in the eigenvalues. Hence, the reachable canonical form is not always well conditioned and must be used with some care.

7.2 STABILIZATION BY STATE FEEDBACK

The state of a dynamical system is a collection of variables that permits prediction of the future evolution of a system given its future inputs. We now explore the idea of designing the dynamics of a system through feedback of the state. We will assume that the system to be controlled is described by a linear state model and has a single input (for simplicity). The feedback control law will be developed step by step using a single idea: the positioning of closed loop eigenvalues in desired locations.

State Space Controller Structure

Figure 7.5 is a diagram of a typical control system using state feedback. The full system consists of the process dynamics, which we take to be linear, the controller elements K and k_f, the reference input (or command signal) r, and process disturbances v. The goal of the feedback controller is to regulate the output of the system y such that it tracks the reference input in the presence of disturbances and also uncertainty in the process dynamics.

An important element of the control design is the performance specification. The simplest performance specification is that of stability: given a constant reference r

and in the absence of any disturbances, we would like the equilibrium point of the system to be asymptotically stable. More sophisticated performance specifications typically involve giving desired properties of the step or frequency response of the system, such as specifying the desired rise time, overshoot, and settling time of the step response. Finally, we are often concerned with the disturbance attenuation properties of the system: to what extent can we experience disturbance inputs v and still hold the output y near the desired value?

Consider a system described by the linear differential equation

$$\frac{dx}{dt} = Ax + Bu, \qquad y = Cx + Du, \tag{7.9}$$

where we have ignored the disturbance signal v for now. Our goal is to drive the output y to a given reference value r and hold it there.

We begin by assuming that all components of the state vector are measured. Since the state at time t contains all the information necessary to predict the future behavior of the system, the most general time-invariant control law is a function of the state and the reference input:

$$u = \alpha(x, r).$$

If the control law is restricted to be linear, it can be written as

$$u = -Kx + k_f r, \tag{7.10}$$

where r is the reference value, assumed for now to be a constant.

This control law corresponds to the structure shown in Figure 7.5. The negative sign is a convention to indicate that negative feedback is the normal situation. The term $k_f r$ represents a feedforward signal from the reference to the control. The closed loop system obtained when the feedback (7.10) is applied to the system (7.9) is given by

$$\frac{dx}{dt} = (A - BK)x + Bk_f r. \tag{7.11}$$

We attempt to determine the feedback gain K so that the closed loop system has the characteristic polynomial

$$p(s) = s^n + p_1 s^{n-1} + \cdots + p_{n-1} s + p_n. \tag{7.12}$$

This control problem is called the *eigenvalue assignment problem* or *pole placement problem* (we will define poles more formally in Chapter 9).

Note that k_f does not affect the stability of the system (which is determined by the eigenvalues of $A - BK$) but does affect the steady-state solution. In particular, the equilibrium point and steady-state output for the closed loop system are given by

$$x_e = -(A - BK)^{-1} Bk_f r, \qquad y_e = Cx_e + Du_e,$$

and hence k_f should be chosen such that $y_e = r$ (the desired output value). Since k_f is a scalar, we can easily solve to show that if $D = 0$ (the most common case),

$$k_f = -1/\big(C(A - BK)^{-1}B\big). \tag{7.13}$$

Notice that k_f is exactly the inverse of the zero frequency gain of the closed loop system. The solution for $D \neq 0$ is left as an exercise.

Using the gains K and k_f, we are thus able to design the dynamics of the closed loop system to satisfy our goal. To illustrate how to construct such a state feedback control law, we begin with a few examples that provide some basic intuition and insights.

Example 7.4 Vehicle steering

In Example 6.13 we derived a normalized linear model for vehicle steering. The dynamics describing the lateral deviation were given by the normalized dynamics

$$A = \begin{pmatrix} 0 & 1 \\ 0 & 0 \end{pmatrix}, \qquad B = \begin{pmatrix} \gamma \\ 1 \end{pmatrix},$$

$$C = \begin{pmatrix} 1 & 0 \end{pmatrix}, \qquad D = 0,$$

where $\gamma = a/b$ is the ratio of the distance between the center of mass and the rear wheel, a, and the wheelbase b. We want to design a controller that stabilizes the dynamics and tracks a given reference value r of the lateral position of the vehicle. To do this we introduce the feedback

$$u = -Kx + k_f r = -k_1 x_1 - k_2 x_2 + k_f r,$$

and the closed loop system becomes

$$\frac{dx}{dt} = (A - BK)x + Bk_f r = \begin{pmatrix} -\gamma k_1 & 1 - \gamma k_2 \\ -k_1 & -k_2 \end{pmatrix} x + \begin{pmatrix} \gamma k_f \\ k_f \end{pmatrix} r,$$

$$y = Cx + Du = \begin{pmatrix} 1 & 0 \end{pmatrix} x. \tag{7.14}$$

The closed loop system has the characteristic polynomial

$$\det (sI - A + BK) = \det \begin{pmatrix} s + \gamma k_1 & \gamma k_2 - 1 \\ k_1 & s + k_2 \end{pmatrix} = s^2 + (\gamma k_1 + k_2)s + k_1.$$

Suppose that we would like to use feedback to design the dynamics of the system to have the characteristic polynomial

$$p(s) = s^2 + 2\zeta_c \omega_c s + \omega_c^2.$$

Comparing this polynomial with the characteristic polynomial of the closed loop system, we see that the feedback gains should be chosen as

$$k_1 = \omega_c^2, \qquad k_2 = 2\zeta_c \omega_c - \gamma \omega_c^2.$$

Equation (7.13) gives $k_f = k_1 = \omega_c^2$, and the control law can be written as

$$u = k_1(r - x_1) - k_2 x_2 = \omega_c^2 (r - x_1) - (2\zeta_c \omega_c - \gamma \omega_c^2) x_2.$$

To find reasonable values of ω_c we have to balance the speed of response with the available control authority. The unit step responses for the closed loop system for different values of the design parameters are shown in Figure 7.6. The effect of

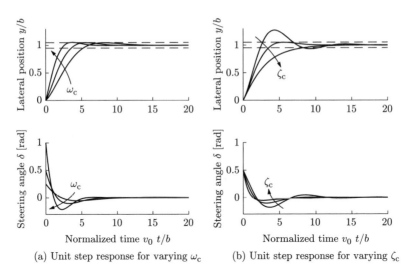

(a) Unit step response for varying ω_{c} (b) Unit step response for varying ζ_{c}

Figure 7.6: State feedback control of a steering system. Unit step responses (from zero initial condition) obtained with controllers designed with $\zeta_{\mathrm{c}} = 0.7$ and $\omega_{\mathrm{c}} = 0.5$, 0.7, and 1 [rad/s] are shown in (a). The dashed lines indicate $\pm 5\%$ deviations from the setpoint. Notice that response speed increases with increasing ω_{c}, but that large ω_{c} also give large initial control actions. Unit step responses obtained with a controller designed with $\omega_{\mathrm{c}} = 0.7$ and $\zeta_{\mathrm{c}} = 0.5$, 0.7, and 1 are shown in (b).

ω_{c} is shown in Figure 7.6a, which shows that the response speed increases with increasing ω_{c}. All responses have overshoot less than 5%, as indicated by the dashed lines, which corresponds to 15 cm assuming a wheelbase $b = 3$ m. The settling times range from 3 to 6 normalized time units, which corresponds to about 2–4 s at $v_0 = 15$ m/s. The effect of ζ_{c} is shown in Figure 7.6b. The response speed and the overshoot increase with decreasing damping.

To select the specific gains to use, we can evaluate how the choice of parameters affects vehicle handling characteristics. For example, a lateral error of 20% of the wheelbase is relatively large and we might choose ω_{c} to exert a relatively large steering angle to correct for such an error. For $\omega_{\mathrm{c}} = 0.7$ and a step input of size 0.2 (in normalized units), Figure 7.6a indicates that the initial steering angle will be 0.1 rad, which is aggressive but not unreasonable at moderate speeds. The value for ζ_{c} can be also be chosen as 0.7, which gives a fast response with approximately 5% overshoot. ∇

The example of the vehicle steering system illustrates how state feedback can be used to set the eigenvalues of a closed loop system to arbitrary values. We see that for this example we can set the eigenvalues to any location. We now show that this is a general property for reachable systems.

State Feedback for Systems in Reachable Canonical Form

The reachable canonical form has the property that the parameters of the system are the coefficients of the characteristic polynomial. It is therefore natural to consider systems in this form when solving the eigenvalue assignment problem.

Consider a system in reachable canonical form, i.e.,

$$
\frac{dz}{dt} = \tilde{A}z + \tilde{B}u = \begin{pmatrix} -a_1 & -a_2 & -a_3 & \cdots & -a_n \\ 1 & 0 & & & \\ & 1 & 0 & & 0 \\ & & \ddots & \ddots & \\ 0 & & & 1 & 0 \end{pmatrix} z + \begin{pmatrix} 1 \\ 0 \\ 0 \\ \vdots \\ 0 \end{pmatrix} u
$$

$$
y = \tilde{C}z = \begin{pmatrix} b_1 & b_2 & \cdots & b_n \end{pmatrix} z.
$$

(7.15)

It follows from equation (7.7) that the open loop system has the characteristic polynomial

$$
\det(sI - A) = s^n + a_1 s^{n-1} + \cdots + a_{n-1}s + a_n.
$$

Before making a formal analysis we can gain some insight by investigating the block diagram of the system shown in Figure 7.4. The characteristic polynomial is given by the parameters a_k in the figure. Notice that the parameter a_k can be changed by feedback from state z_k to the input u. It is thus straightforward to change the coefficients of the characteristic polynomial by state feedback.

Returning to equations, introducing the control law

$$
u = -\tilde{K}z + k_f r = -\tilde{k}_1 z_1 - \tilde{k}_2 z_2 - \cdots - \tilde{k}_n z_n + k_f r,
$$

(7.16)

the closed loop system becomes

$$
\frac{dz}{dt} = \begin{pmatrix} -a_1 - \tilde{k}_1 & -a_2 - \tilde{k}_2 & -a_3 - \tilde{k}_3 & \cdots & -a_n - \tilde{k}_n \\ 1 & 0 & & & \\ & 1 & 0 & & 0 \\ & & \ddots & \ddots & \\ 0 & & & 1 & 0 \end{pmatrix} z + \begin{pmatrix} k_f \\ 0 \\ 0 \\ \vdots \\ 0 \end{pmatrix} r,
$$

$$
y = \begin{pmatrix} b_1 & b_2 & \cdots & b_n \end{pmatrix} z.
$$

(7.17)

The feedback changes the elements of the first row of the A matrix, which corresponds to the parameters of the characteristic polynomial. The closed loop system thus has the characteristic polynomial

$$
s^n + (a_1 + \tilde{k}_1)s^{n-1} + (a_2 + \tilde{k}_2)s^{n-2} + \cdots + (a_{n-1} + \tilde{k}_{n-1})s + a_n + \tilde{k}_n.
$$

Requiring this polynomial to be equal to the desired closed loop polynomial

$$
p(s) = s^n + p_1 s^{n-1} + \cdots + p_{n-1}s + p_n,
$$

we find that the controller gains should be chosen as

$$
\tilde{k}_1 = p_1 - a_1, \qquad \tilde{k}_2 = p_2 - a_2, \qquad \cdots \qquad \tilde{k}_n = p_n - a_n.
$$

This feedback simply replaces the parameters a_i in the system (7.15) by p_i. The feedback gain for a system in reachable canonical form is thus

$$
\tilde{K} = \begin{pmatrix} p_1 - a_1 & p_2 - a_2 & \cdots & p_n - a_n \end{pmatrix}.
$$

(7.18)

To have zero frequency gain equal to unity, we compute the equilibrium point z_e by setting the right hand side of equation (7.17) to zero and then compute the corresponding output. It can be seen that $z_{e,1}, \ldots, z_{e,n-1}$ must all be zero and we are left with

$$(-a_n - \tilde{k}_n)z_{e,n} + k_f r = 0 \quad \text{and} \quad y_e = b_n z_{e,n}.$$

It follows that in order for y_e to be equal to r then the parameter k_f should be chosen as

$$k_f = \frac{a_n + \tilde{k}_n}{b_n} = \frac{p_n}{b_n}. \tag{7.19}$$

Notice that it is essential to know the precise values of parameters a_n and b_n in order to obtain the correct zero frequency gain. The zero frequency gain is thus obtained by precise calibration. This is very different from obtaining the correct steady-state value by integral action, which we shall see in later sections.

Eigenvalue Assignment

We have seen through the examples how feedback can be used to design the dynamics of a system through assignment of its eigenvalues. To solve the problem in the general case, we simply change coordinates so that the system is in reachable canonical form. Consider the system

$$\frac{dx}{dt} = Ax + Bu, \qquad y = Cx + Du. \tag{7.20}$$

We can change the coordinates by a linear transformation $z = Tx$ so that the transformed system is in reachable canonical form (7.15). For such a system the feedback is given by equation (7.16), where the coefficients are given by equation (7.18). Transforming back to the original coordinates gives the control law

$$u = -\tilde{K}z + k_f r = -\tilde{K}Tx + k_f r.$$

The form of the controller is a feedback term $-Kx$ and a feedforward term $k_f r$.

The results obtained can be summarized as follows.

Theorem 7.3 (Eigenvalue assignment by state feedback). *Consider the system given by equation (7.20), with one input and one output. Let $\lambda(s) = s^n + a_1 s^{n-1} + \cdots + a_{n-1}s + a_n$ be the characteristic polynomial of A. If the system is reachable, then there exists a control law*

$$u = -Kx + k_f r$$

that gives a closed loop system with the characteristic polynomial

$$p(s) = s^n + p_1 s^{n-1} + \cdots + p_{n-1}s + p_n$$

and unity zero frequency gain between r and y. The feedback gain is given by

$$K = \tilde{K}T = \begin{pmatrix} p_1 - a_1 & p_2 - a_2 & \cdots & p_n - a_n \end{pmatrix} \tilde{W}_r W_r^{-1}, \tag{7.21}$$

where a_i are the coefficients of the characteristic polynomial of the matrix A and the matrices W_r and $\tilde{W}_r$ are given by

$$W_r = \begin{pmatrix} B & AB & \cdots & A^{n-1}B \end{pmatrix}, \qquad \tilde{W}_r = \begin{pmatrix} 1 & a_1 & a_2 & \cdots & a_{n-1} \\ & 1 & a_1 & \cdots & a_{n-2} \\ & & \ddots & \ddots & \vdots \\ & 0 & & 1 & a_1 \\ & & & & 1 \end{pmatrix}^{-1}.$$

The feedforward gain is given by

$$k_f = -1/\big(C(A - BK)^{-1}B\big).$$

For simple problems, the eigenvalue assignment problem can be solved by introducing the elements k_i of K as unknown variables. We then compute the characteristic polynomial

$$\lambda(s) = \det(sI - A + BK)$$

and equate coefficients of equal powers of s to the coefficients of the desired characteristic polynomial

$$p(s) = s^n + p_1 s^{n-1} + \cdots + p_{n-1}s + p_n.$$

This gives a system of linear equations to determine k_i. The equations can always be solved if the system is reachable, exactly as we did in Example 7.4.

Equation (7.21), which is called Ackermann's formula [3, 4], can be used for numeric computations. It is implemented in the MATLAB function `acker`. The MATLAB function `place` is preferable for systems of high order because it is better conditioned numerically.

Example 7.5 Predator–prey

Consider the problem of regulating the population of an ecosystem by modulating the food supply. We use the predator–prey model introduced in Example 5.16 and described in more detail in Section 4.7. The dynamics for the system are given by

$$\frac{dH}{dt} = (r + u)H\left(1 - \frac{H}{k}\right) - \frac{aHL}{c + H}, \qquad H \geq 0,$$

$$\frac{dL}{dt} = b\frac{aHL}{c + H} - dL, \qquad\qquad L \geq 0.$$

We choose the following nominal parameters for the system, which correspond to the values used in previous simulations:

$$a = 3.2, \qquad b = 0.6, \qquad c = 50,$$
$$d = 0.56, \qquad k = 125 \qquad r = 1.6.$$

We take the parameter r, corresponding to the growth rate for hares, as the input to the system, which we might modulate by controlling a food source for the hares. This is reflected in our model by the term $(r + u)$ in the first equation, where here r represents a constant parameter (not the reference signal) and u represents the controlled modulation. We choose the number of lynxes L as the output of our system.

To control this system, we first linearize the system around the equilibrium point of the system (H_e, L_e), which can be determined numerically to be $x_e \approx (20.6, 29.5)$.

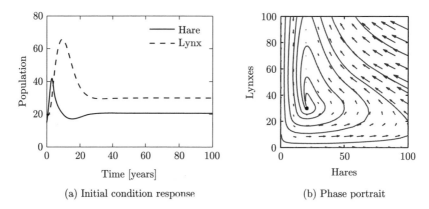

(a) Initial condition response   (b) Phase portrait

Figure 7.7: Simulation results for the controlled predator–prey system. The population of lynxes and hares as a function of time is shown in (a), and a phase portrait for the controlled system is shown in (b). Feedback is used to make the population stable at $H_e = 20.6$ and $L_e = 30$.

This yields a linear dynamical system

$$\frac{d}{dt}\begin{pmatrix} z_1 \\ z_2 \end{pmatrix} = \begin{pmatrix} 0.13 & -0.93 \\ 0.57 & 0 \end{pmatrix}\begin{pmatrix} z_1 \\ z_2 \end{pmatrix} + \begin{pmatrix} 17.2 \\ 0 \end{pmatrix}v, \qquad w = \begin{pmatrix} 0 & 1 \end{pmatrix}\begin{pmatrix} z_1 \\ z_2 \end{pmatrix},$$

where $z_1 = H - H_e$, $z_2 = L - L_e$, and $v = u$. It is easy to check that the system is reachable around the equilibrium point $(z, v) = (0, 0)$, and hence we can assign the eigenvalues of the system using state feedback.

Selecting the eigenvalues of the closed loop system requires balancing the ability to modulate the input against the natural dynamics of the system. This can be done by the process of trial and error or by using some of the more systematic techniques discussed in the remainder of the text. For now, we simply choose the desired closed loop eigenvalues to be at $\lambda = \{-0.1, -0.2\}$. We can then solve for the feedback gains using the techniques described earlier, which results in

$$K = \begin{pmatrix} 0.025 & -0.052 \end{pmatrix}.$$

Finally, we solve for the feedforward gain k_f, using equation (7.13) to obtain $k_f = 0.002$.

Putting these steps together, our control law becomes

$$v = -Kz + k_f L_d,$$

where L_d is the desired number of lynxes. In order to implement the control law, we must rewrite it using the original coordinates for the system, yielding

$$u = u_e - K(x - x_e) + k_f(L_d - y_e)$$

$$= -\begin{pmatrix} 0.025 & -0.052 \end{pmatrix}\begin{pmatrix} H - 20.6 \\ L - 29.5 \end{pmatrix} + 0.002\,(L_d - 29.5).$$

This rule tells us how much we should modulate u as a function of the current number of lynxes and hares in the ecosystem. Figure 7.7a shows a simulation of the resulting closed loop system using the parameters defined above and starting with an initial population of 15 hares and 20 lynxes. Note that the system stabilizes the population of lynxes at the reference value ($L_d = 30$). A phase portrait of the system is given in Figure 7.7b, showing how other initial conditions converge to the stabilized equilibrium population. Notice that the dynamics are very different from the natural dynamics (shown in Figure 4.20). ∇

The results of this section show that we can use state feedback to design the dynamics of a reachable system, under the strong assumption that we can measure all of the states. We shall address the availability of the states in the next chapter, when we consider output feedback and state estimation. In addition, Theorem 7.3, which states that the eigenvalues can be assigned to arbitrary locations, is also highly idealized and assumes that the dynamics of the process are known to high precision. The robustness of state feedback combined with state estimators is considered in Chapter 13 after we have developed the requisite tools.

7.3   DESIGN CONSIDERATIONS

The location of the eigenvalues determines the behavior of the closed loop dynamics, and hence where we place the eigenvalues is the main design decision to be made. As with all other feedback design problems, there are trade-offs among the magnitude of the control inputs, the robustness of the system to perturbations, and the closed loop performance of the system. In this section we examine some of these trade-offs starting with the special case of second-order systems.

Second-Order Systems

One class of systems that occurs frequently in the analysis and design of feedback systems is second-order linear differential equations. Because of their ubiquitous nature, it is useful to apply the concepts of this chapter to that specific class of systems and build more intuition about the relationship between stability and performance.

A canonical second-order system is a differential equation of the form

$$\ddot{q} + 2\zeta\omega_0\dot{q} + \omega_0^2 q = k\omega_0^2 u, \qquad y = q. \tag{7.22}$$

In state space form, this system can be represented as

$$\frac{dx}{dt} = \begin{pmatrix} 0 & \omega_0 \\ -\omega_0 & -2\zeta\omega_0 \end{pmatrix} x + \begin{pmatrix} 0 \\ k\omega_0 \end{pmatrix} u, \qquad y = \begin{pmatrix} 1 & 0 \end{pmatrix} x, \tag{7.23}$$

where $x = (q, \dot{q}/\omega_0)$ represents a normalized choice of states. The eigenvalues of this system are given by

$$\lambda = -\zeta\omega_0 \pm \omega_0\sqrt{(\zeta^2 - 1)},$$

and we see that the system is stable if $\omega_0 > 0$ and $\zeta > 0$. Note that the eigenvalues are complex if $\zeta < 1$ and real otherwise. Equations (7.22) and (7.23) can be used to describe many second-order systems, including damped oscillators, active filters, and flexible structures, as shown in the examples below.

The form of the solution depends on the value of ζ, which is referred to as the *damping ratio* for the system. If $\zeta > 1$, we say that the system is *overdamped*, and the natural response $(u = 0)$ of the system is given by

$$y(t) = \frac{\beta x_{10} + x_{20}}{\beta - \alpha} e^{-\alpha t} - \frac{\alpha x_{10} + x_{20}}{\beta - \alpha} e^{-\beta t},$$

where $\alpha = \omega_0(\zeta + \sqrt{\zeta^2 - 1})$ and $\beta = \omega_0(\zeta - \sqrt{\zeta^2 - 1})$. We see that the response consists of the sum of two exponentially decaying signals. If $\zeta = 1$, then the system is *critically damped* and solution becomes

$$y(t) = e^{-\zeta \omega_0 t}\big(x_{10} + (x_{20} + \zeta \omega_0 x_{10})t\big).$$

Note that this is still asymptotically stable as long as $\omega_0 > 0$, although the second term within the outer parentheses is increasing with time (but more slowly than the decaying exponential that is multiplying it).

Finally, if $0 < \zeta < 1$, then the solution is oscillatory and equation (7.22) is said to be *underdamped*. The natural response of the system is given by

$$y(t) = e^{-\zeta \omega_0 t}\left(x_{10} \cos \omega_{\mathrm{d}} t + \left(\frac{\zeta \omega_0}{\omega_{\mathrm{d}}} x_{10} + \frac{1}{\omega_{\mathrm{d}}} x_{20}\right) \sin \omega_{\mathrm{d}} t\right),$$

where $\omega_{\mathrm{d}} = \omega_0 \sqrt{1 - \zeta^2}$ is called the *damped frequency*. For $\zeta \ll 1$, $\omega_{\mathrm{d}} \approx \omega_0$ defines the oscillation frequency of the solution and ζ gives the damping rate relative to ω_0. The parameter ω_0 is referred to as the *natural frequency* of the system, stemming from the fact that for $\zeta = 0$ the oscillation frequency is given by ω_0.

Because of the simple form of a second-order system, it is possible to solve for the step and frequency responses in analytical form. The solution for the step response depends on the magnitude of ζ:

$$y(t) = \begin{cases} k\left(1 - e^{-\zeta \omega_0 t} \cos \omega_{\mathrm{d}} t - \dfrac{\zeta}{\sqrt{1 - \zeta^2}} e^{-\zeta \omega_0 t} \sin \omega_{\mathrm{d}} t\right), & \text{if } \zeta < 1; \\[2mm] k\left(1 - e^{-\omega_0 t}(1 + \omega_0 t)\right), & \text{if } \zeta = 1; \\[2mm] k\left(1 - \dfrac{1}{2}\left(\dfrac{\zeta}{\sqrt{\zeta^2 - 1}} + 1\right)e^{-\omega_0 t(\zeta - \sqrt{\zeta^2 - 1})}\right. \\[2mm] \qquad \left. + \dfrac{1}{2}\left(\dfrac{\zeta}{\sqrt{\zeta^2 - 1}} - 1\right)e^{-\omega_0 t(\zeta + \sqrt{\zeta^2 - 1})}\right), & \text{if } \zeta > 1, \end{cases} \qquad (7.24)$$

where we have taken $x(0) = 0$. Note that for the lightly damped case $(\zeta < 1)$ we have an oscillatory solution at frequency ω_{d}.

Step responses of systems with $k = 1$ and different values of ζ are shown in Figure 7.8. The shape of the response is determined by ζ, and the speed of the response is determined by ω_0 (included in the time axis scaling): the response is faster if ω_0 is larger.

In addition to the explicit form of the solution, we can also compute the properties of the step response that were defined in Section 6.3. For example, to compute the maximum overshoot for an underdamped system, we rewrite the output as

$$y(t) = k\left(1 - \frac{1}{\sqrt{1 - \zeta^2}} e^{-\zeta \omega_0 t} \sin(\omega_{\mathrm{d}} t + \varphi)\right), \qquad (7.25)$$

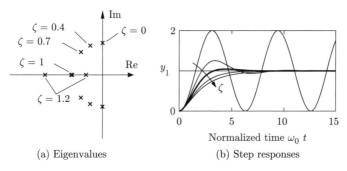

(a) Eigenvalues   (b) Step responses

Figure 7.8: Step response for a second-order system. Normalized step responses for the system (7.23) for $\zeta = 0$, 0.4, 0.7 (thicker), 1, and 1.2. As the damping ratio is increased, the rise time of the system gets longer, but there is less overshoot. The horizontal axis is in scaled units $\omega_0 t$; higher values of ω_0 result in a faster response (rise time and settling time).

Table 7.1: Properties of the step response for a second-order system $\ddot{q} + 2\zeta\omega_0\dot{q} + \omega_0^2 q = k\omega_0^2 u$ with $0 < \zeta \le 1$.

Property	Value	$\zeta = 0.5$	$\zeta = 1/\sqrt{2}$	$\zeta = 1$
Steady-state value	k	k	k	k
Rise time (inverse slope)	$T_r = e^{\varphi/\tan\varphi}/\omega_0$	$1.8/\omega_0$	$2.2/\omega_0$	$2.7/\omega_0$
Overshoot	$M_p = e^{-\pi\zeta/\sqrt{1-\zeta^2}}$	16%	4%	0%
Settling time (2%)	$T_s \approx 4/\zeta\omega_0$	$8.0/\omega_0$	$5.6/\omega_0$	$4.0/\omega_0$

where $\varphi = \arccos\zeta$. The maximum overshoot will occur at the first time in which the derivative of y is zero, at which time the fraction of the final value can be shown to be

$$M_p = e^{-\pi\zeta/\sqrt{1-\zeta^2}}.$$

The rise time is normally defined as the time for the step response to go from $p\%$ of its final value to $(100 - p)\%$. Typical values are $p = 5$ or 10%. An alternative definition is the inverse of the steepest slope: by differentiating equation (7.25) we find after straightforward but tedious calculations that

$$T_r = \frac{1}{\omega_0} e^{\varphi/\tan\varphi}, \qquad \varphi = \arccos\zeta.$$

Similar computations can be done for the other characteristics of a step response. Table 7.1 summarizes these calculations.

The frequency response for a second-order system can also be computed explicitly and is given by

$$M e^{i\theta} = \frac{k\omega_0^2}{(i\omega)^2 + 2\zeta\omega_0(i\omega) + \omega_0^2} = \frac{k\omega_0^2}{\omega_0^2 - \omega^2 + 2i\zeta\omega_0\omega}.$$

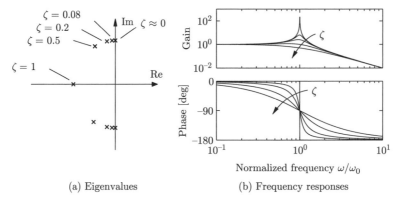

(a) Eigenvalues   (b) Frequency responses

Figure 7.9: Frequency response of a second-order system (7.23). (a) Eigenvalues as a function of ζ. (b) Frequency response as a function of ζ. The upper curve shows the gain ratio M, and the lower curve shows the phase shift θ. For small ζ there is a large peak in the magnitude of the frequency response and a rapid change in phase centered at $\omega = \omega_0$. As ζ is increased, the magnitude of the peak drops and the phase changes more smoothly between $0°$ and $-180°$.

Table 7.2: Properties of the frequency response for a second-order system $\ddot{q} + 2\zeta\omega_0\dot{q} + \omega_0^2 q = k\omega_0^2 u$ with $0 < \zeta \leq 1$.

Property	Value	$\zeta = 0.1$	$\zeta = 0.5$	$\zeta = 1/\sqrt{2}$
Zero frequency gain	M_0	k	k	k
Bandwidth	$\omega_b = \omega_0\sqrt{1 - 2\zeta^2 + \sqrt{(1-2\zeta^2)^2 + 1}}$	$1.54\,\omega_0$	$1.27\,\omega_0$	ω_0
Resonant peak gain	$M_r = \begin{cases} k/(2\zeta\sqrt{1-\zeta^2}) & \zeta \leq \sqrt{2}/2, \\ \text{N/A} & \zeta > \sqrt{2}/2 \end{cases}$	$5\,k$	$1.15\,k$	k
Resonant frequency	$\omega_{mr} = \begin{cases} \omega_0\sqrt{1-2\zeta^2} & \zeta \leq \sqrt{2}/2, \\ 0 & \zeta > \sqrt{2}/2 \end{cases}$	ω_0	$0.707\omega_0$	0

A graphical illustration of the frequency response is given in Figure 7.9. Notice the resonant peak that increases with decreasing ζ. The peak is often characterized by its Q-value, defined as $Q = 1/2\zeta$. The properties of the frequency response for a second-order system are summarized in Table 7.2.

Example 7.6 Drug administration
To illustrate the use of these formulas, consider the two-compartment model for drug administration, described in Section 4.6. The dynamics of the system are

$$\frac{dc}{dt} = \begin{pmatrix} -k_0 - k_1 & k_1 \\ k_2 & -k_2 \end{pmatrix} c + \begin{pmatrix} b_0 \\ 0 \end{pmatrix} u, \qquad y = \begin{pmatrix} 0 & 1 \end{pmatrix} c,$$

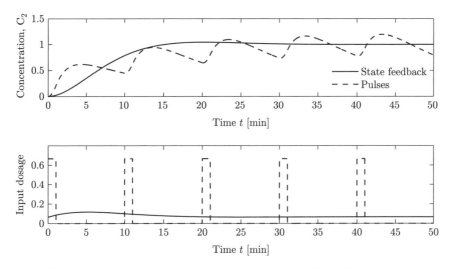

Figure 7.10: Open loop versus closed loop drug administration. Comparison between drug administration using a sequence of doses versus continuously monitoring the concentrations and adjusting the dosage continuously. In each case, the concentration is (approximately) maintained at the desired level, but the closed loop system has substantially less variability in drug concentration.

where c_1 and c_2 are the concentrations of the drug in each compartment, k_0, k_1, k_2, and b_0 are parameters of the system, u is the flow rate of the drug into compartment 1, and y is the concentration of the drug in compartment 2. We assume that we can measure the concentrations of the drug in each compartment, and we would like to design a feedback law to maintain the output at a given reference value r.

We choose $\zeta = 1/\sqrt{2}$ to minimize the overshoot and additionally require the rise time to be $T_r = 10$ min. Using the formulas in Table 7.1, this gives a value for $\omega_0 = 0.22$. We can now compute the gains to place the eigenvalues at this location. Setting $u = -Kx + k_f r$, the closed loop eigenvalues for the system satisfy

$$\lambda(s) = -0.2 \pm 0.096i.$$

Choosing $\tilde{k}_1 = -0.2$ and $\tilde{k}_2 = 0.2$, with $K = (\tilde{k}_1, \tilde{k}_2)$ to avoid confusion with the rates k_i in the dynamics matrix, gives the desired closed loop behavior. Equation (7.13) gives the feedforward gain $k_f = 0.065$. The response of the controller is shown in Figure 7.10 and compared with an open loop strategy involving administering periodic doses of the drug. $\qquad\nabla$

Higher-Order Systems

Our emphasis so far has considered only second-order systems. For higher-order systems, eigenvalue assignment is considerably more difficult, especially when trying to account for the many trade-offs that are present in a feedback design.

One of the other reasons why second-order systems play such an important role in feedback systems is that even for more complicated systems the response is often characterized by the *dominant eigenvalues*. To define these more precisely, consider a stable system with eigenvalues λ_j, $j = 1, \ldots, n$. We say that a complex conjugate pair of eigenvalues λ, λ^* is a *dominant pair* if they are the closest pair to the imaginary axis. In the case when multiple eigenvalues pairs are the same distance to the imaginary axis, a second criterion is to look at the relative damping of the system modes. We define the *damping ratio* for a complex eigenvalue λ as

$$\zeta = \frac{-\operatorname{Re} \lambda}{|\lambda|}.$$

Given multiple complex conjugate pairs with the same real part, the dominant pair will be the set with the lowest damping ratio.

Assuming that a system is stable, the dominant pair of eigenvalues tends to be the most important element of the response. To see this, assume that we have a system in Jordan form with a simple Jordan block corresponding to the dominant pair of eigenvalues:

$$\frac{dz}{dt} = \begin{pmatrix} \lambda & & & & \\ & \lambda^* & & & \\ & & J_2 & & \\ & & & \ddots & \\ & & & & J_k \end{pmatrix} z + Bu, \qquad y = Cz.$$

(Note that the state z may be complex because of the Jordan transformation.) The response of the system will be a linear combination of the responses from each of the individual Jordan subsystems. As we see from Figure 7.8, for $\zeta < 1$ the subsystem with the slowest response is precisely the one with eigenvalues closest to the imaginary axis. Hence, when we add the responses from each of the individual subsystems, it is the dominant pair of eigenvalues that will be the primary factor after the initial transients due to the other terms in the solution die out. While this simple analysis does not always hold (e.g., if some non-dominant terms have larger coefficients because of the particular form of the system), it is often the case that the dominant eigenvalues determine the (step) response of the system.

The only formal requirement for eigenvalue assignment is that the system be reachable. In practice there are many other constraints because the selection of eigenvalues has a strong effect on the magnitude and rate of change of the control signal. Large eigenvalues will in general require large control signals as well as fast changes of the signals. The capability of the actuators will therefore impose constraints on the possible location of closed loop eigenvalues. These issues will be discussed in depth in Chapters 12–14.

We illustrate some of the main ideas using the balance system as an example.

Example 7.7 Balance system
Consider the problem of stabilizing a balance system, whose dynamics were given in Example 7.2. The dynamics are given by

$$A = \begin{pmatrix} 0 & 0 & 1 & 0 \\ 0 & 0 & 0 & 1 \\ 0 & m^2 l^2 g/\mu & -cJ_t/\mu & -\gamma lm/\mu \\ 0 & M_t mgl/\mu & -clm/\mu & -\gamma M_t/\mu \end{pmatrix}, \qquad B = \begin{pmatrix} 0 \\ 0 \\ J_t/\mu \\ lm/\mu \end{pmatrix},$$

where $M_t = M + m$, $J_t = J + ml^2$, $\mu = M_t J_t - m^2 l^2$ and we have left c and γ nonzero. We use the following parameters for the system (corresponding roughly to a human being balanced on a stabilizing cart):

$$M = 10\,\text{kg}, \qquad m = 80\,\text{kg}, \qquad c = 0.1\,\text{N s/m},$$
$$J = 100\,\text{kg m}^2/\text{s}^2, \qquad l = 1\,\text{m}, \qquad \gamma = 0.01\,\text{N m s}, \qquad g = 9.8\,\text{m/s}^2.$$

The eigenvalues of the open loop dynamics are given by $\lambda \approx 0, -0.0011, \pm 2.68$. We have verified already in Example 7.2 that the system is reachable, and hence we can use state feedback to stabilize the system and provide a desired level of performance.

To decide where to place the closed loop eigenvalues, we note that the closed loop dynamics will roughly consist of two components: a set of fast dynamics that stabilize the pendulum in the inverted position and a set of slower dynamics that control the position of the cart. For the fast dynamics, we look to the natural period of the pendulum (in the hanging-down position), which is given by $\omega_0 = \sqrt{mgl/(J + ml^2)} \approx 2.1\,\text{rad/s}$. To provide a fast response we choose a damping ratio of $\zeta = 0.5$ and try to place the first pair of eigenvalues at $\lambda_{1,2} \approx -\zeta \omega_0 \pm i\omega_0 \approx -1 \pm 2i$, where we have used the approximation that $\sqrt{1 - \zeta^2} \approx 1$. For the slow dynamics, we choose the damping ratio to be 0.7 to provide a small overshoot and choose the natural frequency to be 0.5 to give a rise time of approximately 5 s. This gives eigenvalues $\lambda_{3,4} = -0.35 \pm 0.35i$.

The controller consists of feedback on the state and a feedforward gain for the reference input. The feedback gain is given by

$$K = \begin{pmatrix} -15.6 & 1730 & -50.1 & 443 \end{pmatrix},$$

which can be computed using Theorem 7.3 or using the MATLAB place command. The feedforward gain is $k_f = -1/(C(A - BK)^{-1}B) = -15.6$. The step response for the resulting controller (applied to the linearized system) is given in Figure 7.11a. While the step response gives the desired characteristics, the input required (lower left) is excessively large, almost three times the force of gravity at its peak.

To provide a more realistic response, we can redesign the controller to have slower dynamics. We see that the peak of the input force occurs on the fast time scale, and hence we choose to slow this down by approximately a factor of 3, leaving the damping ratio unchanged. We also slow down the second set of eigenvalues, with the intuition that we should move the position of the cart more slowly than we stabilize the pendulum dynamics. Leaving the damping ratio for the slow dynamics unchanged at 0.7 and changing the frequency to 1 (corresponding to a rise time of approximately 10 s), the desired eigenvalues become

$$\lambda = \{-0.33 \pm 0.66i, \ -0.18 \pm 0.18i\}.$$

The performance of the resulting controller is shown in Figure 7.11b. ∇

As we see from this example, it can be difficult to decide where to place the eigenvalues using state feedback. This is one of the principal limitations of this approach, especially for systems of higher dimension. Optimal control, such as the linear quadratic regulator problem discussed in Section 7.5, is one approach that is available. One can also focus on the frequency response for performing the design, which is the subject of Chapters 9–13.

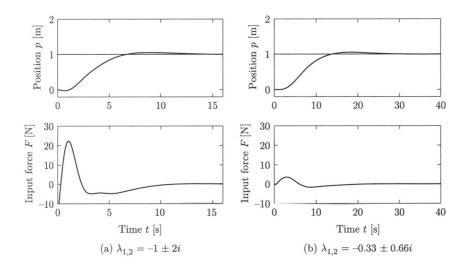

Figure 7.11: State feedback control of a balance system. The step response of a controller designed to give fast performance is shown in (a). Although the response characteristics (upper left) look very good, the input magnitude (lower left) is very large. Also note that the force is negative initially. A less aggressive controller is shown in (b). Here the response time is slowed down, but the input magnitude is much more reasonable. Both step responses are applied to the linearized dynamics.

7.4 INTEGRAL ACTION

Controllers based on state feedback achieve the correct steady-state response to command signals by careful calibration of the gain k_f. However, one of the primary uses of feedback is to allow good performance in the presence of uncertainty and hence requiring that we have an *exact* model of the process is undesirable. An alternative to calibration is to make use of integral feedback, in which the controller uses an integrator to provide zero steady-state error. The basic concept of integral feedback was introduced in Section 1.6 and discussed briefly in Sections 2.3 and 2.4; here we provide a more complete description and analysis.

System Augmentation

The basic approach in integral feedback is to create a state within the controller that computes the integral of the error signal, which is then used as a feedback term. We do this by augmenting the description of the system with a new state z, which is the integral of the difference between the the actual output y and desired output r. The augmented state equations become

$$\frac{d}{dt}\begin{pmatrix} x \\ z \end{pmatrix} = \begin{pmatrix} Ax + Bu \\ y - r \end{pmatrix} = \begin{pmatrix} Ax + Bu \\ Cx - r \end{pmatrix}. \tag{7.26}$$

Note that if we find a controller that stabilizes the system, then we will necessarily have $\dot{z} = 0$ in steady state and hence $y = r$ in steady state.

Given the augmented system, we design a state space controller in the usual fashion, with a control law of the form

$$u = -Kx - k_i z + k_f r, \qquad (7.27)$$

where K is the usual state feedback term, k_i is the integral term, and k_f is used to set the nominal input for the desired steady state. The resulting equilibrium point for the system is given by

$$x_e = -(A - BK)^{-1} B(k_f r - k_i z_e), \qquad Cx_e = r,$$

which comes from setting the right hand side of equation (7.26) to zero and substituting u from equation (7.27). Note that the value of z_e is not specified but rather will automatically settle to the value that makes $\dot{z} = y - r = 0$, which implies that at equilibrium the output will equal the reference value. This holds independently of the specific values of A, B, and K as long as the system is stable (which can be done through appropriate choice of K and k_i).

The final control law is given by

$$u = -Kx - k_i z + k_f r, \qquad \frac{dz}{dt} = y - r,$$

where we have now included the dynamics of the integrator as part of the specification of the controller. This type of control law is known as a *dynamic compensator* since it has its own internal dynamics. The following example illustrates the basic approach.

Example 7.8 Cruise control

Consider the cruise control example introduced in Section 1.5 and considered further in Example 6.11 (see also Section 4.1). The linearized dynamics of the process around an equilibrium point v_e, u_e are given by

$$\frac{dx}{dt} = -ax - b_g \theta + bw, \qquad y = v = x + v_e,$$

where $x = v - v_e$, $w = u - u_e$, m is the mass of the car, and θ is the angle of the road. The constants a, b, and b_g depend on the properties of the car and are given in Example 6.11.

If we augment the system with an integrator, the system dynamics become

$$\frac{dx}{dt} = -ax - b_g \theta + bw, \qquad \frac{dz}{dt} = y - v_r = v_e + x - v_r,$$

or, in state space form,

$$\frac{d}{dt} \begin{pmatrix} x \\ z \end{pmatrix} = \begin{pmatrix} -a & 0 \\ 1 & 0 \end{pmatrix} \begin{pmatrix} x \\ z \end{pmatrix} + \begin{pmatrix} b \\ 0 \end{pmatrix} w + \begin{pmatrix} -b_g \\ 0 \end{pmatrix} \theta + \begin{pmatrix} 0 \\ v_e - v_r \end{pmatrix}.$$

Note that when the system is at equilibrium, we have that $\dot{z} = 0$, which implies that the vehicle speed $v = v_e + x$ should be equal to the desired reference speed v_r. Our controller will be of the form

$$\frac{dz}{dt} = y - v_r, \qquad w = -k_p x - k_i z + k_f v_r,$$

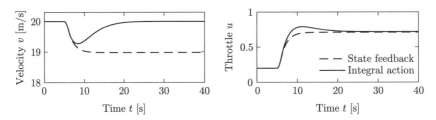

Figure 7.12: Velocity and throttle for a car with cruise control based on state feedback (dashed) and state feedback with integral action (solid). The controller with integral action is able to adjust the throttle to compensate for the effect of the hill and maintain the speed at the reference value of $v_r = 20 \, \text{m/s}$. The controller gains are $k_p = 0.5$ and $k_i = 0.1$.

and the gains k_p, k_i, and k_f will be chosen to stabilize the system and provide the correct input for the reference speed.

Assume that we wish to design the closed loop system to have the characteristic polynomial

$$\lambda(s) = s^2 + a_1 s + a_2.$$

Setting the disturbance $\theta = 0$, the characteristic polynomial of the closed loop system is given by

$$\det\big(sI - (A - BK)\big) = s^2 + (bk_p + a)s + bk_i,$$

and hence we set

$$k_p = \frac{a_1 - a}{b}, \qquad k_i = \frac{a_2}{b}, \qquad k_f = -1/\big(C(A - BK)^{-1}B\big) = \frac{a_1}{b}.$$

The resulting controller stabilizes the system and hence brings $\dot{z} = y - v_r$ to zero, resulting in perfect tracking. Notice that even if we have a small error in the values of the parameters defining the system, as long as the closed loop eigenvalues are still stable, then the tracking error will approach zero. Thus the exact calibration required in our previous approach (using k_f) is not needed here. Indeed, we can even choose $k_f = 0$ and let the feedback controller do all of the work. However, k_f does influence the transient response to reference signals and setting it properly will generally give a more favorable response.

Integral feedback can also be used to compensate for constant disturbances. Figure 7.12 shows the results of a simulation in which the car encounters a hill with angle $\theta = 4°$ at $t = 5 \, \text{s}$. The steady-state values of the throttle for a state feedback controller and a controller with integral action are very close, but the corresponding values of the car velocity are quite different. The reason for this is that the zero frequency gain from throttle to velocity is $-b/a = 130$ is high. The stability of the system is not affected by this external disturbance, and so we once again see that the car's velocity converges to the reference speed. This ability to handle constant disturbances is a general property of controllers with integral feedback (see Exercise 7.15). ∇

Reachability of the Augmented System

Eigenvalue assignment requires that the augmented system (7.26) is reachable. To explore this we compute the reachability matrix of the augmented system:

$$W_r = \begin{pmatrix} B & AB & \cdots & A^n B \\ 0 & CB & \cdots & CA^{n-1}B \end{pmatrix}.$$

To find the conditions for W_r to be of full rank, the matrix will be transformed by making column operations. Let a_k be the coefficients of the characteristic polynomial of the matrix A:

$$\lambda_A(s) = s^n + a_1 s^{n-1} + \cdots + a_{n-1}s + a_n.$$

Multiplying the first column by a_n, the second by a_{n-1}, through multiplication of the $(n$-1)th column by a_1 and then adding these to the last column of the matrix W_r, it follows from the Cayley–Hamilton theorem (Exercise 7.3) that the transformed matrix becomes

$$W_r = \begin{pmatrix} B & AB & \cdots & A^{n-1}B & 0 \\ 0 & CB & \cdots & CA^{n-2}B & b_n \end{pmatrix},$$

where

$$b_n = C(A^{n-1}B + a_1 A^{n-2}B + \cdots + a_{n-1}B). \tag{7.28}$$

If the matrix A is invertible, implying that there are no eigenvalues at the origin, then we can rewrite the formula for b_n as

$$b_n = CA^{-1}(A^n + a_1 A^{n-1} + \cdots + a_{n-1}A)B = -a_n CA^{-1}B,$$

where the final equality follows from a second application of the Cayley–Hamilton theorem. As long as the coefficient $b_n \neq 0$, then the system is reachable and it is possible to assign the eigenvalues of the augmented system to arbitrary values.

We will see in Chapter 9 that the coefficient b_n can be identified with a coefficient of the transfer function

$$G(s) = \frac{b_1 s^{n-1} + b_2 s^{n-2} + \cdots + b_n}{s^n + a_1 s^{n-1} + \cdots + a_n}.$$

The condition for reachability is thus that the original system does not contain a pure derivative in the input/output response.

7.5   LINEAR QUADRATIC REGULATORS

As an alternative to selecting the closed loop eigenvalue locations to accomplish a certain objective, the gains for a state feedback controller can instead be chosen by attempting to optimize a cost function. This can be particularly useful in helping balance the performance of the system with the magnitude of the inputs required to achieve that level of performance.

The linear quadratic regulator (LQR) problem is one of the most common optimal control problems. Given a multi-input linear system

$$\frac{dx}{dt} = Ax + Bu, \qquad x \in \mathbb{R}^n, \, u \in \mathbb{R}^p$$

with initial condition $x(0) = x_0$, we attempt to minimize the quadratic cost function

$$J(x_0) = \int_0^{t_f} \left(x^T Q_x x + u^T Q_u u \right) dt + x^T(t_f) Q_f x(t_f), \tag{7.29}$$

where $Q_x \succeq 0$, $Q_u \succ 0$ and $Q_f \succeq 0$ are symmetric, positive (semi-) definite matrices of the appropriate dimensions. This cost function represents a trade-off between the deviation of the state from the origin and the cost of the control input. By choosing the matrices Q_x, Q_u, and Q_f we can balance the rate of convergence of the solutions with the cost of the control.

The solution to the LQR problem is given by a linear control law of the form

$$u = -Kx, \quad K = Q_u^{-1} B^T S, \tag{7.30}$$

where $S \in \mathbb{R}^{n \times n}$ is a positive definite, symmetric matrix given by

$$-\frac{dS}{dt} = A^T S + SA - SBQ_u^{-1}B^T S + Q_x, \quad S(t_f) = Q_f. \tag{7.31}$$

This differential equation, called the *Riccati differential equation*, is integrated backwards in time starting with $S(t_f) = Q_f$. The minimal cost function, representing the optimal cost, is given by

$$\min_u \int_0^{t_f} \left(x^T Q_x x + u^T Q_u u \right) dt + x^T(t_f) Q_f x(t_f) = x^T(0) S(0) x(0). \tag{7.32}$$

The matrices A, B, Q_x, Q_u, and K may depend on time. A solution to the optimal control problem exists if the Riccati equation has a unique positive solution. The LQR approach is particularly well suited when linearizing around a trajectory, as will be done later in Section 8.5.

The LQR problem is simplified significantly if the time horizon is infinite and all matrices are constants, in which case S is a constant matrix given by the steady-state solution of (7.31):

$$A^T S + SA - SBQ_u^{-1}B^T S + Q_x = 0. \tag{7.33}$$

This equation is called the *algebraic Riccati equation*. If the system is reachable, it can be shown that there is a unique positive definite matrix S satisfying equation (7.33) that makes the closed loop system stable. The feedback gain $K = Q_u^{-1}B^T S$ is then also a constant matrix. The MATLAB command `lqr` returns K, S, and the dynamics matrix $E = A - BK$ of the closed loop system.

A key question in LQR design is how to choose the weights Q_x, Q_u, and Q_f. To guarantee that a solution exists, we must have $Q_x \succeq 0$ and $Q_u \succ 0$. In addition, there are certain "observability" conditions on Q_x that limit its choice. Here we assume $Q_x \succ 0$ to ensure that solutions to the algebraic Riccati equation always exist. To choose specific values for the cost function weights Q_x and Q_u, we must use our knowledge of the system we are trying to control. A particularly simple choice is to use diagonal weights

$$Q_x = \begin{pmatrix} q_1 & & 0 \\ & \ddots & \\ 0 & & q_n \end{pmatrix}, \quad Q_u = \begin{pmatrix} \rho_1 & & 0 \\ & \ddots & \\ 0 & & \rho_n \end{pmatrix}.$$

For this choice of Q_x and Q_u, the individual diagonal elements describe how much each state and input (squared) should contribute to the overall cost. Hence, we can take states that should remain small and attach higher weight values to them.

Similarly, we can penalize an input versus the states and other inputs through choice of the corresponding input weight ρ.

Example 7.9 Vectored thrust aircraft

Consider the original dynamics of the system (3.28), written in state space form as

$$
\frac{dz}{dt} = \begin{pmatrix} z_4 \\ z_5 \\ z_6 \\ -\frac{c}{m} z_4 \\ -g - \frac{c}{m} z_5 \\ 0 \end{pmatrix} + \begin{pmatrix} 0 \\ 0 \\ 0 \\ \frac{F_1}{m}\cos\theta - \frac{F_2}{m}\sin\theta \\ \frac{F_1}{m}\sin\theta + \frac{F_2}{m}\cos\theta \\ \frac{r}{J} F_1 \end{pmatrix}
$$

(see also Example 6.4). The system parameters are $m = 4\,\mathrm{kg}$, $J = 0.0475\,\mathrm{kg\,m^2}$, $r = 0.25\,\mathrm{m}$, $g = 9.8\,\mathrm{m/s^2}$, and $c = 0.05\,\mathrm{N\,s/m}$, which correspond to a scaled model of the system. The equilibrium point for the system is given by $F_1 = 0$, $F_2 = mg$, and $z_e = (x_e, y_e, 0, 0, 0, 0)$. To derive the linearized model near an equilibrium point, we compute the linearization according to equation (6.35):

$$
A = \begin{pmatrix} 0 & 0 & 0 & 1 & 0 & 0 \\ 0 & 0 & 0 & 0 & 1 & 0 \\ 0 & 0 & 0 & 0 & 0 & 1 \\ 0 & 0 & -g & -c/m & 0 & 0 \\ 0 & 0 & 0 & 0 & -c/m & 0 \\ 0 & 0 & 0 & 0 & 0 & 0 \end{pmatrix}, \qquad B = \begin{pmatrix} 0 & 0 \\ 0 & 0 \\ 0 & 0 \\ 1/m & 0 \\ 0 & 1/m \\ r/J & 0 \end{pmatrix},
$$

$$
C = \begin{pmatrix} 1 & 0 & 0 & 0 & 0 & 0 \\ 0 & 1 & 0 & 0 & 0 & 0 \end{pmatrix}, \qquad D = \begin{pmatrix} 0 & 0 \\ 0 & 0 \end{pmatrix}.
$$

Letting $\xi = z - z_e$ and $v = F - F_e$, the linearized system is given by

$$
\frac{d\xi}{dt} = A\xi + Bv, \qquad y = C\xi.
$$

It can be verified that the system is reachable.

To compute a linear quadratic regulator for the system, we write the cost function as

$$
J = \int_0^\infty (\xi^T Q_\xi \xi + v^T Q_v v)\,dt,
$$

where $\xi = z - z_e$ and $v = F - F_e$ again represent the local coordinates around the desired equilibrium point (z_e, F_e). We begin with diagonal matrices for the state and input costs:

$$
Q_\xi = \begin{pmatrix} 1 & 0 & 0 & 0 & 0 & 0 \\ 0 & 1 & 0 & 0 & 0 & 0 \\ 0 & 0 & 1 & 0 & 0 & 0 \\ 0 & 0 & 0 & 1 & 0 & 0 \\ 0 & 0 & 0 & 0 & 1 & 0 \\ 0 & 0 & 0 & 0 & 0 & 1 \end{pmatrix}, \qquad Q_v = \begin{pmatrix} \rho & 0 \\ 0 & \rho \end{pmatrix}.
$$

This gives a control law of the form $v = -K\xi$, which can then be used to derive the control law in terms of the original variables:

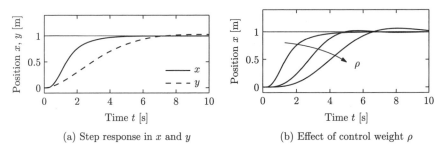

(a) Step response in x and y (b) Effect of control weight ρ

Figure 7.13: Step response for a vectored thrust aircraft with an LQR controller. The plot in (a) shows the x and y positions of the aircraft when it is commanded to move $1\,\mathrm{m}$ in each direction. In (b) the x motion is shown for control weights $\rho = 1$, 10^2, 10^4. A higher weight of the input term in the cost function causes a more sluggish response.

$$F = v + F_{\mathrm{e}} = -K(z - z_{\mathrm{e}}) + F_{\mathrm{e}}.$$

As computed in Example 6.4, the equilibrium points have $F_{\mathrm{e}} = (0, mg)$ and $z_{\mathrm{e}} = (x_{\mathrm{e}}, y_{\mathrm{e}}, 0, 0, 0, 0)$. The response of the controller to a step change in the desired position is shown in Figure 7.13a for $\rho = 1$. The response can be tuned by adjusting the weights in the LQR cost. Figure 7.13b shows the response in the x direction for different choices of the weight ρ. ∇

Linear quadratic regulators can also be designed for discrete-time systems, as illustrated by the following example.

Example 7.10 Web server control
Consider the web server example given in Section 4.4, where a discrete-time model for the system was given. We wish to design a control law that sets the server parameters so that the average server processor load is maintained at a desired level. Since other processes may be running on the server, the web server must adjust its parameters in response to changes in the load.

A block diagram for the control system is shown in Figure 7.14. We focus on the special case where we wish to control only the processor load using both the `KeepAlive` and `MaxClients` parameters. We also include a "disturbance" on the measured load that represents the use of the processing cycles by other processes running on the server. The system has the same basic structure as the generic control system in Figure 7.5, with the variation that the disturbance enters after the process dynamics.

The dynamics of the system are given by a set of difference equations of the form
$$x[k+1] = Ax[k] + Bu[k], \qquad y_{\mathrm{cpu}}[k] = x_{\mathrm{cpu}}[k] + d_{\mathrm{cpu}}[k],$$

where $x = (x_{\mathrm{cpu}}, x_{\mathrm{mem}})$ is the state of the web server, $u = (u_{\mathrm{ka}}, u_{\mathrm{mc}})$ is the input, d_{cpu} is the processing load from other processes on the computer, and y_{cpu} is the total processor load. The matrices $A \in \mathbb{R}^{2\times2}$ and $B \in \mathbb{R}^{2\times2}$ are described in Section 4.4.

We choose our controller to be a feedback controller of the form
$$u = -K \begin{pmatrix} y_{\mathrm{cpu}} \\ x_{\mathrm{mem}} \end{pmatrix} + k_{\mathrm{f}} r_{\mathrm{cpu}},$$

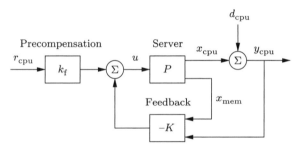

Figure 7.14: Feedback control of a web server. The controller sets the values of the web server parameters based on the difference between the nominal parameters (determined by $k_f r_{\text{cpu}}$) and the current load y_{cpu}. The disturbance d_{cpu} represents the load due to other processes running on the server. Note that the measurement is taken after the disturbance so that we measure the total load on the server.

where r_{cpu} is the desired processor load. Note that we have used the measured processor load y_{cpu} instead of the CPU state x_{cpu} to ensure that we adjust the system operation based on the actual load. (This modification is necessary because of the nonstandard way in which the disturbance enters the process dynamics.)

The feedback gain matrix K can be chosen by any of the methods described in this chapter. Here we use a linear quadratic regulator, with the cost function given by

$$Q_x = \begin{pmatrix} 5 & 0 \\ 0 & 1 \end{pmatrix}, \qquad Q_u = \begin{pmatrix} 1/50^2 & 0 \\ 0 & 1/1000^2 \end{pmatrix}.$$

The cost function for the state Q_x is chosen so that we place more emphasis on the processor load versus the memory usage. The cost function for the inputs Q_u is chosen so as to normalize the two inputs, with a `KeepAlive` timeout of $50\,\text{s}$ having the same weight as a `MaxClients` value of 1000. These values are squared since the cost associated with the inputs is given by $u^T Q_u u$. Using the dynamics in Section 4.4 and the `dlqr` command in MATLAB, the resulting gains become

$$K = \begin{pmatrix} -22.3 & 10.1 \\ 382.7 & 77.7 \end{pmatrix}.$$

As in the case of a continuous-time control system, the feedforward gain k_f is chosen to yield the desired operating point for the system. Setting $x[k+1] = x[k] = x_e$, the steady-state equilibrium point and output for a given reference input r are given by

$$x_e = (A - BK)x_e + Bk_f r, \qquad y_e = Cx_e.$$

This is a matrix equation in which k_f is a column vector that sets the two input values based on the desired reference. Since we have two inputs, we can set both the desired CPU load $y_{\text{cpu,e}}$ and the desired memory usage $x_{\text{mem,e}}$. If we take the desired equilibrium state to be of the form $x_e = (r, 0)$, where we choose the desired value of memory usage to be zero to make as much memory as possible available for other tasks, then we must solve

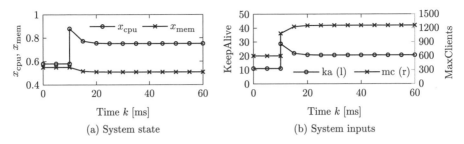

Figure 7.15: Web server with LQR control. The plot in (a) shows the state of the system under a change in external load applied at $k = 10\,\text{ms}$. The corresponding web server parameters (system inputs) are shown in (b). The controller is able to reduce the effect of the disturbance by approximately 40%.

$$\begin{pmatrix} r \\ 0 \end{pmatrix} = (A - BK - I)^{-1} B k_\text{f} \, r.$$

Solving this equation for k_f, we obtain

$$k_\text{f} = \left(\left((A - BK - I)^{-1} B \right) \right)^{-1} \begin{pmatrix} 1 \\ 0 \end{pmatrix} = \begin{pmatrix} 49.3 \\ 539.5 \end{pmatrix}.$$

The dynamics of the closed loop system are illustrated in Figure 7.15. We apply a change in load of $d_\text{cpu} = 0.3$ at time $t = 10\,\text{s}$, forcing the controller to adjust the operation of the server to attempt to maintain the desired load at 0.57. Note that both the KeepAlive and MaxClients parameters are adjusted. Although the load is decreased, it remains approximately 0.2 above the desired steady state. ▽

7.6   FURTHER READING

The importance of state models and state feedback was discussed in the seminal paper by Kalman [134], where the state feedback gain was obtained by solving an optimization problem that minimized a quadratic loss function. The notions of reachability and observability (Chapter 8) are also due to Kalman [136] (see also [100, 139]). Kalman defines controllability and reachability as the ability to reach the origin and an arbitrary state, respectively [138]. Reachability is also used in graph theory as the ability to get from one vertex to another. We note that in most textbooks the term "controllability" is used instead of "reachability," but we prefer the latter term because it is more descriptive of the fundamental property of being able to reach arbitrary states. The result that the eigenvalues of a reachable linear system could be placed in arbitrary positions was first realized by J. Bertram in 1959 [138], who worked in a control group at IBM Research led by Kalman. Bertram's results were based on root-locus analysis; an analytical proof was given in 1960 [209]. Most undergraduate textbooks on control contain material on state space systems, including, for example, Franklin, Powell, and Emami-Naeini [93] and Ogata [195]. Friedland's textbook [94] covers the material in the previous, current, and next chapter in considerable detail, including the topic of optimal control.

EXERCISES

7.1 (Double integrator) Consider the double integrator. Find a piecewise constant control strategy that drives the system from the origin to the state $x = (1, 1)$.

7.2 (Reachability from nonzero initial state) Extend the argument in Section 7.1 to show that if a system is reachable from an initial state of zero, it is reachable from a nonzero initial state.

7.3 (Cayley–Hamilton theorem) Let $A \in \mathbb{R}^{n \times n}$ be a matrix with characteristic polynomial $\lambda(s) = \det(sI - A) = s^n + a_1 s^{n-1} + \cdots + a_{n-1}s + a_n$. Show that the matrix A satisfies

$$\lambda(A) = A^n + a_1 A^{n-1} + \cdots + a_{n-1}A + a_n I = 0,$$

where the zero on the right hand side represents a matrix of elements with all zeros. Use this result to show that A^n can be written in terms of lower order powers of A and hence any matrix polynomial in A can be rewritten using terms of order at most $n - 1$.

7.4 (Unreachable systems) Consider a system with the state x and z described by the equations

$$\frac{dx}{dt} = Ax + Bu, \qquad \frac{dz}{dt} = Az + Bu.$$

If $x(0) = z(0)$ it follows that $x(t) = z(t)$ for all t regardless of the input that is applied. Show that this violates the definition of reachability and further show that the reachability matrix W_r is not full rank.

7.5 (Rear-steered bicycle) A simple model for a bicycle was given by equation (4.5) in Section 4.2. A model for a bicycle with rear-wheel steering is obtained by reversing the sign of the velocity in the model. Determine the conditions under which this systems is reachable and explain any situations in which the system is not reachable.

7.6 (Characteristic polynomial for reachable canonical form) Show that the characteristic polynomial for a system in reachable canonical form is given by equation (7.7) and that

$$\frac{d^n z_k}{dt^n} + a_1 \frac{d^{n-1}z_k}{dt^{n-1}} + \cdots + a_{n-1}\frac{dz_k}{dt} + a_n z_k = \frac{d^{n-k}u}{dt^{n-k}},$$

where z_k is the kth state.

7.7 (Reachability matrix for reachable canonical form) Consider a system in reachable canonical form. Show that the inverse of the reachability matrix is given by

$$\tilde{W}_\mathrm{r}^{-1} = \begin{pmatrix} 1 & a_1 & a_2 & \cdots & a_{n-1} \\ & 1 & a_1 & \cdots & a_{n-2} \\ & & 1 & \ddots & \vdots \\ & 0 & & \ddots & a_1 \\ & & & & 1 \end{pmatrix}.$$

7.8 (Non-maintainable equilibrium points) Consider the normalized model of a pendulum on a cart

$$\frac{d^2x}{dt^2} = u, \qquad \frac{d^2\theta}{dt^2} = -\theta + u,$$

where x is cart position and θ is pendulum angle. Can the angle $\theta = \theta_0$ for $\theta_0 \neq 0$ be maintained?

7.9 (Eigenvalue assignment) Consider the system

$$\frac{dx}{dt} = Ax + Bu = \begin{pmatrix} -1 & 0 \\ 1 & 0 \end{pmatrix} x + \begin{pmatrix} a-1 \\ 1 \end{pmatrix} u,$$

with $a = 1.25$. Design a state feedback that gives $\det(sI - BK) = s^2 + 2\zeta_c\omega_c s + \omega_c^2$, where $\omega_c = 5$, and $\zeta_c = 0.6$.

7.10 (Eigenvalue assignment for unreachable system) Consider the system

$$\frac{dx}{dt} = \begin{pmatrix} 0 & 1 \\ 0 & 0 \end{pmatrix} x + \begin{pmatrix} 1 \\ 0 \end{pmatrix} u, \qquad y = \begin{pmatrix} 1 & 0 \end{pmatrix} x,$$

with the control law

$$u = -k_1 x_1 - k_2 x_2 + k_f r.$$

Compute the rank of the reachability matrix for the system and show that eigenvalues of the system cannot be assigned to arbitrary values.

7.11 (Motor drive) Consider the normalized model of the motor drive in Exercise 3.7. Using the following normalized parameters,

$$J_1 = 10/9, \qquad J_2 = 10, \qquad c = 0.1, \qquad k = 1, \qquad k_I = 1,$$

verify that the eigenvalues of the open loop system are $0, 0, -0.05 \pm i$. Design a state feedback that gives a closed loop system with eigenvalues -2, -1, and $-1 \pm i$. This choice implies that the oscillatory eigenvalues will be well damped and that the eigenvalues at the origin are replaced by eigenvalues on the negative real axis. Simulate the responses of the closed loop system to step changes in the reference signal for θ_2 and a step change in a disturbance torque on the second rotor.

7.12 (Whipple bicycle model) Consider the Whipple bicycle model given by equation (4.8) in Section 4.2. Using the parameters from the companion web site, the model is unstable at the velocity $v_0 = 5\,\text{m/s}$ and the open loop eigenvalues are -1.84, -14.29, and $1.30 \pm 4.60i$. Find the gains of a controller that stabilizes the bicycle and gives closed loop eigenvalues at -2, -10, and $-1 \pm i$. Simulate the response of the system to a step change in the steering reference of $0.002\,\text{rad}$.

7.13 (Dominant eigenvalues) Consider the following two linear systems:

$$\Sigma_1: \begin{aligned} \frac{dx}{dt} &= \begin{pmatrix} -1.1 & -0.1 \\ 1 & 0 \end{pmatrix} x + \begin{pmatrix} 1 \\ 0 \end{pmatrix} u, \\ y &= \begin{pmatrix} 1.01 & 0.11 \end{pmatrix} x, \end{aligned} \qquad \Sigma_2: \begin{aligned} \frac{dx}{dt} &= \begin{pmatrix} -1.1 & -0.1 \\ 1 & 0 \end{pmatrix} x + \begin{pmatrix} 1 \\ 0 \end{pmatrix} u, \\ y &= \begin{pmatrix} 1.1 & 1.01 \end{pmatrix} x. \end{aligned}$$

Show that although both systems have the same eigenvalues, the step responses of the two systems are dominated by different sets of eigenvalues.

7.14 Consider the second-order system

$$\frac{d^2y}{dt^2} + 0.5\frac{dy}{dt} + y = a\frac{du}{dt} + u.$$

Let the initial conditions be zero.

a) Show that the initial slope of the unit step response is a. Discuss what it means when $a < 0$.

b) Show that there are points on the unit step response that are invariant with a. Discuss qualitatively the effect of the parameter a on the solution.

c) Simulate the system and explore the effect of a on the rise time and overshoot.

7.15 (Integral feedback for rejecting constant disturbances) Consider a linear system of the form

$$\frac{dx}{dt} = Ax + Bu + Fd, \qquad y = Cx,$$

where u is a scalar and v is a disturbance that enters the system through a disturbance vector $F \in \mathbb{R}^n$. Assume that the matrix A is invertible and the zero frequency gain $CA^{-1}B$ is nonzero. Show that integral feedback can be used to compensate for a constant disturbance by giving zero steady-state output error even when $d \neq 0$.

7.16 (Bryson's rule) Bryson and Ho [58] have suggested the following method for choosing the matrices Q_x and Q_u in equation (7.29). Start by choosing Q_x and Q_u as diagonal matrices whose elements are the inverses of the squares of the maxima of the corresponding variables. Then modify the elements to obtain a compromise among response time, damping, and control effort. Apply this method to the motor drive in Exercise 7.11. Assume that the largest values of the φ_1 and φ_2 are 1, the largest values of $\dot{\varphi}_1$ and $\dot{\varphi}_2$ are 2, and the largest control signal is 10. Simulate the closed loop system for $\varphi_2(0) = 1$ and all other states are initialized to 0. Explore the effects of different values of the diagonal elements for Q_x and Q_u.

7.17 (LQR proof) Use the Riccati equation (7.31) and the relation

$$x^T(t_f)Q_f x(t_f) - x^T(0)S(0)x(0) =$$
$$\int_0^{t_f} \left(\dot{x}^T(t)S(t)x(t) + x^T \dot{S}(t)x(t) + x^T(t)S(t)\dot{x}(t) \right) dt$$

to show that the cost function for the linear quadratic regulator problem can be written as

$$\int_0^{t_f} \left(x^T(t)Q_x x(t) + u^T(t)Q_u u(t) \right) dt + x^T(t_f)Q_f x(t_f)$$
$$= x^T(0)S(0)x(0) + \int_0^{t_f} \left(u(t) + Q_u^{-1}B^T S(t)x(t) \right)^T Q_u \left(u(t) + Q_u^{-1}B^T S(t)x(t) \right) dt,$$

from which it follows that the control law $u(t) = -Kx(t) = -Q_u^{-1}B^T S(t)x(t)$ is optimal. Does the proof hold when all matrices depend on time?

Chapter Eight

Output Feedback

> One may separate the problem of physical realization into two stages:
> computation of the "best approximation" $\hat{x}(t_1)$ of the state from
> knowledge of $y(t)$ for $t \leq t_1$ and computation of $u(t_1)$ given $\hat{x}(t_1)$.
>
> —R. E. Kalman, "Contributions to the Theory
> of Optimal Control," 1960 [134].

In this chapter we show how to use output feedback to modify the dynamics of
the system through the use of observers. We introduce the concept of observability
and show that if a system is observable, it is possible to recover the state from
measurements of the inputs and outputs to the system. We then show how to
design a controller with feedback from the observer state. A general controller with
two degrees of freedom is obtained by adding feedforward. We illustrate by outlining
a controller for a nonlinear system that also employs gain scheduling.

8.1 OBSERVABILITY

In Section 7.2 of the previous chapter it was shown that it is possible to find a state
feedback law that gives desired closed loop eigenvalues provided that the system is
reachable and that all the states are measured by sensors. For many situations, it
is highly unrealistic to assume that all the states are measured. In this section we
investigate how the state can be estimated by using a mathematical model and a
few measurements. It will be shown that computation of the states can be carried
out by a dynamical system called an *observer*.

Definition of Observability

Consider a system described by a set of differential equations

$$\frac{dx}{dt} = Ax + Bu, \qquad y = Cx + Du, \tag{8.1}$$

where $x \in \mathbb{R}^n$ is the state, $u \in \mathbb{R}^p$ the input, and $y \in \mathbb{R}^q$ the measured output. We
wish to estimate the state of the system from its inputs and outputs, as illustrated
in Figure 8.1. In some situations we will assume that there is only one measured
signal, i.e., that the signal y is a scalar and that C is a (row) vector. This signal
may be corrupted by noise w, although we shall start by considering the noise-free
case. We write $\hat{x}$ for the state estimate given by the observer.

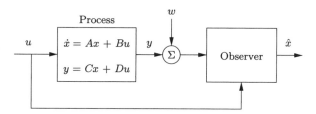

Figure 8.1: Block diagram for an observer. The observer uses the process measurement y (possibly corrupted by noise w) and the input u to estimate the current state of the process, denoted $\hat{x}$.

Definition 8.1 (Observability). A linear system is *observable* if for every $T > 0$ it is possible to determine the state of the system $x(T)$ through measurements of $y(t)$ and $u(t)$ on the interval $[0, T]$.

The definition above holds for nonlinear systems as well, and the results discussed here have extensions to the nonlinear case.

The problem of observability is one that has many important applications, even outside feedback systems. If a system is observable, then there are no "hidden" dynamics inside it; we can understand everything that is going on through observation (over time) of the inputs and outputs. As we shall see, the problem of observability is of significant practical interest because it will determine if a set of sensors is sufficient for controlling a system. Sensors combined with a mathematical model of the system can also be viewed as a "virtual sensor" that gives information about variables that are not measured directly. The process of reconciling signals from many sensors using mathematical models is also called *sensor fusion*.

Testing for Observability

When discussing reachability in the previous chapter, we neglected the output and focused on the state. Similarly, it is convenient here to initially neglect the input and focus on the autonomous system

$$\frac{dx}{dt} = Ax, \qquad y = Cx, \tag{8.2}$$

where $x \in \mathbb{R}^n$ and $y \in \mathbb{R}^q$. We wish to understand when it is possible to determine the state from observations of the output.

The output itself gives the projection of the state onto vectors that are rows of the matrix C. The observability problem can immediately be solved if $n = q$ (number of outputs equals number of states) and the matrix C is invertible. If the matrix is not square and invertible, we can take derivatives of the output to obtain

$$\frac{dy}{dt} = C\frac{dx}{dt} = CAx.$$

From the derivative of the output we thus get the projection of the state on vectors that are rows of the matrix CA. Proceeding in this way, we get at every time t

$$
\begin{pmatrix} y(t) \\ \dot{y}(t) \\ \ddot{y}(t) \\ \vdots \\ y^{(n-1)}(t) \end{pmatrix} = \begin{pmatrix} C \\ CA \\ CA^2 \\ \vdots \\ CA^{n-1} \end{pmatrix} x(t). \tag{8.3}
$$

We thus find that the state at time t can be determined from the output and its derivatives at time t if the *observability matrix*

$$
W_{\mathrm{o}} = \begin{pmatrix} C \\ CA \\ CA^2 \\ \vdots \\ CA^{n-1} \end{pmatrix} \tag{8.4}
$$

has full row rank (n independent rows). As in the case of reachability, it turns out that we need not consider any derivatives higher than $n-1$ (this is an application of the Cayley–Hamilton theorem [Exercise 7.3]).

The calculation can easily be extended to systems with inputs and many measured signals. The state is then given by a linear combination of inputs and outputs and their higher derivatives. The observability criterion is unchanged. We leave this case as an exercise for the reader.

In practice, differentiation of the output can give large errors when there is measurement noise, and therefore the method sketched above is not particularly practical. We will address this issue in more detail in the next section, but for now we have the following basic result.

Theorem 8.1 (Observability rank condition). *A linear system of the form* (8.1) *is observable if and only if the observability matrix W_o is full row rank.*

 Proof. The sufficiency of the observability rank condition follows from the previous analysis. To prove necessity, suppose that the system is observable but W_{o} is not full row rank. Let $v \in \mathbb{R}^n$, $v \neq 0$, be a vector in the null space of W_{o}, so that $W_{\mathrm{o}}v = 0$. (Such a v exists using the fact that the row and column rank of a matrix are always equal.) If we let $x(0) = v$ be the initial condition for the system and choose $u = 0$, then the output is given by $y(t) = Ce^{At}v$. Since e^{At} can be written as a power series in A and since A^n and higher powers can be rewritten in terms of lower powers of A (by the Cayley–Hamilton theorem), it follows that $y(t)$ will be identically zero (the reader should fill in the missing steps). However, if both the input and output of the system are zero, then a valid estimate of the state is $\hat{x} = 0$ for all time, which is clearly incorrect since $x(0) = v \neq 0$. Hence by contradiction we must have that W_{o} is full row rank if the system is observable. $\qquad\square$

Example 8.1 Compartment model

Consider the two-compartment model in Figure 4.18a, but assume that only the concentration in the first compartment can be measured. The system is described by the linear system

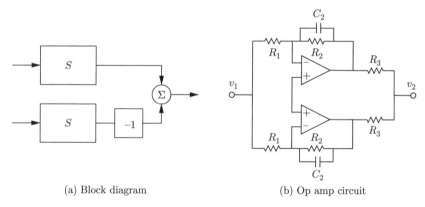

(a) Block diagram   (b) Op amp circuit

Figure 8.2: An unobservable system. Two identical subsystems have outputs that add together to form the overall system output. The individual states of the subsystem cannot be determined since the contributions of each to the output are not distinguishable. The circuit diagram on the right is an example of such a system.

$$\frac{dc}{dt} = \begin{pmatrix} -k_0 - k_1 & k_1 \\ k_2 & -k_2 \end{pmatrix} c + \begin{pmatrix} b_0 \\ 0 \end{pmatrix} u, \qquad y = \begin{pmatrix} 1 & 0 \end{pmatrix} c.$$

The first compartment represents the drug concentration in the blood plasma, and the second compartment the drug concentration in the tissue where it is active. To determine if it is possible to find the concentration in the tissue compartment from a measurement of blood plasma, we investigate the observability of the system by forming the observability matrix

$$W_o = \begin{pmatrix} C \\ CA \end{pmatrix} = \begin{pmatrix} 1 & 0 \\ -k_0 - k_1 & k_1 \end{pmatrix}.$$

The rows are linearly independent if $k_1 \neq 0$, and under this condition it is thus possible to determine the concentration of the drug in the active compartment from measurements of the drug concentration in the blood. $\qquad \nabla$

It is useful to have an understanding of the mechanisms that make a system unobservable. Such a system is shown in Figure 8.2. The system is composed of two identical systems whose outputs are subtracted. It seems intuitively clear that it is not possible to deduce the states from the output since we cannot deduce the individual output contributions from the difference. This can also be seen formally (Exercise 8.2).

Observable Canonical Form

As in the case of reachability, certain canonical forms will be useful in studying observability. A linear single-input, single-output state space system is in *observable canonical form* if its dynamics are given by

$$\frac{dz}{dt} = \begin{pmatrix} -a_1 & 1 & & 0 \\ -a_2 & 0 & \ddots & \\ \vdots & & \ddots & 1 \\ -a_n & 0 & & 0 \end{pmatrix} z + \begin{pmatrix} b_1 \\ b_2 \\ \vdots \\ b_n \end{pmatrix} u,$$

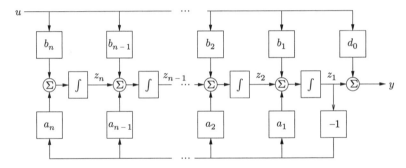

Figure 8.3: Block diagram of a system in observable canonical form. The states of the system are represented by individual integrators whose inputs are a weighted combination of the next integrator in the chain, the first state (rightmost integrator), and the system input. The output is a combination of the first state and the input. Compare with the block diagram of the system in reachable form in Figure 7.4.

$$y = \begin{pmatrix} 1 & 0 & \cdots & 0 \end{pmatrix} z + d_0 \, u.$$

This definition can be extended to systems with many inputs; the only difference is that the vector multiplying u is replaced by a matrix.

Figure 8.3 is a block diagram for a system in observable canonical form. As in the case of reachable canonical form, we see that the coefficients in the system description appear directly in the block diagram. The characteristic polynomial for a system in observable canonical form is

$$\lambda(s) = s^n + a_1 s^{n-1} + \cdots + a_{n-1} s + a_n. \tag{8.5}$$

It is possible to reason about the observability of a system in observable canonical form by studying the block diagram. If the input u and the output y are available, the state z_1 can clearly be computed. Differentiating z_1, we obtain the input to the integrator that generates z_1, and we can now obtain $z_2 = \dot{z}_1 + a_1 z_1 - b_1 u$. Proceeding in this way, we can compute all states. The computation will, however, require that the signals be differentiated.

To check observability more formally, we compute the observability matrix for a system in observable canonical form, which is given by

$$\tilde{W}_o = \begin{pmatrix} 1 & & & & 0 \\ -a_1 & 1 & & & \\ -a_1^2 - a_2 & -a_1 & 1 & & \\ \vdots & \vdots & & \ddots & \\ * & * & & \cdots & 1 \end{pmatrix},$$

where $*$ represents an entry whose exact value is not important. The columns of this matrix are linearly independent (since it is lower triangular), and hence W_o is invertible. A straightforward but tedious calculation shows that the inverse of the observability matrix has a simple form given by

$$\tilde{W}_{\mathrm{o}}^{-1} = \begin{pmatrix} 1 & & & & \\ a_1 & 1 & & & \\ a_2 & a_1 & 1 & & \\ \vdots & \vdots & & \ddots & \\ & & & & 1 \\ a_{n-1} & a_{n-2} & \cdots & a_1 & 1 \end{pmatrix}.$$

As in the case of reachability, it turns out that a system is observable if and only if there exists a transformation T that converts the system into observable canonical form. This is useful for proofs since it lets us assume that a system is in observable canonical form without any loss of generality. The observable canonical form may be poorly conditioned numerically.

8.2   STATE ESTIMATION

Having defined the concept of observability, we now return to the question of how to construct an observer for a system. We will look for observers that can be represented as a linear dynamical system that takes the inputs and outputs of the system we are observing and produces an estimate of the system's state. That is, we wish to construct a dynamical system of the form

$$\frac{d\hat{x}}{dt} = F\hat{x} + Gu + Hy,$$

where u and y are the input and output of the original system and $\hat{x} \in \mathbb{R}^n$ is an estimate of the state with the property that $\hat{x}(t) \to x(t)$ as $t \to \infty$.

The Observer

We consider the system in equation (8.1) with D set to zero to simplify the exposition:

$$\frac{dx}{dt} = Ax + Bu, \qquad y = Cx. \tag{8.6}$$

We can attempt to determine the state simply by simulating the equations with the correct input. An estimate of the state is then given by

$$\frac{d\hat{x}}{dt} = A\hat{x} + Bu. \tag{8.7}$$

To find the properties of this estimate, introduce the estimation error $\tilde{x} = x - \hat{x}$. It follows from equations (8.6) and (8.7) that

$$\frac{d\tilde{x}}{dt} = A\tilde{x}.$$

If the dynamics matrix A has all its eigenvalues in the left half-plane, the error $\tilde{x}$ will go to zero, and hence equation (8.7) is a dynamical system whose output converges to the state of the system (8.6). However, the convergence might be slower than desired.

The observer given by equation (8.7) uses only the process input u; the measured signal does not appear in the equation. We must also require that the system be stable, and essentially our estimator converges because the transient dynamics of both the observer and the estimator are going to zero. This is not very useful in a control design context since we want to have our estimate converge quickly to a nonzero state so that we can make use of it in our controller. We will therefore attempt to modify the observer so that the output is used and its convergence properties can be designed to be fast relative to the system's dynamics. This version will also work for unstable systems.

Consider the observer

$$\frac{d\hat{x}}{dt} = A\hat{x} + Bu + L(y - C\hat{x}). \tag{8.8}$$

This can be considered as a generalization of equation (8.7). Feedback from the measured output is provided by adding the term $L(y - C\hat{x})$, which is proportional to the difference between the observed output and the output predicted by the observer. It follows from equations (8.6) and (8.8) that

$$\frac{d\tilde{x}}{dt} = (A - LC)\tilde{x}.$$

If the matrix L can be chosen in such a way that the matrix $A - LC$ has eigenvalues with negative real parts, the error $\tilde{x}$ will go to zero. The convergence rate is determined by an appropriate selection of the eigenvalues.

Notice the similarity between the problems of finding a state feedback and finding the observer. State feedback design by eigenvalue assignment is equivalent to finding a matrix K so that $A - BK$ has given eigenvalues. Designing an observer with prescribed eigenvalues is equivalent to finding a matrix L so that $A - LC$ has given eigenvalues. Since the eigenvalues of a matrix and its transpose are the same we can establish the following equivalences:

$$A \leftrightarrow A^T, \qquad B \leftrightarrow C^T, \qquad K \leftrightarrow L^T, \qquad W_{\mathrm{r}} \leftrightarrow W_{\mathrm{o}}^T. \tag{8.9}$$

The observer design problem is the *dual* of the state feedback design problem. Using the results of Theorem 7.3, we get the following theorem on observer design.

Theorem 8.2 (Observer design by eigenvalue assignment). *Consider the system given by*

$$\frac{dx}{dt} = Ax + Bu, \qquad y = Cx, \tag{8.10}$$

with one input and one output. Let $\lambda(s) = s^n + a_1 s^{n-1} + \cdots + a_{n-1}s + a_n$ be the characteristic polynomial for A. If the system is observable, then the dynamical system

$$\frac{d\hat{x}}{dt} = A\hat{x} + Bu + L(y - C\hat{x}) \tag{8.11}$$

is an observer for the system, with L chosen as

$$L = W_o^{-1} \widetilde{W}_o \begin{pmatrix} p_1 - a_1 \\ p_2 - a_2 \\ \vdots \\ p_n - a_n \end{pmatrix} \tag{8.12}$$

and the matrices W_o and $\widetilde{W}_o$ given by

$$W_o = \begin{pmatrix} C \\ CA \\ \vdots \\ CA^{n-1} \end{pmatrix}, \qquad \tilde{W}_o = \begin{pmatrix} 1 & & & & \\ a_1 & 1 & & 0 & \\ a_2 & a_1 & 1 & & \\ \vdots & \vdots & & \ddots & \\ & & & & 1 \\ a_{n-1} & a_{n-2} & \cdots & a_1 & 1 \end{pmatrix}^{-1}.$$

The resulting observer error $\tilde{x} = x - \hat{x}$ is governed by a differential equation having the characteristic polynomial

$$p(s) = s^n + p_1 s^{n-1} + \cdots + p_n.$$

The dynamical system (8.11) is called an *observer* for (the states of) the system (8.10) because it will generate an approximation of the states of the system from its inputs and outputs. This form of an observer is a much more useful form than the one given by pure differentiation in equation (8.3).

Example 8.2 Compartment model

Consider the compartment model in Example 8.1, which is characterized by the matrices

$$A = \begin{pmatrix} -k_0 - k_1 & k_1 \\ k_2 & -k_2 \end{pmatrix}, \qquad B = \begin{pmatrix} b_0 \\ 0 \end{pmatrix}, \qquad C = \begin{pmatrix} 1 & 0 \end{pmatrix}.$$

The observability matrix was computed in Example 8.1, where we concluded that the system was observable if $k_1 \neq 0$. The dynamics matrix has the characteristic polynomial

$$\lambda(s) = \det \begin{pmatrix} s + k_0 + k_1 & -k_1 \\ -k_2 & s + k_2 \end{pmatrix} = s^2 + (k_0 + k_1 + k_2)s + k_0 k_2.$$

Letting the desired characteristic polynomial of the observer be $s^2 + p_1 s + p_2$, equation (8.12) gives the observer gain

$$L = \begin{pmatrix} 1 & 0 \\ -k_0 - k_1 & k_1 \end{pmatrix}^{-1} \begin{pmatrix} 1 & 0 \\ k_0 + k_1 + k_2 & 1 \end{pmatrix}^{-1} \begin{pmatrix} p_1 - k_0 - k_1 - k_2 \\ p_2 - k_0 k_2 \end{pmatrix}$$

$$= \begin{pmatrix} p_1 - k_0 - k_1 - k_2 \\ (p_2 - p_1 k_2 + k_1 k_2 + k_2^2)/k_1 \end{pmatrix}.$$

Notice that the observability condition $k_1 \neq 0$ is essential. The behavior of the observer is illustrated by the simulation in Figure 8.4b. Notice how the observed concentrations approach the true concentrations.   ▽

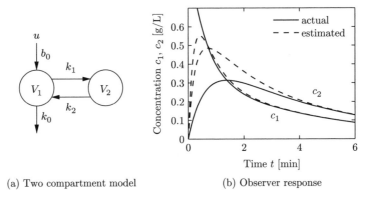

(a) Two compartment model   (b) Observer response

Figure 8.4: Observer for a two compartment system. A two compartment model is shown on the left. The observer measures the input concentration u and output concentration $y = c_1$ to determine the compartment concentrations, shown on the right. The true concentrations are shown by solid lines and the estimates generated by the observer by dashed lines.

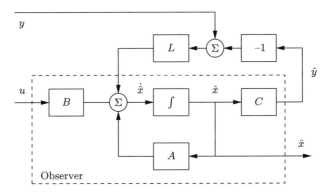

Figure 8.5: Block diagram of an observer. The observer takes the signals y and u as inputs and produces an estimate x. Notice that the observer contains a copy of the process model that is driven by $y - \hat{y}$ through the observer gain L.

The observer is a dynamical system whose inputs are the process input u and the process output y. The rate of change of the estimate is composed of two terms. One term, $A\hat{x} + Bu$, is the rate of change computed from the model with $\hat{x}$ substituted for x. The other term, $L(y - \hat{y})$, is proportional to the difference $e = y - \hat{y}$ between measured output y and its estimate $\hat{y} = C\hat{x}$. The observer gain L is a matrix that determines how the error e is weighted and distributed among the state estimates. The observer thus combines measurements with a dynamical model of the system. A block diagram of the observer is shown in Figure 8.5.

Computing the Observer Gain

For simple low-order problems it is convenient to introduce the elements of the observer gain L as unknown parameters and solve for the values required to give the desired characteristic polynomial, as illustrated in the following example.

Example 8.3 Vehicle steering

The normalized linear model for vehicle steering derived in Examples 6.13 and 7.4 gives the following state space model dynamics relating lateral path deviation y to steering angle u:

$$\frac{dx}{dt} = \begin{pmatrix} 0 & 1 \\ 0 & 0 \end{pmatrix} x + \begin{pmatrix} \gamma \\ 1 \end{pmatrix} u, \qquad y = \begin{pmatrix} 1 & 0 \end{pmatrix} x. \tag{8.13}$$

Recall that the state x_1 represents the lateral path deviation and that x_2 represents the turning rate. We will now derive an observer that uses the system model to determine the turning rate from the measured path deviation.

The observability matrix is

$$W_{\mathrm{o}} = \begin{pmatrix} 1 & 0 \\ 0 & 1 \end{pmatrix},$$

i.e., the identity matrix. The system is thus observable, and the eigenvalue assignment problem can be solved. We have

$$A - LC = \begin{pmatrix} -l_1 & 1 \\ -l_2 & 0 \end{pmatrix},$$

which has the characteristic polynomial

$$\det\left(sI - A + LC\right) = \det \begin{pmatrix} s + l_1 & -1 \\ l_2 & s \end{pmatrix} = s^2 + l_1 s + l_2.$$

Assuming that we want to have an observer with the characteristic polynomial

$$s^2 + p_1 s + p_2 = s^2 + 2\zeta_{\mathrm{o}}\omega_{\mathrm{o}}s + \omega_{\mathrm{o}}^2,$$

the observer gains should be chosen as

$$l_1 = p_1 = 2\zeta_{\mathrm{o}}\omega_{\mathrm{o}}, \qquad l_2 = p_2 = \omega_{\mathrm{o}}^2.$$

The observer is then

$$\frac{d\hat{x}}{dt} = A\hat{x} + Bu + L(y - C\hat{x}) = \begin{pmatrix} 0 & 1 \\ 0 & 0 \end{pmatrix} \hat{x} + \begin{pmatrix} \gamma \\ 1 \end{pmatrix} u + \begin{pmatrix} l_1 \\ l_2 \end{pmatrix} (y - \hat{x}_1).$$

A simulation of the observer for a vehicle driving on a curvy road is shown in Figure 8.6. Figure 8.6a shows the trajectory of the vehicle on the road, as viewed from above. The response of the observer is shown in Figure 8.6b, where time is normalized to the vehicle length. We see that the observer error settles in about 4 vehicle lengths. ∇

To compute the observer gains for systems of high order we have to use numerical calculations. The duality between the design of a state feedback and the design of an observer means that the computer algorithms for state feedback can also be used for the observer design; we simply use the transpose of the dynamics matrix and the output matrix. The MATLAB command $\mathtt{acker}$, which essentially is a direct implementation of the calculations given in Theorem 8.2, can be used for systems

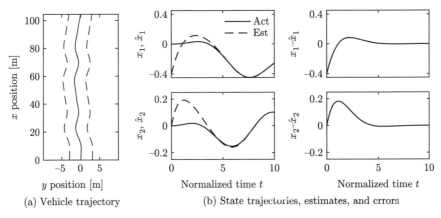

(a) Vehicle trajectory   (b) State trajectories, estimates, and errors

Figure 8.6: Simulation of an observer for a vehicle driving on a curvy road. (a) The vehicle trajectory, as viewed from above, with the lane boundaries shown as dashed lines. (b) The response of the observer with an initial position error. The plots on the left show the lateral deviation x_1 and the lateral velocity x_2 with solid lines and their estimates $\hat{x}_1$ and $\hat{x}_2$ with dashed lines. The plots on the right show the estimation errors. The parameters used to design the estimator were $\omega_o = 1$ and $\zeta_o = 0.7$.

with one output. The MATLAB command `place` can be used for systems with many outputs. It is also better conditioned numerically.

Requirements on a control system typically involve fast response to reference inputs and disturbances at the same time as avoiding amplification of noise. Choosing a fast observer gives fast convergence but the observer gains will be high and the estimated state will be sensitive to measurement noise. If noise characteristics are known it is possible to find the best compromise, as will be discussed in Section 8.4, the observer is then called a *Kalman filter*.

8.3   CONTROL USING ESTIMATED STATE

In this section we will consider a state space system of the form

$$\frac{dx}{dt} = Ax + Bu, \qquad y = Cx. \tag{8.14}$$

We wish to design a feedback controller for the system where only the output is measured. Notice that we have assumed that there is no direct term in the system ($D = 0$), which is often a realistic assumption. The presence of a direct term in combination with a controller having proportional action creates an algebraic loop, which will be discussed in Section 9.4. The problem can still be solved even if there is a direct term, but the calculations are more complicated.

As before, we will assume that u and y are scalars. We also assume that the system is reachable and observable. In Chapter 7 we found a feedback of the form

$$u = -Kx + k_f r$$

for the case that all states could be measured, and in Section 8.2 we developed an observer that can generate estimates of the state $\hat{x}$ based on inputs and outputs.

In this section we will combine the ideas of these sections to find a feedback that gives desired closed loop eigenvalues for systems where only outputs are available for feedback.

If all states are not measurable, it seems reasonable to try the feedback

$$u = -K\hat{x} + k_{\mathrm{f}}r, \tag{8.15}$$

where $\hat{x}$ is the output of an observer of the state, i.e.,

$$\frac{d\hat{x}}{dt} = A\hat{x} + Bu + L(y - C\hat{x}). \tag{8.16}$$

It is not clear that such a combination will have the desired effect. To explore this, note that since the system (8.14) and the observer (8.16) are both of state dimension n, the closed loop system has state dimension $2n$ with state $(x, \hat{x})$. The evolution of the states is described by equations (8.14)–(8.16). To analyze the closed loop system, we change coordinates and replace the estimated state variable $\hat{x}$ by the estimation error

$$\tilde{x} = x - \hat{x}. \tag{8.17}$$

Subtraction of equation (8.16) from equation (8.14) gives

$$\frac{d\tilde{x}}{dt} = Ax - A\hat{x} - L(Cx - C\hat{x}) = A\tilde{x} - LC\tilde{x} = (A - LC)\tilde{x}.$$

Returning to the process dynamics, introducing u from equation (8.15) into equation (8.14) and using equation (8.17) to eliminate $\hat{x}$ gives

$$\frac{dx}{dt} = Ax + Bu = Ax - BK\hat{x} + Bk_{\mathrm{f}}r = Ax - BK(x - \tilde{x}) + Bk_{\mathrm{f}}r$$
$$= (A - BK)x + BK\tilde{x} + Bk_{\mathrm{f}}r.$$

The closed loop system is thus governed by

$$\frac{d}{dt}\begin{pmatrix} x \\ \tilde{x} \end{pmatrix} = \begin{pmatrix} A - BK & BK \\ 0 & A - LC \end{pmatrix}\begin{pmatrix} x \\ \tilde{x} \end{pmatrix} + \begin{pmatrix} Bk_{\mathrm{f}} \\ 0 \end{pmatrix}r. \tag{8.18}$$

Notice that the state $\tilde{x}$, representing the observer error, is not affected by the reference signal r. This is desirable since we do not want the reference signal to generate observer errors.

Since the dynamics matrix is block diagonal, we find that the characteristic polynomial of the closed loop system is

$$\lambda(s) = \det(sI - A + BK)\det(sI - A + LC).$$

This polynomial is a product of two terms: the characteristic polynomial of the closed loop system obtained with state feedback $\det(sI - A + BK)$ and the characteristic polynomial of the observer $\det(sI - A + LC)$. The design procedure thus separates into two subproblems: design of a state feedback and design of an observer. The feedback (8.15) that was motivated heuristically therefore provides an elegant solution to the eigenvalue assignment problem for output feedback. The result is summarized as follows.

Theorem 8.3 (Eigenvalue assignment by output feedback). *Consider the system*

$$\frac{dx}{dt} = Ax + Bu, \qquad y = Cx.$$

The controller described by

$$\frac{d\hat{x}}{dt} = A\hat{x} + Bu + L(y - C\hat{x}) = (A - BK - LC)\hat{x} + Bk_f r + Ly,$$

$$u = -K\hat{x} + k_f r$$

gives a closed loop system with the characteristic polynomial

$$\lambda(s) = \det{(sI - A + BK)}\det{(sI - A + LC)}.$$

This polynomial can be assigned arbitrary roots if the system is reachable and observable.

The controller has a strong intuitive appeal: it can be thought of as being composed of two parts: state feedback and an observer. The controller is now a dynamic compensator with internal state dynamics generated by the observer. The control action makes use of feedback from the estimated states $\hat{x}$. The feedback gain K can be computed as if all state variables can be measured, and it depends only on A and B. The observer gain L depends only on A and C. The property that the eigenvalue assignment for output feedback can be separated into an eigenvalue assignment for a state feedback and an observer is called the *separation principle*.

A block diagram of the controller is shown in Figure 8.7. Notice that the controller contains a dynamical model of the plant. This is called the *internal model principle*: the controller contains a model of the process being controlled.

Design of control systems involves a balance between achieving high performance while maintaining adequate robustness in the presence of uncertainties. It is not obvious how such properties are reflected in the closed loop eigenvalues. It is therefore important to evaluate the design for example by plotting time responses to get more insight into the properties of the design. Additional discussion is presented in Section 14.5, where we consider the robustness of eigenvalue assignment (pole placement) design and also give some design rules.

Example 8.4 Vehicle steering

Consider again the normalized linear model for vehicle steering in Example 7.4. The dynamics relating the steering angle u to the lateral path deviation y are given by the state space model (8.13). Combining the state feedback derived in Example 7.4 with the observer determined in Example 8.3, we find that the controller is given by

$$\frac{d\hat{x}}{dt} = A\hat{x} + Bu + L(y - C\hat{x}) = \begin{pmatrix} 0 & 1 \\ 0 & 0 \end{pmatrix}\hat{x} + \begin{pmatrix} \gamma \\ 1 \end{pmatrix}u + \begin{pmatrix} l_1 \\ l_2 \end{pmatrix}(y - \hat{x}_1),$$

$$u = -K\hat{x} + k_f r = k_1(r - \hat{x}_1) - k_2\hat{x}_2.$$

Elimination of the variable u gives

$$\frac{d\hat{x}}{dt} = (A - BK - LC)\hat{x} + Ly + Bk_f r$$

$$= \begin{pmatrix} -l_1 - \gamma k_1 & 1 - \gamma k_2 \\ -k_1 - l_2 & -k_2 \end{pmatrix}\hat{x} + \begin{pmatrix} l_1 \\ l_2 \end{pmatrix}y + \begin{pmatrix} \gamma \\ 1 \end{pmatrix}k_1 r,$$

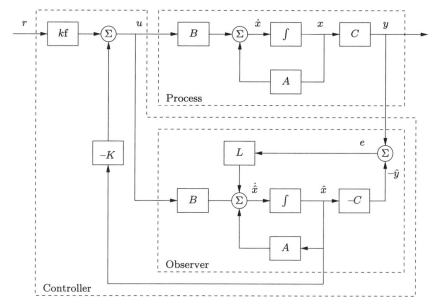

Figure 8.7: Block diagram of an observer-based control system. The observer uses the measured output y and the input u to construct an estimate of the state. This estimate is used by a state feedback controller to generate the corrective input. The controller consists of the observer and the state feedback; the observer is identical to that in Figure 8.5.

where we have set $k_f = k_1$ as described in Example 7.4. The controller is a dynamical system of second order, with two inputs y and r and one output u. Figure 8.8 shows a simulation of the system when the vehicle is driven along a curvy road. Since we are using a normalized model, the length unit is the vehicle length and the time unit is the time it takes to travel one vehicle length. The estimator is initialized with all states equal to zero but the real system has an initial lateral position of 0.8. The figures show that the estimates converge quickly to their true values. The vehicle roughly tracks the desired path, but there are errors because the road is curving. The tracking error can be improved by introducing feedforward (Section 8.5). ∇

Kalman's Decomposition of a Linear System

In this chapter and the previous one we have seen that two fundamental properties of a linear input/output system are reachability and observability. It turns out that these two properties can be used to classify the dynamics of a system. The key result is Kalman's decomposition theorem, which says that a linear system can be divided into four subsystems: Σ_{ro} which is reachable and observable, $\Sigma_{r\bar{o}}$ which is reachable but not observable, $\Sigma_{\bar{r}o}$ which is not reachable but is observable, and $\Sigma_{\bar{r}\bar{o}}$ which is neither reachable nor observable.

We will first consider this in the special case of systems with one input and one output, and where the matrix A has distinct eigenvalues. In this case we can find a set of coordinates such that the A matrix is diagonal and, with some additional reordering of the states, the system can be written as

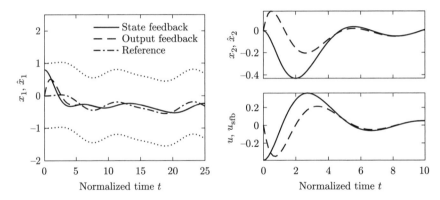

Figure 8.8: Simulation of a vehicle driving on a curvy road with a controller based on state feedback and an observer. The left plot shows the lane boundaries (dotted), the vehicle position (solid), and its estimate (dashed), the upper right plot shows the velocity (solid) and its estimate (dashed), and the lower right plot shows the control signal using state feedback (solid) and the control signal using the estimated state (dashed).

$$
\frac{dx}{dt} = \begin{pmatrix} A_{\mathrm{ro}} & 0 & 0 & 0 \\ 0 & A_{\mathrm{r\bar{o}}} & 0 & 0 \\ 0 & 0 & A_{\bar{\mathrm{r}}\mathrm{o}} & 0 \\ 0 & 0 & 0 & A_{\bar{\mathrm{r}\bar{\mathrm{o}}}} \end{pmatrix} x + \begin{pmatrix} B_{\mathrm{ro}} \\ B_{\mathrm{r\bar{o}}} \\ 0 \\ 0 \end{pmatrix} u,
$$

$$
y = \begin{pmatrix} C_{\mathrm{ro}} & 0 & C_{\bar{\mathrm{r}}\mathrm{o}} & 0 \end{pmatrix} x + Du.
$$

(8.19)

All states x_k such that $B_k \neq 0$ are reachable, and all states such that $C_k \neq 0$ are observable. If we set the initial state to zero (or equivalently look at the steady-state response if A is stable), the states given by $x_{\bar{\mathrm{r}}\mathrm{o}}$ and $x_{\bar{\mathrm{r}\bar{\mathrm{o}}}}$ will be zero and $x_{\mathrm{r\bar{o}}}$ does not affect the output. Hence the output y can be determined from the system

$$
\frac{dx_{\mathrm{ro}}}{dt} = A_{\mathrm{ro}}x_{\mathrm{ro}} + B_{\mathrm{ro}}u, \qquad y = C_{\mathrm{ro}}x_{\mathrm{ro}} + Du.
$$

Thus from the input/output point of view, it is only the reachable and observable dynamics that matter. A block diagram of the system illustrating this property is given in Figure 8.9a.

The general case of the Kalman decomposition is more complicated and requires some additional linear algebra; see the original paper by Kalman, Ho, and Narendra [139]. The key result is that the state space can still be decomposed into four parts, but there will be additional coupling so that the equations have the form

$$
\frac{dx}{dt} = \begin{pmatrix} A_{\mathrm{ro}} & 0 & * & 0 \\ * & A_{\mathrm{r\bar{o}}} & * & * \\ 0 & 0 & A_{\bar{\mathrm{r}}\mathrm{o}} & 0 \\ 0 & 0 & * & A_{\bar{\mathrm{r}\bar{\mathrm{o}}}} \end{pmatrix} x + \begin{pmatrix} B_{\mathrm{ro}} \\ B_{\mathrm{r\bar{o}}} \\ 0 \\ 0 \end{pmatrix} u,
$$

$$
y = \begin{pmatrix} C_{\mathrm{ro}} & 0 & C_{\bar{\mathrm{r}}\mathrm{o}} & 0 \end{pmatrix} x,
$$

(8.20)

where $*$ denotes block matrices of appropriate dimensions. If $x_{\bar{\mathrm{r}}\mathrm{o}}(0) = 0$ then the input/output response of the system is given by

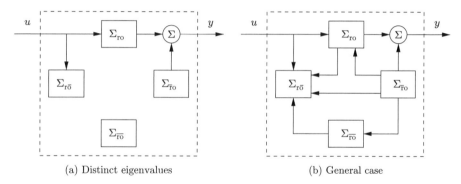

(a) Distinct eigenvalues   (b) General case

Figure 8.9: Kalman's decomposition of a linear system. The decomposition in (a) is for a system with distinct eigenvalues and the one in (b) is the general case. The system is broken into four subsystems, representing the various combinations of reachable and observable states. The input/output relationship only depends on the subset of states that are both reachable and observable.

$$\frac{dx_{\mathrm{ro}}}{dt} = A_{\mathrm{ro}}x_{\mathrm{ro}} + B_{\mathrm{ro}}u, \qquad y = C_{\mathrm{ro}}x_{\mathrm{ro}} + Du, \qquad (8.21)$$

which are the dynamics of the reachable and observable subsystem Σ_{ro}. A block diagram of the system is shown in Figure 8.9b.

The following example illustrates Kalman's decomposition.

Example 8.5 System and controller with feedback from observer states
Consider the system

$$\frac{dx}{dt} = Ax + Bu, \qquad y = Cx.$$

The following controller, based on feedback from the observer state, was given in Theorem 8.3:

$$\frac{d\hat{x}}{dt} = A\hat{x} + Bu + L(y - C\hat{x}), \qquad u = -K\hat{x} + k_{\mathrm{f}}r.$$

Introducing the states x and $\tilde{x} = x - \hat{x}$, the closed loop system can be written as

$$\frac{d}{dt}\begin{pmatrix} x \\ \tilde{x} \end{pmatrix} = \begin{pmatrix} A - BK & BK \\ 0 & A - LC \end{pmatrix} \begin{pmatrix} x \\ \tilde{x} \end{pmatrix} + \begin{pmatrix} Bk_{\mathrm{f}} \\ 0 \end{pmatrix} r, \qquad y = \begin{pmatrix} C & 0 \end{pmatrix} \begin{pmatrix} x \\ \tilde{x} \end{pmatrix},$$

which is a Kalman decomposition like the one shown in Figure 8.9b with only two subsystems Σ_{ro} and $\Sigma_{\bar{\mathrm{r}}\mathrm{o}}$. The subsystem Σ_{ro}, with state x, is reachable and observable, and the subsystem $\Sigma_{\bar{\mathrm{r}}\mathrm{o}}$, with state $\tilde{x}$, is not reachable but observable. It is natural that the state $\tilde{x}$ is not reachable from the reference signal r because it would not make sense to design a system where changes in the reference signal could generate observer errors. The relationship between the reference r and the output y is given by

$$\frac{dx}{dt} = (A - BK)x + Bk_{\mathrm{f}}r, \qquad y = Cx,$$

which is the same relationship as for a system with full state feedback.   ∇

8.4   KALMAN FILTERING

One of the principal uses of observers in practice is to estimate the state of a system in the presence of *noisy* measurements. We have not yet treated noise in our analysis, and a full treatment of stochastic dynamical systems is beyond the scope of this text. In this section, we present a brief introduction to the use of stochastic systems analysis for constructing observers. We work primarily in discrete time to avoid some of the complications associated with continuous-time random processes and to keep the mathematical prerequisites to a minimum. This section assumes basic knowledge of random variables and stochastic processes; see Kumar and Varaiya [155] or Åström [17] for the required material.

Discrete-Time Systems

Consider a discrete-time linear system with dynamics

$$x[k+1] = Ax[k] + Bu[k] + v[k], \qquad y[k] = Cx[k] + w[k], \tag{8.22}$$

where $v[k]$ and $w[k]$ are Gaussian white noise processes satisfying

$$\mathbb{E}(v[k]) = 0, \qquad\qquad\qquad \mathbb{E}(w[k]) = 0,$$

$$\mathbb{E}(v[k]v^T[j]) = \begin{cases} 0 & \text{if } k \neq j, \\ R_v & \text{if } k = j, \end{cases} \qquad \mathbb{E}(w[k]w^T[j]) = \begin{cases} 0 & \text{if } k \neq j, \\ R_w & \text{if } k = j, \end{cases} \tag{8.23}$$

$$\mathbb{E}(v[k]w^T[j]) = 0.$$

$\mathbb{E}(v[k])$ represents the expected value of $v[k]$ and $\mathbb{E}(v[k]v^T[j])$ is the covariance matrix. The matrices R_v and R_w are the covariance matrices for the process disturbance v and measurement noise w. (R_v is allowed to be singular if the disturbances do not affect all states.) We assume that the initial condition is also modeled as a Gaussian random variable with

$$\mathbb{E}(x[0]) = x_0, \qquad \mathbb{E}\left((x[0] - x_0)(x[0] - x_0)^T\right) = P_0. \tag{8.24}$$

We would like to find an estimate $\hat{x}[k]$ that minimizes the mean square error

$$P[k] = \mathbb{E}\left((x[k] - \hat{x}[k])(x[k] - \hat{x}[k])^T\right) \tag{8.25}$$

given the measurements $\{y(\kappa) : 0 \leq \kappa \leq k\}$. We consider an observer in the same basic form as derived previously:

$$\hat{x}[k+1] = A\hat{x}[k] + Bu[k] + L[k](y[k] - C\hat{x}[k]). \tag{8.26}$$

The following theorem summarizes the main result.

Theorem 8.4 (Kalman, 1961). *Consider a random process $x[k]$ with dynamics given by equation (8.22) and noise processes and initial conditions described by equations (8.23) and (8.24). The observer gain L that minimizes the mean square error is given by*

$$L[k] = AP[k]C^T(R_w + CP[k]C^T)^{-1},$$

where

$$P[k+1] = (A - LC)P[k](A - LC)^T + R_v + LR_wL^T,$$
$$P[0] = \mathbb{E}\left((x[0] = x_0)(x[0] - x_0)^T\right). \tag{8.27}$$

Before we prove this result, we reflect on its form and function. First, note that the Kalman filter has the form of a *recursive* filter: given the mean square error $P[k] = \mathbb{E}((x[k] - \hat{x}[k])(x[k] - \hat{x}[k])^T)$ at time k, we can compute how the estimate and error *change*. Thus we do not need to keep track of old values of the output. Furthermore, the Kalman filter gives the estimate $\hat{x}[k]$ *and* the error covariance $P[k]$, so we can see how reliable the estimate is. It can also be shown that the Kalman filter extracts the maximum possible information about output data. If we form the residual between the measured output and the estimated output,

$$e[k] = y[k] - C\hat{x}[k],$$

we can show that for the Kalman filter the covariance matrix is

$$R_e(j, k) = \mathbb{E}(e[j]e^T[k]) = W[k]\delta_{jk}, \qquad \delta_{jk} = \begin{cases} 1 & \text{if } j = k, \\ 0 & \text{if } j \neq k. \end{cases}$$

In other words, the error is a white noise process, so there is no remaining dynamic information content in the error.

The Kalman filter is extremely versatile and can be used even if the process, noise, or disturbances are time-varying. When the system is time-invariant and *if* $P[k]$ converges, then the observer gain is constant:

$$L = APC^T(R_w + CPC^T),$$

where P satisfies

$$P = APA^T + R_v - APC^T(R_w + CPC^T)^{-1}CPA^T.$$

We see that the optimal gain depends on both the process noise and the measurement noise, but in a nontrivial way. Like the use of LQR to choose state feedback gains, the Kalman filter permits a systematic derivation of the observer gains given a description of the noise processes. The solution for the constant gain case is solved by the `dlqe` command in MATLAB.

Proof of theorem. We wish to minimize the mean square of the error $\mathbb{E}((x[k] - \hat{x}[k])(x[k] - \hat{x}[k])^T)$. We will define this quantity as $P[k]$ and then show that it satisfies the recursion given in equation (8.27). By definition,

$$\begin{aligned} P[k+1] &= \mathbb{E}\left((x[k+1] - \hat{x}[k+1])(x[k+1] - \hat{x}[k+1])^T\right) \\ &= (A - LC)P[k](A - LC)^T + R_v + LR_wL^T \\ &= AP[k]A^T + R_v - AP[k]C^TL^T - LCP[k]A^T \\ &\quad + L(R_w + CP[k]C^T)L^T. \end{aligned}$$

Letting $R_\epsilon = (R_w + CP[k]C^T)$, we have

$$\begin{aligned} P[k+1] &= AP[k]A^T + R_v - AP[k]C^TL^T - LCP[k]A^T + LR_\epsilon L^T \\ &= AP[k]A^T + R_v + \left(L - AP[k]C^TR_\epsilon^{-1}\right)R_\epsilon\left(L - AP[k]C^TR_\epsilon^{-1}\right)^T \\ &\quad - AP[k]C^TR_\epsilon^{-1}CP^T[k]A^T. \end{aligned}$$

To minimize this expression, we choose $L = AP[k]C^T R_\epsilon^{-1}$, and the theorem is proved.   □

Continuous-Time Systems

The Kalman filter can also be applied to continuous-time stochastic processes. The mathematical derivation of this result requires more sophisticated tools, but the final form of the estimator is relatively straightforward.

Consider a continuous stochastic system

$$
\frac{dx}{dt} = Ax + Bu + v, \qquad \mathbb{E}(v(s)v^T(t)) = R_v\delta(t-s),
$$

$$
y = Cx + w, \qquad \mathbb{E}(w(s)w^T(t)) = R_w\delta(t-s), \tag{8.28}
$$

where $\delta(\tau)$ is the unit impulse function, and the initial value is Gaussian with mean x_0 and covariance $P_0 = \mathbb{E}((x(0) - x_0)(x(0) - x_0)^T)$. Assume that the disturbance v and noise w are zero mean and Gaussian (but not necessarily time-invariant):

$$
\text{pdf}(v) = \frac{1}{\sqrt[n]{2\pi}\sqrt{\det R_v}}e^{-\frac{1}{2}v^T R_v^{-1}v}, \quad \text{pdf}(w) = \frac{1}{\sqrt[n]{2\pi}\sqrt{\det R_w}}e^{-\frac{1}{2}w^T R_w^{-1}w}. \tag{8.29}
$$

The model (8.28) is very general. We can model the dynamics both of the process and disturbances, as illustrated by the following example.

Example 8.6 Modeling a noisy sinusoidal disturbance

Consider a process whose dynamics are described by

$$
\frac{dx}{dt} = x + u + v, \qquad y = x + w.
$$

The disturbance v is a noisy sinusoidal disturbance with frequency ω_0 and w is white measurement noise. We model the oscillatory load disturbance as $v = z_1$, where

$$
\frac{d}{dt}\begin{pmatrix} z_1 \\ z_2 \end{pmatrix} = \begin{pmatrix} -0.01\omega_0 & \omega_0 \\ -\omega_0 & -0.01\omega_0 \end{pmatrix}\begin{pmatrix} z_1 \\ z_2 \end{pmatrix} + \begin{pmatrix} 0 \\ \omega_0 \end{pmatrix}e,
$$

and e is zero mean white noise with covariance function $r\delta(t)$.

Augmenting the system state with the states of the noise model by introducing the new state $\xi = (x, z_1, z_2)$, we obtain the model

$$
\frac{d\xi}{dt} = \begin{pmatrix} 1 & 1 & 0 \\ 0 & -0.01\omega_0 & \omega_0 \\ 0 & -\omega_0 & -0.01\omega_0 \end{pmatrix}\xi + \begin{pmatrix} 1 \\ 0 \\ 0 \end{pmatrix}u + v, \quad y = \begin{pmatrix} 1 & 0 & 0 \end{pmatrix}\xi + w,
$$

where v is white Gaussian noise with zero mean and the covariance $R_v\delta(t)$ with

$$
R_v = \begin{pmatrix} 0 & 0 & 0 \\ 0 & 0 & 0 \\ 0 & 0 & \omega_0^2 r \end{pmatrix}.
$$

The model is in the standard form given by equations (8.28) and (8.29).   ∇

We will now return to the filtering problem. Specifically, we wish to find the estimate $\hat{x}(t)$ that minimizes the mean square error $P(t) = \mathbb{E}((x(t) - \hat{x}(t))(x(t) - \hat{x}(t))^T)$ given $\{y(\tau) : 0 \leq \tau \leq t\}$.

Theorem 8.5 (Kalman–Bucy, 1961). *The optimal estimator has the form of a linear observer*

$$\frac{d\hat{x}}{dt} = A\hat{x} + Bu + L(y - C\hat{x}), \qquad \hat{x}(0) = \mathbb{E}(x(0)),$$

where $L = PC^T R_w^{-1}$ and $P = \mathbb{E}((x(t) - \hat{x}(t))(x(t) - \hat{x}(t))^T)$ and satisfies

$$\frac{dP}{dt} = AP + PA^T - PC^T R_w^{-1} CP + R_v, \quad P(0) = \mathbb{E}\left((x(0) - x_0)(x(0) - x_0)^T\right). \tag{8.30}$$

All matrices A, B, C, R_v, R_w, P, and L can be time varying. The essential condition is that the Riccati equation (8.30) has a unique positive solution.

As in the discrete case, when the system is time-invariant and if $P(t)$ converges, the observer gain $L = PC^T R_w^{-1}$ is constant and P is the solution to

$$AP + PA^T - PC^T R_w^{-1} CP + R_v = 0, \tag{8.31}$$

which is called the *algebraic Riccati equation*.

Notice that there are a strong similarities between the Riccati equations (8.30) and (8.31) for the Kalman filtering problem and the corresponding equations (7.31) and (7.33) for the linear quadratic regulator (LQR). We have the equivalences

$$A \leftrightarrow A^T, \quad B \leftrightarrow C^T, \quad K \leftrightarrow L^T, \quad P \leftrightarrow S, \quad Q_x \leftrightarrow R_v, \quad Q_u \leftrightarrow R_w, \tag{8.32}$$

which we can compare with equation (8.9). The MATLAB command `kalman` can be used to compute optimal filter gains.

Example 8.7 Vectored thrust aircraft
The dynamics for a vectored thrust aircraft were considered in Examples 3.12 and 7.9. We consider the (linearized) lateral dynamics of the system, consisting of the subsystems whose states are given by $z = (x, \theta, \dot{x}, \dot{\theta})$. The dynamics of the linearized system can be obtained from Example 7.9 by extracting only the relevant states and outputs, giving

$$A = \begin{pmatrix} 0 & 0 & 1 & 0 \\ 0 & 0 & 0 & 1 \\ 0 & -g & -c/m & 0 \\ 0 & 0 & 0 & 0 \end{pmatrix}, \qquad B = \begin{pmatrix} 0 \\ 0 \\ 0 \\ r/J \end{pmatrix}, \qquad C = \begin{pmatrix} 0 & 0 & 0 & 1 \end{pmatrix},$$

where the linearized state $\xi = z - z_e$ represents the system state linearized around the equilibrium point z_e. To design a Kalman filter for the system, we must include a description of the process disturbances and the sensor noise. We thus augment the system to have the form

$$\frac{d\xi}{dt} = A\xi + Bu + Fv, \qquad y = C\xi + w,$$

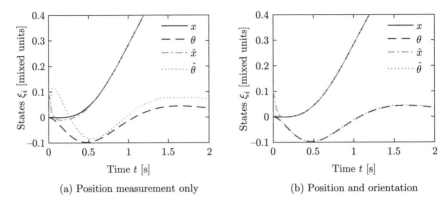

(a) Position measurement only   (b) Position and orientation

Figure 8.10: Kalman filter response for a (linearized) vectored thrust aircraft with disturbances and noise during the initial portion of a step response. In the first design (a) only the lateral position of the aircraft is measured. Adding a direct measurement of the roll angle produces a much better observer (b). The initial estimator state for both simulations is $(0.1, 0.0175, 0.01, 0)$ and the controller gains are $K = (-1, 7.9, -1.6, 2.1)$ and $k_f = -1$.

where F represents the structure of the disturbances (including the effects of nonlinearities that we have ignored in the linearization), v represents the disturbance source (modeled as zero mean, Gaussian white noise), and w represents that measurement noise (also zero mean, Gaussian, and white).

For this example, we choose F as the identity matrix and choose disturbances v, $i = 1, \ldots, n$, to be independent random variables with covariance matrix elements given by $R_{ii} = 0.1$, $R_{ij} = 0$, $i \neq j$. The sensor noise is a single random variable that we model as white noise having covariance $R_w = 10^{-4}$. Using the same parameters as before, the resulting Kalman gain is given by

$$L = PC^T R_w^{-1} = \begin{pmatrix} 37.0 \\ -46.9 \\ 185 \\ -31.6 \end{pmatrix} \quad \text{where} \quad AP + PA^T - PC^T R_w^{-1} CP + R_v = 0.$$

The performance of the estimator is shown in Figure 8.10a. We see that while the estimator roughly tracks the system state, it contains significant overshoot in the state estimate and has significant error in the estimate for θ even after 2 seconds, which can lead to poor performance in a closed loop setting.

To improve the performance of the estimator, we explore the impact of adding a new output measurement. Suppose that instead of measuring just the output position x, we also measure the orientation of the aircraft θ. The output becomes

$$y = \begin{bmatrix} 1 & 0 & 0 & 0 \\ 0 & 1 & 0 & 0 \end{bmatrix} \xi + \begin{pmatrix} w_1 \\ w_2 \end{pmatrix},$$

and if we assume that w_1 and w_2 are independent white noise sources each with covariance $R_{w_i} = 10^{-4}$, then the optimal estimator gain matrix becomes

$$L = \begin{pmatrix} 32.6 & -0.150 \\ -0.150 & 32.6 \\ 32.7 & -9.79 \\ -0.0033 & 31.6 \end{pmatrix}.$$

These gains provide good immunity to noise and high performance, as illustrated in Figure 8.10b. ∇

Linear Quadratic Gaussian Control (LQG)

In Section 7.5 we considered optimization of the criterion (7.29) when the control $u(t)$ could be a function of the state $x(t)$. We will now explore the same problem for the stochastic system (8.28) where the control $u(t)$ is a function of the output $y(t)$.

Consider the system given by equation (8.28) where the initial state is Gaussian with mean x_0 and covariance P_0 and the disturbances v and w are characterized by equation (8.29). Assume that the requirement can be captured by the cost function

$$J = \min_u \mathbb{E} \left(\int_0^{t_f} (x^T Q_x x + u^T Q_u u) \, dt + x^T(t_f) Q_f x(t_f) \right), \tag{8.33}$$

where we minimize over all controls such that $u(t)$ is a function of all measurements $y(\tau)$, $0 \leq \tau \leq t$ obtained up to time t.

The optimal control law is simply $u(t) = -K\hat{x}(t)$ where $K = SBQ_u^{-1}$ and S is the solution of the Riccati equation (7.31) (for the linear quadratic regulator) and $\hat{x}(t)$ is given by the Kalman filter (Theorem 8.5). The solution of the problem can thus be separated into a deterministic control problem (LQR) and an optimal filtering problem. This remarkable result is also known as the *separation principle*, as mentioned briefly in Section 8.3.

The minimum cost function is

$$\min J = x_0^T S(0) x_0 + \text{Tr}\, (S(0) P_0) + \int_0^{t_f} \text{Tr}\, (R_v S) \, dt + \int_0^{t_f} \text{Tr}\, (L^T Q_u L P) \, dt,$$

where Tr is the trace of a matrix. The first two terms represent the cost of the mean x_0 and covariance P_0 of the initial state, the third term represents the cost due to the load disturbance, and the last term represents the cost of prediction. Notice that the models we have used do not have a direct term in the output. The separation theorem does not hold in this case because the nature of the disturbances is then influenced by the feedback.

8.5 STATE SPACE CONTROLLER DESIGN

State estimators and state feedback are important components of a controller. In this section, we will add feedforward to arrive at a general controller structure that appears in many places in control theory and is the heart of most modern control systems. We will also briefly sketch how computers can be used to implement a controller based on output feedback.

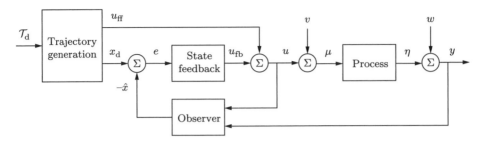

Figure 8.11: Block diagram of a controller based on a structure with two degrees of freedom that combines feedback and feedforward. The controller consists of a trajectory generator, state feedback, and an observer. The trajectory generation subsystem computes a feedforward command u_{ff} along with the desired state x_{d}. The state feedback controller uses the estimated state and desired state to compute a corrective input u_{fb}.

Two Degree-of-Freedom Controller Architecture

In this chapter and the previous one we have emphasized feedback as a mechanism for minimizing tracking error; reference values were introduced simply by adding them to the state feedback through a gain k_{f}. A more sophisticated way of doing this is shown by the block diagram in Figure 8.11, where the controller consists of three parts: an observer that computes estimates of the states based on a model and measured process inputs and outputs, a state feedback, and a trajectory generator that computes the desired behavior of all states x_{d} and a feedforward signal u_{ff}. Under the ideal conditions of no disturbances and no modeling errors the signal u_{ff} generates the desired behavior x_{d} when applied to the process. The signals x_{d} and u_{ff} are generated from the task description $\mathcal{T}_{\mathrm{d}}$, which can represent different types of command signals depending on the application. In simple cases the task description is simply the reference signal r, and x_{d} and u_{ff} are generated by sending r through linear systems. For motion control problems, such as vehicle steering and robotics, the task description consists of the coordinates of a number of points (waypoints) that the vehicle should pass. In other situations the task description could be to transition from one state to another while optimizing some criterion.

To get some insight into the behavior of the system, consider the case when there are no disturbances and the system is in equilibrium with a constant reference signal and with the observer state $\hat{x}$ equal to the process state x. When the reference signal is changed, the signals u_{ff} and x_{d} will change. The observer tracks the state perfectly because the initial state was correct. The estimated state $\hat{x}$ is thus equal to the desired state x_{d}, and the feedback signal $u_{\mathrm{fb}} = K(x_{\mathrm{d}} - \hat{x})$ will also be zero. All action is thus created by the signals from the trajectory generator. If there are some disturbances or some modeling errors, the feedback signal will attempt to correct the situation.

This controller is said to have *two degrees of freedom* because the responses to reference signals and disturbances are decoupled. Disturbance responses are governed by the observer and the state feedback, while the response to command signals is governed by the trajectory generator (feedforward).

Feedforward Design and Trajectory Generation

We will now discuss design of controllers with the architecture shown in Figure 8.11. For an analytic description we start with the full nonlinear dynamics of the process

$$\frac{dx}{dt} = f(x, u), \qquad y = h(x, u). \tag{8.34}$$

A *feasible trajectory* for the system (8.34) is a pair $(x_d(t), u_{ff}(t))$ that satisfies the differential equation and generates the desired trajectory:

$$\frac{dx_d(t)}{dt} = f\big(x_d(t), u_{ff}(t)\big), \qquad r(t) = h\big(x_d(t), u_{ff}(t)\big).$$

The problem of finding a feasible trajectory for a system is called the *trajectory generation* problem, with x_d representing the desired state for the (nominal) system and u_{ff} representing the desired input or the feedforward control. If we can find a feasible trajectory for the system, we can search for controllers of the form $u = \alpha(x, x_d, u_{ff})$ that track the desired reference trajectory.

In many applications, it is possible to attach a cost function to trajectories that describe how well they balance trajectory tracking with other factors, such as the magnitude of the inputs required. In such applications, it is natural to ask that we find the *optimal* controller with respect to some cost function:

$$\min_{u(\cdot)} \int_0^T L(x, u)\, dt + V\big(x(T)\big),$$

subject to the constraint

$$\dot{x} = f(x, u), \qquad x \in \mathbb{R}^n,\ u \in \mathbb{R}^p.$$

Abstractly, this is a constrained optimization problem where we seek a feasible trajectory $(x_d(t), u_{ff}(t))$ that minimizes the cost function. Depending on the form of the dynamics, this problem can be quite complex to solve, but there are good numerical packages for solving such problems, including handling constraints on the range of inputs as well as the allowable values of the state.

In some situations we can simplify the approach of generating feasible trajectories by exploiting the structure of the system. The next example illustrates one such approach.

Example 8.8 Vehicle steering

To illustrate how we can use a two degree-of-freedom design to improve the performance of the system, consider the problem of steering a car to change lanes on a road, as illustrated in Figure 8.12a.

We use the non-normalized form of the dynamics, which were derived in Example 3.11. As shown in Exercise 3.6, using the center of the rear wheels as the reference $(\alpha = 0)$ the dynamics can be written as

$$\frac{dx}{dt} = v \cos\theta, \qquad \frac{dy}{dt} = v \sin\theta, \qquad \frac{d\theta}{dt} = \frac{v}{b} \tan\delta,$$

where v is the forward velocity of the vehicle, θ is the heading angle, and δ is the steering angle. To generate a trajectory for the system, we note that we can solve for the states and inputs of the system given $x(t)$, $y(t)$ by solving the following sets of equations:

$$\dot{x} = v \cos\theta, \qquad \ddot{x} = \dot{v} \cos\theta - v\dot{\theta} \sin\theta,$$

$$\dot{y} = v \sin\theta, \qquad \ddot{y} = \dot{v} \sin\theta + v\dot{\theta} \cos\theta, \tag{8.35}$$

$$\dot{\theta} = (v/b) \tan\delta.$$

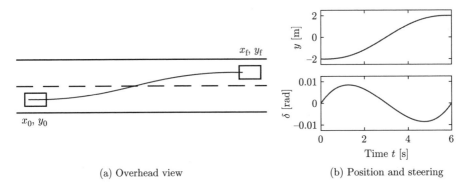

(a) Overhead view   (b) Position and steering

Figure 8.12: Trajectory generation for changing lanes. We wish to change from the right lane to the left lane over a distance of 90 m in 6 s. The planned trajectory in the xy plane is shown in (a) and the lateral position y and the steering angle δ over the maneuver time interval are shown in (b).

This set of five equations has five unknowns (θ, $\dot{\theta}$, v, $\dot{v}$ and δ) that can be solved using trigonometry and linear algebra given the path variables $x(t)$, $y(t)$ and their time derivatives. It follows that we can compute a feasible state trajectory for the system given any path $x(t)$, $y(t)$. (This special property of a system is known as *differential flatness* and is described in more detail below.)

To find a trajectory from an initial state (x_0, y_0, θ_0) to a final state (x_f, y_f, θ_f) at a time T, we look for a path $x(t), y(t)$ that satisfies

$$
\begin{aligned}
x(0) &= x_0, & x(T) &= x_f, \\
y(0) &= y_0, & y(T) &= y_f, \\
\dot{x}(0) \sin\theta_0 - \dot{y}(0) \cos\theta_0 &= 0, & \dot{x}(T) \sin\theta_f - \dot{y}(T) \cos\theta_f &= 0, \\
\dot{y}(0) \sin\theta_0 + \dot{x}(0) \cos\theta_0 &= v_0, & \dot{y}(T) \sin\theta_f + \dot{x}(T) \cos\theta_f &= v_f,
\end{aligned}
\tag{8.36}
$$

where v_0 is the initial velocity and v_f is the final velocity along the trajectory. One such trajectory can be found by choosing $x(t)$ and $y(t)$ to have the form

$$
x_d(t) = \alpha_0 + \alpha_1 t + \alpha_2 t^2 + \alpha_3 t^3, \qquad y_d(t) = \beta_0 + \beta_1 t + \beta_2 t^2 + \beta_3 t^3.
$$

Substituting these equations into equation (8.36), we are left with a set of linear equations that can be solved for α_i, β_i, $i = 0, 1, 2, 3$. This gives a feasible trajectory for the system by using equation (8.35) to solve for θ_d, v_d, and δ_d.

Figure 8.12b shows a sample trajectory generated by a set of higher-order equations that also set the initial and final steering angle to zero. Notice that the feedforward input is different from zero, allowing the controller to command a steering angle that executes the turn in the absence of errors.   ▽

The concept of differential flatness that we exploited in the previous example is a fairly general one and can be applied to many interesting trajectory generation problems. A nonlinear system (8.34) is *differentially flat* if there exists a flat output z such that the state x and the input u can be expressed as functions of the flat output z and a finite number of its derivatives:

$$
x = \beta(z, \dot{z}, \ldots, z^{(m)}), \qquad u = \gamma(z, \dot{z}, \ldots, z^{(m)}).
\tag{8.37}
$$

The number of flat outputs is always equal to the number of system inputs. The vehicle steering model is differentially flat with the position of the rear wheels as the flat output.

A broad class of systems that is differentially flat is the class of reachable linear systems. For the linear system given in equation (7.6), which is in reachable canonical form, we have

$$z_1 = z_n^{(n-1)}, \quad z_2 = z_n^{(n-2)}, \quad \ldots, \quad z_{n-1} = \dot{z}_n,$$

$$u = z_n^{(n)} + a_1 z_n^{(n-1)} + a_2 z_n^{(n-2)} + \cdots + a_n z_n,$$

and the nth component z_n of the state vector is thus a flat output. Since any reachable system can be transformed to reachable canonical form, it follows that every reachable linear system is differentially flat.

Note that no differential equations need to be integrated in order to compute the feasible trajectories for a differentially flat system (unlike optimal control methods, which often involve parameterizing the *input* and then solving the differential equations). The practical implication is that nominal trajectories and inputs that satisfy the equations of motion for a differentially flat system can be computed efficiently. The concept of differential flatness is described in more detail in the review article by Fliess et al. [88].

Disturbance Modeling and State Augmentation

We often have some information about load disturbances: they can be unknown constants, drifting with unknown rates, sinusoidal with known or unknown frequency, or stochastic signals. This information can be used by modeling the disturbances by differential equations and augmenting the process state with the disturbance states as was done in Section 7.4 and Example 8.6. We illustrate with a simple example.

Example 8.9 Integral action by state augmentation

Consider the system (8.1) and assume that there is a constant but unknown disturbance z acting additively on the process input. The system and the disturbance can then be modeled by augmenting the state x with z. An unknown constant can be modeled by the differential equation $dz/dt = 0$ and we obtain the following model for the process and its environment:

$$\frac{d}{dt} \begin{pmatrix} x \\ z \end{pmatrix} = \begin{pmatrix} A & B \\ 0 & 0 \end{pmatrix} \begin{pmatrix} x \\ z \end{pmatrix} + \begin{pmatrix} B \\ 0 \end{pmatrix} u, \quad y = \begin{pmatrix} C & 0 \end{pmatrix} \begin{pmatrix} x \\ z \end{pmatrix}.$$

Notice that the disturbance state z is not reachable from u, but because the disturbance enters at the process input it can be attenuated by the control law

$$u = -K\hat{x} - \hat{z}, \tag{8.38}$$

where $\hat{x}$ and $\hat{z}$ are estimates of the state x and the disturbance z. The estimated disturbance can be obtained from the observer:

$$\frac{d\hat{x}}{dt} = A\hat{x} + B\hat{z} + Bu + L_x(y - C\hat{x}), \quad \frac{d\hat{z}}{dt} = L_z(y - C\hat{x}).$$

Integrating the last equation and inserting the expression for $\hat{z}$ in the control law (8.38) gives

$$u = -K\hat{x} - L_z \int_0^t (y(\tau) - C\hat{x}(\tau))d\tau,$$

which is a state feedback controller with integral action. Notice that the integral action is created through estimation of a disturbance state. $\qquad\qquad \nabla$

The idea of the example can be extended to many types of disturbances and we emphasized that much can be gained from modeling a process and its environment (disturbances acting on the process and measurement noise).

Feedback Design and Gain Scheduling

We now assume that the trajectory generator is able to compute a desired trajectory (x_d, u_{ff}) that satisfies the dynamics (8.34) and satisfies $r = h(x_d, u_{ff})$. To design the feedback controller, we construct the error system. Let $\xi = x - x_d$ and $u_{fb} = u - u_{ff}$ and compute the dynamics for the error:

$$\dot{\xi} = \dot{x} - \dot{x}_d = f(x, u) - f(x_d, u_{ff})$$
$$= f(\xi + x_d, v + u_{ff}) - f(x_d, u_{ff}) =: F(\xi, v, x_d(t), u_{ff}(t)).$$

For trajectory tracking, we can assume that ξ is small (if our controller is doing a good job), and so we can linearize around $\xi = 0$:

$$\frac{d\xi}{dt} \approx A(t)\xi + B(t)v, \quad h(x, u) \approx C(t)x(t)$$

$$A(t) = \left.\frac{\partial F}{\partial \xi}\right|_{(x_d(t), u_{ff}(t))}, \quad B(t) = \left.\frac{\partial F}{\partial v}\right|_{(x_d(t), u_{ff}(t))}, \quad C(t) = \left.\frac{\partial h}{\partial \xi}\right|_{(x_d(t), u_{ff}(t))}.$$

In general, this system is time-varying. Note that ξ corresponds to $-e$ in Figure 8.11 due to the convention of using negative feedback in the block diagram. We can now proceed to use LQR to compute the time-varying feedback gain $K(t) = Q_u^{-1}(t)B^T(t)S(t)$ by solving the Riccati differential equation (7.31) and the Kalman filter gain $L(t) = P(t)C^T(t)R_w^{-1}(t)$, where $P(t)$ is obtained by solving the Riccati equation (8.30).

Assume now that x_d and u_{ff} are either constant or slowly varying (with respect to the process dynamics). It is often the case that $A(t)$, $B(t)$, and $C(t)$ depend only on x_d, in which case it is convenient to write $A(t) = A(x_d)$, $B(t) = B(x_d)$, and $C(t) = C(x_d)$. This allows us to consider just the linear system given by $A(x_d)$, $B(x_d)$, and $C(x_d)$. If we design a state feedback controller $K(x_d)$ for each x_d, then we can regulate the system using the feedback

$$u_{fb} = -K(x_d)\xi.$$

Substituting back the definitions of ξ and u_{fb}, our controller becomes

$$u = u_{fb} + u_{ff} = -K(x_d)(x - x_d) + u_{ff}.$$

This form of controller is called a *gain scheduled* linear controller with *feedforward* u_{ff}.

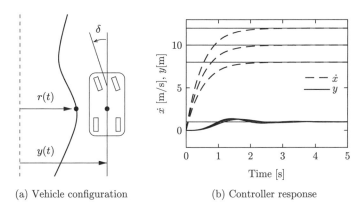

(a) Vehicle configuration   (b) Controller response

Figure 8.13: Vehicle steering using gain scheduling. (a) Vehicle configuration consists of the x, y position of the vehicle, its angle with respect to the road, and the steering wheel angle. (b) Step responses for the vehicle lateral position (solid) and forward velocity (dashed). Gain scheduling is used to set the feedback controller gains for the different forward velocities.

Example 8.10 Steering control with velocity scheduling

Consider the problem of controlling the motion of a automobile so that it follows a given trajectory on the ground, as shown in Figure 8.13a. We use the model derived in Example 8.8. A simple feasible trajectory for the system is to follow a straight line in the x direction at lateral position y_r and fixed velocity v_r. This corresponds to a desired state $x_d = (v_r t, y_r, 0)$ and nominal input $u_{ff} = (v_r, 0)$. Note that (x_d, u_{ff}) is not an equilibrium point for the full system, but it does satisfy the equations of motion.

Linearizing the system about the desired trajectory, we obtain

$$A_d = \frac{\partial f}{\partial x}\bigg|_{(x_d, u_{ff})} = \begin{pmatrix} 0 & 0 & -\sin\theta \\ 0 & 0 & \cos\theta \\ 0 & 0 & 0 \end{pmatrix}\bigg|_{(x_d, u_{ff})} = \begin{pmatrix} 0 & 0 & 0 \\ 0 & 0 & 1 \\ 0 & 0 & 0 \end{pmatrix},$$

$$B_d = \frac{\partial f}{\partial u}\bigg|_{(x_d, u_{ff})} = \begin{pmatrix} 1 & 0 \\ 0 & 0 \\ 0 & v_r/l \end{pmatrix}.$$

We form the error dynamics by setting $e = x - x_d$ and $w = u - u_{ff}$:

$$\frac{de_x}{dt} = w_1, \qquad \frac{de_y}{dt} = e_\theta, \qquad \frac{de_\theta}{dt} = \frac{v_r}{l} w_2.$$

We see that the first state is decoupled from the second two states and hence we can design a controller by treating these two subsystems separately. Suppose that we wish to place the closed loop eigenvalues of the longitudinal dynamics (e_x) at λ_1 and place the closed loop eigenvalues of the lateral dynamics (e_y, e_θ) at the roots of the polynomial equation $s^2 + a_1 s + a_2 = 0$. This can be accomplished by setting

$$w_1 = -\lambda_1 e_x, \qquad w_2 = \frac{l}{v_r}(a_1 e_y + a_2 e_\theta).$$

Note that the gain l/v_r depends on the velocity v_r (or equivalently on the nominal input u_{ff}), giving us a gain scheduled controller.

In the original inputs and state coordinates, the controller has the form

$$
\begin{pmatrix} v \\ \delta \end{pmatrix} = - \underbrace{\begin{bmatrix} \lambda_1 & 0 & 0 \\ 0 & \dfrac{a_1 l}{v_r} & \dfrac{a_2 l}{v_r} \end{bmatrix}}_{K_d} \underbrace{\begin{pmatrix} x - v_r t \\ y - y_r \\ \theta \end{pmatrix}}_{e} + \underbrace{\begin{pmatrix} v_r \\ 0 \end{pmatrix}}_{u_{ff}}.
$$

The form of the controller shows that at low speeds the gains in the steering angle will be high, meaning that we must turn the wheel harder to achieve the same effect. As the speed increases, the gains become smaller. This matches the usual experience that at high speed a very small amount of actuation is required to control the lateral position of a car. Note that the gains go to infinity when the vehicle is stopped ($v_r = 0$), corresponding to the fact that the system is not reachable at this point.

Figure 8.13b shows the response of the controller to a step change in lateral position at three different reference speeds. Notice that the rate of the response is constant, independent of the reference speed, reflecting the fact that the gain scheduled controllers each set the closed loop eigenvalues to the same values. $\quad \nabla$

Nonlinear Estimation

Finally, we briefly comment on the observer represented in Figure 8.11 for the case where the process dynamics are not necessarily linear. Since we are now considering a nonlinear system that operates over a wide range of the state space, it is desirable to use full nonlinear dynamics for the prediction portion of the observer. This can then be combined with a linear correction term, so that the observer has the form:

$$
\frac{d\hat{x}}{dt} = f(\hat{x}, u) + L(\hat{x})(y - h(\hat{x})).
$$

The estimator gain $L(\hat{x})$ is the observer gain obtained by linearizing the system around the currently estimated state. This form of the observer is known as an *extended Kalman filter* and has proved to be a very effective means of estimating the state of a nonlinear system.

The combination of trajectory generation, trajectory tracking, and nonlinear estimation provides a means for state space control of nonlinear systems. There are many ways to generate the feedforward signal, and there are also many different ways to compute the feedback gain K and the observer gain L. Note that once again the internal model principle applies: the overall controller contains a model of the system to be controlled and its environment through the observer.

Computer Implementation

The controllers obtained so far have been described by ordinary differential equations. They can be implemented directly using analog components, whether electronic circuits, hydraulic valves, or other physical devices. Since in modern engineering applications most controllers are implemented using computers, we will briefly discuss how this can be done.

A computer-controlled system typically operates periodically: every cycle, signals from the sensors are sampled and converted to digital form by an

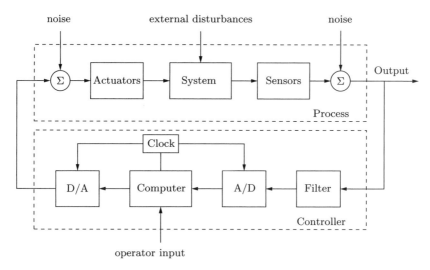

Figure 8.14: Components of a computer-controlled system. The controller consists of analog-to-digital (A/D) and digital-to-analog (D/A) converters, as well as a computer that implements the control algorithm. A system clock controls the operation of the controller, synchronizing the A/D, D/A, and computing processes. The operator input is also fed to the computer as an external input.

analog-to-digital (A/D) converter, the control signal is computed and the resulting output is converted to analog form for the actuators, as shown in Figure 8.14. To illustrate the main principles of how to implement feedback in this environment, we consider the controller described by equations (8.15) and (8.16), i.e.,

$$\frac{d\hat{x}}{dt} = A\hat{x} + Bu + L(y - C\hat{x}), \qquad u = -K\hat{x} + k_f r.$$

The second equation consists only of additions and multiplications and can thus be implemented directly on a computer. The first equation can be implemented by approximating the derivative by a difference

$$\frac{d\hat{x}}{dt} \approx \frac{\hat{x}(t_{k+1}) - \hat{x}(t_k)}{h} = A\hat{x}(t_k) + Bu(t_k) + L\big(y(t_k) - C\hat{x}(t_k)\big),$$

where t_k are the sampling instants and $h = t_{k+1} - t_k$ is the sampling period. Rewriting the equation to isolate $\hat{x}(t_{k+1})$, we get the difference equation

$$\hat{x}(t_{k+1}) = \hat{x}(t_k) + h\big(A\hat{x}(t_k) + Bu(t_k) + L\big(y(t_k) - C\hat{x}(t_k)\big)\big). \qquad (8.39)$$

The calculation of the estimated state at time t_{k+1} requires only addition and multiplication and can easily be done by a computer. A section of pseudocode for the program that performs this calculation is as follows.

```
% Control algorithm - main loop
r = adin(ch1)                    % read reference
y = adin(ch2)                    % get process output
(xd, uff) = trajgen(r, t)        % generate feedforward
u = K*(xd - xhat) + uff          % compute control variable
```

```
daout(ch1, u)                    % set analog output
xhat = xhat + h*(A*x+B*u+L*(y-C*x))  % update state estimate
```

The program runs periodically at a fixed sampling period h. Notice that the number of computations between reading the analog input and setting the analog output has been minimized by updating the state after the analog output has been set. The program has an array of states xhat that represents the state estimate. The choice of sampling period requires some care.

There are more sophisticated ways of approximating a differential equation by a difference equation. If the control signal is constant between the sampling instants, it is possible to obtain exact equations; see [22].

There are several practical issues that also must be dealt with. For example, it is necessary to filter measured signals before they are sampled so that the filtered signal has little frequency content above $f_s/2$ (the Nyquist frequency), where $f_s = 1/h$ is the sampling frequency. This avoids a phenomenon known as *aliasing*. If controllers with integral action are used, it is also necessary to provide protection so that the integral does not become too large when the actuator saturates. This issue, called *integrator windup*, is studied in more detail in Chapter 11. Care must also be taken so that parameter changes do not cause disturbances.

8.6   FURTHER READING

The notion of observability is due to Kalman [136] and, combined with the dual notion of reachability, it was a major stepping stone toward establishing state space control theory beginning in the 1960s. The observer first appeared as the Kalman filter in the paper by Kalman [135] for the discrete-time case and Kalman and Bucy [137] for the continuous-time case. Kalman also conjectured that the controller for output feedback could be obtained by combining a state feedback with a Kalman filter; see the quote in the beginning of this chapter. This result, which is known as the *separation theorem*, is mathematically subtle. Attempts of proof were made by Josep and Tou [132] and Gunckel and Franklin [112], but a rigorous proof was given by Georgiou and Lindquist [99] in 2013. The combined result is known as the linear quadratic Gaussian control theory; a compact treatment is given in the books by Anderson and Moore [9], Åström [17], and Lindquist and Picci [165]. It was also shown that solutions to robust control problems had a similar structure but with different ways of computing observer and state feedback gains [77]. The importance of systems with two degrees of freedom that combine feedback and feedforward was emphasized by Horowitz [121]. The controller structure discussed in Section 8.5 is based on these ideas. The particular form in Figure 8.11 appeared in [22], where computer implementation of the controller was discussed in detail. The hypothesis that motion control in humans is based on a combination of feedback and feedforward was proposed by Ito in 1970 [127]. Differentially flat systems were originally studied by Fliess et al. [87]; they are very useful for trajectory generation.

EXERCISES

8.1 (Coordinate transformations) Consider a system under a coordinate transformation $z = Tx$, where $T \in \mathbb{R}^{n \times n}$ is an invertible matrix. Show that the observability matrix for the transformed system is given by $\widetilde{W}_o = W_o T^{-1}$ and hence observability is independent of the choice of coordinates.

8.2 Show that the system depicted in Figure 8.2 is not observable.

8.3 (Multi-input, multi-output observability) Consider the multi-input, multi-output system given by

$$\frac{dx}{dt} = Ax + Bu, \qquad y = Cx,$$

where $x \in \mathbb{R}^n$, $u \in \mathbb{R}^p$, and $y \in \mathbb{R}^q$. Show that the states can be determined from the input u and the output y and their derivatives if the observability matrix W_o given by equation (8.4) has n independent rows.

8.4 (Observable canonical form) Show that if a system is observable, then there exists a change of coordinates $z = Tx$ that puts the transformed system into observable canonical form.

8.5 (Bicycle dynamics) The linearized model for a bicycle is given in equation (4.5), which has the form

$$J\frac{d^2\varphi}{dt^2} - \frac{Dv_0}{b}\frac{d\delta}{dt} = mgh\varphi + \frac{mv_0^2 h}{b}\delta,$$

where φ is the tilt of the bicycle and δ is the steering angle. Give conditions under which the system is observable and explain any special situations where it loses observability.

8.6 (Observer design by eigenvalue assignment) Consider the system

$$\frac{dx}{dt} = Ax = \begin{pmatrix} -1 & 0 \\ 1 & 0 \end{pmatrix} x + \begin{pmatrix} a-1 \\ 1 \end{pmatrix} u, \qquad y = Cx = \begin{pmatrix} 0 & 1 \end{pmatrix} x.$$

Design an observer such that $\det(sI - LC) = s^2 + 2\zeta_o\omega_o s + \omega_o^2$ with values $\omega_o = 10$ and $\zeta_o = 0.6$.

8.7 (Vectored thrust aircraft) The lateral dynamics of the vectored thrust aircraft example described in Example 7.9 can be obtained by considering the motion described by the states $z = (x, \theta, \dot{x}, \dot{\theta})$. Construct an estimator for these dynamics by setting the eigenvalues of the observer into a *Butterworth pattern* with $\lambda_{bw} = -3.83 \pm 9.24i$, $-9.24 \pm 3.83i$. Using this estimator combined with the state space controller computed in Example 7.9, plot the step response of the closed loop system.

8.8 (Observer for Teorell's compartment model) Teorell's compartment model, shown in Figure 4.17, has the following state space representation:

$$\frac{dx}{dt} = \begin{pmatrix} -k_1 & 0 & 0 & 0 & 0 \\ k_1 & -k_2-k_4 & 0 & k_3 & 0 \\ 0 & k_4 & 0 & 0 & 0 \\ 0 & k_2 & 0 & -k_3-k_5 & 0 \\ 0 & 0 & 0 & k_5 & 0 \end{pmatrix} x + \begin{pmatrix} 1 \\ 0 \\ 0 \\ 0 \\ 0 \end{pmatrix} u,$$

where representative parameters are $k_1 = 0.02$, $k_2 = 0.1$, $k_3 = 0.05$, $k_4 = k_5 = 0.005$. The concentration of a drug that is active in compartment 5 is measured in the bloodstream (compartment 2). Determine the compartments that are observable from measurement of concentration in the bloodstream and design an estimator for these concentrations base on eigenvalue assignment. Choose the closed loop eigenvalues -0.03, -0.05, and -0.1. Simulate the system when the input is a pulse injection.

8.9 (Whipple bicycle model) Consider the Whipple bicycle model given by equation (4.8) in Section 4.2. A state feedback for the system was designed in Exercise 7.12. Design an observer and an output feedback for the system.

8.10 (Kalman decomposition) Consider a linear system characterized by the matrices

$$A = \begin{pmatrix} -2 & 1 & -1 & 2 \\ 1 & -3 & 0 & 2 \\ 1 & 1 & -4 & 2 \\ 0 & 1 & -1 & -1 \end{pmatrix}, \quad B = \begin{pmatrix} 2 \\ 2 \\ 2 \\ 1 \end{pmatrix}, \quad C = \begin{pmatrix} 0 & 1 & -1 & 0 \end{pmatrix}, \quad D = 0.$$

Construct a Kalman decomposition for the system. (Hint: Try to diagonalize.)

8.11 (Kalman filtering for a first-order system) Consider the system

$$\frac{dx}{dt} = ax + v, \qquad y = cx + w$$

where all variables are scalar. The signals v and w are uncorrelated white noise disturbances with zero mean values and covariance functions

$$\mathbb{E}(v(s)v^T(t)) = r_v \delta(t - s), \quad \mathbb{E}(w(s)w^T(t)) = r_w \delta(t - s).$$

The initial condition is Gaussian with mean value x_0 and covariance P_0. Determine the Kalman filter for the system and analyze what happens for large t.

8.12 (Vertical alignment) In navigation systems it is important to align a system to the vertical. This can be accomplished by measuring the vertical acceleration and controlling the platform so that the measured acceleration is zero. A simplified one-dimensional version of the problem can be modeled by

$$\frac{d\varphi}{dt} = u, \qquad u = -ky, \qquad y = \varphi + w,$$

where φ is the alignment error, u the control signal, y the measured signal, and w the measurement noise, which is assumed to be white noise with zero mean and covariance function $\mathbb{E}(w(s)w^T(t)) = r_w \delta(t - s)$. The initial misalignment is assumed to be a random variable with zero mean and the covariance P_0. Determine a time-varying gain $k(t)$ such that the error goes to zero as fast as possible. Compare this with a constant gain.

Chapter Nine

Transfer Functions

> The typical regulator system can frequently be described, in essentials, by differential equations of no more than perhaps the second, third, or fourth order. ... In contrast, the order of the set of differential equations describing the typical negative feedback amplifier used in telephony is likely to be very much greater. As a matter of idle curiosity, I once counted to find out what the order of the set of equations in an amplifier I had just designed would have been, if I had worked with the differential equations directly. It turned out to be 55.
>
> —Hendrik Bode, 1960 [52].

This chapter introduces the concept of the *transfer function*, which is a compact description of the input/output relation for a linear time-invariant system. We show how to obtain transfer functions analytically and experimentally. Combining transfer functions with block diagrams gives a powerful algebraic method to analyze linear systems with many blocks. The transfer function allows new interpretations of system dynamics. We also introduce the Bode plot, a powerful graphical representation of the transfer function that was introduced by Bode to analyze and design feedback amplifiers.

9.1 FREQUENCY DOMAIN MODELING

Figure 9.1 is a block diagram for a typical control system, consisting of a process to be controlled and a controller that combines feedback and feedforward. We saw in the previous two chapters how to analyze and design such systems using state space descriptions of the blocks. As mentioned in Chapter 3, an alternative approach is to focus on the input/output characteristics of the system. Since it is the inputs and outputs that are used to connect the systems, one could expect that this point of view would allow an understanding of the overall behavior of the system. Transfer functions are the main tool in implementing this approach for linear systems.

The basic idea of the transfer function comes from looking at the frequency response of a system. Suppose that we have an input signal that is periodic. Then we can decompose this signal into the sum of a set of sines and cosines,

$$u(t) = \sum_{k=0}^{\infty} a_k \sin(k\omega_{\mathrm{f}} t) + b_k \cos(k\omega_{\mathrm{f}} t),$$

where ω_{f} is the fundamental frequency of the periodic input. As we saw in Section 6.3, the input $u(t)$ generates corresponding sine and cosine outputs (in steady

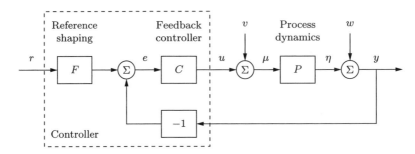

Figure 9.1: A block diagram for a feedback control system. The reference signal r is fed through a reference shaping block, which generates a signal that is compared with the output y to form the error e. The control signal u is generated by the controller, which has the error as the input. The load disturbance v and the measurement noise w are external signals.

state), with possibly shifted magnitude and phase. The gain and phase at each frequency are determined by the frequency response given in equation (6.24):

$$G(i\omega) = C(i\omega I - A)^{-1}B + D, \tag{9.1}$$

where we set $\omega = k\omega_{\mathrm{f}}$ for each $k = 1, \ldots, \infty$. We can thus use the steady-state frequency response $G(i\omega)$ and superposition to compute the steady-state response for any periodic signal.

The transfer function generalizes this notion to allow a broader class of input signals besides periodic ones. As we shall see in the next section, the transfer function represents the response of the system to an *exponential input*, $u = e^{st}$. It turns out that the form of the transfer function is precisely the same as that of equation (9.1). This should not be surprising since we derived equation (9.1) by writing sinusoids as sums of complex exponentials. The transfer function can also be introduced as the ratio of the Laplace transforms of the output and the input when the initial state is zero, although one does not have to understand the details of Laplace transforms in order to make use of transfer functions.

Modeling a system through its response to sinusoidal and exponential signals is known as *frequency domain modeling*. This terminology stems from the fact that we represent the dynamics of the system in terms of the generalized frequency s rather than the time domain variable t. The transfer function provides a complete representation of a linear system in the frequency domain.

The power of transfer functions is that they provide a particularly convenient representation in manipulating and analyzing complex linear feedback systems. As we shall see, there are graphical representations of transfer functions (Bode and Nyquist plots) that capture interesting properties of the underlying dynamics. Transfer functions also make it possible to express the changes in a system because of modeling error, which is essential when considering sensitivity to process variations of the sort discussed in Chapter 13. More specifically, using transfer functions it is possible to analyze what happens when dynamical models are approximated by static models or when high-order models are approximated by low-order models. One consequence is that we can introduce concepts that express the degree of stability of a system.

While many of the concepts for state space modeling and analysis apply directly to nonlinear systems, frequency domain analysis applies primarily to linear systems. The notions of gain and phase can, however, be generalized to nonlinear systems and, in particular, propagation of sinusoidal signals through a nonlinear system can approximately be captured by an analog of the frequency response called the describing function. These extensions of frequency response will be discussed in Section 10.5.

9.2   DETERMINING THE TRANSFER FUNCTION

As we have seen in previous chapters, the input/output dynamics of a linear system have two components: the initial condition response and the forced response, which depends on the system input. The forced response can be characterized by the transfer function. In this section we will compute transfer functions for general linear time-invariant systems. Transfer functions will also be determined for systems with time delays and systems described by partial differential equations, for which the transfer functions obtained are then transcendental functions of a complex variable.

Transmission of Exponential Signals

To formally compute the transfer function of a system, we will make use of a special type of signal, called an *exponential signal,* of the form e^{st}, where $s = \sigma + i\omega$ is a complex number. Exponential signals play an important role in linear systems. They appear in the solution of differential equations and in the impulse response of linear systems, and many signals can be represented as exponentials or sums of exponentials. For example, a constant signal is simply $e^{\alpha t}$ with $\alpha = 0$. Using Euler's formula, damped sine and cosine signals can be represented by

$$e^{(\sigma + i\omega)t} = e^{\sigma t} e^{i\omega t} = e^{\sigma t}(\cos \omega t + i \sin \omega t),$$

where $\sigma < 0$ determines the decay rate. Figure 9.2 gives examples of signals that can be represented by complex exponentials; many other signals can be represented by linear combinations of these signals.

As in the case of the sinusoidal signals we considered in Section 6.3, we will allow complex-valued signals in the derivation that follows, although in practice we always add together combinations of signals that result in real-valued functions.

To find the transfer function for the state space system

$$\frac{dx}{dt} = Ax + Bu, \qquad y = Cx + Du, \tag{9.2}$$

we let the input be the exponential signal $u(t) = e^{st}$ and assume that $s \notin \lambda(A)$. The state is then given by

$$x(t) = e^{At}x(0) + \int_0^t e^{A(t-\tau)} B e^{s\tau} \, d\tau = e^{At}x(0) + e^{At}(sI - A)^{-1}\left(e^{(sI-A)t} - I\right)B.$$

The output y of equation (9.2) then becomes

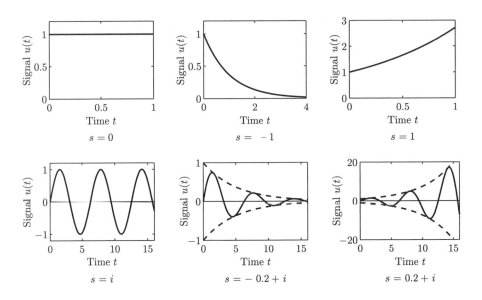

Figure 9.2: Examples of exponential signals. The top row corresponds to exponential signals with a real exponent, and the bottom row corresponds to those with complex exponents. The dashed line in the last two cases denotes the bounding envelope for the oscillatory signals. In each case, if the real part of the exponent is negative then the signal decays, while if the real part is positive then it grows.

$$y(t) = Cx(t) + Du(t)$$

$$= \underbrace{Ce^{At}x(0)}_{\substack{\text{initial condition} \\ \text{response}}} + \underbrace{\left(C(sI - A)^{-1}B + D\right)e^{st} - Ce^{At}(sI - A)^{-1}B}_{\text{input response}} \tag{9.3}$$

$$= \underbrace{Ce^{At}\left(x(0) - (sI - A)^{-1}B\right)}_{\text{transient response}} + \underbrace{\left(C(sI - A)^{-1}B + D\right)e^{st}}_{\text{pure exponential response } y_{\mathrm{p}}},$$

and the *transfer function* from u to y of the system (9.2) is the coefficient of the term e^{st}, hence

$$G(s) = C(sI - A)^{-1}B + D. \tag{9.4}$$

Compare this with the definition of frequency response given by equations (6.23) and (6.24).

An important point in the derivation of the transfer function is the fact that we have restricted s so that $s \neq \lambda_j(A)$, the eigenvalues of A. At those values of s, we see that the response (9.3) of the system is singular (since $sI - A$ then is not invertible). The transfer function can, however, be extended to all values of s by analytic continuation.

To give some insight we will now discuss the structure of equation (9.3). We first notice that the output $y(t)$ can be separated into two terms in two different ways, as is indicated by braces in the equation.

The response of the system to initial conditions is $Ce^{At}x(0)$. Recall that e^{At} can be written in terms of the eigenvalues of A (using the Jordan form in the case of repeated eigenvalues), and hence the transient response is a linear combination of terms of the form $p_j(t)e^{\lambda_j t}$, where λ_j are eigenvalues of A and $p_j(t)$ is a polynomial whose degree is less than the multiplicity of the eigenvalue (Exercise 9.1).

The response to the input $u(t) = e^{st}$ contains a mixture of terms $p_j(t)e^{\lambda_j t}$ and the exponential function

$$y_{\mathrm{p}}(t) = \big(C(sI - A)^{-1}B + D\big)e^{st} = G(s)e^{st}, \qquad (9.5)$$

which is a *particular solution* to the differential equation (9.2). We call equation (9.5) the *pure exponential solution* because it has only one exponential e^{st}. It follows from equation (9.3) that the output $y(t)$ is equal to the pure exponential solution $y_{\mathrm{p}}(t)$ if the initial condition is chosen as

$$x(0) = (sI - A)^{-1}B. \qquad (9.6)$$

If the system (9.2) is asymptotically stable, then $e^{At} \to 0$ as $t \to \infty$. If in addition the input $u(t)$ is a constant $u(t) = e^{0 \cdot t}$ or a sinusoid $u(t) = e^{i\omega t}$ then the response converges to a constant or sinusoidal *steady-state solution* (as shown in equation (6.23)).

To simplify manipulation of the equations describing linear time-invariant systems, we introduce $\mathcal{E}$ as the class of time functions that can be created from combinations of signals of the form $X(s)e^{st}$, where the parameter s is a complex variable and $X(s)$ is a complex function (vector valued if needed). It follows from equations (9.3) and (9.4) that if a system with transfer function $G(s)$ has the input $u \in \mathcal{E}$ then there is a particular solution $y \in \mathcal{E}$ that satisfies the dynamics of the system. This solution is the actual response of the system if the initial condition is chosen as equation (9.6). Since the transfer function of a system is given by the pure exponential response, we can derive transfer functions using exponential signals, and we will use the notation

$$y = G_{yu}\, u, \qquad (9.7)$$

where G_{yu} is the transfer function for the linear input/output system taking u to y. Mathematically, it is important to remember that this notation assumes the use of combinations of exponential signals. We will also often drop the subscripts on G and just write $y = Gu$ when the meaning is clear from context.

Example 9.1 Damped oscillator
Consider the response of a damped linear oscillator, whose state space dynamics were studied in Section 7.3:

$$\frac{dx}{dt} = \begin{pmatrix} 0 & \omega_0 \\ -\omega_0 & -2\zeta\omega_0 \end{pmatrix} x + \begin{pmatrix} 0 \\ k\omega_0 \end{pmatrix} u, \qquad y = \begin{pmatrix} 1 & 0 \end{pmatrix} x. \qquad (9.8)$$

This system is asymptotically stable if $\zeta > 0$, and so we can look at the steady-state response to an input $u = e^{st}$:

$$G_{yu}(s) = C(sI - A)^{-1}B = \begin{pmatrix} 1 & 0 \end{pmatrix} \begin{pmatrix} s & -\omega_0 \\ \omega_0 & s + 2\zeta\omega_0 \end{pmatrix}^{-1} \begin{pmatrix} 0 \\ k\omega_0 \end{pmatrix}$$

$$= \begin{pmatrix} 1 & 0 \end{pmatrix} \left(\frac{1}{s^2 + 2\zeta\omega_0 s + \omega_0^2} \begin{pmatrix} s + 2\zeta\omega_0 & -\omega_0 \\ \omega_0 & s \end{pmatrix} \right) \begin{pmatrix} 0 \\ k\omega_0 \end{pmatrix} \qquad (9.9)$$

$$= \frac{k\omega_0^2}{s^2 + 2\zeta\omega_0 s + \omega_0^2}.$$

The steady-state response to a step input is obtained by setting $s = 0$, which gives

$$u = 1 \qquad \Longrightarrow \qquad y = G_{yu}(0)u = k.$$

If we wish to compute the steady-state response to a sinusoid, we write

$$u = \sin \omega t = \frac{1}{2} \left(ie^{-i\omega t} - ie^{i\omega t} \right) \qquad \Longrightarrow \qquad y = \frac{1}{2} \left(iG_{yu}(-i\omega)e^{-i\omega t} - iG_{yu}(i\omega)e^{i\omega t} \right).$$

We can now write $G(i\omega)$ in terms of its magnitude and phase,

$$G(i\omega) = \frac{k\omega_0^2}{-\omega^2 + (2\zeta\omega_0\omega)i + \omega_0^2} = Me^{i\theta},$$

where the magnitude (or gain) M and phase θ are given by

$$M = \frac{k\omega_0^2}{\sqrt{(\omega_0^2 - \omega^2)^2 + (2\zeta\omega_0\omega)^2}}, \qquad \frac{\sin\theta}{\cos\theta} = \frac{-2\zeta\omega_0\omega}{\omega_0^2 - \omega^2}.$$

We can also make use of the fact that $G(-i\omega)$ is given by its complex conjugate $G^*(i\omega)$, and it follows that $G(-i\omega) = Me^{-i\theta}$. Substituting these expressions into our output equation, we obtain

$$y = \frac{1}{2} \left(i(Me^{-i\theta})e^{-i\omega t} - i(Me^{i\theta})e^{i\omega t} \right)$$

$$= M \cdot \frac{1}{2} \left(ie^{-i(\omega t + \theta)} - ie^{i(\omega t + \theta)} \right) = M\sin(\omega t + \theta).$$

The responses to other signals can be computed by writing the input as an appropriate combination of exponential responses and using linearity. $\qquad \nabla$

Example 9.2 Operational amplifier circuit
To further illustrate the use of exponential signals, we consider the operational amplifier circuit described in Section 4.3 and reproduced in Figure 9.3a. The model in Section 4.3 is a simplification because the linear behavior of the amplifier is modeled as a constant gain. In reality there are significant dynamics in the amplifier, and the static model $v_{\text{out}} = -kv$ (equation (4.11)) should therefore be replaced by a dynamical model $v_{\text{out}} = -Gv$. A simple transfer function is

$$G(s) = \frac{ak}{s + a}. \qquad (9.10)$$

These dynamics correspond to a first-order system with time constant $1/a$. The parameter k is called the *open loop gain*, and the product ak is called the

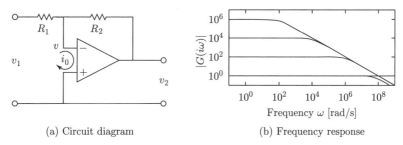

(a) Circuit diagram   (b) Frequency response

Figure 9.3: Stable amplifier based on negative feedback around an operational amplifier. The circuit diagram on the left shows a typical amplifier with low-frequency gain R_2/R_1. If we model the dynamic response of the op amp as $G(s) = ak/(s+a)$, then the gain falls off at frequency $\omega = aR_1k/R_2$, as shown in the gain curves on the right. The frequency response is computed for $k = 10^7$, $a = 10\,\text{rad/s}$, $R_2 = 10^6\,\Omega$, and $R_1 = 1$, 10^2, 10^4, and $10^6\,\Omega$.

gain-bandwidth product; typical values for these parameters are $k = 10^7$ and $ak = 10^7$–10^9 rad/s.

If the input v_1 is an exponential signal e^{st}, then there are solutions where all signals in the circuit are exponentials, $v, v_1, v_2 \in \mathcal{E}$, since all of the elements of the circuit are modeled as being linear. The equations describing the system can then be manipulated algebraically.

Assuming that the current into the amplifier is zero, as is done in Section 4.3, the currents through the resistors R_1 and R_2 are the same, hence

$$\frac{v_1 - v}{R_1} = \frac{v - v_2}{R_2}, \quad \text{or} \quad (R_1 + R_2)v = R_2v_1 + R_1v_2.$$

Combining the above equation with the open loop dynamics of the operational amplifier (9.10), which can be written as $v_2 = -Gv$ in the simplified notation (9.7), gives the following model for the closed loop system:

$$(R_1 + R_2)v = R_2v_1 + R_1v_2, \qquad v_2 = -Gv, \qquad v, v_1, v_2 \in \mathcal{E}. \qquad (9.11)$$

Eliminating v between these equations yields

$$v_2 = \frac{-R_2G}{R_1 + R_2 + R_1G}v_1 = \frac{-R_2ak}{R_1ak + (R_1 + R_2)(s+a)}v_1,$$

and the transfer function of the closed loop system is

$$G_{v_2v_1} = \frac{-R_2ak}{R_1ak + (R_1 + R_2)(s+a)}. \qquad (9.12)$$

The low-frequency gain is obtained by setting $s = 0$, hence

$$G_{v_2v_1}(0) = \frac{-kR_2}{(k+1)R_1 + R_2} \approx -\frac{R_2}{R_1},$$

which is the result given by equation (4.12) in Section 4.3. The bandwidth of the amplifier circuit is

$$\omega_{\mathrm{b}} = a\frac{R_1(k+1)+R_2}{R_1+R_2} \approx a\frac{R_1 k}{R_2} \quad \text{for} \quad k \gg 1,$$

where the approximation holds for $R_2/R_1 \gg 1$. The gain of the closed loop system drops off at high frequencies as $R_2 ak/(\omega(R_1+R_2))$. The frequency response of the transfer function is shown in Figure 9.3b for $k=10^7$, $a=10\,\mathrm{rad/s}$, $R_2=10^6\,\Omega$, and $R_1=1$, 10^2, 10^4, and $10^6\,\Omega$.

Note that in solving this example, we bypassed explicitly writing the signals as $v = V(s)e^{st}$ and instead worked directly with v, assuming it was an exponential. This shortcut is handy in solving problems of this sort and when manipulating block diagrams. A comparison with Section 4.3, where we make the same calculation when $G(s)$ is a constant, shows analysis of systems using transfer functions is as easy as using static systems. The calculations are the same if the resistances R_1 and R_2 are replaced by impedances, as discussed further in Example 9.3. ∇

Transfer Functions for Linear Differential Equations

Consider a linear system described by the controlled differential equation

$$\frac{d^n y}{dt^n} + a_1\frac{d^{n-1}y}{dt^{n-1}} + \cdots + a_n y = b_0\frac{d^m u}{dt^m} + b_1\frac{d^{m-1}u}{dt^{m-1}} + \cdots + b_m u, \qquad (9.13)$$

where u is the input and y is the output. Notice that here we have generalized our system description from Section 3.2 to allow both the input and its derivatives to appear. This type of description arises in many applications, as described briefly in Chapter 2 and Section 3.2; bicycle dynamics and AFM modeling are two specific examples.

To determine the transfer function of the system (9.13), let the input be $u(t) = e^{st}$. Since the system is linear, there is an output of the system that is also an exponential function $y(t) = y_0 e^{st}$. Inserting the signals into equation (9.13), we find

$$(s^n + a_1 s^{n-1} + \cdots + a_n)y_0 e^{st} = (b_0 s^m + b_1 s^{m-1} \cdots + b_m)e^{st},$$

and the response of the system can be completely described by two polynomials

$$a(s) = s^n + a_1 s^{n-1} + \cdots + a_n, \qquad b(s) = b_0 s^m + b_1 s^{m-1} + \cdots + b_m. \qquad (9.14)$$

The polynomial $a(s)$ is the characteristic polynomial of the ordinary differential equation. If $a(s) \neq 0$, it follows that

$$y(t) = y_0 e^{st} = \frac{b(s)}{a(s)}e^{st}. \qquad (9.15)$$

The transfer function of the system (9.13) is thus the rational function

$$G(s) = \frac{b(s)}{a(s)} = \frac{b_0 s^m + b_1 s^{m-1} + \cdots + b_m}{s^n + a_1 s^{n-1} + \cdots + a_n}, \qquad (9.16)$$

where the polynomials $a(s)$ and $b(s)$ are given by equation (9.14). Notice that the transfer function for the system (9.13) can be obtained by inspection since the coefficients of $a(s)$ and $b(s)$ are precisely the coefficients of the derivatives of u and y. The *poles* and the *zeros* of the transfer function are the roots of the polynomials

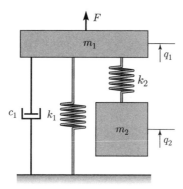

Figure 9.4: A vibration damper. Vibrations of the mass m_1 can be damped by providing it with an auxiliary mass m_2, attached to m_1 by a spring with stiffness k_2. The parameters m_2 and k_2 are chosen so that the frequency $\sqrt{k_2/m_2}$ matches the frequency of the vibration.

$a(s)$ and $b(s)$. The properties of the system are determined by the poles and zeros of the transfer function, as we shall see in the examples that follow and shall explore in more detail in Section 9.5.

Example 9.3 Electrical circuit elements

Modeling of electrical circuits is a common use of transfer functions. Consider, for example, a resistor modeled by Ohm's law $V = IR$, where V is the voltage across the resistor, I is the current through the resistor, and R is the resistance value. If we consider current to be the input and voltage to be the output, the resistor has the transfer function $Z(s) = R$, which is also called the *generalized impedance* of the circuit element.

Next we consider an inductor whose input/output characteristic is given by

$$L\frac{dI}{dt} = V.$$

Letting the current be $I(t) = e^{st}$, we find that the voltage is $V(t) = Lse^{st}$ and the transfer function of an inductor is thus $Z(s) = Ls$. A capacitor is characterized by

$$C\frac{dV}{dt} = I,$$

and a similar analysis gives a transfer function from current to voltage of $Z(s) = 1/(Cs)$. Using transfer functions, complex electrical circuits can be analyzed algebraically by using the generalized impedance $Z(s)$ just as one would use the resistance value in a resistor network. ∇

Example 9.4 Vibration damper

Damping vibrations is a common engineering problem. A schematic diagram of a vibration damper is shown in Figure 9.4. To analyze the system we use Newton's equations for the two masses:

$$m_1\ddot{q}_1 + c_1\dot{q}_1 + k_1q_1 + k_2(q_1 - q_2) = F, \qquad m_2\ddot{q}_2 + k_2(q_2 - q_1) = 0.$$

To determine the transfer function from the force F to the position q_1 of the mass m_1 we first find particular exponential solutions:

$$(m_1 s^2 + c_1 s + k_1)q_1 + k_2(q_1 - q_2) = F, \qquad m_2 s^2 q_2 + k_2(q_2 - q_1) = 0. \qquad (9.17)$$

We solve q_2 from the second expression,

$$q_2 = \frac{k_2}{m_2 s^2 + k_2} q_1,$$

and insert this into the first expression to obtain

$$(m_1 s^2 + c_1 s + k_1)q_1 + k_2 \left(1 - \frac{k_2}{m_2 s^2 + k_2}\right) q_1 = F,$$

and hence

$$\left((m_1 s^2 + c_1 s + k_1 + k_2)(m_2 s^2 + k_2) - k_2^2\right) q_1 = (m_2 s^2 + k_2)F.$$

Expanding the expression gives the transfer function

$$G_{q_1 F}(s) = \frac{m_2 s^2 + k_2}{m_1 m_2 s^4 + m_2 c_1 s^3 + (m_1 k_2 + m_2(k_1 + k_2))s^2 + k_2 c_1 s + k_1 k_2}$$

from the disturbance force F to the position q_1 of the mass m_1. The transfer function has a zero at $s = \pm i\sqrt{k_2/m_2}$, which means that transmission of sinusoidal signals with this frequency are blocked (this blocking property will be discussed in Section 9.5). $\qquad \nabla$

As the examples above illustrate, transfer functions provide a simple representation for linear input/output systems. Transfer functions for some common linear time-invariant systems are given in Table 9.1. Transfer functions of a form similar to equation (9.13) can also be constructed for systems with many inputs and many outputs.

Time Delays and Partial Differential Equations

Although we have focused thus far on ordinary differential equations, transfer functions can also be used for other types of linear systems. We illustrate this using time delays and systems described by partial differential equations.

Example 9.5 Time delay
Time delays appear in many systems: typical examples are delays in nerve propagation, communication systems, and mass transport. A system with a time delay has the input/output relation

$$y(t) = u(t - \tau). \qquad (9.18)$$

To obtain the corresponding transfer function we let the input be $u(t) = e^{st}$, and the output is then

$$y(t) = u(t - \tau) = e^{s(t-\tau)} = e^{-s\tau} e^{st} = e^{-s\tau} u(t).$$

Table 9.1: Transfer functions for some common linear time-invariant systems.

Type	System	Transfer Function
Integrator	$\dot{y} = u$	$\dfrac{1}{s}$
Differentiator	$y = \dot{u}$	s
First-order system	$\dot{y} + ay = u$	$\dfrac{1}{s + a}$
Double integrator	$\ddot{y} = u$	$\dfrac{1}{s^2}$
Damped oscillator	$\ddot{y} + 2\zeta\omega_0\dot{y} + \omega_0^2 y = u$	$\dfrac{1}{s^2 + 2\zeta\omega_0 s + \omega_0^2}$
State space system	$\dot{x} = Ax + Bu, \; y = Cx + Du$	$C(sI - A)^{-1}B + D$
PID controller	$y = k_{\mathrm{p}}u + k_{\mathrm{d}}\dot{u} + k_{\mathrm{i}}\int u$	$k_{\mathrm{p}} + k_{\mathrm{d}}s + \dfrac{k_{\mathrm{i}}}{s}$
Time delay	$y(t) = u(t - \tau)$	$e^{-\tau s}$

We find that the transfer function of a time delay is thus $G(s) = e^{-s\tau}$, which is not a rational function. ∇

Example 9.6 Heat propagation

Consider the problem of one-dimensional heat propagation in a semi-infinite metal rod. Assume that the input is the temperature at one end and that the output is the temperature at a point along the rod. Let $\theta(x, t)$ be the temperature at position x and time t. With a proper choice of length scales and units, heat propagation is described by the partial differential equation

$$\frac{\partial \theta}{\partial t} = \frac{\partial^2 \theta}{\partial^2 x}, \qquad y(t) = \theta(1, t), \tag{9.19}$$

and the point of interest can be assumed to have $x = 1$. The boundary condition for the partial differential equation is

$$\theta(0, t) = u(t).$$

To determine the transfer function we choose the input as $u(t) = e^{st}$. Assume that there is a solution to the partial differential equation of the form $\theta(x, t) = \psi(x)e^{st}$ and insert this into equation (9.19) to obtain

$$s\psi(x) = \frac{d^2\psi}{dx^2},$$

with boundary condition $\psi(0) = 1$. This ordinary differential equation (with independent variable x) has the solution

$$\psi(x) = Ae^{x\sqrt{s}} + Be^{-x\sqrt{s}}.$$

Since the temperature of the rod is bounded we have $A = 0$, the boundary condition gives $B = 1$, and the solution is then

$$y(t) = \theta(1, t) = \psi(1)e^{st} = e^{-\sqrt{s}}e^{st} = e^{-\sqrt{s}}u(t).$$

The system thus has the transfer function $G(s) = e^{-\sqrt{s}}$. As in the case of a time delay, the transfer function is not a rational function. ∇

State Space Realizations of Transfer Functions

We have seen in equation (9.4) how to compute the transfer function for a given state space control system. The inverse problem, computing a state space control system for a given transfer function, is known as the *realization problem*. Given a transfer function $G(s)$, we say that a state space system with matrices A, B, C, and D is a (state space) *realization* of $G(s)$ if $G(s) = C(sI - A)^{-1}B + D$. We explore here some of the properties of realizations of transfer functions, starting with the question of uniqueness.

As we saw in Section 6.3, it is possible to choose a different set of coordinates for the state space of a linear system and still preserve the input/output response. In other words, the matrices A, B, C, and D in the state space equations (9.2) depend on the choice of coordinate system used for the states, but since the transfer function relates input to outputs, it should be invariant to coordinate changes in the state space. Repeating the analysis in Chapter 6, consider a model (9.2) and introduce new coordinates z by the transformation $z = Tx$, where T is a nonsingular matrix. The system is then described by

$$\frac{dz}{dt} = T(Ax + Bu) = TAT^{-1}z + TBu =: \tilde{A}z + \tilde{B}u,$$

$$y = Cx + Du = CT^{-1}z + Du =: \tilde{C}z + Du.$$

This system has the same form as equation (9.2), but the matrices A, B, and C are different:

$$\tilde{A} = TAT^{-1}, \qquad \tilde{B} = TB, \qquad \tilde{C} = CT^{-1}. \tag{9.20}$$

Computing the transfer function of the transformed model, we get

$$\tilde{G}(s) = \tilde{C}(sI - \tilde{A})^{-1}\tilde{B} + D = CT^{-1}(sI - TAT^{-1})^{-1}TB + D$$

$$= C\left(T^{-1}(sI - TAT^{-1})T\right)^{-1}B + D = C(sI - A)^{-1}B + D = G(s),$$

which is identical to the transfer function (9.4) computed from the system description (9.2). The transfer function is thus invariant to changes of the coordinates in the state space.

One consequence of this coordinate invariance is that it is not possible for there to be a *unique* state space realization for a given transfer function. Given any one realization, we can compute another realization by simply changing coordinates using any invertible matrix T. Note, however, that the dimension of the state space realization is not changed by this transformation. It therefore makes sense to talk about a *minimal realization*, in which the number of states is as small as possible. For a transfer function $G(s) = b(s)/a(s)$ with denominator $a(s)$ of degree n, it can be shown that there is always a realization with n states, given by a state space system in reachable canonical form (7.6). In general, a minimal realization will always have at most n states. However, the degree may be lower if there are pole/zero cancellations, as illustrated by the following example.

Example 9.7 Cancellation of poles and zeros
Consider the system

$$\frac{dx}{dt} = \begin{pmatrix} -3 & 1 \\ -2 & 0 \end{pmatrix} x + \begin{pmatrix} 1 \\ 1 \end{pmatrix} u, \qquad y = \begin{pmatrix} 1 & 0 \end{pmatrix} x.$$

Equation (9.4) gives the following transfer function

$$G(s) = \begin{pmatrix} 1 & 0 \end{pmatrix} \begin{pmatrix} s+3 & -1 \\ 2 & s \end{pmatrix}^{-1} \begin{pmatrix} 1 \\ 1 \end{pmatrix} = \frac{1}{s^2 + 3s + 2} \begin{pmatrix} 1 & 0 \end{pmatrix} \begin{pmatrix} s & 1 \\ -2 & s+3 \end{pmatrix} \begin{pmatrix} 1 \\ 1 \end{pmatrix}$$

$$= \frac{s+1}{s^2 + 3s + 2} = \frac{s+1}{(s+1)(s+2)} = \frac{1}{s+2}.$$

Even though the original state space system was of second order, the transfer function is a first-order rational function. The reason is that the factor $s+1$ has been canceled when computing the transfer function. Cancellation of poles and zeros is related to lack of reachability and observability. In this particular case the reachability matrix

$$W_{\mathrm{r}} = \begin{pmatrix} B & AB \end{pmatrix} = \begin{pmatrix} 1 & -2 \\ 1 & -2 \end{pmatrix}$$

has rank 1 and the system is not reachable. Notice that it was shown in Section 8.3 that the transfer function is given by the reachable and observable subsystem Σ_{ro} in the Kalman decomposition of a linear system, which in this case is of first order. $\qquad\qquad \nabla$

The general approach to understand realizations (and minimal realizations) is to make use of the Kalman decomposition in Section 8.3. We see from the structure of equation (8.20) that the input/output response of a linear control system is determined solely by the reachable and observable subsystem Σ_{ro}. When a system lacks reachability and observability, this shows up as cancellation of poles and zeros in the transfer function computed from the full system matrices.

Cancellation of poles and zeros was controversial for a long time, which was manifested in rules for manipulating transfer functions: do not cancel factors with roots in the right half-plane. Special algebraic methods were also developed to do block diagram algebra. Kalman's decomposition, which clarifies that the transfer function only represents part of the dynamics, gives clear insight into what is happening. These issues are discussed in more detail in Section 9.5.

The results of this section can also be extended to the case of multi-input, multi-output (MIMO) systems. The transfer function $G(s)$ for a single-input, single-output given by equation (9.4) is a function of complex variables, $G: \mathbb{C} \to \mathbb{C}$. For systems with p inputs and q outputs the transfer function is matrix-valued, $G: \mathbb{C} \to \mathbb{C}^{q \times p}$. The techniques described above can be generalized to this case, but the notion of a (minimal) realization becomes substantially more complicated.

Table 9.2: Laplace transforms for some common signals.

Signal $u(t)$	Laplace transform $U(s)$	Signal $u(t)$	Laplace transform $U(s)$
$S(t)$ [unit step]	$\dfrac{1}{s}$	$\delta(t)$ [impulse]	1
$\sin(at)$	$\dfrac{a}{s^2 + a^2}$	$\cos(at)$	$\dfrac{s}{s^2 + a^2}$
$e^{-\alpha t}\sin(at)$	$\dfrac{a}{(s + \alpha)^2 + a^2}$	$e^{-\alpha t}\cos(at)$	$\dfrac{s + \alpha}{(s + \alpha)^2 + a^2}$

9.3   LAPLACE TRANSFORMS

The traditional way to derive the transfer function for a linear, time-invariant, input/output system is to make use of Laplace transforms. The Laplace transform method was particularly important before the advent of computers, since it provided a practical way to compute the response of a system to a given input. Today, we compute the response of a linear (or nonlinear) system to complex inputs using numerical simulation, and the Laplace transform is no longer needed for this purpose. It is however, still useful to gain insight into the response of linear systems.

In this section, we provide a brief introduction to the use of Laplace transforms and their connections with transfer functions. Only a few elementary properties of Laplace transforms are needed for basic control applications; students who are not familiar with them can safely skip this section. A good reference for the mathematical material in this section is the classic book by Widder [251] or the more modern treatments available in standard textbooks on signals and systems [162, 198].

Consider a function $f(t)$, $f : \mathbb{R}^+ \to \mathbb{R}$, that is integrable and grows no faster than $e^{s_0 t}$ for some finite $s_0 \in \mathbb{R}$ and large t. The Laplace transform maps f to a function $F = \mathcal{L}f : \mathbb{C} \to \mathbb{C}$ of a complex variable. It is defined by

$$F(s) = \int_0^\infty e^{-st} f(t)\, dt, \quad \operatorname{Re} s > s_0. \tag{9.21}$$

Using this formula, it is possible to compute the Laplace transform of some common functions; see Table 9.2.

The Laplace transform has some properties that makes it well suited to deal with linear systems. First we observe that the transform itself is linear because

$$\mathcal{L}(af + bg) = \int_0^\infty e^{-st}(af(t) + bg(t))\, dt$$

$$= a \int_0^\infty e^{-st} f(t)\, dt + b \int_0^\infty e^{-st} g(t)\, dt = a\mathcal{L}f + b\mathcal{L}g. \tag{9.22}$$

Using linearity we can compute the Laplace transform of combinations of simple inputs, such as those that make up the set of exponential signals $\mathcal{E}$ introduced earlier.

Next we will calculate the Laplace transform of the integral of a function. Using integration by parts, we get

$$\mathcal{L} \int_0^t f(\tau)\, d\tau = \int_0^\infty \left(e^{-st} \int_0^t f(\tau)\, d\tau \right) dt$$

$$= -\frac{e^{-st}}{s} \int_0^t f(\tau)\, d\tau \Big|_0^\infty + \int_0^\infty \frac{e^{-s\tau}}{s} f(\tau)\, d\tau = \frac{1}{s} \int_0^\infty e^{-s\tau} f(\tau)\, d\tau,$$

hence

$$\mathcal{L} \int_0^t f(\tau)\, d\tau = \frac{1}{s} \mathcal{L} f = \frac{1}{s} F(s). \qquad (9.23)$$

Integration of a time function thus corresponds to division of the corresponding Laplace transform by s.

Since integration corresponds to division by s, we can expect that differentiation corresponds to multiplication by s. This is not quite true as we will see by calculating the Laplace transform of the derivative of a function. We have

$$\mathcal{L} \frac{df}{dt} = \int_0^\infty e^{-st} f'(t)\, dt = e^{-st} f(t) \Big|_0^\infty + s \int_0^\infty e^{-st} f(t)\, dt = -f(0) + s\mathcal{L}f,$$

where the second equality is obtained using integration by parts. We thus obtain

$$\mathcal{L} \frac{df}{dt} = s\mathcal{L}f - f(0) = sF(s) - f(0). \qquad (9.24)$$

Notice the appearance of the initial value $f(0)$ of the function. The formula (9.24) is particularly simple if the initial conditions are zero, because if $f(0) = 0$ it follows that differentiation of a function corresponds to multiplication of the transform by s.

Using Laplace transforms the transfer function for a linear time-invariant system can be defined as the ratio of the transform of the input and the output, when the transforms are computed under the assumption that all initial conditions are zero. We will now illustrate how Laplace transforms can be used to compute transfer functions.

Example 9.8 Transfer function of state space model
Consider the state space system described by equation (9.2). Taking Laplace transforms gives

$$sX(s) - x(0) = AX(s) + BU(s), \qquad Y(s) = CX(s) + DU(s).$$

Elimination of $X(s)$ gives

$$X(s) = (sI - A)^{-1} x(0) + (sI - A)^{-1} BU(s). \qquad (9.25)$$

When the initial condition $x(0)$ is zero we have

$$X(s) = (sI - A)^{-1} BU(s), \qquad Y(s) = \left(C(sI - A)^{-1} B + D \right) U(s),$$

and the transfer function is given by $G(s) = C(sI - A)^{-1}B + D$ (compare with equation (9.4)). ∇

Example 9.9 Transfer functions and impulse response

Consider a linear time-invariant system with zero initial state. We saw in Section 6.3 that the relation between the input u and the output y is given by the convolution integral

$$y(t) = \int_0^\infty h(t-\tau)u(\tau)\,d\tau,$$

where $h(t)$ is the impulse response for the system (assumed causal). Taking the Laplace transform of this expression and using the fact that $h(t') = 0$ for $t' = t - \tau < 0$ gives

$$Y(s) = \int_0^\infty e^{-st}y(t)\,dt = \int_0^\infty e^{-st}\int_0^\infty h(t-\tau)u(\tau)\,d\tau\,dt$$

$$= \int_0^\infty \int_0^t e^{-s(t-\tau)}e^{-s\tau}h(t-\tau)u(\tau)\,d\tau\,dt$$

$$= \int_0^\infty \int_0^\infty e^{-st'}h(t')e^{-s\tau}u(\tau)\,d\tau\,dt'$$

$$= \int_0^\infty e^{-st}h(t)\,dt \int_0^\infty e^{-s\tau}u(\tau)\,d\tau = H(s)U(s).$$

Thus, the input/output response is given by $Y(s) = H(s)U(s)$, where H, U, and Y are the Laplace transforms of h, u, and y.

The system theoretic interpretation is that the Laplace transform of the output of a linear system is a product of two terms, the Laplace transform of the input $U(s)$ and the Laplace transform of the impulse response of the system $H(s)$. A mathematical interpretation is that the Laplace transform of a convolution is the product of the transforms of the functions that are convolved. The fact that the formula $Y(s) = H(s)U(s)$ is much simpler than a convolution is one reason why Laplace transforms have traditionally been popular in engineering.   ▽

A variety of theorems are available using Laplace transforms that are useful in a control systems setting. The *initial value theorem* states that

$$\lim_{t \to 0} f(t) = \lim_{s \to \infty} sF(s).$$

Using this theorem and the fact that a step input has Laplace transform $1/s$, we can compute the initial value of signals in a control system in response to step inputs. For example, if G_{ur} represents that transfer function between the reference r and control input u, then the step response will have the property that

$$u(0) = \lim_{t \to 0} u(t) = \lim_{s \to \infty} sU(s) = \lim_{s \to \infty} s \cdot G_{ur}(s) \cdot \frac{1}{s} = G_{ur}(\infty).$$

Similarly, the *final value theorem* states that

$$\lim_{t \to \infty} f(t) = \lim_{s \to 0} sF(s),$$

and this can be used to show that for a step input $r(t)$ we have $\lim_{t \to \infty} y(t) = G_{yr}(0)$.

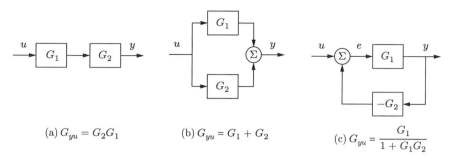

(a) $G_{yu} = G_2 G_1$ (b) $G_{yu} = G_1 + G_2$ (c) $G_{yu} = \dfrac{G_1}{1 + G_1 G_2}$

Figure 9.5: Interconnections of linear systems. Series (a), parallel (b), and feedback (c) connections are shown. The transfer functions for the composite systems can be derived by algebraic manipulations assuming exponential functions for all signals.

9.4   BLOCK DIAGRAMS AND TRANSFER FUNCTIONS

The combination of block diagrams and transfer functions is a powerful way to represent control systems. Transfer functions relating different signals in the system can be derived by purely algebraic manipulations of the transfer functions of the blocks using *block diagram algebra*. Outputs resulting from several input signals can be derived using *superposition*. To show how this can be done, we will begin with simple combinations of systems. We will assume that all signals are exponential signals $\mathcal{E}$ and we will use the compact notation $y = Gu$ for the output $y \in \mathcal{E}$ of a linear time-invariant system with the input $u \in \mathcal{E}$ and the transfer function G (see equation (9.7) and recall its interpretation).

Consider a system that is a cascade combination of systems with the transfer functions $G_1(s)$ and $G_2(s)$, as shown in Figure 9.5a. Let the input of the system be $u \in \mathcal{E}$. The output of the first block is then $G_1 u \in \mathcal{E}$, which is also the input to the second system. The output of the second system is then

$$y = G_2(G_1 u) = (G_2 G_1)u. \tag{9.26}$$

The transfer function of the series connection is thus $G = G_2 G_1$, i.e., the product of the transfer functions. The order of the individual transfer functions is due to the fact that we place the input signal on the right-hand side of this expression, hence we first multiply by G_1 and then by G_2. Unfortunately, this has the opposite ordering from the diagrams that we use, where we typically have the signal flow from left to right, so one needs to be careful. The ordering is important if either G_1 or G_2 is a vector-valued transfer function, as we shall see in some examples.

Consider next a parallel connection of systems with the transfer functions G_1 and G_2, as shown in Figure 9.5b, and assume that all signals are exponential signals. The outputs of the first and second systems are simply $G_1 u$ and $G_2 u$ and the output of the parallel connection is

$$y = G_1 u + G_2 u = (G_1 + G_2)u.$$

The transfer function for a parallel connection is thus $G = G_1 + G_2$.

Finally, consider a feedback connection of systems with the transfer functions G_1 and G_2, as shown in Figure 9.5c. Writing the relations between the signals for

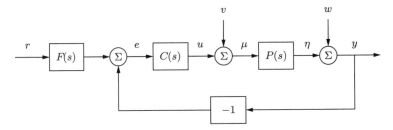

Figure 9.6: Block diagram of a feedback system. The inputs to the system are the reference signal r, the process disturbance v, and the measurement noise w. The remaining signals in the system can all be chosen as possible outputs, and transfer functions can be used to relate the system inputs to the other labeled signals.

the different blocks and the summation unit, we find

$$y = G_1 e, \qquad e = u - G_2 y. \tag{9.27}$$

Elimination of e gives

$$y = G_1(u - G_2 y) \quad \Longrightarrow \quad (1 + G_1 G_2)y = G_1 u \quad \Longrightarrow \quad y = \frac{G_1}{1 + G_1 G_2} u.$$

The transfer function of the feedback connection is thus

$$G = \frac{G_1}{1 + G_1 G_2}. \tag{9.28}$$

These three basic interconnections can be used as the basis for computing transfer functions for more complicated systems.

Control System Transfer Functions

Consider the system in Figure 9.6, which was given at the beginning of the chapter. The system has three blocks representing a process P, a feedback controller C, and a feedforward controller F. Together, C and F define the *control law* for the system. There are three external signals: the reference (or command) signal r, the load disturbance v, and the measurement noise w. A typical problem is to determine how the error e is related to the signals r, v, and w.

To derive the transfer functions we are interested in, we assume that all signals are exponential signals $\mathcal{E}$ and we write the relations between the signals for each block in the system block diagram. Assume for example that we are interested in the control error e. The summation point and the block $F(s)$ gives

$$e = Fr - y.$$

The signal y is the sum of w and η, where η is the output of the process $P(s)$:

$$y = w + \eta, \qquad \eta = P(v + u), \qquad u = Ce.$$

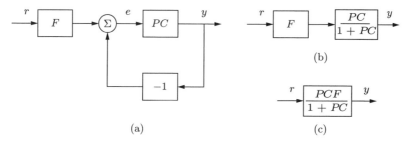

Figure 9.7: Example of block diagram algebra. The results from multiplying the process and controller transfer functions (from Figure 9.6) are shown in (a). Replacing the feedback loop with its transfer function equivalent yields (b), and finally multiplying the two remaining blocks gives the reference to output representation in (c).

Combining these equations gives

$$e = Fr - y = Fr - (w + \eta) = Fr - \big(w + P(v + u)\big)$$
$$= Fr - \big(w + P(v + Ce)\big),$$

and hence

$$e = Fr - w - Pv - PCe.$$

Finally, solving this equation for e gives

$$e = \frac{F}{1 + PC}\, r - \frac{1}{1 + PC}\, w - \frac{P}{1 + PC}\, v = G_{er} r + G_{ew} w + G_{ev} v, \qquad (9.29)$$

and the error is thus the sum of three terms, depending on the reference r, the measurement noise w, and the load disturbance v. The functions

$$G_{er} = \frac{F}{1 + PC}, \qquad G_{ew} = \frac{-1}{1 + PC}, \qquad G_{ev} = \frac{-P}{1 + PC} \qquad (9.30)$$

are transfer functions from reference r, noise w, and disturbance v to the error e. Equation (9.29) can also be obtained by computing the outputs for each input and using superposition.

We can also derive transfer functions by manipulating the block diagrams directly, as illustrated in Figure 9.7. Suppose we wish to compute the transfer function between the reference r and the output y. We begin by combining the process and controller blocks in Figure 9.6 to obtain the diagram in Figure 9.7a. We can now eliminate the feedback loop using the algebra for a feedback interconnection (Figure 9.7b) and then use the series interconnection rule to obtain

$$G_{yr} = \frac{PCF}{1 + PC}. \qquad (9.31)$$

Similar manipulations can be used to obtain the other transfer functions (Exercise 9.10).

The above analysis illustrates an effective way to manipulate the equations to obtain the relations between inputs and outputs in a feedback system. The general idea is to start with the variable of interest and to trace variables backwards around the feedback loop. With some practice, equations (9.29) and (9.30) can be written directly by inspection of the block diagram. Notice, for example, that all terms in equation (9.30) have the same denominator and that the numerators are the products of the blocks that one passes through when going directly from input to output (ignoring the feedback). This type of rule can be used to compute transfer functions by inspection, although for systems with multiple feedback loops it can be tricky to compute them without writing down the algebra explicitly.

We can also use block diagram algebra to obtain insights about state space controllers. Consider a state space controller that uses an observer, such as the one shown in Figure 8.7. The process model is

$$\frac{dx}{dt} = Ax + Bu, \qquad y = Cx,$$

and the controller (8.15) is given by

$$u = -K\hat{x} + k_f r, \tag{9.32}$$

where $\hat{x}$ is the output of a state observer (8.16) given by

$$\frac{d\hat{x}}{dt} = A\hat{x} + Bu + L(y - C\hat{x}). \tag{9.33}$$

The controller is a system with one output u and two inputs, the reference r and the measured signal y. Using transfer functions and exponential signals it can be represented as

$$u = G_{ur}r + G_{uy}y. \tag{9.34}$$

The transfer function G_{uy} from y to u describes the feedback action and G_{ur} from r to u describes the feedforward action. We call these *open loop* transfer functions because they represent the relationships between the signals without considering the dynamics of the process (e.g., removing P from the system description or cutting the loop at the process input or output).

To derive the controller transfer functions we rewrite equation (9.33) as

$$\frac{d\hat{x}}{dt} = (A - BK - LC)\hat{x} + Bk_f r + Ly. \tag{9.35}$$

Letting $\hat{x}$, r, and y be exponential signals, the above equations give

$$u = -K\hat{x} + k_f r, \qquad (sI - (A - BK - LC))\hat{x} = Bk_f r + Ly,$$

and we find that the controller transfer functions in equation (9.34) are

$$G_{ur} = k_f - K(sI - A + BK + LC)^{-1}Bk_f,$$
$$G_{uy} = -K(sI - A + BK + LC)^{-1}L. \tag{9.36}$$

We illustrate with an example.

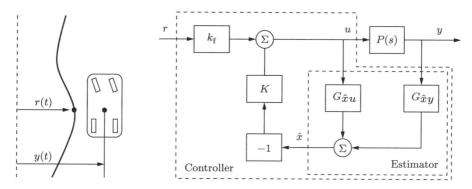

Figure 9.8: Block diagram for a steering control system. The control system is designed to maintain the lateral position of the vehicle along a reference curve (left). The structure of the control system is shown on the right as a block diagram of transfer functions. The estimator consists of two components that compute the estimated state $\hat{x}$ from the combination of the input u and output y of the process. The estimated state is fed through a state feedback controller and combined with a feedforward gain obtain the commanded steering angle u.

Example 9.10 Vehicle steering
Consider the linearized model for vehicle steering introduced in Example 6.13. In Examples 7.4 and 8.3 we designed a state feedback controller and state estimator for the system. A block diagram for the resulting control system is given in Figure 9.8. Note that we have split the estimator into two components, $G_{\hat{x}u}(s)$ and $G_{\hat{x}y}(s)$, corresponding to its inputs u and y. To compute these transfer functions we use equation (9.33) and the expressions for A, B, C, and L from Example 8.3, hence

$$G_{\hat{x}u}(s) = \begin{pmatrix} \dfrac{\gamma s + 1}{s^2 + l_1 s + l_2} \\ \dfrac{s + l_1 - \gamma l_2}{s^2 + l_1 s + l_2} \end{pmatrix}, \qquad G_{\hat{x}y}(s) = \begin{pmatrix} \dfrac{l_1 s + l_2}{s^2 + l_1 s + l_2} \\ \dfrac{l_2 s}{s^2 + l_1 s + l_2} \end{pmatrix},$$

where l_1 and l_2 are the observer gains and γ is the scaled position of the center of mass from the rear wheels. Applying block diagram algebra to the controller in Figure 9.8 we obtain

$$G_{ur}(s) = \frac{k_f}{1 + K G_{\hat{x}u}(s)} = \frac{k_f(s^2 + l_1 s + l_2)}{s^2 + s(\gamma k_1 + k_2 + l_1) + k_1 + l_2 + k_2 l_1 - \gamma k_2 l_2},$$

and

$$G_{uy}(s) = \frac{-K G_{\hat{x}y}(s)}{1 + K G_{\hat{x}u}(s)} = \frac{s(k_1 l_1 + k_2 l_2) + k_1 l_2}{s^2 + s(\gamma k_1 + k_2 + l_1) + k_1 + l_2 + k_2 l_1 - \gamma k_2 l_2},$$

where k_1 and k_2 are the state feedback gains and k_f is the feedforward gain. The last equalities are obtained applying block diagram algebra to Figure 9.8, but can also be obtained by applying equation (9.36).

To compute the closed loop transfer function G_{yr} from reference r to output y, we begin by deriving the transfer function for the process $P(s)$. We can compute this directly from the state space description, which was given in Example 6.13. Using that description, we have

$$P(s) = G_{yu}(s) = C(sI - A)^{-1}B + D = \begin{pmatrix} 1 & 0 \end{pmatrix} \begin{pmatrix} s & -1 \\ 0 & s \end{pmatrix}^{-1} \begin{pmatrix} \gamma \\ 1 \end{pmatrix} = \frac{\gamma s + 1}{s^2}.$$

The transfer function for the full closed loop system between the input r and the output y is then given by

$$G_{yr} = \frac{P(s)G_{ur}(s)}{1 - P(s)G_{uy}(s)} = \frac{k_{\mathrm{f}}(\gamma s + 1)}{s^2 + (k_1\gamma + k_2)s + k_1}.$$

(The unusual sign in the denominator of the middle expression occurs because G_{ur} is in the feedback path and incorporates the -1 gain element.) ∇

Note that in the previous example the observer gains l_1 and l_2 do not appear in the transfer function G_{yr}. This is true in general, as follows from Figure 8.9b in Section 8.3.

We also note that a control system using an observer should be implemented as the multivariable system (9.35), which is of order n. It should not be implemented using two separate transfer functions, as described in equation (9.34), because the controller would then be of order $2n$, and there will be unobservable modes.

Algebraic Loops

When analyzing or simulating a system described by a block diagram, it is necessary to form the differential equations that describe the complete system. In many cases the equations can be obtained by combining the differential equations that describe each subsystem and substituting variables. This simple procedure cannot be used when there are closed loops of subsystems that all have a direct connection between inputs and outputs, known as an *algebraic loop*.

To see what can happen, consider a system with two blocks, a first-order nonlinear system,

$$\frac{dx}{dt} = f(x, u), \qquad y = h(x), \tag{9.37}$$

and a proportional controller described by $u = -ky$. There is no direct term since the function h does not depend on u. In that case we can obtain the equation for the closed loop system simply by replacing u by $-ky$ in equation (9.37) to give

$$\frac{dx}{dt} = f(x, -ky), \qquad y = h(x).$$

Such a procedure can easily be automated using simple formula manipulation.

The situation is more complicated if there is a direct term. If $y = h(x, u)$, then replacing u by $-ky$ gives

$$\frac{dx}{dt} = f(x, -ky), \qquad y = h(x, -ky).$$

To obtain a differential equation for x, the algebraic equation $y = h(x, -ky)$ must be solved to give $y = \alpha(x)$, which in general is a complicated task.

When algebraic loops are present, it is necessary to solve algebraic equations to obtain the differential equations for the complete system. Resolving algebraic loops is a nontrivial problem because it requires the symbolic solution of algebraic equations. Most block diagram-oriented modeling languages cannot handle algebraic loops, and they simply give a diagnosis that such loops are present. In the era of analog computing, algebraic loops were eliminated by introducing fast dynamics between the loops. This created differential equations with fast and slow modes that are difficult to solve numerically. Advanced modeling languages like Modelica use several sophisticated methods to resolve algebraic loops.

9.5   ZERO FREQUENCY GAIN, POLES, AND ZEROS

The transfer function has many useful interpretations and the features of a transfer function are often associated with important system properties. Three of the most important features are the gain and the locations of the poles and zeros.

Zero Frequency Gain

The *zero frequency gain* of a system is given by the magnitude of the transfer function at $s = 0$. It represents the ratio of the steady-state value of the output with respect to a step input (which can be represented as $u = e^{st}$ with $s = 0$). For a state space system, we computed the zero frequency gain in equation (6.22):

$$G(0) = D - CA^{-1}B.$$

For a system modeled as the linear differential equation

$$\frac{d^n y}{dt^n} + a_1 \frac{d^{n-1} y}{dt^{n-1}} + \cdots + a_n y = b_0 \frac{d^m u}{dt^m} + b_1 \frac{d^{m-1} u}{dt^{m-1}} + \cdots + b_m u,$$

if we assume that the input u and output y are constants y_0 and u_0, then we find that $a_n y_0 = b_m u_0$, and the zero frequency gain is

$$G(0) = \frac{y_0}{u_0} = \frac{b_m}{a_n}. \tag{9.38}$$

Poles and Zeros

Next consider a linear system with the rational transfer function

$$G(s) = \frac{b(s)}{a(s)}.$$

The roots of the polynomial $a(s)$ are called the *poles* of the system, and the roots of $b(s)$ are called the *zeros* of the system. If p is a pole, it follows that $y(t) = e^{pt}$ is a solution of equation (9.13) with $u = 0$ (the solution to the homogeneous equation). A pole p corresponds to a *mode* of the system with corresponding modal solution e^{pt}. The unforced motion of the system after an arbitrary excitation is a weighted sum of modes.

Zeros have a different interpretation. Since the pure exponential output corresponding to the input $u(t) = e^{st}$ with $a(s) \neq 0$ is $G(s)e^{st}$, it follows that the pure exponential output is zero if $b(s) = 0$. Zeros of the transfer function thus *block transmission* of the corresponding exponential signals.

The difference between the number of poles and zeros $n_{pe} = n - m$ is called the *pole excess* (also sometimes referred to as the *relative degree*). A rational transfer function is called *proper* if $n_{pe} \geq 0$ and *strictly proper* if $n_{pe} > 0$.

Effective use of zeros can be seen in integral control. To obtain a closed loop system where a constant disturbance does not create a steady-state error, the controller is designed so that the transfer function from disturbance to control error has a zero at the origin. Vibration dampers are another example where the system is designed so that the transfer function from disturbance force to motion has a zero at the frequency we want to damp (Example 9.4).

For a state space system with transfer function $G(s) = C(sI - A)^{-1}B + D$, the poles of the transfer function are the eigenvalues of the matrix A in the state space model. One easy way to see this is to notice that the value of $G(s)$ is unbounded when s is an eigenvalue of a system since this is precisely the set of points where the characteristic polynomial $\lambda(s) = \det(sI - A) = 0$ (and hence $sI - A$ is noninvertible). It follows that the poles of a state space system depend only on the matrix A, which represents the intrinsic dynamics of the system. We say that a transfer function is *stable* if all of its poles have negative real part.

To find the zeros of a state space system, we observe that the zeros are complex numbers s such that the input $u(t) = U_0 e^{st}$ gives zero output. Inserting the pure exponential response $x(t) = X_0 e^{st}$ and setting $y(t) = 0$ in equation (9.2) gives

$$se^{st}x_0 = AX_0 e^{st} + BU_0 e^{st} \qquad 0 = Ce^{st}X_0 + De^{st}U_0,$$

which can be written as

$$\begin{pmatrix} A - sI & B \\ C & D \end{pmatrix} \begin{pmatrix} X_0 \\ U_0 \end{pmatrix} e^{st} = 0.$$

This equation has a solution with nonzero X_0, U_0 only if the matrix on the left does not have full column rank. The zeros are thus the values s such that the matrix

$$\begin{pmatrix} A - sI & B \\ C & D \end{pmatrix} \tag{9.39}$$

loses rank.

Since the zeros depend on A, B, C, and D, they therefore depend on how the inputs and outputs are coupled to the states. Notice in particular that if the matrix B has full row rank, then the matrix in equation (9.39) has n linearly independent rows for all values of s. Similarly there are n linearly independent columns if the matrix C has full column rank. This implies that systems where the matrix B or C is square and full rank do not have zeros. In particular it means that a system has no zeros if it is fully actuated (each state can be controlled independently) or if the full state is measured.

A convenient way to view the poles and zeros of a transfer function is through a *pole zero diagram*, as shown in Figure 9.9. In this diagram, each pole is marked with a cross, and each zero with a circle. If there are multiple poles or zeros at a fixed location, these are often indicated with overlapping crosses or circles (or other

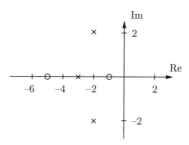

Figure 9.9: A pole zero diagram for a transfer function with zeros at -5 and -1 and poles at -3 and $-2 \pm 2j$. The circles represent the locations of the zeros, and the crosses the locations of the poles.

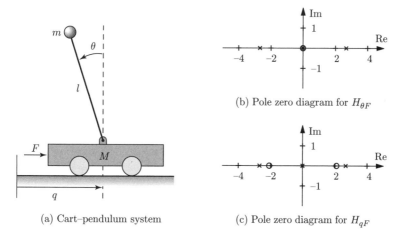

(a) Cart–pendulum system   (b) Pole zero diagram for $H_{\theta F}$

(c) Pole zero diagram for H_{qF}

Figure 9.10: Poles and zeros for a balance system. The balance system (a) can be modeled around its vertical equilibrium point by a fourth order linear system. The poles and zeros for the transfer functions $H_{\theta F}$ and H_{qF} are shown in (b) and (c), respectively.

annotations). Poles in the left half-plane correspond to stable modes of the system, and poles in the right half-plane correspond to unstable modes. We thus call a pole in the left half-plane a *stable pole* and a pole in the right half-plane an *unstable pole*. A similar terminology is used for zeros, even though the zeros do not directly relate to stability or instability of the system. Notice that the gain must also be given to have a complete description of the transfer function.

Example 9.11 Balance system
Consider the dynamics for a balance system, shown in Figure 9.10. The transfer function for a balance system can be derived directly from the second-order equations, given in Example 3.2:

$$M_t \frac{d^2 q}{dt^2} - ml \frac{d^2 \theta}{dt^2} \cos\theta + c \frac{dq}{dt} + ml \sin\theta \left(\frac{d\theta}{dt}\right)^2 = F,$$

$$-ml \cos\theta \frac{d^2 q}{dt^2} + J_t \frac{d^2 \theta}{dt^2} + \gamma \frac{d\theta}{dt} - mgl \sin\theta = 0.$$

If we assume that θ and $\dot{\theta}$ are small, we can approximate this nonlinear system by a set of linear second-order differential equations,

$$M_t \frac{d^2 q}{dt^2} - ml \frac{d^2 \theta}{dt^2} + c \frac{dq}{dt} = F,$$

$$-ml \frac{d^2 q}{dt^2} + J_t \frac{d^2 \theta}{dt^2} + \gamma \frac{d\theta}{dt} - mgl\theta = 0.$$

If we let F be an exponential signal, the resulting response satisfies

$$M_t s^2 q - ml s^2 \theta + cs\,q = F,$$

$$J_t s^2 \theta - ml s^2 q + \gamma s\,\theta - mgl\,\theta = 0,$$

where all signals are exponential signals. The resulting transfer functions for the position of the cart and the orientation of the pendulum are given by solving for q and θ in terms of F to obtain

$$H_{\theta F}(s) = \frac{mls}{(M_t J_t - m^2 l^2)s^3 + (\gamma M_t + c J_t)s^2 + (c\gamma - M_t mgl)s - mglc},$$

$$H_{qF}(s) = \frac{J_t s^2 + \gamma s - mgl}{(M_t J_t - m^2 l^2)s^4 + (\gamma M_t + c J_t)s^3 + (c\gamma - M_t mgl)s^2 - mglcs},$$

where each of the coefficients is positive. The pole zero diagrams for these two transfer functions are shown in Figure 9.10 using the parameters from Example 7.7.

If we assume the damping is small and set $c = 0$ and $\gamma = 0$, we obtain

$$H_{\theta F}(s) = \frac{ml}{(M_t J_t - m^2 l^2)s^2 - M_t mgl},$$

$$H_{qF}(s) = \frac{J_t s^2 - mgl}{s^2\big((M_t J_t - m^2 l^2)s^2 - M_t mgl\big)}.$$

This gives nonzero poles and zeros at

$$p = \pm\sqrt{\frac{mgl M_t}{M_t J_t - m^2 l^2}} \approx \pm 2.68, \qquad z = \pm\sqrt{\frac{mgl}{J_t}} \approx \pm 2.09.$$

We see that these are quite close to the pole and zero locations in Figure 9.10. ∇

Pole/Zero Cancellations

Because transfer functions are often polynomials in s, it can sometimes happen that the numerator and denominator have a common factor, which can be canceled. Sometimes these cancellations are simply algebraic simplifications, but in other situations they can mask potential fragilities in the model. In particular, if a pole/zero cancellation occurs because terms in separate blocks just happen to coincide, the cancellation may not occur if one of the systems is slightly perturbed. In some situations this can result in severe differences between the expected behavior and the actual behavior.

Consider the block diagram in Figure 9.6 with $F = 1$ (no feedforward compensation) and let C and P be given by

$$C(s) = \frac{n_c(s)}{d_c(s)}, \qquad P(s) = \frac{n_p(s)}{d_p(s)}.$$

The transfer function from r to e is then given by

$$G_{er}(s) = \frac{1}{1 + PC} = \frac{d_c(s)d_p(s)}{d_c(s)d_p(s) + n_c(s)n_p(s)}.$$

If there are common factors in the numerator and denominator polynomials, then these terms can be factored out and eliminated from both the numerator and denominator. For example, if the controller has a zero at $s = -a$ and the process has a pole at $s = -a$, then we will have

$$G_{er}(s) = \frac{(s+a)d_c(s)d_p'(s)}{(s+a)d_c(s)d_p'(s) + (s+a)n_c'(s)n_p(s)} = \frac{d_c(s)d_p'(s)}{d_c(s)d_p'(s) + n_c'(s)n_p(s)},$$

where $n_c'(s)$ and $d_p'(s)$ represent the relevant polynomials with the term $s + a$ factored out. We see that the $s + a$ term does not appear in the transfer function G_{er}.

Suppose instead that we compute the transfer function from v to e, which represents the effect of a disturbance on the error between the reference and the output. This transfer function is given by

$$G_{ev}(s) = -\frac{d_c(s)n_p(s)}{(s+a)d_c(s)d_p'(s) + (s+a)n_c'(s)n_p(s)}.$$

Notice that if $a < 0$, then the pole is in the right half-plane and the transfer function G_{ev} is *unstable*. Hence, even though the transfer function from r to e appears to be okay (assuming a perfect pole/zero cancellation), the transfer function from v to e can exhibit unbounded behavior. This unwanted behavior is typical of an *unstable pole/zero cancellation*.

As noted at the end of Section 9.2, the cancellation of a pole with a zero can be understood in terms of the state space representation of the systems. Reachability or observability is lost when there are cancellations of poles and zeros (Example 9.7 and Exercise 9.14) and the transfer function depends only on the dynamics in the reachable and observable subsystem Σ_{ro}.

Example 9.12 Cruise control

A cruise control system can be modeled by the block diagram in Figure 9.6, where y is the vehicle velocity, r the desired velocity, v the slope of the road, and u the throttle. Furthermore $F(s) = 1$, and the input/output response from throttle to velocity for the linearized model for a car has the transfer function $P(s) = b/(s+a)$. A simple (but not necessarily good) way to design a PI controller is to choose the parameters of the PI controller as $k_i = ak_p$. The controller transfer function is then $C(s) = k_p + k_i/s = k_p(s+a)/s$. It has a zero at $s = -k_i/k_p = -a$ that cancels the process pole at $s = -a$. We have $P(s)C(s) = bk_p/s$ giving the transfer function from reference to vehicle velocity as $G_{yr}(s) = bk_p/(s + bk_p)$, and control design is then simply a matter of choosing the gain k_p. The closed loop system dynamics are of first order with the time constant $1/(bk_p)$. Notice that the canceled pole $1/a$ is much slower than the other pole.

Figure 9.11 shows the velocity error when the car encounters an increase in the road slope. A comparison with the controller used in Figure 4.3b (reproduced in

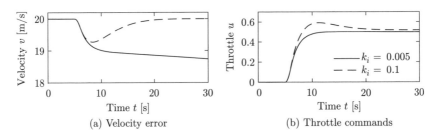

(a) Velocity error   (b) Throttle commands

Figure 9.11: Car with PI cruise control encountering a sloping road. The velocity error is shown on the left and the throttle is shown on the right. Results for a PI controller with $k_\mathrm{p} = 0.5$ and $k_\mathrm{i} = 0.005$ are shown by solid lines, and for a controller with $k_\mathrm{p} = 0.5$ and $k_\mathrm{i} = 0.1$ are shown by dashed lines. Compare with Figure 4.3b.

dashed curves) shows that the controller based on pole/zero cancellation has very poor performance. The velocity error is larger, and it takes a long time to settle.

Notice that the control signal remains practically constant after $t = 15$ even if the error is large after that time. To understand what happens we will analyze the system. The parameters of the system are $a = 0.01$ and $b = 1.32$, and the controller parameters are $k_\mathrm{p} = 0.5$ and $k_\mathrm{i} = 0.005$. The closed loop time constant is $1/(bk_\mathrm{p}) = 1.5\,\mathrm{s}$, and we would expect that the error would settle in about $6\,\mathrm{s}$ (4 time constants). The transfer functions from road slope to velocity and control signals are

$$G_{yv}(s) = \frac{b_\mathrm{g}s}{(s+a)(s+bk_\mathrm{p})}, \qquad G_{uv}(s) = \frac{bk_\mathrm{p}}{s+bk_\mathrm{p}}.$$

Notice that the slow canceled mode $s = -a = -0.01$ appears in G_{yv} but not in G_{uv}. The reason why the control signal remains constant is that the controller has a zero at $s = -0.01$, which cancels the slowly decaying process mode. Note also that the error would diverge if the canceled pole was unstable. ∇

The lesson we can learn from this example is that it is a bad idea to try to cancel unstable or slow process poles. A more detailed discussion of pole/zero cancellations and their impact on robustness is given in Section 14.5.

9.6   THE BODE PLOT

The frequency response of a linear system can be computed from its transfer function by setting $s = i\omega$, corresponding to a complex exponential

$$u(t) = e^{i\omega t} = \cos(\omega t) + i\sin(\omega t).$$

The resulting output has the form

$$y(t) = G(i\omega)e^{i\omega t} = Me^{i(\omega t+\theta)} = M\cos(\omega t + \theta) + iM\sin(\omega t + \theta),$$

where M and θ are the gain and phase of G:

$$M = |G(i\omega)|, \qquad \theta = \arctan\frac{\mathrm{Im}\,G(i\omega)}{\mathrm{Re}\,G(i\omega)}.$$

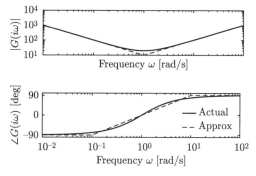

Figure 9.12: Bode plot of the transfer function $C(s) = 20 + \frac{10}{s} + 10s = 10\frac{(s+1)^2}{s}$, corresponding to an ideal PID controller. The upper plot is the gain curve and the lower plot is the phase curve. The dashed lines show straight-line approximations of the gain curve and the corresponding phase curve.

The gain and phase of G are also called the *magnitude* and *argument* of G, terms that come from the theory of complex variables.

It follows from linearity that the response to a single sinusoid ($\sin(\omega t)$ or $\cos(\omega t)$) is amplified by M and phase-shifted by θ. It will often be convenient to represent the phase in degrees rather than radians. We will use the notation $\angle G(i\omega)$ for the phase in degrees and $\arg G(i\omega)$ for the phase in radians. In addition, while we always take $\arg G(i\omega)$ to be in the range $(-\pi, \pi]$, we will take $\angle G(i\omega)$ to be continuous, so that it can take on values outside the range of $-180°$ to $180°$.

The frequency response $G(i\omega)$ can thus be represented by two curves: the gain curve and the phase curve. The *gain curve* gives $|G(i\omega)|$ as a function of frequency ω and the *phase curve* gives $\angle G(i\omega)$. One particularly useful way of drawing these curves is to use a log/log scale for the gain curve and a log/linear scale for the phase curve. This type of plot is called a *Bode plot* and is shown in Figure 9.12.

Sketching and Interpreting Bode Plots

Part of the popularity of Bode plots is that they are easy to sketch and interpret. Since the frequency scale is logarithmic, they cover the behavior of a linear system over a wide frequency range.

Consider a transfer function that is a rational function of the form

$$G(s) = \frac{b_1(s)b_2(s)}{a_1(s)a_2(s)}.$$

We have

$$\log|G(s)| = \log|b_1(s)| + \log|b_2(s)| - \log|a_1(s)| - \log|a_2(s)|,$$

and hence we can compute the gain curve by simply adding and subtracting gains corresponding to terms in the numerator and denominator. Similarly,

$$\angle G(s) = \angle b_1(s) + \angle b_2(s) - \angle a_1(s) - \angle a_2(s),$$

and so the phase curve can be determined in an analogous fashion. Since a polynomial can be written as a product of terms of the type

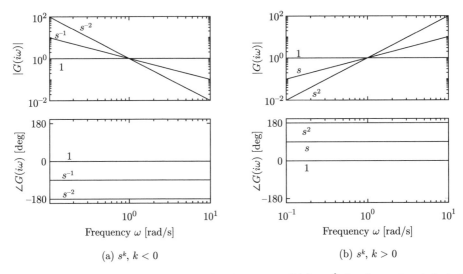

Figure 9.13: Bode plots of the transfer functions $G(s) = s^k$ for $k = -2, -1, 0, 1, 2$. On a log-log scale, the gain curve is a straight line with slope k. The phase curves for the transfer functions are constants, with phase equal to $k \times 90°$.

$$k, \quad s, \quad s + a, \quad s^2 + 2\zeta\omega_0 s + \omega_0^2,$$

it suffices to be able to sketch Bode diagrams for these terms. The Bode plot of a complex system is then obtained by adding the gains and phases of the terms.

The function $G(s) = s^k$ is a simple transfer function, with the important special cases of $k = 1$ corresponding to a differentiator and $k = -1$ to an integrator. The gain and phase of the term are given by

$$\log |G(i\omega)| = k \times \log \omega, \qquad \angle G(i\omega) = k \times 90°.$$

The gain curve is thus a straight line with slope k, and the phase curve is a constant at $k \times 90°$. The case when $k = 1$ corresponds to a differentiator and has slope 1 with phase 90°. The case when $k = -1$ corresponds to an integrator and has slope -1 with phase $-90°$. Bode plots of the various powers of k are shown in Figure 9.13.

Consider next the transfer function of a first-order system, given by

$$G(s) = \frac{a}{s + a}, \qquad a > 0.$$

We have

$$|G(s)| = \frac{|a|}{|s + a|}, \qquad \angle G(s) = \angle(a) - \angle(s + a),$$

and hence

$$\log |G(i\omega)| = \log a - \frac{1}{2} \log (\omega^2 + a^2), \qquad \angle G(i\omega) = -\frac{180}{\pi} \arctan \frac{\omega}{a}.$$

The Bode plot is shown in Figure 9.14a, with the magnitude normalized by the zero frequency gain. Both the gain curve and the phase curve can be approximated by

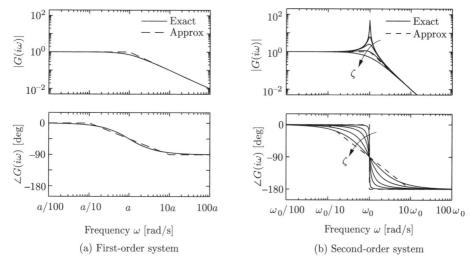

Figure 9.14: Bode plots for first- and second-order systems. (a) The first-order system $G(s) = a/(s+a)$ can be approximated by asymptotic curves (dashed) in both the gain and the frequency, with the breakpoint in the gain curve at $\omega = a$ and the phase decreasing by $90°$ over a factor of 100 in frequency. (b) The second-order system $G(s) = \omega_0^2/(s^2 + 2\zeta\omega_0 s + \omega_0^2)$ has a peak at frequency ω_0 and then a slope of -2 beyond the peak; the phase decreases from $0°$ to $-180°$. The height of the peak and the rate of change of phase depend on the damping ratio ζ ($\zeta = 0.02$, 0.1, 0.2, 0.5, and 1.0 shown).

the following straight lines

$$\log|G(i\omega)| \approx \begin{cases} 0 & \text{if } \omega < a, \\ \log a - \log \omega & \text{if } \omega > a, \end{cases}$$

$$\angle G(i\omega) \approx \begin{cases} 0 & \text{if } \omega < a/10, \\ -45 - 45(\log \omega - \log a,) & \text{if } a/10 < \omega < 10a, \\ -90 & \text{if } \omega > 10a. \end{cases}$$

The approximate gain curve consists of a horizontal line up to frequency $\omega = a$, called the *breakpoint* or *corner frequency*, after which the curve is a line of slope -1 (on a log-log scale). The phase curve is zero up to frequency $a/10$ and then decreases linearly by $45°$/decade up to frequency $10a$, at which point it remains constant at $-90°$. Notice that a first-order system behaves like a constant for low frequencies and like an integrator for high frequencies; compare with the Bode plot in Figure 9.13.

Finally, consider the transfer function for a second-order system,

$$G(s) = \frac{\omega_0^2}{s^2 + 2\omega_0 \zeta s + \omega_0^2},$$

with $0 < \zeta < 1$, for which we have

$$\log |G(i\omega)| = 2\log \omega_0 - \frac{1}{2}\log \left(\omega^4 + 2\omega_0^2\omega^2(2\zeta^2 - 1) + \omega_0^4\right),$$

$$\angle G(i\omega) = -\frac{180}{\pi}\arctan\frac{2\zeta\omega_0\omega}{\omega_0^2 - \omega^2}.$$

The gain curve has an asymptote with zero slope for $\omega \ll \omega_0$. For large values of ω the gain curve has an asymptote with slope -2. The largest gain $Q = \max_\omega |G(i\omega)| \approx 1/(2\zeta)$, called the Q-value, is obtained for $\omega \approx \omega_0$. The phase is zero for low frequencies and approaches $-180°$ for large frequencies. The curves can be approximated with the following piecewise linear expressions

$$\log |G(i\omega)| \approx \begin{cases} 0 & \text{if } \omega \ll \omega_0, \\ 2\log \omega_0 - 2\log \omega & \text{if } \omega \gg \omega_0, \end{cases}$$

$$\angle G(i\omega) \approx \begin{cases} 0 & \text{if } \omega \ll \omega_0, \\ -180 & \text{if } \omega \gg \omega_0. \end{cases}$$

The Bode plot is shown in Figure 9.14b. Note that the asymptotic approximation is poor near $\omega = \omega_0$ and that the Bode plot depends strongly on ζ near this frequency.

Given the Bode plots of the basic functions, we can now sketch the frequency response for a more general system. The following example illustrates the basic idea.

Example 9.13 Asymptotic approximation for a transfer function
Consider the transfer function given by

$$G(s) = \frac{k(s+b)}{(s+a)(s^2 + 2\zeta\omega_0 s + \omega_0^2)}, \qquad a \ll b \ll \omega_0.$$

The Bode plot for this transfer function appears in Figure 9.15, with the complete transfer function shown as a solid curve and the asymptotic approximation shown as a dashed curve.

We begin with the gain curve. At low frequency, the magnitude is given by

$$G(0) = \frac{kb}{a\omega_0^2}.$$

When we reach $\omega = a$, the effect of the pole begins and the gain decreases with slope -1. At $\omega = b$, the zero comes into play and we increase the slope by 1, leaving the asymptote with net slope 0. This slope is used until the effect of the second-order pole is seen at $\omega = \omega_0$, at which point the asymptote changes to slope -2. We see that the gain curve is fairly accurate except in the region of the peak due to the second-order pole (indicating that for this case ζ is reasonably small).

The phase curve is more complicated since the effect of the phase stretches out much further. The effect of the pole begins at $\omega = a/10$, at which point we change from phase 0 to a slope of $-45°$/decade. The zero begins to affect the phase at $\omega = b/10$, producing a flat section in the phase. At $\omega = 10a$ the phase contribution from the pole ends, and we are left with a slope of $+45°$/decade (from the zero). At the location of the second-order pole, $s \approx i\omega_0$, we get a jump in phase of $-180°$. Finally, at $\omega = 10b$ the phase contribution of the zero ends, and we are left with a phase of -180 degrees. We see that the straight-line approximation for the phase is not quite as accurate as it was for the gain curve, but it does capture the basic features of the phase changes as a function of frequency. ∇

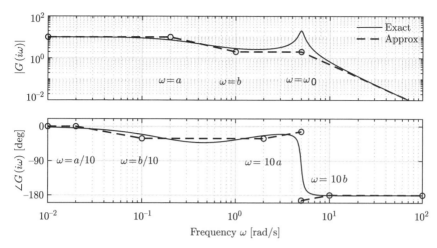

Figure 9.15: Asymptotic approximation to a Bode plot. The solid curve is the Bode plot for the transfer function $G(s) = k(s+b)/(s+a)(s^2 + 2\zeta\omega_0 s + \omega_0^2)$, where $a \ll b \ll \omega_0$. Each segment in the gain and phase curves represents a separate portion of the approximation, where either a pole or a zero begins to have effect. Each segment of the approximation is a straight line between these points at a slope given by the rules for computing the effects of poles and zeros.

Poles and Zeros in the Right Half-Plane

The gain curve of a transfer function remains the same if a pole or a zero of a transfer function is shifted from the left half-plane to the right half-plane by mirror imaging in the imaginary axis. The phase will, however, change significantly as is illustrated by the following example.

Example 9.14 Transfer function with a zero in the right half-plane
Consider the transfer functions

$$G(s) = \frac{s+1}{(s+0.1)(s+10)}, \qquad G_{\mathrm{rhpp}}(s) = \frac{s+1}{(s-0.1)(s+10)},$$

and

$$G_{\mathrm{rhpz}}(s) = \frac{-s+1}{(s+0.1)(s+10)}.$$

The transfer functions G and G_{rhpp} have the zero at $s = -1$ and the pole at $s = -10$ in common, while G has the pole at $s = -0.1$ but G_{rhpp} has the pole at $s = 0.1$. Similarly, the transfer functions G and G_{rhpz} have the same poles, but G has the zero at $s = -1$ while G_{rhpz} has the zero at $s = 1$. Notice that all transfer functions have the same gain curves but that the phase curves differ significantly, as shown in Figure 9.16. Notice in particular that the transfer functions G_{rhpp} and G_{rhpz} have much larger phase lags than G. $\qquad\qquad \nabla$

A time delay, which has the transfer function $G(s) = e^{-s\tau}$, is an even more striking example of a change in phase than a right half-plane zero. Since $|G(i\omega\tau)| = |e^{-i\omega\tau}| = 1$ the gain curve is constant but the phase is $\angle G(i\omega\tau) = -180\,\omega\tau/\pi$, which

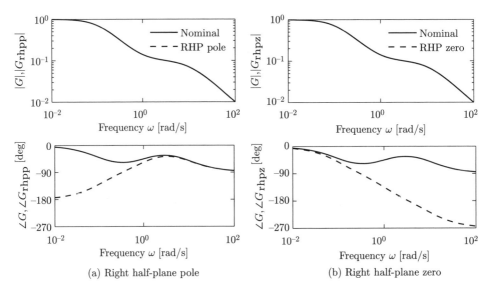

(a) Right half-plane pole   (b) Right half-plane zero

Figure 9.16: Effect of a right half-plane pole and a right half-plane zero on the Bode plot. The curves for G, which has all poles and zeros in the right half-plane, are shown in solid lines and the curves for G_{rhpp} and G_{rhpz} are shown as dashed curves. (a) Bode plots for the transfer functions G and G_{rhpp}, which have a pole at $s = -10$ and a zero at $s = -1$, but G has a pole at $s = -0.1$ while G_{rhpp} has a corresponding pole at $s = 0.1$. (b) Bode plots for the transfer functions G and G_{rhpz}, which have the same poles at $s- = 0.1$ and $s = -10$, while G has a zero at $s = -1$ and G_{rhpz} has a zero $s = 1$.

has a large negative value for large ω. Time delays are in this respect similar to right half-plane zeros. Intuitively it seems reasonable that extra phase will cause difficulties for control since there is a delay between applying an input and seeing its effect. Poles and zeros in the right half-plane and time delay will indeed limit the achievable control performance, as will be discussed in detail in Section 10.4 and Chapter 14.

System Insights from the Bode Plot

The Bode plot gives a quick overview of a system. The plot covers wide ranges in amplitude and frequency because of the logarithmic scales. Since many useful signals can be decomposed into a sum of sinusoids, it is possible to visualize the behavior of a system for different frequency ranges. The system can be viewed as a filter that can change the amplitude (and phase) of the input signals according to the frequency response. For example, if there are frequency ranges where the gain curve has constant slope and the phase is close to zero, the action of the system for signals with these frequencies can be interpreted as a pure gain. Similarly, for frequencies where the slope is $+1$ and the phase close to $90°$, the action of the system can be interpreted as a differentiator.

Three common types of frequency responses are shown in Figure 9.17. The system in Figure 9.17a is called a *low-pass filter* because the gain is constant for low frequencies and drops for high frequencies. Notice that the phase is zero for low

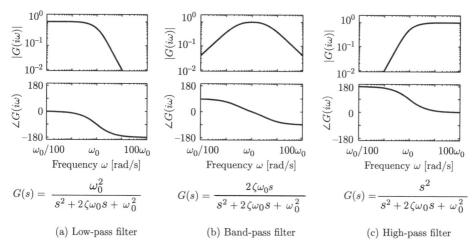

$$G(s) = \frac{\omega_0^2}{s^2 + 2\zeta\omega_0 s + \omega_0^2}$$

$$G(s) = \frac{2\zeta\omega_0 s}{s^2 + 2\zeta\omega_0 s + \omega_0^2}$$

$$G(s) = \frac{s^2}{s^2 + 2\zeta\omega_0 s + \omega_0^2}$$

(a) Low-pass filter

(b) Band-pass filter

(c) High-pass filter

Figure 9.17: Bode plots for low-pass, band-pass, and high-pass filters. The upper plots are the gain curves and the lower plots are the phase curves. Each system passes frequencies in a different range and attenuates frequencies outside of that range.

frequencies and $-180°$ for high frequencies. The systems in Figures 9.17b and 9.17c are called a *band-pass filter* and a *high-pass filter* for similar reasons.

To illustrate how different system behaviors can be read from the Bode plots we consider the band-pass filter in Figure 9.17b. For frequencies around $\omega = \omega_0$, the signal is passed through with no change in gain. However, for frequencies well below or well above ω_0, the signal is attenuated. The phase of the signal is also affected by the filter, as shown in the phase curve. For frequencies below $\omega_0/100$ there is a phase lead of $90°$, and for frequencies above $100\omega_0$ there is a phase lag of $90°$. These actions correspond to differentiation and integration of the signal in these frequency ranges.

The intuition captured in the Bode plot can also be related to the transfer function: the approximations of $G(s)$ for small and large s capture the propagation of slow and fast signals, respectively, as illustrated in the following examples.

Example 9.15 Spring–mass system
Consider a spring–mass system with input u (force) and output q (position), whose dynamics satisfy the second-order differential equation

$$m\ddot{q} + c\dot{q} + kq = u.$$

The system has the transfer function

$$G(s) = \frac{1}{ms^2 + cs + k},$$

and the Bode plot is shown in Figure 9.18. For small s we have $G(s) \approx 1/k$. The corresponding input/output relation is $q = (1/k)u$, which implies that for low-frequency inputs, the system behaves like a spring driven by a force. For large s we have $G(s) \approx 1/(ms^2)$. The corresponding differential equation is $m\ddot{q} = u$ and the system thus behaves like mass driven by a force (a double integrator). ∇

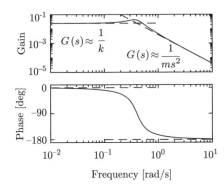

Figure 9.18: Bode plot for a spring–mass system. At low frequency the system behaves like a spring with $G(s) \approx 1/k$ and at high frequency the system behaves like a pure mass with $G(s) \approx 1/(ms^2)$.

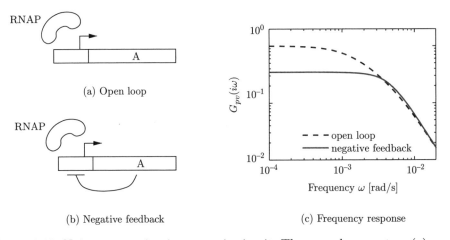

(a) Open loop

(b) Negative feedback

(c) Frequency response

Figure 9.19: Noise attenuation in a genetic circuit. The open loop system (a) consists of a constitutive promoter, while the closed loop circuit (b) is self-regulated with negative feedback (repressor). The frequency response for each circuit is shown in (c).

Example 9.16 Transcriptional regulation

Consider a genetic circuit consisting of a single gene. We wish to study the response of the protein concentration to fluctuations in the mRNA dynamics. We consider two cases: a "constitutive" promoter (no regulation) and self-repression (negative feedback), illustrated in Figure 9.19. The dynamics of the system are given by

$$\frac{dm}{dt} = \alpha(p) - \delta m + v, \qquad \frac{dp}{dt} = \kappa m - \gamma p,$$

where v is a disturbance term that affects mRNA transcription.

For the case of no feedback we have $\alpha(p) = \alpha_0$, and when $v = 0$ the system has an equilibrium point at $m_e = \alpha_0/\delta$, $p_e = \kappa\alpha_0/(\gamma\delta)$. The open loop transfer function from v to p is given by

$$G_{pv}^{\text{ol}}(s) = \frac{\kappa}{(s+\delta)(s+\gamma)}.$$

For the case of negative regulation, we have

$$\alpha(p) = \frac{\alpha_1}{1+kp^n} + \alpha_0,$$

and the equilibrium points satisfy

$$m_e = \frac{\gamma}{\kappa} p_e, \qquad \frac{\alpha_1}{1+kp_e^n} + \alpha_0 = \delta m_e = \frac{\delta\gamma}{\kappa} p_e.$$

The transfer function can be obtained by linearization around the equilibrium point and can be shown to be

$$G_{pv}^{\text{cl}}(s) = \frac{\kappa}{(s+\delta)(s+\gamma)+\kappa\sigma}, \qquad \sigma = \frac{n\alpha_1 k p_e^{n-1}}{(1+kp_e^n)^2}.$$

Figure 9.19c shows the frequency response for the two circuits. We see that the feedback circuit attenuates the response of the system to disturbances with low-frequency content but slightly amplifies disturbances at high frequency (compared to the open loop system). ∇

Determining Transfer Functions Experimentally

The transfer function of a system provides a summary of the input/output response and is very useful for analysis and design. We can often build an input/output model for a given application by directly measuring the frequency response and fitting a transfer function to it. To do so, we perturb the input to the system using a sinusoidal signal at a fixed frequency. When steady state is reached, the amplitude ratio and the phase lag give the frequency response for the excitation frequency. The complete frequency response is obtained by sweeping over a range of frequencies.

By using correlation techniques it is possible to determine the frequency response very accurately, and an analytic transfer function can be obtained from the frequency response by curve fitting. The success of this approach has led to instruments and software that automate this process, called *spectrum analyzers*. We illustrate the basic concept through two examples.

Example 9.17 Atomic force microscope
To illustrate the utility of spectrum analysis, we consider the dynamics of the atomic force microscope, described in Section 4.5. Experimental determination of the frequency response is particularly attractive for this system because its dynamics are very fast and hence experiments can be done quickly. A typical example is given in Figure 9.20, which shows an experimentally determined frequency response (solid line). In this case the frequency response was obtained in less than a second. The transfer function

$$G(s) = \frac{k\omega_2^2\omega_3^2\omega_5^2(s^2+2\zeta_1\omega_1 s+\omega_1^2)(s^2+2\zeta_4\omega_4 s+\omega_4^2)e^{-s\tau}}{\omega_1^2\omega_4^2(s^2+2\zeta_2\omega_2 s+\omega_2^2)(s^2+2\zeta_3\omega_3 s+\omega_3^2)(s^2+2\zeta_5\omega_5 s+\omega_5^2)},$$

with $\omega_i = 2\pi f_i$, $k = 5$,

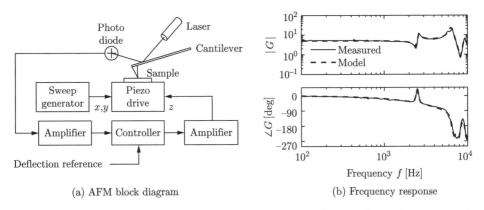

(a) AFM block diagram   (b) Frequency response

Figure 9.20: Frequency response of a preloaded piezoelectric drive for an atomic force microscope. The Bode plot shows the response of the measured transfer function (solid) and the fitted transfer function (dashed).

$$f_1 = 2.4\,\text{kHz}, \quad f_2 = 2.6\,\text{kHz}, \quad f_3 = 6.5\,\text{kHz}, \quad f_4 = 8.3\,\text{kHz}, \quad f_5 = 9.3\,\text{kHz},$$

$$\zeta_1 = 0.03, \qquad \zeta_2 = 0.03, \qquad \zeta_3 = 0.042, \qquad \zeta_4 = 0.025, \qquad \zeta_5 = 0.032,$$

and $\tau = 10^{-4}\,$s, was fitted to the data (dashed line). The frequencies ω_1 and ω_4 associated with the zeros are located where the gain curve has minima, and the frequencies ω_2, ω_3, and ω_5 associated with the poles are located where the gain curve has local maxima. The relative damping ratios are adjusted to give a good fit to maxima and minima. When a good fit to the gain curve is obtained, the time delay is adjusted to give a good fit to the phase curve. The piezo drive is preloaded, and a simple model of its dynamics is derived in Exercise 4.6. The pole at 2.55 kHz corresponds to a "trampoline" mode; the other resonances are higher modes.   ∇

Example 9.18 Pupillary light reflex dynamics

The human eye is an organ that is easily accessible for experiments. It has a control system that adjusts the pupil opening to regulate the light intensity at the retina.

This control system was explored extensively by Stark in the 1960s [227]. To determine the dynamics, light intensity on the eye was varied sinusoidally and the pupil opening was measured. A fundamental difficulty is that the closed loop system is insensitive to internal system parameters, so analysis of a closed loop system thus gives little information about the internal properties of the system. Stark used a clever experimental technique that allowed him to investigate both open and closed loop dynamics. He excited the system by varying the intensity of a light beam focused on the eye and measured pupil area, as illustrated in Figure 9.21. By using a wide light beam that covers the whole pupil, the measurement gives the closed loop dynamics. The open loop dynamics were obtained by using a narrow beam, which is small enough that it is not influenced by the pupil opening. The result of one experiment for determining open loop dynamics is given in Figure 9.22. Fitting a transfer function to the gain curve gives a good fit for $G(s) = 0.17/(1 + 0.08s)^3$. This curve gives a poor fit to the phase curve as shown by the dashed curve in Figure 9.22. The fit to the phase curve is improved by adding a 0.2 s time delay, which leaves the gain curve unchanged while substantially modifying the phase

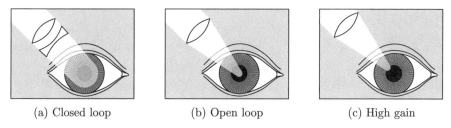

(a) Closed loop   (b) Open loop   (c) High gain

Figure 9.21: Light stimulation of the eye. In (a) the light beam is so large that it always covers the whole pupil, giving closed loop dynamics. In (b) the light is focused into a beam which is so narrow that it is not influenced by the pupil opening, giving open loop dynamics. In (c) the light beam is focused on the edge of the pupil opening, which has the effect of increasing the gain of the system since small changes in the pupil opening have a large effect on the amount of light entering the eye. From Stark [227].

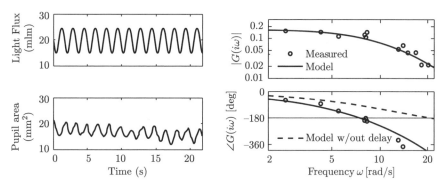

Figure 9.22: Sample curves from an open loop frequency response of the eye (left) and a Bode plot for the open loop dynamics (right). The solid curve shows a fit of the data using a third-order transfer function with 0.2 s time delay. The dashed curve in the Bode plot is the phase of the system without time delay, showing that the delay is needed to properly capture the phase. (Figure redrawn from the data of Stark [227].)

curve. The final fit gives the model

$$G(s) = \frac{0.17}{(1 + 0.08s)^3} e^{-0.2s}.$$

The Bode plot of this is shown with solid curves in Figure 9.22. Modeling of the pupillary reflex from first principles is discussed in detail in [141].   ∇

Notice that for both the AFM drive and pupillary dynamics it is not easy to derive appropriate models from first principles. In practice, it is often fruitful to use a combination of analytical modeling and experimental identification of parameters. Experimental determination of frequency response is less attractive for systems with slow dynamics because the experiment takes a long time.

9.7   FURTHER READING

The idea of characterizing a linear system by its steady-state response to sinusoids was introduced by Fourier in his investigation of heat conduction in solids [90]. Much later, it was used by the electrical engineer Steinmetz who introduced the $i\omega$ method for analyzing electrical circuits. Transfer functions were introduced via the Laplace transform by Gardner and Barnes [98], who also used them to calculate the response of linear systems. The Laplace transform was very important in the early phase of control because it made it possible to find transients via tables (see, e.g., [130]). Combined with block diagrams and transfer functions, Laplace transforms provided powerful techniques for dealing with complex systems. Calculation of responses based on Laplace transforms is less important today, when responses of linear systems can easily be generated using computers. The frequency response of a system can also be measured directly using a frequency response analyzer. There are many excellent books on the use of Laplace transforms and transfer functions for modeling and analysis of linear input/output systems. Traditional texts on control such as [73], [93] and [195] are representative examples. Pole/zero cancellation was one of the mysteries of early control theory. It is clear that common factors can be canceled in a rational function, but cancellations have system theoretical consequences that were not clearly understood until Kalman's decomposition of a linear system was introduced [139]. In the following chapters, we will use transfer functions extensively to analyze stability and to describe model uncertainty.

EXERCISES

9.1 Consider the system

$$\frac{dx}{dt} = ax + u.$$

Compute the exponential response of the system and use this to derive the transfer function from u to x. Show that when $s = a$, a pole of the transfer function, the response to the exponential input $u(t) = e^{st}$ is $x(t) = e^{at}x(0) + te^{at}$.

9.2 Let $G(s)$ be the transfer function for a linear system. Show that if we apply an input $u(t) = A\sin(\omega t)$, then the steady-state output is given by $y(t) = |G(i\omega)|A\sin(\omega t + \arg G(i\omega))$. (Hint: start by showing that the real part of a complex number is a linear operation and then use this fact.)

9.3 (Inverted pendulum) A model for an inverted pendulum was introduced in Example 3.3. Neglecting damping and linearizing the pendulum around the upright position gives a linear system characterized by the matrices

$$A = \begin{pmatrix} 0 & 1 \\ mgl/J_t & 0 \end{pmatrix}, \quad B = \begin{pmatrix} 0 \\ 1/J_t \end{pmatrix}, \quad C = \begin{pmatrix} 1 & 0 \end{pmatrix}, \quad D = 0.$$

Determine the transfer function of the system.

9.4 (Operational amplifier) Consider the operational amplifier described in Section 4.3 and analyzed in Example 9.2. An analog implementation of a PI controller can be constructed using an op amp by replacing the resistor R_2 with a resistor and capacitor in series, as shown in Figure 4.10. The resulting transfer function of the circuit is given by

$$H(s) = -\left(R_2 + \frac{1}{Cs}\right) \cdot \left(\frac{kCs}{((k+1)R_1C + R_2C)s + 1}\right),$$

where k is the gain of the op amp, R_1 and R_2 are the resistances in the compensation network and C is the capacitance.

a) Sketch the Bode plot for the system under the assumption that $k \gg R_2 > R_1$. You should label the key features in your plot, including the gain and phase at low frequency, the slopes of the gain curve, the frequencies at which the gain changes slope, etc.

b) Suppose now that we include some dynamics in the amplifier, as outlined in Example 9.2. This would involve replacing the gain k with the transfer function

$$G(s) = \frac{ak}{s + a}.$$

Compute the resulting transfer function for the system (i.e., replace k with $G(s)$) and find the poles and zeros assuming the following parameter values

$$\frac{R_2}{R_1} = 100, \qquad k = 10^6, \qquad R_2C = 1, \qquad a = 100.$$

c) Sketch the Bode plot for the transfer function in part (b) using straight line approximations and compare this to the exact plot of the transfer function (using MATLAB). Make sure to label the important features in your plot.

9.5 (Delay differential equation) Consider a system described by

$$\frac{dx}{dt} = -x(t) + u(t - \tau).$$

Derive the transfer function for the system.

9.6 (Congestion control) Consider the congestion control model described in Section 4.4. Let w represent the individual window size for a set of N identical sources, q represent the end-to-end probability of a dropped packet, b represent the number of packets in the router's buffer, and p represent the probability that a packet is dropped by the router. We write $\bar{w} = Nw$ to represent the total number of packets being received from all N sources. Show that the linearized model can be described by the transfer functions

$$G_{b\bar{w}}(s) = \frac{e^{-\tau^{\mathrm{f}}s}}{\tau_{\mathrm{e}}^{\mathrm{P}}s + e^{-\tau^{\mathrm{f}}s}}, \qquad G_{\bar{w}q}(s) = \frac{N}{q_{\mathrm{e}}(\tau_{\mathrm{e}}^{\mathrm{P}}s + q_{\mathrm{e}}w_{\mathrm{e}})},$$

$$G_{qp}(s) = e^{-\tau^{\mathrm{b}}s}, \qquad\qquad G_{pb}(s) = \rho e^{-\tau_{\mathrm{e}}^{\mathrm{P}}s},$$

where $(w_{\mathrm{e}}, b_{\mathrm{e}})$ is the equilibrium point for the system, $\tau_{\mathrm{e}}^{\mathrm{P}}$ is the router processing time, and τ^{f} and τ^{b} are the forward and backward propagation times.

9.7 (Transfer function for state space system) Consider the linear state space system

$$\frac{dx}{dt} = Ax + Bu, \qquad y = Cx.$$

a) Show that the transfer function is

$$G(s) = \frac{b_1 s^{n-1} + b_2 s^{n-2} + \cdots + b_n}{s^n + a_1 s^{n-1} + \cdots + a_n},$$

where the coefficients for the numerator polynomial are linear combinations of the *Markov parameters* $CA^i B$, $i = 0, \ldots, n-1$:

$$b_1 = CB, \quad b_2 = CAB + a_1 CB, \quad \ldots, \quad b_n = CA^{n-1}B + a_1 CA^{n-2}B + \cdots + a_{n-1}CB$$

and $\lambda(s) = s^n + a_1 s^{n-1} + \cdots + a_n$ is the characteristic polynomial for A.

b) Compute the transfer function for a linear system in reachable canonical form and show that it matches the transfer function given above.

9.8 Consider linear time-invariant systems with the control matrices

(a) $\quad A = \begin{pmatrix} -1 & 0 \\ 0 & -2 \end{pmatrix}, \qquad B = \begin{pmatrix} 2 \\ 1 \end{pmatrix}, \qquad C = \begin{pmatrix} 1 & -1 \end{pmatrix}, \qquad D = 0,$

(b) $\quad A = \begin{pmatrix} -3 & 1 \\ -2 & 0 \end{pmatrix}, \qquad B = \begin{pmatrix} 1 \\ 3 \end{pmatrix}, \qquad C = \begin{pmatrix} 1 & 0 \end{pmatrix}, \qquad D = 0,$

(c) $\quad A = \begin{pmatrix} -3 & -2 \\ 1 & 0 \end{pmatrix}, \qquad B = \begin{pmatrix} 1 \\ 0 \end{pmatrix}, \qquad C = \begin{pmatrix} 1 & 3 \end{pmatrix}, \qquad D = 0.$

Show that all systems have the transfer function $G(s) = \dfrac{s+3}{(s+1)(s+2)}$.

 9.9 (Kalman decomposition) Show that the transfer function of a system depends only on the dynamics in the reachable and observable subspace of the Kalman decomposition.

9.10 Using block diagram algebra, show that the transfer functions from v to y and w to y in Figure 9.6 are given by

$$G_{yv} = \frac{P}{1 + PC}, \qquad G_{yw} = \frac{1}{1 + PC}.$$

9.11 (Vectored thrust aircraft) Consider the lateral dynamics of a vectored thrust aircraft as described in Example 3.12. Show that the dynamics can be described using the following block diagram:

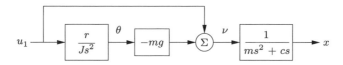

Use this block diagram to compute the transfer functions from u_1 to θ and x and show that they satisfy

$$H_{\theta u_1} = \frac{r}{Js^2}, \qquad H_{xu_1} = \frac{Js^2 - mgr}{Js^2(ms^2 + cs)}.$$

9.12 (Vehicle suspension [116]) Active and passive damping are used in cars to give a smooth ride on a bumpy road. A schematic diagram of a car with a damping system in shown in the following figure.

(Porter Class I race car driven by Todd Cuffaro)

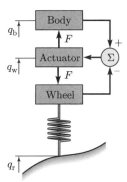

This model is called a *quarter car model*, and the car is approximated with two masses, one representing one fourth of the car body and the other a wheel. The actuator exerts a force F between the wheel and the body based on feedback from the distance between the body and the center of the wheel (the *rattle space*).

Let q_b, q_w, and q_r represent the heights of body, wheel, and road measured from their equilibrium points. A simple model of the system is given by Newton's equations for the body and the wheel,

$$m_b \ddot{q}_b = F, \qquad m_w \ddot{q}_w = -F + k_t(q_r - q_w),$$

where m_b is a quarter of the body mass, m_w is the effective mass of the wheel including brakes and part of the suspension system (the *unsprung mass*) and k_t is the tire stiffness. For a conventional damper consisting of a spring and a damper, we have $F = k(q_w - q_b) + c(\dot{q}_w - \dot{q}_b)$. For an active damper the force F can be more general and can also depend on riding conditions. Rider comfort can be characterized by the transfer function G_{aq_r} from road height q_r to body acceleration $a = \ddot{q}_b$. Show that this transfer function has the property $G_{aq_r}(i\omega_t) = k_t/m_b$, where $\omega_t = \sqrt{k_t/m_w}$ (the *tire hop frequency*). The equation implies that there are fundamental limits to the comfort that can be achieved with any damper.

9.13 (Solutions corresponding to poles and zeros) Consider the differential equation

$$\frac{d^n y}{dt^n} + a_1 \frac{d^{n-1}y}{dt^{n-1}} + \cdots + a_n y = b_1 \frac{d^{n-1}u}{dt^{n-1}} + b_2 \frac{d^{n-2}u}{dt^{n-2}} + \cdots + b_n u.$$

a) Let λ be a root of the characteristic equation

$$s^n + a_1 s^{n-1} + \cdots + a_n = 0.$$

Show that if $u(t) = 0$, the differential equation has the solution $y(t) = e^{\lambda t}$.

b) Let κ be a zero of the polynomial

$$b(s) = b_1 s^{n-1} + b_2 s^{n-2} + \cdots + b_n.$$

Show that if the input is $u(t) = e^{\kappa t}$, then there is a solution to the differential equation that is identically zero.

 9.14 (Pole/zero cancellation) Consider a closed loop system of the form of Figure 9.6, with $F = 1$ and P and C having a pole/zero cancellation. Show that if each system is written in state space form, the resulting closed loop system is not reachable and not observable.

9.15 (Inverted pendulum with PD control) Consider the normalized inverted pendulum system, whose transfer function is given by $P(s) = 1/(s^2 - 1)$ (Exercise 9.3). A proportional-derivative control law for this system has transfer function $C(s) = k_\mathrm{p} + k_\mathrm{d} s$ (see Table 9.1). Suppose that we choose $C(s) = \alpha(s - 1)$. Compute the closed loop dynamics and show that the system has good tracking of reference signals but does not have good disturbance rejection properties.

Chapter Ten

Frequency Domain Analysis

> Mr. Black proposed a negative feedback repeater and proved by tests that it possessed the advantages which he had predicted for it. In particular, its gain was constant to a high degree, and it was linear enough so that spurious signals caused by the interaction of the various channels could be kept within permissible limits. For best results the feedback factor $\mu\beta$ had to be numerically much larger than unity. The possibility of stability with a feedback factor larger than unity was puzzling.
>
> —Harry Nyquist, "The Regeneration Theory," 1956 [193].

In this chapter we study how the stability and robustness of closed loop systems can be determined by investigating how sinusoidal signals of different frequencies propagate around the feedback loop. This technique allows us to reason about the closed loop behavior of a system through the frequency domain properties of the *open loop* transfer function. The Nyquist stability theorem is a key result that provides a way to analyze stability and introduce measures of degrees of stability.

10.1 THE LOOP TRANSFER FUNCTION

Understanding how the behavior of a closed loop system is influenced by the properties of its open loop dynamics is tricky. Indeed, as the quote from Nyquist above illustrates, the behavior of feedback systems can often be puzzling. However, using the mathematical framework of transfer functions provides an elegant way to reason about such systems, which we call *loop analysis*.

The basic idea of loop analysis is to trace how a sinusoidal signal propagates in the feedback loop and explore the resulting stability by investigating if the propagated signal grows or decays. This is easy to do because the transmission of sinusoidal signals through a linear dynamical system is characterized by the frequency response of the system. The key result is the Nyquist stability theorem, which provides a great deal of insight regarding the stability of a system. Unlike proving stability with Lyapunov functions, studied in Chapter 5, the Nyquist criterion allows us to determine more than just whether a system is stable or unstable. It provides a measure of the degree of stability through the definition of stability margins. The Nyquist criterion also indicates how an unstable system should be changed to make it stable, which we shall study in detail in Chapters 11–13.

Consider the system in Figure 10.1a. The traditional way to determine if the closed loop system is stable is to investigate if the closed loop characteristic polynomial has all its roots in the left half-plane. If the process and the controller

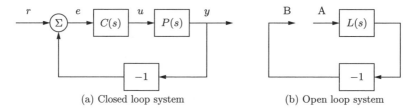

(a) Closed loop system   (b) Open loop system

Figure 10.1: The loop transfer function. The stability of the feedback system (a) can be determined by tracing signals around the loop. Letting $L = PC$ represent the loop transfer function, we break the loop in (b) and ask whether a signal injected at the point A has the same magnitude and phase when it reaches point B.

have rational transfer functions $P(s) = n_p(s)/d_p(s)$ and $C(s) = n_c(s)/d_c(s)$, then the closed loop system has the transfer function

$$G_{yr}(s) = \frac{PC}{1 + PC} = \frac{n_p(s)n_c(s)}{d_p(s)d_c(s) + n_p(s)n_c(s)},$$

and the characteristic polynomial is

$$\lambda(s) = d_p(s)d_c(s) + n_p(s)n_c(s).$$

To check stability, we simply compute the roots of the characteristic polynomial and verify that they each have negative real part. This approach is straightforward but it gives little guidance for design: it is not easy to tell how the controller should be modified to make an unstable system stable.

Nyquist's idea was to first investigate conditions under which oscillations can occur in a feedback loop. To study this, we introduce the *loop transfer function* $L(s) = P(s)C(s)$, which is the transfer function obtained by breaking the feedback loop, as shown in Figure 10.1b. The loop transfer function is simply the transfer function from the input at position A to the output at position B multiplied by -1 (to account for the usual convention of negative feedback).

Assume that a sinusoid of frequency ω_0 is injected at point A. In steady state the signal at point B will also be a sinusoid with the frequency ω_0. It seems reasonable that an oscillation can be maintained if the signal at B has the same amplitude and phase as the injected signal because we can then disconnect the injected signal and connect A to B. Tracing signals around the loop, we find that the signals at A and B are identical if there is a frequency ω_0 such that

$$L(i\omega_0) = -1, \tag{10.1}$$

which then provides a condition for maintaining an oscillation. The condition in equation (10.1) implies that the frequency response goes through the value -1, which is called the *critical point*. Letting ω_c represent a frequency at which $\angle L(i\omega_c) = 180°$, we can further reason that the system is stable if $|L(i\omega_c)| < 1$, since the signal at point B will have smaller amplitude than the injected signal. This is essentially true, but there are several subtleties that require a proper mathematical analysis, leading to Nyquist's stability criterion. Before discussing the details we give an example of calculating the loop transfer function.

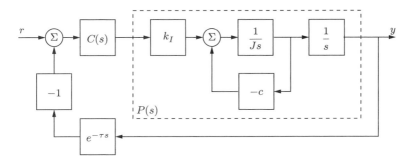

Figure 10.2: Block diagram of a DC motor control system with a short delay in the sensed position of the motor.

Example 10.1 Electric motor with proportional controller and delay

Consider a simple direct current electric motor with inertia J and damping (or back EMF) c. We wish to control the position of the motor using a feedback controller, and we consider the case where there is a small delay in the measurement of the motor position (a common case for controllers implemented on a computer with a fixed sampling rate). A block diagram for the motor with a controller $C(s)$ is shown in Figure 10.2. Using block diagram algebra, the process dynamics can be shown to be

$$P(s) = \frac{k_I}{Js^2 + cs}.$$

We now use a proportional controller of the form

$$C(s) = k_{\mathrm{p}}.$$

The loop transfer function for the control system is given by

$$L(s) = P(s)C(s)e^{-\tau s} = \frac{k_I k_{\mathrm{p}}}{Js^2 + cs}e^{-\tau s},$$

where τ is the delay in sensing of the motor position. We wish to understand under which conditions the closed loop system is stable.

The condition for oscillation is given by equation (10.1), which requires that the phase of the loop transfer function must be 180° at some frequency ω_0. Examining the loop transfer function we see that if $\tau = 0$ (no delay) then for s near 0 the phase of $L(s)$ will be 90° while for large s the phase of $L(s)$ will approach 180°. Since the gain of the system decreases as s increases, it is not possible for the condition of oscillation to be met in the case of no delay (the gain will always be less than 1 at arbitrarily high frequency).

When there is a small delay in the system, however, it is possible that we might get oscillations in the closed loop system. Suppose that ω_0 represents the frequency at which the magnitude of $L(i\omega)$ is equal to 1 (the specific value of ω_0 will depend on the parameters of the motor and the controller). Notice that the magnitude of the loop transfer function is not affected by the delay, but the phase increases as τ increases. In particular, if we let θ_0 be the phase of the undelayed system at frequency ω_0, then a time delay of $\tau_c = (\pi + \theta_0)/\omega_0$ will cause $L(i\omega_0)$ to be equal to -1. This means that as signals traverse the feedback loop, they can return in phase with the original signal and an oscillation may result.

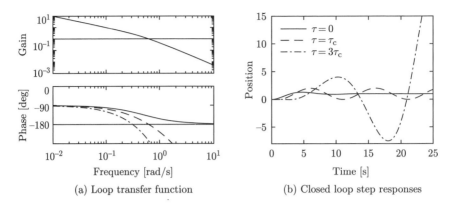

(a) Loop transfer function   (b) Closed loop step responses

Figure 10.3: Loop transfer function and step response for the DC motor control system. The system parameters are $k_I = 1$, $J = 2$, $c = 1$ and the controller parameters are $k_\mathrm{p} = 1$ and $\tau = 0$, 1, and 3.

Figure 10.3 shows three controllers that result in stable, oscillatory, and unstable closed loop performance, depending on the amount of delay in the system. The instability is caused by the fact that the disturbance signals that propagate around the feedback loop can be in phase with the original disturbance due to the delay. If the gain around the loop is greater than or equal to one, this can lead to instability.   ∇

One of the powerful concepts embedded in Nyquist's approach to stability analysis is that it allows us to study the stability of the feedback system by looking at properties of the loop transfer function $L = PC$. The advantage of doing this is that it is easy to see how the controller should be chosen to obtain a desired loop transfer function. For example, if we change the gain of the controller, the loop transfer function will be scaled accordingly and the critical point can be avoided. A simple way to stabilize an unstable system is thus to reduce the gain or to otherwise modify the controller so that the critical point -1 is avoided. Different ways to do this, called loop shaping, will be developed and discussed in Chapter 12.

10.2   THE NYQUIST CRITERION

In this section we present Nyquist's criterion for determining the stability of a feedback system through analysis of the loop transfer function. We begin by introducing a convenient graphical tool, the Nyquist plot, and show how it can be used to ascertain stability.

The Nyquist Plot

We saw in the previous chapter that the dynamics of a linear system can be represented by its frequency response and graphically illustrated by a Bode plot. To study the stability of a system, we will make use of a different representation of the frequency response called a *Nyquist plot*. The Nyquist plot of the loop transfer function $L(s)$ is formed by tracing $s \in \mathbb{C}$ around the *Nyquist contour*, consisting of the imaginary axis combined with an arc at infinity connecting the endpoints of the

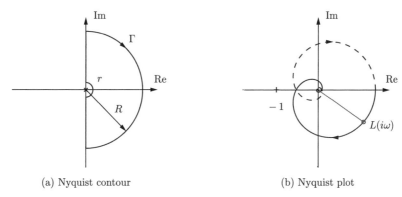

(a) Nyquist contour   (b) Nyquist plot

Figure 10.4: The Nyquist contour and the Nyquist plot. (a) The Nyquist contour Γ encloses the right half-plane, with a small semicircle around any poles of $L(s)$ at the origin or on the imaginary axis (illustrated here at the origin) and an arc whose radius R extends towards infinity. (b) The Nyquist plot is the image of the loop transfer function $L(s)$ when s traverses Γ in the clockwise direction. The solid curve corresponds to $\omega > 0$, and the dashed curve to $\omega < 0$. The gain and phase at the frequency ω are $g = |L(i\omega)|$ and $\varphi = \angle L(i\omega)$. The curve is generated for $L(s) = 1.4\,e^{-s}/(s+1)^2$.

imaginary axis. This contour, sometimes called the "Nyquist D contour" is denoted as $\Gamma \subset \mathbb{C}$ and is illustrated in Figure 10.4a. The image of $L(s)$ when s traverses Γ gives a closed curve in the complex plane and is referred to as the Nyquist plot for $L(s)$, as shown in Figure 10.4b. Note that if the transfer function $L(s)$ goes to zero as s gets large (the usual case), then the portion of the contour "at infinity" maps to the origin. Furthermore, the portion of the plot corresponding to $\omega < 0$, shown in dashed lines in Figure 10.4b, is the mirror image of the portion with $\omega > 0$.

There is a subtlety in the Nyquist plot when the loop transfer function has poles on the imaginary axis because the gain is infinite at the poles. To solve this problem, we modify the contour Γ to include small deviations that avoid any poles on the imaginary axis, as illustrated in Figure 10.4a (assuming a pole of $L(s)$ at the origin). The deviation consists of a small semicircle to the right of the imaginary axis pole location. Formally the contour Γ is defined as

$$\Gamma = \lim_{\substack{r \to 0 \\ R \to \infty}} \left(-iR, -ir\right) \cup \left\{re^{i\theta} : \theta \in \left[-\tfrac{\pi}{2}, \tfrac{\pi}{2}\right]\right\} \cup \left(ir, iR\right) \cup \left\{Re^{-i\theta} : \theta \in \left[-\tfrac{\pi}{2}, \tfrac{\pi}{2}\right]\right\}$$

(10.2)

for the case with a pole at the origin.

We now state the Nyquist criterion for the special case where the loop transfer function $L(s)$ has no poles in the right half-plane and no poles on the imaginary axis except possibly at the origin.

Theorem 10.1 (Simplified Nyquist criterion). *Let $L(s)$ be the loop transfer function for a negative feedback system (as shown in Figure 10.1a) and assume that L has no poles in the closed right half-plane ($\text{Re}\,s \geq 0$) except possibly at the origin. Then the closed loop system $G_{cl}(s) = L(s)/(1 + L(s))$ is stable if and only if the image of L along the closed contour Γ given by equation (10.2) has no net encirclements of the critical point $s = -1$.*

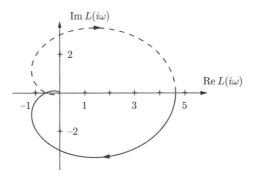

Figure 10.5: Nyquist plot for a third-order transfer function $L(s)$. The Nyquist plot consists of a trace of the loop transfer function $L(s) = 1/(s+a)^3$ with $a = 0.6$. The solid line represents the portion of the transfer function along the positive imaginary axis, and the dashed line the negative imaginary axis. The outer arc of the Nyquist contour Γ maps to the origin.

The following conceptual procedure can be used to determine that there are no net encirclements. Fix a pin at the critical point $s = -1$, orthogonal to the plane. Attach a string with one end at the critical point and the other on the Nyquist plot. Let the end of the string attached to the Nyquist curve traverse the whole curve. There are no encirclements if the string does not wind up on the pin when the curve is encircled. The number of encirclements is called the *winding number*.

Example 10.2 Nyquist plot for a third-order system
Consider a third-order transfer function

$$L(s) = \frac{1}{(s+a)^3}.$$

To compute the Nyquist plot we start by evaluating points on the imaginary axis $s = i\omega$, which yields

$$L(i\omega) = \frac{1}{(i\omega + a)^3} = \frac{(a - i\omega)^3}{(a^2 + \omega^2)^3} = \frac{a^3 - 3a\omega^2}{(a^2 + \omega^2)^3} + i\frac{\omega^3 - 3a^2\omega}{(a^2 + \omega^2)^3}.$$

This is plotted in the complex plane in Figure 10.5, with the points corresponding to $\omega > 0$ drawn as a solid line and $\omega < 0$ as a dashed line. Notice that these curves are mirror images of each other.

To complete the Nyquist plot, we compute $L(s)$ for s on the outer arc of the Nyquist contour. This arc has the form $s = Re^{-i\theta}$ for $\theta \in [-\pi/2, \pi/2]$ and $R \to \infty$. This gives

$$L(Re^{-i\theta}) = \frac{1}{(Re^{-i\theta} + a)^3} \to 0 \quad \text{as} \quad R \to \infty.$$

Thus the outer arc of the Nyquist contour Γ maps to the origin on the Nyquist plot. ∇

An alternative to computing the Nyquist plot explicitly is to determine the plot from the frequency response (Bode plot), which gives the Nyquist curve for $s = i\omega$, $\omega > 0$. We start by plotting $L(i\omega)$ from $\omega = 0$ to $\omega = \infty$, which can be read off

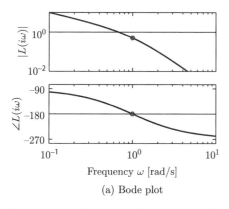

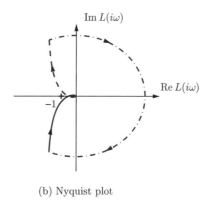

(a) Bode plot   (b) Nyquist plot

Figure 10.6: Sketching Nyquist and Bode plots. The loop transfer function is $L(s) = 1/(s(s+1)^2)$. The frequency response (a) can be used to construct the Nyquist plot (b). The large semicircle is the map of the small semicircle of the Nyquist contour around the pole at the origin. The closed loop is stable because the Nyquist curve does not encircle the critical point. The point where the phase is $-180°$ is marked with a circle in the Bode plot.

from the magnitude and phase of the transfer function. We then plot $L(Re^{i\theta})$ with $\theta \in [\pi/2, 0]$ and $R \to \infty$, which goes to zero if the high-frequency gain of $L(i\omega)$ goes to zero (if and only if $L(s)$ is strictly proper). The remaining parts of the plot can be determined by taking the mirror image of the curve thus far (normally plotted using a dashed line). The plot can then be labeled with arrows corresponding to a clockwise traversal around the Nyquist contour (the same direction in which the first portion of the curve was plotted).

Example 10.3 Nyquist criterion for a third-order system with a pole at the origin
Consider the transfer function

$$L(s) = \frac{k}{s(s+1)^2},$$

where the gain has the nominal value $k = 1$. The Bode plot is shown in Figure 10.6a. The system has a single pole at $s = 0$ and a double pole at $s = -1$. The gain curve of the Bode plot thus has the slope -1 for low frequencies, and at the double pole $s = 1$ the slope changes to -3. For small s we have $L \approx k/s$, which means that the low-frequency asymptote intersects the unit gain line at $\omega = k$. The phase curve starts at $-90°$ for low frequencies, it is $-180°$ at the breakpoint $\omega = 1$, and it is $-270°$ at high frequencies.

Having obtained the Bode plot, we can now sketch the Nyquist plot, shown in Figure 10.6b. It starts with a phase of $-90°$ for low frequencies, intersects the negative real axis at the breakpoint $\omega = 1$ where $L(i) = -0.5$ and goes to zero along the imaginary axis for high frequencies. The small half-circle of the Nyquist contour at the origin is mapped onto a large circle enclosing the right half-plane. The Nyquist curve does not encircle the critical point $s = -1$, and it follows from the simplified Nyquist criterion that the closed loop system is stable. Since $L(i) = -k/2$, we find the closed loop system becomes unstable if the gain is increased to $k = 2$ or beyond.   ▽

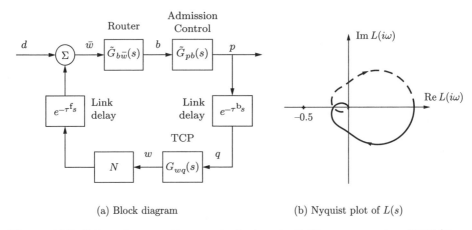

(a) Block diagram   (b) Nyquist plot of $L(s)$

Figure 10.7: Internet congestion control. A set of N sources using TCP/Reno send messages through a single router with admission control (a). Link delays are included for the forward and backward directions. The Nyquist plot for the loop transfer function is shown in (b).

The Nyquist criterion does not require that $|L(i\omega_c)| < 1$ for all ω_c corresponding to a crossing of the negative real axis. Rather, it says that the number of encirclements must be zero, allowing for the possibility that the Nyquist curve could cross the negative real axis and cross back at magnitudes greater than 1. The fact that it was possible to have high feedback gains surprised the early designers of feedback amplifiers, as mentioned in the quote in the beginning of this chapter.

One advantage of the Nyquist criterion is that it tells us how a system is influenced by changes of the controller parameters. For example, it is very easy to visualize what happens when the gain is changed since this just scales the Nyquist curve.

Example 10.4 Congestion control

Consider the Internet congestion control system described in Section 4.4. Suppose we have N identical sources and a disturbance d representing an external data source, as shown in Figure 10.7a. We let w represent the individual window size for a source, q represent the end-to-end probability of a dropped packet, b represent the number of packets in the router's buffer, and p represent the probability that a packet is dropped by the router. We write $\bar{w}$ for the total number of packets being received from all N sources. We also include forward and backward propagation delays between the router and the senders.

To analyze the stability of the system, we use the transfer functions computed in Exercise 9.6:

$$\tilde{G}_{b\bar{w}}(s) = \frac{1}{\tau_e^{\mathrm{P}} s + e^{-\tau^f s}}, \qquad G_{wq}(s) = -\frac{1}{q_e(\tau_e^{\mathrm{P}} s + q_e w_e)}, \qquad \tilde{G}_{pb}(s) = \rho,$$

where (w_e, b_e) is the equilibrium point for the system, N is the number of sources, τ_e^{P} is the steady-state round-trip time, and τ^f and τ^b are the forward and backward propagation times. We use $\tilde{G}_{b\bar{w}}$ and $\tilde{G}_{qp}$ to represent the transfer functions with the forward and backward time delays removed since this is accounted for as separate blocks in Figure 10.7b. Similarly, $G_{wq} = G_{\bar{w}q}/N$ since we have pulled out the multiplier N as a separate block as well.

The loop transfer function is given by

$$L(s) = \rho \cdot \frac{N}{\tau_e^P s + e^{-\tau^f s}} \cdot \frac{1}{q_e(\tau_e^P s + q_e w_e)} e^{-\tau_e^t s},$$

where $\tau^t = \tau^P + \tau^f + \tau^b$ is the total round trip delay time. Using the fact that $w_e = b_e/N = \tau_e^P c/N$ and $q_e = 2/(2 + w_e^2) \approx 2/w_e^2 = 2N^3/(\tau_e^P c)^2$ from equation (4.17), we can show that

$$L(s) = \rho \cdot \frac{N}{\tau_e^P s + e^{-\tau^f s}} \cdot \frac{c^3(\tau_e^P)^3}{2N^2(c(\tau_e^P)^2 s + 2N)} e^{-\tau_e^t s}.$$

Note that we have chosen the sign of $L(s)$ to use the same sign convention as in Figure 10.1b.

The Nyquist plot for the loop transfer function is shown in Figure 10.7b. To obtain an analytic stability criterion we can approximate the transfer function close to the intersection with the negative real axis, which occurs at the "phase crossover" frequency ω_{pc}. The second factor is stable if $\tau_e^P > \tau^f$ and has fast dynamics, so we approximate it by its zero frequency gain N. The third factor has slow dynamics (it can be shown that $2N \ll c(\tau_e^P)^2 \omega_{pc}$), and we can approximate it by an integrator. We thus obtain the following approximation of the loop transfer function around the frequency ω_{pc}:

$$L(s) \approx \rho \cdot N \cdot \frac{c^3(\tau_e^P)^3}{2N^2 c(\tau_e^P)^2 s} e^{-\tau_e^t s} = \frac{\rho c^2 \tau_e^P}{2Ns} e^{-\tau_e^t s}.$$

The integrator has a phase lag of $\pi/2$ and the transfer function $L(s)$ has the phase crossover frequency $\omega_{pc} = \pi/(2\tau_e^P)$. A necessary condition for stability is thus $|L(i\omega_{pc})| < 1$, which gives the condition

$$\frac{\rho c^2(\tau_e^P)^2}{\pi N} < 1.$$

Using the Nyquist criterion, the closed loop system will be unstable if this quantity is greater than 1. In particular, for a fixed processing time τ_e^P, the system will become unstable as the link capacity c increases. This indicates that the TCP protocol may not be scalable to high-capacity networks, as pointed out by Low *et al.* [168]. Exercise 10.3 provides some ideas of how this might be overcome. ∇

The General Nyquist Criterion

Theorem 10.1 requires that $L(s)$ has no poles in the closed right half-plane, except possibly at the origin. In some situations this is not the case and we need a more general result. This requires some results from the theory of complex variables, for which the reader can consult Ahlfors [6]. Since some precision is needed in stating Nyquist's criterion properly, we will use a more mathematical style of presentation. We also follow the mathematical convention of counting encirclements in the counterclockwise direction for the remainder of this section. The key result is the following theorem about functions of complex variables.

Theorem 10.2 (Principle of variation of the argument). *Let Γ be a closed contour in the complex plane and let D represent the interior of Γ. Assume the function*

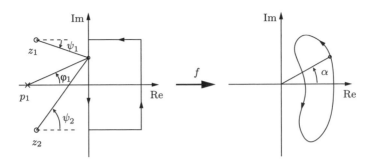

Figure 10.8: Graphical proof of the principle of the variation of the argument.

$f : \mathbb{C} \to \mathbb{C}$ *is analytic on* Γ *and* D *except at a finite number of poles and zeros in* D. *Then the winding number* $n_{\mathrm{w},\Gamma}(f(s))$ *of the function* $f(s)$ *as* s *traverses the contour* Γ *in the counterclockwise direction is given by*

$$n_{\mathrm{w},\Gamma}(f(s)) = \frac{1}{2\pi} \Delta \arg_{\Gamma} f(s) = \frac{1}{2\pi i} \int_{\Gamma} \frac{f'(s)}{f(s)} ds = n_{\mathrm{z},D} - n_{\mathrm{p},D},$$

where $\Delta \arg_{\Gamma}$ *is the net variation in the angle when* s *traverses the contour* Γ *in the counterclockwise direction,* $n_{\mathrm{z},D}$ *is the number of zeros of* $f(s)$ *in* D, *and* $n_{\mathrm{p},D}$ *is the number of poles of* $f(s)$ *in* D. *Poles and zeros of multiplicity* m *are counted* m *times.*

To understand why the principle of variation of the argument is true, we keep track of how the argument (angle) of a function varies as we traverse a closed contour. Figure 10.8 illustrates the basic idea. Consider a function $f : \mathbb{C} \to \mathbb{C}$ of the form

$$f(s) = \frac{(s - z_1) \cdots (s - z_m)}{(s - p_1) \cdots (s - p_n)}, \tag{10.3}$$

where z_i are zeros and p_i are poles. We can rewrite the factors in this function by keeping track of the distance and angle to each pole and zero:

$$f(s) = \frac{r_1 e^{i\psi_1} \cdots r_m e^{i\psi_m}}{\rho_1 e^{i\theta_1} \cdots \rho_m e^{i\theta_n}}.$$

The argument (angle) of $f(s)$ at any given value of s can be computed by adding the contributions for the zeros and subtracting the contributions from the poles,

$$\arg(f(s)) = \sum_{i=1}^{m} \psi_i - \sum_{i=1}^{n} \theta_i.$$

We now consider what happens if we traverse a closed loop contour Γ. If all of the poles and zeros in $f(s)$ are outside of the contour, then the net contribution to the angle from terms in the numerator and denominator will be zero since there is no way for the angle to "accumulate." Thus the contribution from each individual zero and pole will integrate to zero as we traverse the contour. If, however, the zero or pole is inside the contour Γ, then the net change in angle as we transverse the contour will be 2π for terms in the numerator (zeros) or -2π for terms in the denominator (poles). Thus the net change in the angle as we traverse the contour is given by $2\pi(n_{z,D} - n_{p,D})$, where $n_{z,D}$ is the number of zeros inside the contour and $n_{p,D}$ is the number of poles inside the contour.

Formal proof. Assume that $s = a$ is a zero of multiplicity m. In the neighborhood of $s = a$ we have

$$f(s) = (s-a)^m g(s),$$

where the function g is analytic and different from zero. The ratio of the derivative of f to itself is then given by

$$\frac{f'(s)}{f(s)} = \frac{m}{s-a} + \frac{g'(s)}{g(s)},$$

and the second term is analytic at $s = a$. The function f'/f thus has a single pole at $s = a$ with the residue m. The sum of the residues at the zeros of this function is $n_{z,D}$. Similarly, we find that the sum of the residues for the poles is $-n_{p,D}$, and hence

$$n_{z,D} - n_{p,D} = \frac{1}{2\pi i} \int_\Gamma \frac{f'(s)}{f(s)}\, ds = \frac{1}{2\pi i} \int_\Gamma \frac{d}{ds} \log f(s)\, ds = \frac{1}{2\pi i} \Delta \arg_\Gamma \log f(s),$$

where $\Delta \arg_\Gamma$ again denotes the variation along the contour Γ. We have

$$\log f(s) = \log |f(s)| + i \arg f(s),$$

and since the variation of $|f(s)|$ around a closed contour is zero it follows that

$$\Delta \arg_\Gamma \log f(s) = i\Delta \arg_\Gamma \arg f(s),$$

and the theorem is proved. □

This theorem is useful in determining the number of poles and zeros of a function of a complex variable in a given region. By choosing an appropriate closed region D with boundary Γ, we can determine the difference between the number of zeros and poles through computation of the winding number.

Theorem 10.2 can be used to obtain a general version of Nyquist's stability theorem by choosing Γ as the Nyquist contour shown in Figure 10.4a, which encloses the right half-plane. To construct the contour, we start with part of the imaginary axis $-iR \le s \le iR$ and a semicircle to the right with radius R. If the function f has poles on the imaginary axis, we introduce small semicircles with radii r to the right of the poles as shown in the figure to avoid crossing through a singularity. The Nyquist contour is obtained by selecting R large enough and r small enough so that all open loop right half-plane poles are enclosed.

Note that Γ has orientation *opposite* that shown in Figure 10.4a. The convention in engineering is to traverse the Nyquist contour in the clockwise direction since this corresponds to increasing frequency moving upwards along the imaginary axis, which makes it easy to sketch the Nyquist contour from a Bode plot. In mathematics it is customary to define the winding number for a curve with respect to a point so that it is positive when the contour is traversed counterclockwise. This difference does not matter as long as we use the same convention for orientation when traversing the Nyquist contour and computing the winding number.

To use the principle of variation of the argument (Theorem 10.2) to obtain an improved stability criterion we apply it to the function $f(s) = 1 + L(s)$, where $L(s)$ is the loop transfer function of a closed loop system with negative feedback. The generalized Nyquist criterion is given by the following theorem.

Theorem 10.3 (General Nyquist criterion). *Consider a closed loop system with loop transfer function $L(s)$ that has $n_{p,rhp}$ poles in the region enclosed by the Nyquist contour Γ. Let $n_{w,\Gamma}(1 + L(s))$ be the winding number of $f(s) = 1 + L(s)$ when s traverses Γ in the counterclockwise direction. Assume that $1 + L(i\omega) \neq 0$ for all ω on Γ and that $n_{w,\Gamma}(1 + L(s)) + n_{p,rhp} = 0$. Then the closed loop system has no poles in the closed right half-plane and it is thus stable.*

Proof. The proof follows directly from the principle of variation of the argument, Theorem 10.2. The closed loop poles of the system are the zeros of the function $f(s) = 1 + L(s)$. It follows from the assumptions that the function $f(s)$ has no zeros on the contour Γ. To find the zeros in the right half-plane, we investigate the winding number of the function $f(s) = 1 + L(s)$ as s moves along the Nyquist contour Γ in the *counterclockwise* direction. The winding number n_w can be determined from the Nyquist plot. A direct application of Theorem 10.2 shows that since $n_{w,\Gamma}(1 + L(s)) + n_{p,rhp}(L(s)) = 0$, then $f(s)$ has no zeros in the right half-plane. Since the image of $1 + L(s)$ is a shifted version of $L(s)$, we usually express the Nyquist criterion as net encirclements of the -1 point by the image of $L(s)$. □

The condition that $1 + L(i\omega) \neq 0$ on Γ implies that the Nyquist curve does not go through the critical point -1 for any frequency. The condition that $n_{w,\Gamma}(1 + L(s)) + n_{p,rhp}(L(s)) = 0$, which is called *the winding number condition*, implies that the Nyquist curve encircles the critical point as many times as the loop transfer function $L(s)$ has poles in the right half-plane.

As noted above, in practice the Nyquist criterion is most often applied by traversing the Nyquist contour in the *clockwise* direction, since this corresponds to tracing out the Nyquist curve from $\omega = 0$ to ∞, which can be read off from the Bode plot. In this case, the number of net encirclements of the -1 point must also be counted in the *clockwise* direction. If we let P be the number of unstable poles in the loop transfer function, N be the number of clockwise encirclements of the point -1, and Z be the number of unstable stable zeros of $1 + L$ (and hence the number of unstable poles of the closed loop) then the following relation holds:

$$Z = N + P.$$

Note also than when using small semicircles of radius r to avoid poles on the imaginary axis, these generate a section of the Nyquist curve with large magnitude, requiring care in computing the winding number.

Example 10.5 Stabilized inverted pendulum
The linearized dynamics of a normalized inverted pendulum can be represented by the transfer function $P(s) = 1/(s^2 - 1)$, where the input is acceleration of the pivot and the output is the pendulum angle θ, as shown in Figure 10.9 (Exercise 9.3). We attempt to stabilize the pendulum with a proportional-derivative (PD) controller having the transfer function $C(s) = k(s + 2)$. The loop transfer function is

$$L(s) = \frac{k(s + 2)}{s^2 - 1}.$$

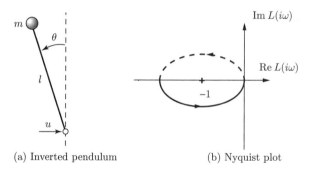

(a) Inverted pendulum   (b) Nyquist plot

Figure 10.9: PD control of an inverted pendulum. (a) The system consists of a mass that is balanced by applying a force at the pivot point. A proportional-derivative controller with transfer function $C(s) = k(s+2)$ is used to command u based on θ. (b) A Nyquist plot of the loop transfer function for gain $k = 1$. There is one counterclockwise encirclement of the critical point, giving $N = -1$ clockwise encirclements.

The Nyquist plot of the loop transfer function is shown in Figure 10.9b. We have $L(0) = -2k$ and $L(\infty) = 0$. If $k > 0.5$, the Nyquist curve encircles the critical point $s = -1$ in the counterclockwise direction when the Nyquist contour γ is encircled in the clockwise direction. The number of encirclements is thus $N = -1$. Since the loop transfer function has one pole in the right half-plane $(P = 1)$, we find that $Z = N + P = 0$ and the system is thus stable for $k > 0.5$. If $k < 0.5$, there is no encirclement and the closed loop will have one pole in the right half-plane. Notice that the system is unstable for small gains but stable for high gains. ∇

Conditional Stability

An unstable system can often be stabilized simply by reducing the loop gain. However, as Example 10.5 illustrates, there are situations where a system can be stabilized by *increasing* the gain. This was first encountered by electrical engineers in the design of feedback amplifiers, who coined the term *conditional stability*. The problem was actually a strong motivation for Nyquist to develop his theory. The following example further illustrates this concept.

Example 10.6 Conditional stability for a third-order system

Consider a feedback system with the loop transfer function

$$L(s) = \frac{3k(s+6)^2}{s(s+1)^2}. \tag{10.4}$$

The Nyquist plot of the loop transfer function is shown in Figure 10.10 for $k = 1$. Notice that the Nyquist curve intersects the negative real axis twice. The first intersection occurs at $L = -12$ for $\omega = 2$ and the second at $L = -4.5$ for $\omega = 3$. The intuitive argument based on signal tracing around the loop in Figure 10.1b is misleading in this case. Injection of a sinusoid with frequency 2 rad/s and amplitude 1 at A gives, in steady state, an oscillation at B that is in phase with the input and has amplitude 12. Intuitively it seems unlikely that closing of the loop will result in a stable system. Evaluating the winding number for the Nyquist plot in Figure 10.10 shows that the winding number is zero and the system is thus shown to

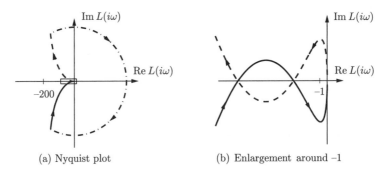

(a) Nyquist plot   (b) Enlargement around –1

Figure 10.10: Nyquist curve for the loop transfer function $L(s) = \left(3(s+6)^2\right)/\left(s(s+1)^2\right)$. The plot in (b) is an enlargement of the box around the origin of the plot in (a). The Nyquist curve intersects the negative real axis twice but has no net encirclements of -1.

be stable by using the version of Nyquist's stability criterion in Theorem 10.3. More specifically, the closed loop system is stable for any $k > 2/9$. It becomes unstable if the gain is reduced to $1/12 < k < 2/9$, and it will be stable again for gains less than $1/12$. ∇

10.3   STABILITY MARGINS

In practice it is not enough that a system is stable. There must also be some margins of stability that describe how far from instability the system is and its robustness to perturbations. Stability is captured by Nyquist's criterion, which says that the loop transfer $L(s)$ function should avoid the critical point -1, while satisfying a winding number condition. Stability margins express how well the Nyquist curve of the loop transfer avoids the critical point. The shortest distance s_{m} of the Nyquist curve to the critical point is a natural criterion, which is called the *stability margin*. It is illustrated in Figure 10.11a, where we have plotted the portion of the curve corresponding to $\omega > 0$. A stability margin s_{m} means that the Nyquist curve of the loop transfer function is outside a circle around the critical point with radius s_{m}.

Other margins are based the influence of the controller on the Nyquist curve. An increase in controller gain expands the Nyquist plot radially. An increase in the phase of the controller turns the Nyquist plot clockwise. Hence from the Nyquist plot we can easily pick off the amount of gain or phase that can be added without causing the system to become unstable.

The *gain margin* g_{m} of a closed loop system is defined as the smallest multiplier of the loop gain that makes the system unstable. It is also the inverse of the distance between the origin and the point between -1 and 0 where the loop transfer function crosses the negative real axis. If there are several crossings the gain margin is defined by the intersection that is closest to the critical point. Let this point be $L(i\omega_{\mathrm{pc}})$, where ω_{pc} represents this frequency, called the *phase crossover frequency*. The gain margin for the system is then

$$g_{\mathrm{m}} = \frac{1}{|L(i\omega_{\mathrm{pc}})|}. \qquad (10.5)$$

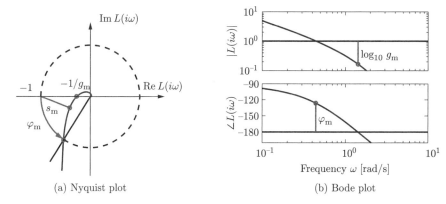

(a) Nyquist plot   (b) Bode plot

Figure 10.11: Stability margins for a third-order loop transfer function $L(s)$. The Nyquist plot (a) shows the stability margin, s_m, the gain margin g_m, and the phase margin φ_m. The stability margin s_m is the shortest distance to the critical point -1. The gain margin corresponds to the smallest increase in gain that creates an encirclement, and the phase margin is the smallest change in phase that creates an encirclement. The Bode plot (b) shows the gain and phase margins.

This number can be obtained directly from the Nyquist plots as shown in Figure 10.11a.

The *phase margin* is the amount of phase lag required to reach the stability limit. Let ω_gc be the *gain crossover frequency*, the frequency where the loop transfer function $L(i\omega_\mathrm{pc})$ intersects the unit half-circle below the real axis. The phase margin is then

$$\varphi_\mathrm{m} = 180° + \angle\, L(i\omega_\mathrm{gc}). \qquad (10.6)$$

As with the gain margin, this number can be obtained from the Nyquist plots as shown in Figure 10.11a. If the Nyquist curve intersects the half-circle many times, the phase margin is defined by the intersection that is closest to the critical point.

The gain and phase margins can also be determined from the Bode plot of the loop transfer function, as illustrated in Figure 10.11b. To find the gain margin we first find the phase crossover frequency ω_pc where the phase is $-180°$. The gain margin is the inverse of the gain at that frequency. To determine the phase margin we first determine the gain crossover frequency ω_gc, i.e., the frequency where the gain of the loop transfer function is 1. The phase margin is the phase of the loop transfer function at that frequency plus $180°$. Figure 10.11b illustrates how the margins are found in the Bode plot of the loop transfer function. The margins are not always easy to determine from the Bode plot if the loop transfer function intersects the lines $|G(i\omega)| = 1$ or $\angle G(i\omega) = -180° \pm n \cdot 360°$ many times. In these cases, the Nyquist plot should be used instead.

The gain and phase margins are classical robustness measures that have been used for a long time in control system design. They were particularly attractive because design was often based on the Bode plot of the loop transfer function. The gain and phase margins are related to the stability margin through the inequalities

$$g_\mathrm{m} \geq \frac{1}{1 - s_\mathrm{m}}, \qquad \varphi_\mathrm{m} \geq 2\arcsin(s_\mathrm{m}/2), \qquad (10.7)$$

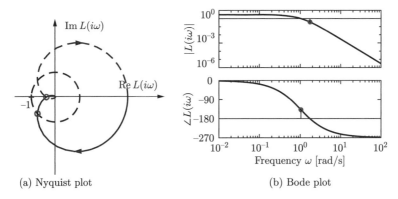

(a) Nyquist plot   (b) Bode plot

Figure 10.12: Stability margins for a third-order transfer function. The Nyquist plot (a) allows the gain, phase, and stability margins to be determined by measuring the distances of relevant features. The gain and phase margins can also be read off of the Bode plot (b).

which follow from Figure 10.12 and the fact that s_m is less than the distance $d = 2\sin(\varphi_m/2)$ from the critical point -1 to the point defining the gain crossover frequency.

A drawback with the stability margin s_m is that it does not have a natural representation in the Bode plot of the loop transfer function. It can be shown that the peak magnitude M_s of the closed loop transfer function $1/(1 + P(s)C(s))$ is related to the stability margin through the formula $s_m = 1/M_s$, as will be discussed in Chapter 13 together with more general robustness measures. A drawback with gain and phase margins is that both have to be given to guarantee that the Nyquist curve is not close to the critical point. It is also difficult to represent the winding number in the Bode plot. In general, it is best to use the Nyquist plot to check stability since this provides more complete information than the Bode plot.

Example 10.7 Stability margins for a third-order system
Consider a loop transfer function $L(s) = 3/(s+1)^3$. The Nyquist and Bode plots are shown in Figure 10.12. To compute the gain, phase, and stability margins, we can use the Nyquist plot shown in Figure 10.12a. This yields the following values:

$$g_m = 2.67, \qquad \varphi_m = 41.7°, \qquad s_m = 0.464.$$

The gain and phase margins can also be determined from the Bode plot.   ∇

Even if both the gain and phase margins are reasonable, the system may still not be robust, as is illustrated by the following example.

Example 10.8 Good gain and phase margins but poor stability margins
Consider a system with the loop transfer function

$$L(s) = \frac{0.38(s^2 + 0.1s + 0.55)}{s(s+1)(s^2 + 0.06s + 0.5)}.$$

A numerical calculation gives the gain margin as $g_m = 266$, and the phase margin is 70°. These values indicate that the system is robust, but the Nyquist curve is still close to the critical point, as shown in Figure 10.13a. The stability margin is

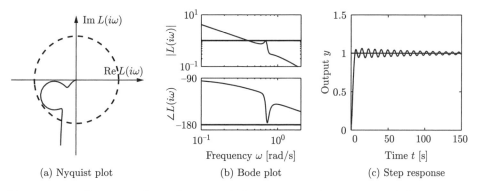

(a) Nyquist plot   (b) Bode plot   (c) Step response

Figure 10.13: System with good gain and phase margins but a poor stability margin. The Nyquist plot (a) and Bode plot (b) of the loop transfer function and step response (c) for a system with good gain and phase margins but with a poor stability margin. The Nyquist plot shows only the portion of the curve corresponding to $\omega > 0$.

$s_\mathrm{m} = 0.27$, which is very low. The closed loop system has two resonant modes, one with damping ratio $\zeta = 0.81$ and the other with $\zeta = 0.014$. The step response of the system is highly oscillatory, as shown in Figure 10.13c. ∇

When designing feedback systems, it will often be useful to define the robustness of the system using gain, phase, and stability margins. These numbers tell us how much the system can vary from our nominal model and still be stable. Reasonable values of the margins are phase margin $\varphi_\mathrm{m} = 30°$–$60°$, gain margin $g_\mathrm{m} = 2$–5, and stability margin $s_\mathrm{m} = 0.5$–0.8.

There are also other stability measures, such as the *delay margin*, which is the smallest time delay required to make the system unstable. For loop transfer functions that decay quickly, the delay margin is closely related to the phase margin, but for systems where the gain curve of the loop transfer function has several peaks at high frequencies, the delay margin is a more relevant measure.

Example 10.9 Nanopositioning system for an atomic force microscope
Consider the system for horizontal positioning of the sample in an atomic force microscope, described in more detail in Section 4.5. The system has oscillatory dynamics, and a simple model is a spring–mass system with low damping. The normalized transfer function is given by

$$P(s) = \frac{\omega_0^2}{s^2 + 2\zeta\omega_0 s + \omega_0^2},\tag{10.8}$$

where the damping ratio typically is a very small number, e.g., $\zeta = 0.1$.

We will start with a controller that has only integral action. The resulting loop transfer function is

$$L(s) = \frac{k_\mathrm{i}\omega_0^2}{s(s^2 + 2\zeta\omega_0 s + \omega_0^2)},$$

where k_i is the gain of the controller. Nyquist and Bode plots of the loop transfer function are shown in Figure 10.14. Notice that the part of the Nyquist curve that is close to the critical point -1 is approximately circular.

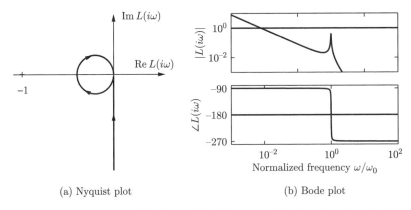

(a) Nyquist plot   (b) Bode plot

Figure 10.14: Nyquist and Bode plots of the loop transfer function for the AFM system (10.8) with an integral controller. The frequency in the Bode plot is normalized by ω_0. The parameters are $\zeta = 0.01$ and $k_i = 0.008$.

From the Bode plot in Figure 10.14b, we see that the phase crossover frequency is $\omega_{\mathrm{pc}} = \omega_0$, which will be independent of the gain k_i. Evaluating the loop transfer function at this frequency, we have $L(i\omega_0) = -k_i/(2\zeta\omega_0)$, which means that the stability margin is $s_m = 1 - k_i/(2\zeta\omega_0)$. To have a desired stability margin of s_m the integral gain should be chosen as

$$k_i = 2\zeta\omega_0(1 - s_m).$$

Figure 10.14 shows Nyquist and Bode plots for the system with gain margin $g_m = 2.5$ and stability margin $s_m = 0.6$. The gain curve in the Bode plot is almost a straight line for low frequencies and has a resonant peak at $\omega = \omega_0$. The gain crossover frequency is approximately equal to k_i and the phase decreases monotonically from $-90°$ to $-270°$: it is equal to $-180°$ at $\omega = \omega_0$. The gain curve can be shifted vertically by changing k_i: increasing k_i shifts the gain curve upward and increases the gain crossover frequency. ∇

10.4   BODE'S RELATIONS AND MINIMUM PHASE SYSTEMS

An analysis of Bode plots reveals that there appears to be a relation between the gain curve and the phase curve. Consider, for example, the Bode plots for the differentiator and the integrator (shown in Figure 9.13). For the differentiator the slope is $+1$ and the phase is a constant $\pi/2$ radians. For the integrator the slope is -1 and the phase is $-\pi/2$. For the first-order system $G(s) = s + a$, the amplitude curve has the slope 0 for small frequencies and the slope $+1$ for high frequencies, and the phase is 0 for low frequencies and $\pi/2$ for high frequencies.

Bode investigated the relations between the gain and phase curves in his plot and he found that for a special class of systems there was indeed a relation between gain and phase. These systems do not have time delays or poles and zeros in the right half-plane, and in addition they have the property that $\log|G(s)|/s$ goes to zero as $s \to \infty$ for $\mathrm{Re}\, s \geq 0$. Bode called these systems *minimum phase systems* because they have the smallest phase lag of all systems with the same gain curve. For minimum phase systems the phase is uniquely given by the shape of the gain curve and vice versa:

$$\arg G(i\omega_0) = \frac{\pi}{2} \int_0^\infty f(\omega) \frac{d \log |G(i\omega)|}{d \log \omega} \frac{d\omega}{\omega} \approx \frac{\pi}{2} \frac{d \log |G(i\omega)|}{d \log \omega}\bigg|_{\omega=\omega_0}, \qquad (10.9)$$

where f is the weighting kernel

$$f(\omega) = \frac{2}{\pi^2} \log \left| \frac{\omega + \omega_0}{\omega - \omega_0} \right| \quad \text{and} \quad \int_0^\infty f(\omega) \frac{d\omega}{\omega} = 1. \qquad (10.10)$$

The phase curve for a minimum phase system is thus a weighted average of the derivative of the gain curve. Notice that since $|G(s)| = |-G(s)|$ and $\angle(-G(s)) = \angle G(s) - 180°$, the sign of the minimum phase $G(s)$ must also be chosen properly. We assume that the sign is always chosen so that $\angle G(s) > \angle(-G(s))$.

We illustrate Bode's relation (10.9) with an example.

Example 10.10 Phase of $G(s) = s^n$
For the transfer function $G(s) = s^n$ we have that $\log G(s) = n \log s$ and hence $d \log G(s)/d \log s = n$. Equation (10.9) then gives

$$\arg G(i\omega_0) = \frac{\pi}{2} \int_0^\infty f(\omega) \frac{d \log |G(i\omega)|}{d \log \omega} \frac{d\omega}{\omega} = \frac{\pi}{2} \int_0^\infty n f(\omega) \frac{d\omega}{\omega} = n \frac{\pi}{2},$$

where the last equality follows from equation (10.10). If the gain curve has constant slope n, the phase curve is a horizontal line $\arg G(i\omega) = n\pi/2$. ▽

We will now give a few examples of transfer functions that are not minimum phase transfer functions. The transfer function of a time delay of τ units is $G(s) = e^{-s\tau}$. This transfer function has unit gain $|G(i\omega)| = 1$, and the phase is $\arg G(i\omega) = -\omega\tau$. The corresponding minimum phase system with unit gain has the transfer function $G(s) = 1$. The time delay thus has an additional phase lag of $\omega\tau$. Notice that the phase lag increases linearly with frequency. Figure 10.15a shows the Bode plot of the transfer function. (Because we use a log scale for frequency, the phase falls off exponentially in the plot.)

Consider a system with the transfer function $G(s) = (a - s)/(a + s)$ with $a > 0$, which has a zero $s = a$ in the right half-plane. The transfer function has unit gain $|G(i\omega)| = 1$, and the phase is $\arg G(i\omega) = -2 \arctan(\omega/a)$. The corresponding minimum phase system with unit gain has the transfer function $G(s) = 1$. Figure 10.15b shows the Bode plot of the transfer function. A similar analysis of the transfer function $G(s) = (s + a)/(s - a)$ with $a > 0$, which has a pole in the right half-plane, shows that its phase is $\arg G(i\omega) = -2 \arctan(a/\omega)$. The Bode plot is shown in Figure 10.15c.

The presence of poles and zeros in the right half-plane imposes severe limits on the achievable performance as will be discussed in Chapter 14. Dynamics of this type should be avoided by redesign of the system. While the poles are intrinsic properties of the system and they do not depend on sensors and actuators, the zeros depend on how inputs and outputs of a system are coupled to the states. Zeros can thus be changed by moving sensors and actuators or by introducing new sensors and actuators. Non-minimum phase systems are unfortunately quite common in practice.

The following example shows that difficulties can arise in the response of non-minimum phase systems.

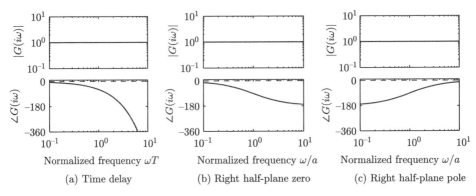

Figure 10.15: Bode plots of systems that are not minimum phase. (a) Time delay $G(s) = e^{-sT}$, (b) system with a right half-plane (RHP) zero $G(s) = (a - s)/(a + s)$, and (c) system with right half-plane pole $G(s) = (s + a)/(s - a)$. The corresponding minimum phase system has the transfer function $G(s) = 1$ in all cases; the phase curves for that system are shown as dashed lines.

Example 10.11 Vehicle steering

The vehicle steering model considered in Examples 6.13 and 9.10 has different properties depending on whether we are driving forward or in reverse. The non-normalized transfer function from steering angle to lateral position for the simple vehicle model is

$$P(s) = \frac{av_0 s + v_0^2}{bs^2},$$

where v_0 is the velocity of the vehicle and $a, b > 0$ (see Example 6.13). The transfer function has a zero at $s = v_0/a$. In normal (forward) driving this zero is in the left half-plane, but it is in the right half-plane when driving in reverse, $v_0 < 0$. The unit step response is

$$y(t) = \frac{av_0 t}{b} + \frac{v_0^2 t^2}{2b}.$$

The lateral position thus begins to respond immediately to a steering command as an integrator. For reverse steering v_0 is negative and the initial response is in the wrong direction, a behavior that is representative for non-minimum phase systems (called an *inverse response*).

Figure 10.16 shows the step response for forward and reverse driving. The parameters are $a = 1.5\,\text{m}$, $b = 3\,\text{m}$, $v_0 = 2\,\text{m/s}$ for forward driving, and $v_0 = -2\,\text{m/s}$ for reverse driving. Thus when driving in reverse there is an initial motion of the center of mass in the *opposite* direction and there is a delay before the car begins to move in the desired manner.

The position of the zero v_0/a depends on the location of the sensor. In our calculation we have assumed that the sensor is at the center of mass. The zero in the transfer function disappears if the sensor is located at the rear wheel. Thus if we look at the center of the rear wheels instead of the center of mass, the inverse response is not present and the resulting input/output behavior is simplified.

The formulas for the unit step response $y(t)$ and the transfer $P(s)$ give an interesting insight between the time and frequency domains. The behavior of the step response for small t, $y(t) \approx av_0 t/b$ is related to the high frequency property of the transfer function $P(s) \approx av_0/(bs)$ and the behavior of the step response for large

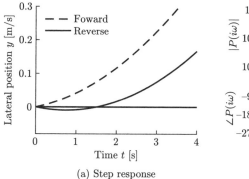
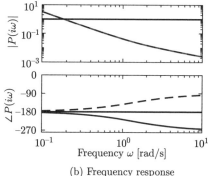

(a) Step response   (b) Frequency response

Figure 10.16: Vehicle steering for driving in reverse. (a) Step responses from steer-ing angle to lateral translation for a simple kinematics model when driving for-ward (dashed) and reverse (solid). With rear-wheel steering the center of mass first moves in the wrong direction and the overall response with rear-wheel steering is significantly delayed compared with that for front-wheel steering. (b) Frequency response for driving forward (dashed) and reverse (solid). Notice that the gain curves are identical, but the phase curve for driving in reverse has non-minimum phase.

t is related to the low frequency properties of the transfer function. This linkage can be made more formal through the use of the initial value theorem, discussed at the end of Section 9.3 ∇

10.5   GENERALIZED NOTIONS OF GAIN AND PHASE

A key idea in frequency domain analysis is to trace the behavior of sinusoidal signals through a system. The concepts of gain and phase represented by the transfer function are strongly intuitive because they describe amplitude and phase relations between input and output. In this section we will see how to extend the concepts of gain and phase to more general systems, including some nonlinear systems. We will also show that there are analogs of Nyquist's stability criterion if signals are approximately sinusoidal.

System Gain and Passivity

We begin by considering the case of a static linear system $y = Au$, where A is a matrix whose elements are complex numbers. The matrix does not have to be square. Let the inputs and outputs be vectors whose elements are complex numbers and use the Euclidean norm

$$\|u\| = \sqrt{\Sigma |u_i|^2}. \tag{10.11}$$

The norm of the output is

$$\|y\|^2 = u^* A^* A u,$$

where $*$ denotes the complex conjugate transpose. The matrix $A^* A$ is symmetric and positive semidefinite, and the right-hand side is a quadratic form. The square roots of the eigenvalues of the matrix $A^* A$ are all real, and we have

$$\|y\|^2 \leq \bar{\lambda}(A^*A)\|u\|^2,$$

where $\bar{\lambda}$ denotes the largest eigenvalue. The gain of the system can then be defined as the maximum ratio of the output to the input over all possible inputs:

$$\gamma = \max_u \frac{\|y\|}{\|u\|} = \sqrt{\bar{\lambda}(A^*A)}. \tag{10.12}$$

The square roots of the eigenvalues of the matrix A^*A are called the *singular values* of the matrix A, and the largest singular value is denoted by $\bar{\sigma}(A)$.

To generalize this to the case of an input/output dynamical system, we need to think of the inputs and outputs not as vectors of real numbers but as vectors of *signals*. For simplicity, consider first the case of scalar signals and let the signal space L_2 be square-integrable functions with the norm

$$\|u\|_2 = \sqrt{\int_0^\infty |u|^2(\tau)\, d\tau}.$$

This definition can be generalized to vector signals by replacing the absolute value with the vector norm (10.11). We can now formally define the *gain* of a system taking inputs $u \in L_2$ and producing outputs $y \in L_2$ as

$$\gamma = \sup_{u \in L_2} \frac{\|y\|_2}{\|u\|_2}, \tag{10.13}$$

where sup is the *supremum*, defined as the smallest number that is larger or equal to its argument. The reason for using the supremum is that the maximum may not be defined for $u \in L_2$. This definition of the system gain is quite general and can even be used for some classes of nonlinear systems, though one needs to be careful about how initial conditions and global nonlinearities are handled.

This generalized notion of gain can be used to define the concept of input/output stability for a system. Roughly speaking, a system is called bounded input/bounded output (BIBO) stable if a bounded input gives a bounded output for all initial states. A system is called input to state stable (ISS) if $\|x(t)\| \leq \beta(\|x(0)\|) + \gamma(\|u\|)$ where β and γ are monotonically increasing functions that vanish at the origin.

The norm (10.13) has some nice properties in the case of linear systems. In particular, given a single-input, single-output stable linear system with transfer function $G(s)$, it can be shown that the norm of the system is given by

$$\gamma = \sup_\omega |G(i\omega)| =: \|G\|_\infty. \tag{10.14}$$

In other words, the gain of the system corresponds to the peak value of the frequency response. This corresponds to our intuition that an input produces the largest output when we are at the resonant frequencies of the system. $\|G\|_\infty$ is called the *infinity norm* of the transfer function $G(s)$.

This notion of gain can be generalized to the multi-input, multi-output case as well. For a linear multivariable system with a transfer function matrix $G(s)$ we can define the gain as

$$\gamma = \|G\|_\infty = \sup_\omega \bar{\sigma}(G(i\omega)). \tag{10.15}$$

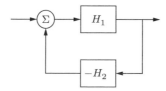

Figure 10.17: Block diagram of feedback connection of two general nonlinear systems H_1 and H_2.

Thus we can combine the idea of the gain of a matrix with the idea of the gain of a linear system by looking at the maximum singular value over all frequencies.

In addition to generalizing the system gain, it is also possible to make generalizations of the concept of phase. The angle between two vectors can be defined by the equation

$$\langle u, y \rangle = \|u\|\|y\| \cos(\varphi), \tag{10.16}$$

where the left argument denotes the scalar product. If systems are defined in such a way that we have norms of signals and a scalar product between signals we can use equation (10.16) to define the phase between two signals. For square-integrable inputs and outputs we have the scalar product

$$\langle u, y \rangle = \int_0^\infty u(\tau)y(\tau)\, d\tau,$$

and the *phase* φ between the signals u and y can now be defined through equation (10.16).

Systems where the phase between inputs and outputs is 90° or less for all inputs are called *passive systems*. Systems where the phase is strictly less than 90° are called *strictly passive*.

Extensions of the Nyquist Criterion

There are many extensions of the Nyquist's criterion, and we briefly sketch a few of them here. For linear systems it follows from Nyquist's theorem that the closed loop is stable if the gain of the loop transfer function is less than 1 for all frequencies. Since we have a notion of gain for nonlinear systems given by equation (10.13), we can extend this case of the Nyquist criterion to nonlinear systems:

Theorem 10.4 (Small gain theorem). *Consider the closed loop system shown in Figure 10.17, where H_1 and H_2 are input/output stable systems and the signal spaces and initial conditions are properly defined. Let the gains of the systems H_1 and H_2 be γ_1 and γ_2. Then the closed loop system is input/output stable if $\gamma_1\gamma_2 < 1$, and the gain of the closed loop system is*

$$\gamma = \frac{\gamma_1}{1 - \gamma_1\gamma_2}.$$

Another extension of the Nyquist criterion to nonlinear systems can be obtained by investigating the phase shift of the nonlinear systems. Consider again the system in Figure 10.17. It follows from the Nyquist criterion that if the blocks H_1 and H_2 are linear transfer functions, then the closed loop system is stable if the phase of

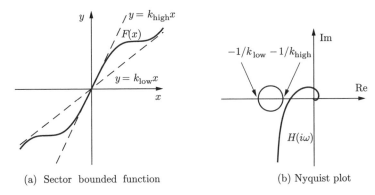

(a) Sector bounded function   (b) Nyquist plot

Figure 10.18: Stability using the circle criterion. For a feedback system with a sector-bounded nonlinearity (a), the Nyquist plot (b) must stay outside of a circle defined by $-1/k_{\mathrm{low}} \leq x \leq -1/k_{\mathrm{high}}$.

$H_1 H_2$ is always less than 180°. A generalization of this to nonlinear systems is that the closed loop system is stable if both H_1 and H_2 are passive and if one of them is strictly passive. This result is called the *passivity theorem*.

A final useful extension of the Nyquist criterion applies to the system in Figure 10.18 where H_1 is a linear system with transfer function $H(s)$ and the nonlinear block H_2 is a static nonlinearity described by a function $F(x)$ that is *sector-bounded*

$$k_{\mathrm{low}}\, x \leq F(x) \leq k_{\mathrm{high}}\, x. \tag{10.17}$$

The following theorem allows us to reason about the stability of such a system.

Theorem 10.5 (Circle criterion). *Consider a negative feedback system consisting of a linear system with transfer function $H(s)$ and a static nonlinearity defined by a function $F(x)$ satisfying the sector bound (10.17). The closed loop system is stable if the Nyquist curve of $H(i\omega)$ is outside a circle with diameter $-1/k_{low} \leq x \leq -1/k_{high}$ and the encirclement condition is satisfied.*

The extensions of Nyquist's criterion that we have discussed are powerful and easy to apply, and we will use them later to in Chapter 13. Details, proofs, and applications are found in [144].

Describing Functions

For special nonlinear systems like the one shown in Figure 10.19a, which consists of a feedback connection between a linear system and a static nonlinearity, it is possible to obtain a generalization of Nyquist's stability criterion based on the idea of *describing functions*. Following the approach of the Nyquist stability condition, we will investigate the conditions for maintaining an oscillation in the system. If the linear subsystem has low-pass character, its output is approximately sinusoidal even if its input is highly irregular. The condition for oscillation can then be found by exploring the propagation of a sinusoid that corresponds to the first harmonic.

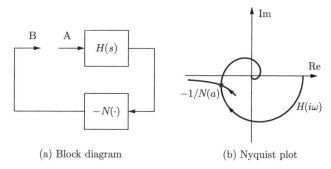

(a) Block diagram   (b) Nyquist plot

Figure 10.19: Describing function analysis. A feedback connection between a static nonlinearity and a linear system is shown in (a). The linear system is characterized by its transfer function $H(s)$, which depends on frequency, and the nonlinearity by its describing function $N(a)$, which depends on the amplitude a of its input. The Nyquist plot of $H(i\omega)$ and the plot of the $-1/N(a)$ are shown in (b). The intersection of the curves represents a possible limit cycle.

To carry out this analysis, we have to analyze how a sinusoidal signal propagates through a static nonlinear system. In particular we investigate how the first harmonic of the output of the nonlinearity is related to its (sinusoidal) input. Letting $F(x)$ represent the nonlinear function, we expand $F(e^{i\omega t})$ in terms of its harmonics:

$$F(ae^{i\omega t}) = \sum_{n=0}^{\infty} M_n(a)e^{i(n\omega t + \varphi_n(a))},$$

where $M_n(a)$ and $\varphi_n(a)$ represent the gain and phase of the nth harmonic, which depend on the input amplitude since the function $F(x)$ is nonlinear. We define the describing function to be the complex gain of the first harmonic:

$$N(a) = M_1(a)e^{i\varphi_1(a)}. \tag{10.18}$$

The function can also be computed by assuming that the input is a sinusoid and using the first term in the Fourier series of the resulting output.

Neglecting higher harmonics and arguing as we did when deriving Nyquist's stability criterion, we find that an oscillation can be maintained if

$$H(i\omega)N(a) = -1. \tag{10.19}$$

This equation means that if we inject a sinusoid of amplitude a at A in Figure 10.19a, the same signal will appear at B and an oscillation can be maintained by connecting the points. Equation (10.19) gives two conditions for finding the frequency ω of the oscillation and its amplitude a: the phase of $H(i\omega)N(a)$ must be 180° and its magnitude must be unity. A convenient way to solve the equation is to plot $H(i\omega)$ and $-1/N(a)$ on the same diagram as shown in Figure 10.19b. The diagram is similar to the Nyquist plot where the critical point -1 is replaced by the curve $-1/N(a)$ and a ranges from 0 to ∞. The intersection of the curves gives the amplitude a and frequency ω of the predicted oscillation.

It is possible to define describing functions for types of inputs other than sinusoids. Describing function analysis is a simple method, but it is approximate because

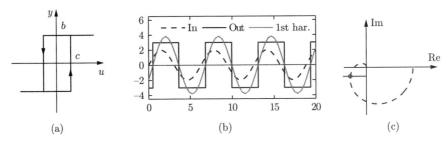

(a)   (b)   (c)

Figure 10.20: Describing function analysis for a relay with hysteresis. The input/output relation of the hysteresis is shown in (a) and the input with amplitude $a = 2$, the output, and its first harmonic are shown in (b). The Nyquist plots of the transfer function $H(s) = (s+1)^{-4}$ and the negative of the inverse describing function for the relay with $b = 3$ and $c = 1$ are shown in (c).

it assumes that higher harmonics can be neglected. Excellent treatments of describing function techniques can be found in the texts by Atherton [25] and Graham and McRuer [108]. The following example illustrates its use.

Example 10.12 Relay with hysteresis

Consider a linear system with a nonlinearity consisting of a relay with hysteresis. The output has amplitude b and the relay switches when the input is $\pm c$, as shown in Figure 10.20a. Assuming that the input is $u = a\sin(\omega t)$, we find that the output is zero if $a \le c$, and if $a > c$ the output is a square wave with amplitude b that switches at times $\omega t = \arcsin(c/a) + n\pi$. The first harmonic is then $y(t) = (4b/\pi)\sin(\omega t - \alpha)$, where $\sin\alpha = c/a$. For $a > c$ the describing function and its inverse are

$$N(a) = \frac{4b}{a\pi}\left(\sqrt{1 - \frac{c^2}{a^2}} - i\frac{c}{a}\right), \qquad \frac{1}{N(a)} = \frac{\pi\sqrt{a^2 - c^2}}{4b} + i\frac{\pi c}{4b},$$

where the inverse is obtained after simple calculations. Figure 10.20b shows the response of the relay to a sinusoidal input with the first harmonic of the output shown as a dashed line. Describing function analysis is illustrated in Figure 10.20c, which shows the Nyquist plot of the transfer function $H(s) = 2/(s+1)^4$ (dashed line) and the negative inverse describing function of a relay with $b = 1$ and $c = 0.5$ (solid line). The curves intersect for $a = 1$ and $\omega = 0.77$ rad/s, indicating the amplitude and frequency for a possible oscillation if the process and the relay are connected in a a feedback loop.   ▽

It follows from the example that the describing function for a relay without hysteresis is $N(a) = 4b/(a\pi)$ and $-1/N(a)$ is thus the negative real axis. For a saturation function, $-1/N(a)$ is the part of the negative real axis from $-\infty$ to -1.

10.6   FURTHER READING

Nyquist's original paper giving his now famous stability criterion was published in the *Bell Systems Technical Journal* in 1932 [192]. More accessible versions are found in the book [33], which also includes other interesting early papers on control.

Nyquist's paper is also reprinted in an IEEE collection of seminal papers on control [28]. Nyquist used $+1$ as the critical point, but Bode changed it to -1, which is now the standard notation. Interesting perspectives on early developments are given by Black [46], Bode [52], and Bennett [35]. Nyquist did a direct calculation based on his insight into the propagation of sinusoidal signals through systems; he did not use results from the theory of complex functions. The idea that a short proof can be given by using the principle of variation of the argument is presented in the delightful book by MacColl [171]. Bode made extensive use of complex function theory in his book [51], which laid the foundation for frequency response analysis where the notion of minimum phase was treated in detail. A good source for complex function theory is the classic by Ahlfors [6].

The extensions of Nyquist's criterion to a closed loop system that is composed of a linear system and a static nonlinearity has received significant attention. An extensive treatment of the passivity and small gain theorems and describing functions is given in the book by Khalil [144]. Describing functions for many nonlinearities are given in the books by Atherton [25] and Graham and McRuer [108]. Frequency response analysis was a key element in the emergence of control theory as described in the early texts by James et al. [130], Brown and Campbell [57], and Oldenburger [196], and it became one of the cornerstones of early control theory. Frequency response methods underwent a resurgence when robust control emerged in the 1980s, as will be discussed in Chapter 13.

EXERCISES

10.1 (Operational amplifier loop transfer function) Consider the operational amplifier circuit shown here, where Z_1 and Z_2 are generalized impedances and the open loop amplifier is modeled by the transfer function $G(s)$.

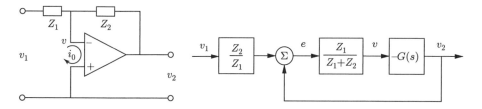

Show that the system can be modeled as the block diagram on the right, with loop transfer function $L = Z_1 G/(Z_1 + Z_2)$ and feedforward transfer function $F = Z_1/(Z_1 + Z_2)$.

10.2 (Atomic force microscope) The dynamics of the tapping mode of an atomic force microscope are dominated by the damping of the cantilever vibrations and the system that averages the vibrations. Modeling the cantilever as a spring–mass system with low damping, we find that the amplitude of the vibrations decays as $\exp(-\zeta\omega_0 t)$, where ζ is the damping ratio and ω_0 is the undamped natural frequency of the cantilever. The cantilever dynamics can thus be modeled by the transfer function

$$G(s) = \frac{a}{s+a},$$

where $a = \zeta\omega_0$. The averaging process can be modeled by the input/output relation

$$y(t) = \frac{1}{\tau} \int_{t-\tau}^{t} u(v)\, dv,$$

where the averaging time is a multiple n of the period of the oscillation $2\pi/\omega$. The dynamics of the piezo scanner can be neglected in the first approximation because they are typically much faster than a. A simple model for the complete system is thus given by the transfer function

$$P(s) = \frac{a(1 - e^{-s\tau})}{s\tau(s+a)}.$$

Plot the Nyquist curve of the system and determine the gain of a proportional controller that brings the system to the boundary of stability.

10.3 (Congestion control in overload conditions) A strongly simplified flow model of a TCP loop under overload conditions is given by the loop transfer function

$$L(s) = \frac{k}{s} e^{-s\tau},$$

where the queuing dynamics are modeled by an integrator, the TCP window control is a time delay τ, and the controller is simply a proportional controller. A major difficulty is that the time delay may change significantly during the operation of the system. Show that if we can measure the time delay, it is possible to choose a gain that gives a stability margin of $s_m \geq 0.6$ for all time delays τ.

10.4 (Heat conduction) A simple model for heat conduction in a solid is given by the transfer function

$$P(s) = k e^{-\sqrt{s}}.$$

Sketch the Nyquist plot of the system. Determine the frequency where the phase of the process is $-180°$ and compute the gain at that frequency. Show that the gain required to bring the system to the stability boundary is $k = e^{\pi}$.

10.5 (Stability margins for second-order systems) A process whose dynamics is described by a double integrator is controlled by an ideal PD controller with the transfer function $C(s) = k_d s + k_p$, where the gains are $k_d = 2\zeta\omega_0$ and $k_p = \omega_0^2$. Calculate and plot the gain, phase, and stability margins as a function ζ.

10.6 (Unity gain operational amplifier) Consider an op amp circuit with $Z_1 = Z_2$ that gives a closed loop system with nominally unit gain. Let the transfer function of the operational amplifier be

$$G(s) = \frac{k a_1 a_2}{(s+a)(s+a_1)(s+a_2)},$$

where $a_1, a_2 \gg a$. Show that the condition for oscillation is $k < a_1 + a_2$ and compute the gain margin of the system. Hint: Assume $a = 0$.

10.7 (Vehicle steering) Consider the linearized model for vehicle steering with a controller based on state feedback discussed in Example 8.4. The transfer functions for the process and controller are given by

$$P(s) = \frac{\gamma s + 1}{s^2}, \quad C(s) = \frac{s(k_1 l_1 + k_2 l_2) + k_1 l_2}{s^2 + s(\gamma k_1 + k_2 + l_1) + k_1 + l_2 + k_2 l_1 - \gamma k_2 l_2},$$

as computed in Example 9.10. Let the process parameter be $\gamma = 0.5$ and assume that the state feedback gains are $k_1 = 0.5$ and $k_2 = 0.75$ and that the observer gains are $l_1 = 1.4$ and $l_2 = 1$. Compute the stability margins numerically.

10.8 (Vectored thrust aircraft)   Consider the state space controller designed for the vectored thrust aircraft in Examples 7.9 and 8.7. The controller consists of two components: an optimal estimator to compute the state of the system from the output and a state feedback compensator that computes the input given the (estimated) state. Compute the loop transfer function for the system and determine the gain, phase, and stability margins for the closed loop dynamics.

10.9 (Kalman's inequality) Consider the linear system (7.20). Let $u = -Kx$ be a state feedback control law obtained by solving the linear quadratic regulator problem. Prove the inequality

$$\left(I + L(-i\omega) \right)^T Q_u \left(I + L(i\omega) \right) \geq Q_u,$$

where
$$K = Q_u^{-1} B^T S, \qquad L(s) = K(sI - A)^{-1} B.$$

(Hint: Use the Riccati equation (7.33), add and subtract the terms sS, multiply with $B^T(sI + A)^{-T}$ from the left and $(sI - A)^{-1}B$ from the right.)
   For single-input single-output systems this result implies that the Nyquist plot of the loop transfer function has the property $|1 + L(i\omega)| \geq 1$, from which it follows that the phase margin for a linear quadratic regulator is always greater than $60°$.

10.10 (Bode's formula) Consider Bode's formula (10.9) for the relation between gain and phase for a transfer function that has all its singularities in the left half-plane. Plot the weighting function and make an assessment of the frequencies where the approximation $\arg G \approx (\pi/2) d \log |G| / d \log \omega$ is valid.

10.11 (Padé approximation to a time delay) Consider the transfer functions

$$G(s) = e^{-s\tau}, \qquad G_1(s) = \frac{1 - s\tau/2}{1 + s\tau/2}. \tag{10.20}$$

Show that the minimum phase properties of the transfer functions are similar for frequencies $\omega < 1/\tau$. A long time delay τ is thus equivalent to a small right half-plane zero. The approximation $G_1(s)$ in equation (10.20) is called a first-order *Padé approximation*.

10.12 (Inverse response) Consider a system whose input/output response is modeled by $G(s) = 6(-s + 1)/(s^2 + 5s + 6)$, which has a zero in the right half-plane. Compute the step response for the system, and show that the output goes in the wrong direction initially, which is also referred to as an *inverse response*. Compare the response to a minimum phase system by replacing the zero at $s = 1$ with a zero at $s = -1$.

10.13 (Circle criterion) Consider the system in Figure 10.17, where H_1 is a linear system with the transfer function $H(s)$ and H_2 is a static nonlinearity $F(x)$ with the property $xF(x) \geq 0$. Use the circle criterion to prove that the closed loop system is stable if $H(s)$ is strictly passive.

10.14 (Describing function analysis) Consider the system with the block diagram shown on the left.

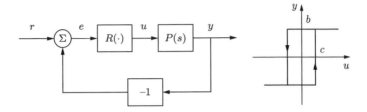

The block R is a relay with hysteresis whose input/output response is shown on the right and the process transfer function is $P(s) = e^{-s\tau}/s$. Use describing function analysis to determine frequency and amplitude of possible limit cycles. Simulate the system and compare with the results of the describing function analysis.

10.15 (Describing functions) Consider the saturation function

$$y = \mathrm{sat}(x) = \begin{cases} -1 & \text{if } x \leq 1, \\ x & \text{if } -1 < x \leq 1, \\ 1 & \text{if } x > 1. \end{cases}$$

Show that the describing function is

$$N(a) = \begin{cases} x & \text{if } |x| \leq 1, \\ \dfrac{2}{\pi}\left(\arcsin\dfrac{1}{x} + \dfrac{1}{x}\sqrt{1 - \dfrac{1}{x^2}}\right) & \text{if } |x| > 1. \end{cases}$$

Chapter Eleven

PID Control

Based on a survey of over eleven thousand controllers in the refining, chemicals and pulp and paper industries, 97% of regulatory controllers utilize a PID feedback control algorithm.

—L. Desborough and R. Miller, 2002 [71].

Proportional-integral-derivative (PID) control is by far the most common way of using feedback in engineering systems. In this chapter we present the basic properties of PID control and the methods for choosing the parameters of the controllers. We also analyze the effects of actuator saturation, an important feature of many feedback systems, and describe methods for compensating for it. Finally, we discuss the implementation of PID controllers as an example of how to implement feedback control systems using analog or digital computation.

11.1   BASIC CONTROL FUNCTIONS

The PID controller was introduced in Section 1.6, where Figure 1.15 illustrates that control action is composed of three terms: the proportional term (P), which depends on the present error; the integral term (I), which depends on past errors; and the derivative term (D), which depends on anticipated future errors. A major difference between a PID controller and an advanced controller based on feedback from estimated states (see Section 8.5) is that the observer-based controller predicts the future state of the system using a mathematical model, while the PID controller makes use of linear extrapolation of the measured output. A PI controller does not make use of any prediction of the future state of the system.

A survey of controllers for more than 100 boiler-turbine units in the Guangdong Province in China is a typical illustration of the prevalence of PID-based control: 94.4% of all controllers were PI, 3.7% PID, and 1.9% used advanced control [235]. The reasons why derivative action is used in only 4% of all controllers are that the benefits of prediction are significant primarily for processes that permit large controller gains. For many systems, prediction by linear extrapolation can generate large undesired control signals because measurement noise is amplified. In addition care must be taken to find a proper prediction horizon. Temperature control is a typical case where derivative action can be beneficial: sensors have low noise levels and controllers can have high gain.

PID control appears in simple dedicated systems and in large factories with thousands of controllers: as stand-alone controllers, as elements of hierarchical, distributed control systems, and as components of embedded systems. Advanced control systems are implemented as hierarchical systems, where high-level controllers

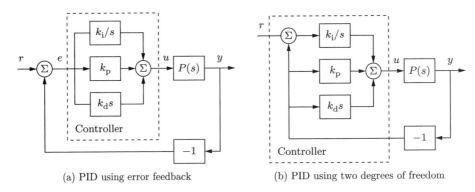

(a) PID using error feedback   (b) PID using two degrees of freedom

Figure 11.1: Block diagrams of closed loop systems with ideal PID controllers. Both controllers have one output, the control signal u. The controller in (a), which is based on error feedback, has one input, the control error $e = r - y$. For this controller proportional, integral, and derivative action acts on the error $e = r - y$. The two degree-of-freedom controller in (b) has two inputs, the reference r and the process output y. Integral action acts on the error, but proportional and derivative action act on only the process output y.

give setpoints to PID controllers in a lower layer. The PID controllers are directly connected to the sensors and actuators of the process. The importance of PID controllers thus has not decreased with the adoption of advanced control methods, because the performance of the system depends critically on the behavior of the PID controllers [71]. There is also growing evidence that PID control appears in biological systems [259].

Block diagrams of closed loop systems with PID controllers are shown in Figure 11.1. The command signal r is called the reference signal in regulation problems or the *setpoint* in the literature of PID control. The control signal u for the system in Figure 11.1a is formed entirely from the error e; there is no feedforward term (which would correspond to $k_f r$ in the state feedback case). A common alternative in which proportional and derivative action do not act on the reference is shown in Figure 11.1b; combinations of the schemes will be discussed in Section 11.5.

The input/output relation for an ideal PID controller with error feedback is

$$u = k_p e + k_i \int_0^t e(\tau)\, d\tau + k_d \frac{de}{dt} = k_p \left(e + \frac{1}{T_i} \int_0^t e(\tau)\, d\tau + T_d \frac{de}{dt} \right). \qquad (11.1)$$

The control action is thus the sum of three terms: proportional feedback, the integral term, and derivative action. For this reason PID controllers were originally called *three-term controllers*.

The controller parameters are the proportional gain k_p, the integral gain k_i, and the derivative gain k_d. The gain k_p is sometimes expressed in terms of the *proportional band*, defined as $PB = 100/k_p$. A proportional band of 10% thus implies that the controller operates linearly for only 10% of the span of the measured signal. The controller can also be parameterized with the time constants $T_i = k_p/k_i$ and $T_d = k_d/k_p$, called the integral time (constant) and the derivative time (constant). The parameters T_i and T_d have dimensions of time and can naturally be related to the time constants of the controller.

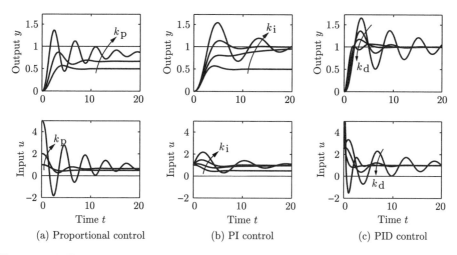

(a) Proportional control   (b) PI control   (c) PID control

Figure 11.2: Responses to step changes in the reference value for a system with a proportional controller (a), PI controller (b), and PID controller (c). The process has the transfer function $P(s) = 1/(s+1)^3$, the proportional controller has parameters $k_p = 1$, 2, and 5, the PI controller has parameters $k_p = 1$, $k_i = 0$, 0.2, 0.5, and 1, and the PID controller has parameters $k_p = 2.5$, $k_i = 1.5$, and $k_d = 0$, 1, 2, and 4.

The controller (11.1) is an idealized representation. It is a useful abstraction for understanding the PID controller, but several modifications must be made to obtain a controller that is practically useful. Before discussing these practical issues we will develop some intuition about PID control.

We start by considering pure proportional feedback. Figure 11.2a shows the responses of the process output to a unit step in the reference value for a system with pure proportional control at different gain settings. In the absence of a feedforward term, the output never reaches the reference, and hence we are left with nonzero steady-state error. Letting the process transfer function be $P(s)$, with proportional feedback we have $C(s) = k_p$ and the transfer function from reference to error is

$$G_{er}(s) = \frac{1}{1 + C(s)P(s)} = \frac{1}{1 + k_p P(s)}. \qquad (11.2)$$

Assuming that the closed loop is stable, the steady-state error for a unit step is

$$G_{er}(0) = \frac{1}{1 + C(0)P(0)} = \frac{1}{1 + k_p P(0)}.$$

For the system in Figure 11.2a with gains $k_p = 1$, 2, and 5, the steady-state error is 0.5, 0.33, and 0.17. The error decreases with increasing gain, but the system also becomes more oscillatory. The system becomes unstable for $k_p = 8$. Notice in the figure that the initial value of the control signal equals the controller gain.

To avoid having a steady-state error, the proportional term can be changed to

$$u(t) = k_p e(t) + u_{ff}, \qquad (11.3)$$

where u_{ff} is a feedforward term that is adjusted to give the desired steady-state value. If the reference value r is constant and we choose $u_{ff} = r/P(0) = k_f r$, then

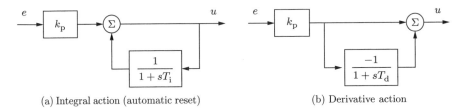

(a) Integral action (automatic reset)   (b) Derivative action

Figure 11.3: Implementation of integral and derivative action. The block diagram in (a) shows how integral action is implemented using *positive feedback* with a first-order system, sometimes called automatic reset. The block diagram in (b) shows how derivative action can be implemented by taking differences between a static system and a first-order system.

the steady-state output will be exactly equal to the reference value, as it was in the state space case, provided that there are no disturbances. However, this requires exact knowledge of the zero frequency gain $P(0)$, which is usually not available. The parameter u_{ff}, called the *reset* value, was adjusted manually in early controllers. Another alternative to avoid a steady-state error is to multiply the reference by $1 + k_{\mathrm{p}}P(0)$, but this also requires precise knowledge of $P(0)$.

As we saw in Section 7.4, integral action guarantees that the process output agrees with the reference in steady state and provides an alternative to the feedforward term. Since this result is so important, we will provide a general proof. Consider the controller given by equation (11.1) with $k_{\mathrm{i}} \neq 0$. Assume that $u(t)$ and $e(t)$ converge to steady-state values $u = u_0$ and $e = e_0$. It then follows from equation (11.1) that

$$u_0 = k_{\mathrm{p}}e_0 + k_{\mathrm{i}} \lim_{t \to \infty} \int_0^t e(t)dt.$$

The limit of the right hand side is not finite unless $e(t)$ goes to zero, which implies that $e_0 = 0$. We can thus conclude that integral control has the property that if a steady state exists, the error will always be zero. This property is sometimes called the *magic of integral action*. Notice that we have not assumed that the process is linear or time-invariant. We have, however, assumed that there is an equilibrium point. It is much better to achieve zero steady-state error by integral action than by feedforward, which requires a precise knowledge of process parameters.

The effect of integral action can also be understood from frequency domain analysis. The transfer function of the PID controller is

$$C(s) = k_{\mathrm{p}} + \frac{k_{\mathrm{i}}}{s} + k_{\mathrm{d}}s. \tag{11.4}$$

The controller has infinite gain at zero frequency ($C(0) = \infty$), and it then follows from equation (11.2) that $G_{er}(0) = 0$, which implies that there is no steady-state error for a step input.

Integral action can also be viewed as a method for generating the feedforward term u_{ff} in the proportional controller (11.3) automatically. This is shown in Figure 11.3a, where the controller output is low-pass filtered and fed back with positive gain. This implementation, called *automatic reset*, was one of the early inventions of integral control (it was much easier to implement a low-pass filter than to implement an integrator). The transfer function of the system in Figure 11.3a is obtained by block diagram algebra: we have

$$G_{ue} = k_\mathrm{p} \frac{1 + sT_\mathrm{i}}{sT_\mathrm{i}} = k_\mathrm{p} + \frac{k_\mathrm{p}}{sT_\mathrm{i}},$$

which is the transfer function for a PI controller.

The properties of integral action are illustrated in Figure 11.2b for a step input. The proportional gain is constant, $k_\mathrm{p} = 1$, and the integral gains are $k_\mathrm{i} = 0$, 0.2, 0.5, and 1. The case $k_\mathrm{i} = 0$ corresponds to pure proportional control, with a steady-state error of 50%. The steady-state error is eliminated when integral gain action is used. The response creeps slowly toward the reference for small values of k_i and converges more quickly for larger integral gains, but the system also becomes more oscillatory.

The integral gain k_i is a useful measure for attenuation of load disturbances. Consider a closed loop system under PID control, like the one in Figure 11.1. Assume that the system is stable and initially at rest with all signals being zero. Apply a unit step load disturbance at the process input. After a transient, the process output goes to zero and the controller output settles at a value that compensates for the disturbance. Since $e(t)$ goes to zero as $t \to \infty$, it follows from equation (11.1) that

$$u(\infty) = k_\mathrm{i} \int_0^\infty e(t)dt.$$

The *integrated error*, IE, for a unit step load disturbance $\mathrm{IE} = \int_0^\infty e(t)dt$ is thus inversely proportional to the integral gain k_i and hence serves as a measure of the effectiveness of disturbance attenuation. A large gain k_i attenuates disturbances effectively, but too large a gain gives oscillatory behavior, poor robustness, and possibly instability.

We now return to the general PID controller and consider the effect of derivative action. Recall that the original motivation for derivative feedback was to provide predictive or anticipatory action. Notice that the combination of the proportional and the derivative terms can be written as

$$u = k_\mathrm{p}e + k_\mathrm{d}\frac{de}{dt} = k_\mathrm{p}\left(e + T_\mathrm{d}\frac{de}{dt}\right) =: k_\mathrm{p}e_\mathrm{p},$$

where $e_\mathrm{p}(t)$ can be interpreted as a prediction of the error at time $t + T_\mathrm{d}$ by linear extrapolation. The prediction time $T_\mathrm{d} = k_\mathrm{d}/k_\mathrm{p}$ is the *derivative time constant*.

Derivative action can be implemented by taking the difference between the signal and its low-pass filtered version as shown in Figure 11.3b. The transfer function for the system is

$$G_{ue}(s) = k_\mathrm{p}\left(1 - \frac{1}{1 + sT_\mathrm{d}}\right) = k_\mathrm{p}\frac{sT_\mathrm{d}}{1 + sT_\mathrm{d}} = \frac{k_\mathrm{d}s}{1 + sT_\mathrm{d}}. \tag{11.5}$$

The transfer function $G_{ue}(s)$ approximates a derivative for low frequencies because for $|s| \ll 1/T_\mathrm{d}$ we have $G(s) \approx k_\mathrm{p}T_\mathrm{d}s = k_\mathrm{d}s$. The transfer function G_{ue} acts like a differentiator for signals with low frequencies and as a constant gain k_p for high-frequency signals, so we can regard this as a filtered derivative.

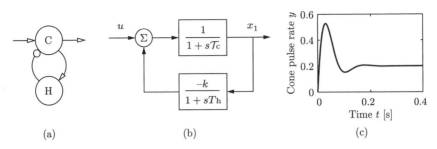

(a)   (b)   (c)

Figure 11.4: Schematic diagram of cone photoreceptors (C) and horizontal cells (H) in the retina. In the schematic diagram in (a), excitatory feedback is indicated by arrows and inhibitory feedback by circles. A block diagram is shown in (b) and the step response in (c).

Figure 11.2c illustrates the effect of derivative action: the system is oscillatory when no derivative action is used, and it becomes more damped as the derivative gain is increased. When the input is a step, the controller output generated by the derivative term will be an impulse. This is clearly visible in Figure 11.2c. The impulse can be avoided by using the controller configuration shown in Figure 11.1b.

Although PID control was developed in the context of engineering applications, it also appears in nature. Disturbance attenuation by feedback in biological systems is often called *adaptation*. A typical example is the pupillary reflex discussed in Example 9.18, where it is said that the eye adapts to changing light intensity. Analogously, feedback with integral action is called perfect adaptation [259]. In biological systems proportional, integral, and derivative action are generated by combining subsystems with dynamical behavior, similar to what is done in engineering systems. For example, PI action can be generated by the interaction of several hormones [81].

Example 11.1 PD action in the retina
The response of cone photoreceptors in the retina is an example where proportional and derivative action is generated by a combination of cones and horizontal cells. The cones are the primary receptors stimulated by light, which in turn stimulate the horizontal cells, and the horizontal cells give inhibitory (negative) feedback to the cones. A schematic diagram of the system is shown in Figure 11.4a. The system can be modeled by ordinary differential equations by representing neuron signals as continuous variables representing the average pulse rate. In [256] it is shown that the system can be represented by the differential equations

$$\frac{dx_1}{dt} = \frac{1}{T_c}(-x_1 - kx_2 + u), \qquad \frac{dx_2}{dt} = \frac{1}{T_h}(x_1 - x_2),$$

where u is the light intensity and x_1 and x_2 are the average pulse rates from the cones and the horizontal cells. A block diagram of the system is shown in Figure 11.4b. The step response of the system given in Figure 11.4c shows that the system has a large initial response followed by a lower, constant steady-state response typical of proportional and derivative action. The parameters used in the simulation are $k = 4$, $T_c = 0.025$, and $T_h = 0.08$. ∇

11.2   SIMPLE CONTROLLERS FOR COMPLEX SYSTEMS

Many of the design methods discussed in previous chapters have the property that the complexity of the controller is a direct reflection of the complexity of the model. When designing controllers by output feedback in Chapter 8, we found for single-input, single-output systems that the order of the controller was the same as the order of the model, possibly one order higher if integral action was required. Applying these design methods to PID control requires that the models must be of first or second order.

Low-order models can be obtained from first principles. Any stable system can be modeled by a static system if its inputs are sufficiently slow. Similarly a first-order model is sufficient if the storage of mass, momentum, or energy can be captured by only one variable; typical examples are the velocity of a car on a road, the angular velocity of a stiff rotational system, the level in a tank, and the concentration in a volume with good mixing. System dynamics are of second order if the storage of mass, energy, and momentum can be captured by two state variables; typical examples are the position and velocity of a car on the road, the orientation and angular velocity of satellites, the levels in two connected tanks, and the concentrations in two-compartment models. A wide range of techniques for model reduction are also available. In this section we will focus on design techniques where we simplify the models to capture the essential properties that are needed for PID design.

We begin by analyzing the case of integral control. Any stable system can be controlled by an integral controller provided that the requirements on the closed loop system are modest. To design a controller we approximate the transfer function of the process by a constant $K = P(0)$, which will be reasonable for any stable system at sufficiently low frequencies. The loop transfer function under integral control then becomes Kk_i/s, and the closed loop characteristic polynomial is simply $s + Kk_i$. Specifying performance by the desired time constant T_{cl} of the closed loop system, we find that the integral gain can be chosen as $k_i = 1/(T_{cl}P(0))$.

This simplified analysis requires that T_{cl} be sufficiently large that the process transfer function can indeed be approximated by a constant. A reasonable criterion is that $T_{cl} > T_{ar}$, where $T_{ar} = -P'(0)/P(0)$ is known as the *average residence time* of the open loop system.

To obtain controllers with higher performance we approximate the process dynamics by a first-order system (rather than a constant):

$$P(s) \approx \frac{P(0)}{1 + sT_{ar}}.$$

A reasonable design criterion is to obtain a step response with small overshoot and reasonable response time. An integral controller with gain

$$k_i = \frac{1}{2P(0)T_{ar}}, \tag{11.6}$$

gives the loop transfer function

$$L(s) = P(s)C(s) \approx \frac{P(0)}{1 + sT_{ar}} \frac{k_i}{s} = \frac{1}{2sT_{ar}(1 + sT_{ar})},$$

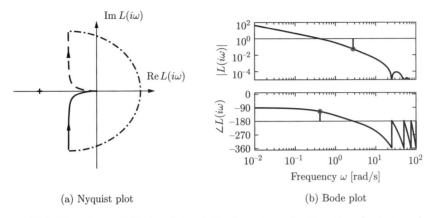

(a) Nyquist plot   (b) Bode plot

Figure 11.5: Nyquist and Bode plots of the loop transfer function for integral control of an AFM in tapping mode. The integrating controller gives good robustness properties based on a simple analysis. At high frequencies the Nyquist plot has an infinite number of small loops with decreasing amplitude in the left half-plane. These loops are not visible in the Nyquist plot but they show up clearly in the Bode plot.

and the closed loop poles become $s = (-0.5 \pm 0.5i)/T_{\mathrm{ar}}$. Using the approximations in Table 7.1 on page 231, we see that this controller has $\omega_0 = 1/(T_{ar}\sqrt{2})$, which gives a rise time of $3.1\,T_{\mathrm{ar}}$, a settling time of $7.9\,T_{\mathrm{ar}}$, and overshoot of 4%.

Example 11.2 Integral control of AFM in tapping mode
A simplified model of the dynamics of the vertical motion of an atomic force microscope in tapping mode was discussed in Exercise 10.2. The transfer function for the system dynamics is
$$P(s) = \frac{a(1 - e^{-s\tau})}{s\tau(s + a)},$$
where $a = \zeta\omega_0$, $\tau = 2\pi n/\omega_0$, and the gain has been normalized to 1. This transfer function is unusual since there is a time-delay term in the numerator.

To design a controller, we focus on the low-frequency dynamics of the system. We have $P(0) = 1$ and $P'(0) = -\tau/2 - 1/a = -(2 + a\tau)/(2a)$. For low frequencies the loop transfer function can then be approximated by
$$L(s) \approx \frac{k_{\mathrm{i}}(P(0) + sP'(0))}{s} = k_{\mathrm{i}}P'(0) + \frac{k_{\mathrm{i}}P(0)}{s}.$$

Using the design rule (11.6) we set $k_{\mathrm{i}} = -1/(2P'(0))$, Nyquist and Bode plots for the resulting loop transfer function are shown in Figure 11.5. We see that the controller provides good performance at low frequency and has good stability margins. Note that even though the system dynamics include a time-delay term, we were able to obtain good performance using a simple integral controller and a simple set of calculations. ∇

Another approach to designing simple controllers is to use the gains of the controller to set the location of the closed loop poles. PI controllers give two gains with which to tune the closed loop dynamics, and for simple models the closed loop poles can be set using these gains.

Consider a first-order system with the transfer function

$$P(s) = \frac{b}{s+a}.$$

With a PI controller the closed loop system has the characteristic polynomial

$$s(s+a) + bk_{\mathrm{p}}s + bk_{\mathrm{i}} = s^2 + (a + bk_{\mathrm{p}})s + bk_{\mathrm{i}}.$$

The closed loop poles can thus be assigned arbitrary values by proper choice of the controller gains k_{p} and k_{i}. Requiring that the closed loop system have the characteristic polynomial

$$p(s) = s^2 + a_1 s + a_2,$$

we find that the controller parameters are

$$k_{\mathrm{p}} = \frac{a_1 - a}{b}, \qquad k_{\mathrm{i}} = \frac{a_2}{b}. \tag{11.7}$$

If we require a response of the closed loop system that is slower than that of the open loop system, a reasonable choice is $a_1 = a + \alpha$ and $a_2 = \alpha a$, where $\alpha < a$ determines the closed loop response. If a response faster than that of the open loop system is required, a possible choice is $a_1 = 2\zeta_{\mathrm{c}}\omega_{\mathrm{c}}$ and $a_2 = \omega_{\mathrm{c}}^2$, where ω_{c} and ζ_{c} are the undamped natural frequency and damping ratio of the dominant mode.

The choice of ω_{c} has a significant impact on the robustness of the system and will be discussed in Section 14.5. An upper limit to ω_{c} is given by highest frequency where the model is valid. Large values of ω_{c} will require fast control actions, and actuators may saturate if the value is too large. A first-order model is unlikely to represent the true dynamics for high frequencies.

Example 11.3 Cruise control using PI feedback
Consider the problem of maintaining the speed of a car as it goes up a hill. In Example 6.11 we found that there was little difference between the linear and nonlinear models when investigating PI control, provided that the throttle did not reach the saturation limits. A simple linear model of a car was given in Example 6.11:

$$\frac{d(v - v_{\mathrm{e}})}{dt} = -a(v - v_{\mathrm{e}}) - b_{\mathrm{g}}(\theta - \theta_{\mathrm{e}}) + b(u - u_{\mathrm{e}}), \tag{11.8}$$

where v is the velocity of the car, u is the input to the engine (throttle), and θ is the slope of the hill. The parameters were $a = 0.01$, $b = 1.32$, $b_{\mathrm{g}} = 9.8$, $v_{\mathrm{e}} = 20$, $\theta_{\mathrm{e}} = 0$, and $u_{\mathrm{e}} = 0.1687$. This model will be used to find suitable parameters of a vehicle speed controller. The transfer function from throttle to velocity is a first-order system. Since the open loop dynamics are quite slow ($1/a \approx 100\,\mathrm{s}$), it is natural to specify a faster closed loop system by requiring that the closed loop system be of second order with damping ratio ζ_{c} and undamped natural frequency ω_{c}. The controller gains are given by equation (11.7).

Figure 11.6 shows the velocity and the throttle for a car that initially moves on a horizontal road and encounters a hill with a slope of $4°$ at time $t = 5\,\mathrm{s}$. To design a PI controller we choose $\zeta_{\mathrm{c}} = 1$ to obtain a response without overshoot, as shown in Figure 11.6a. The choice of ω_{c} is a compromise between response speed and control actions: a large value gives a fast response, but it requires fast control action. The trade-off is illustrated in Figure 11.6b. The largest velocity error

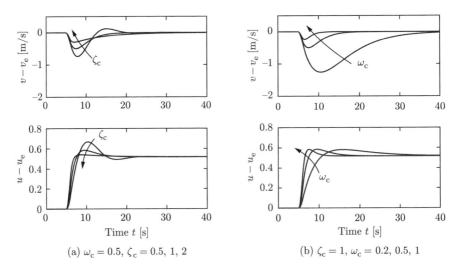

(a) $\omega_c = 0.5$, $\zeta_c = 0.5$, 1, 2   (b) $\zeta_c = 1$, $\omega_c = 0.2$, 0.5, 1

Figure 11.6: Cruise control using PI feedback. The step responses for the error and input illustrate the effect of parameters ζ_c and ω_c on the response of a car with cruise control. The slope of the road changes linearly from $0°$ to $4°$ between $t = 5$ and 6 s. (a) Responses for $\omega_c = 0.5$ and $\zeta_c = 0.5$, 1, and 2. Choosing $\zeta_c \geq 1$ gives no overshoot in the velocity v. (b) Responses for $\zeta_c = 1$ and $\omega_c = 0.2$, 0.5, and 1.0.

decreases with increasing ω_c, but the control signal also changes more rapidly. In the simple model (11.8) it was assumed that the force responds instantaneously to throttle commands. For rapid changes there may be additional dynamics that have to be accounted for. There are also physical limits to the rate of change of the force, which also restricts the admissible value of ω_c. A reasonable choice of ω_c is in the range 0.5–1.0. Notice in Figure 11.6 that even with $\omega_c = 0.2$ the largest velocity error is only about 1.3 m/s.   ∇

A PI controller can also be used for a process with second-order dynamics, but there will be restrictions on the possible locations of the closed loop poles. Using a PID controller, it is possible to control a system of second order in such a way that the closed loop poles have arbitrary locations (Exercise 11.2).

Instead of finding a low-order model and designing controllers for them, we can also use a high-order model and attempt to place only a few dominant poles. An integral controller has one parameter, and it is possible to position one pole. To see this, consider a process with the transfer function $P(s)$. The loop transfer function with an integral controller is $L(s) = k_i P(s)/s$. The roots of the closed loop characteristic polynomial are the roots of $s + k_i P(s) = 0$. Requiring that $s = -a$ be a root, the controller gain should be chosen as $k_i = a/P(-a)$. The pole $s = -a$ will be a dominant closed loop pole if a is smaller than the magnitude of the other closed loop process poles. A similar approach can be applied to PI and PID controllers (Exercise 11.3).

11.3   PID TUNING

Users of control systems are frequently faced with the task of adjusting the controller parameters to obtain a desired behavior. There are many different ways to do this. One approach is to go through the conventional steps of modeling and control

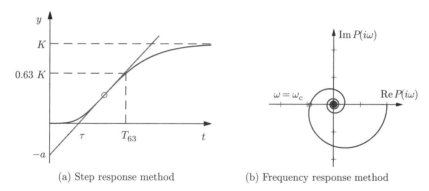

(a) Step response method   (b) Frequency response method

Figure 11.7: Ziegler–Nichols step and frequency response experiments. The open loop unit step response in (a) is characterized by the parameters a and τ. The frequency response method (b) characterizes process dynamics by the point where the Nyquist curve of the process transfer function first intersects the negative real axis and the frequency ω_c where this occurs.

design as described in the previous section. A typical process may have thousands of PID controllers. Since the PID controller has so few parameters a number of special empirical methods have been developed for direct adjustment of the controller parameters.

Ziegler–Nichols' Tuning

The first tuning rules were developed by Ziegler and Nichols [266] in the 1940s. Their idea was to perform a simple experiment on the process and extract features of process dynamics in the time and frequency domains.

The time domain method is based on a measurement of part of the open loop unit step response of the process, as shown in Figure 11.7a. The step response is measured by a *bump test*. The process is first brought to steady state, the input is then changed by a suitable amount, and finally the output is measured and scaled to correspond to a unit step input. Ziegler and Nichols characterized the step response by only two parameters a and τ, which are the intercepts of the steepest tangent of the step response with the coordinate axes. The parameter τ is an approximation of the time delay of the system and a/τ is the steepest slope of the step response. Notice that it is not necessary to wait until steady state is reached to be able to determine the parameters; it suffices to wait until the response has had an inflection point. The suggested controller parameters are given in Table 11.1. They were obtained by extensive simulation of a range of representative processes. A controller was tuned manually for each process, and an attempt was then made to correlate the controller parameters with a and τ.

In the frequency domain method, a controller is connected to the process, the integral and derivative gains are set to zero, and the proportional gain is increased until the system starts to oscillate. The critical value k_c of the proportional gain is observed together with the period of oscillation T_c. It follows from Nyquist's stability criterion that the Nyquist contour for the loop transfer function $L = k_c P(s)$ passes through the critical point at the frequency $\omega_c = 2\pi/T_c$. The experiment thus gives the point on the Nyquist curve of the process transfer function $P(s)$ where the phase lag is 180°, as shown in Figure 11.7b. The suggested controller parameters are then given by Table 11.1b.

Table 11.1: Original Ziegler–Nichols tuning rules. (a) The step response method gives the parameters in terms of the intercept a and the apparent time delay τ. (b) The frequency response method gives controller parameters in terms of *critical gain* k_c and *critical period* T_c.

Type	k_p	T_i	T_d
P	$1/a$		
PI	$0.9/a$	$\tau/0.3$	
PID	$1.2/a$	$\tau/0.5$	0.5τ

(a) Step response method

Type	k_p	T_i	T_d
P	$0.5k_c$		
PI	$0.45k_c$	$T_c/1.2$	
PID	$0.6k_c$	$T_c/2$	$T_c/8$

(b) Frequency response method

The Ziegler–Nichols methods had a huge impact when they were introduced in the 1940s. The rules were simple to use and gave initial conditions for manual tuning. The ideas were adopted by manufacturers of controllers for routine use. The Ziegler–Nichols tuning rules unfortunately have two severe drawbacks: too little process information is used, and the closed loop systems that are obtained lack robustness.

Tuning Based on the FOTD Model

The Ziegler–Nichols methods use only two parameters to characterize process dynamics, a and τ for the step response method and k_c and T_c for the frequency domain method. Tuning of PID controllers can be improved if we characterize the process by more parameters. The first-order and time-delay (FOTD) model

$$P(s) = \frac{K}{1+sT}e^{-\tau s}, \qquad \tau_n = \frac{\tau}{T+\tau}, \tag{11.9}$$

is commonly used to approximate the step response of systems with essentially monotone step responses. The parameter τ_n, which has values between 0 and 1, is called the *relative time delay* or the *normalized time delay*. The dynamics are characterized as being *lag dominated* if τ_n is close to zero, *delay dominated* if τ_n is close to one, and *balanced* for intermediate values.

The parameters of the FOTD model can be determined from a bump test as indicated in Figure 11.7a. The zero frequency gain K is the steady-state value of the unit step response. The time delay τ is the intercept of the steepest tangent with the time axis, as in the Ziegler–Nichols method. The time T_{63} is the time where the output has reached 63% of its steady-state value and T is then given by $T = T_{63} - \tau$. Notice that it takes a longer time to find an FOTD model than the Ziegler–Nichols model (a and τ) because to determine K it is necessary to wait until the steady state has been reached.

There are many versions of improved tuning rules for the model (11.9). As an illustration we give the following rules for PI control, based on [19]:

$$k_p = \frac{0.15\tau + 0.35T}{K\tau} \left(\frac{0.9T}{K\tau} \right), \qquad k_i = \frac{0.46\tau + 0.02T}{K\tau^2} \left(\frac{0.27T}{K\tau^2} \right), \tag{11.10a}$$

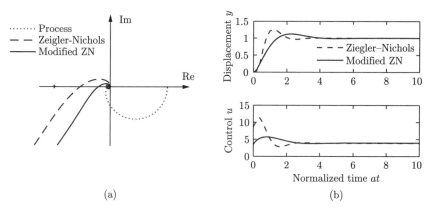

Figure 11.8: PI control of an AFM in tapping mode. Nyquist plots (a) and step responses (b) for PI control of the vertical motion of an atomic force microscope in tapping mode. Results with Ziegler–Nichols tuning are shown by dashed lines, and modified Ziegler–Nichols tuning is shown by solid lines. The Nyquist plot of the process transfer function is shown by dotted lines.

$$k_\mathrm{p} = 0.16k_\mathrm{c} \quad \left(0.45k_\mathrm{c}\right), \qquad k_\mathrm{i} = \frac{0.16k_\mathrm{c} + 0.72/K}{T_\mathrm{c}} \quad \left(\frac{0.54k_\mathrm{c}}{T_\mathrm{c}}\right). \quad (11.10\mathrm{b})$$

The values for the Ziegler–Nichols rule from Table 11.1 are given in parentheses. Notice that the improved formulas typically give lower controller gains than the original Ziegler–Nichols method.

Example 11.4 Atomic force microscope in tapping mode

A simplified model of the dynamics of the vertical motion of an atomic force microscope in tapping mode was discussed in Example 11.2. The transfer function is normalized by choosing $1/a$ as the time unit, yielding

$$P(s) = \frac{1 - e^{-sT_n}}{sT_n(s+1)},$$

where $T_n = 2n\pi a/\omega_0 = 2n\pi\zeta$. The Nyquist plot of $P(s)$ is shown as a dotted line in Figure 11.8a for $\zeta = 0.002$ and $n = 20$. The first intersection with the real axis occurs at $\mathrm{Re}\, s = -0.0461$ for $\omega_\mathrm{c} = 13.1$. The critical gain is thus $k_\mathrm{c} = 21.7$ and the critical period is $T_\mathrm{c} = 0.48$. Using the Ziegler–Nichols tuning rule, we find the parameters $k_\mathrm{p} = 8.67$ and $k_\mathrm{i} = 22.6$ ($T_\mathrm{i} = 0.384$) for a PI controller. With this controller the stability margin is $s_\mathrm{m} = 0.31$, which is quite small. The step response of the controller is shown using dashed lines in Figure 11.8. Notice in particular that there is a large overshoot in the control signal.

The modified Ziegler–Nichols rule (11.10b) gives the controller parameters $k_\mathrm{p} = 3.47$ and $k_\mathrm{i} = 8.73$ ($T_\mathrm{i} = 0.397$) and the stability margin becomes $s_\mathrm{m} = 0.61$. The step response with this controller is shown using solid lines in Figure 11.8. A comparison of the responses obtained with the original Ziegler–Nichols rule shows that the overshoot has been reduced. Notice that the control signal reaches its steady-state value almost instantaneously. It follows from Example 11.2 that a pure integral controller has the normalized gain $k_\mathrm{i} = 1/(2 + T_n) = 0.44$, which is more than an order of magnitude smaller than the integral gain of the PI controller. ∇

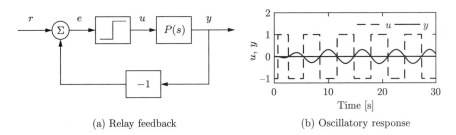

(a) Relay feedback   (b) Oscillatory response

Figure 11.9: Block diagram of a process with relay feedback (a) and typical signals
(b). The process output y is a solid line, and the relay output u is a dashed line.
Notice that the signals u and y have opposite phases.

The tuning rules based on the FOTD model work well for PI controllers. Derivative action has little effect on processes with delay-dominated dynamics, but can
give substantial performance for processes with lag-dominated dynamics. Tuning
of PID controllers for processes with lag-dominated dynamics cannot, however, be
based on the FOTD model; see [19].

Relay Feedback

The Ziegler–Nichols frequency response method increases the gain of a proportional
controller until oscillation to determine the critical gain k_c and the corresponding
critical period T_c or, equivalently, the point where the Nyquist curve intersects the
negative real axis. One way to obtain this information automatically is to connect
the process in a feedback loop with a nonlinear element having a relay function
as shown in Figure 11.9a. For many systems there will then be an oscillation, as
shown in Figure 11.9b, where the relay output u is a square wave and the process
output y is close to a sinusoid. Moreover, the fundamental sinusoidal components
of the input and the output are 180° out of phase, which means that the system
oscillates with the critical period T_c. Notice that an oscillation with constant period
is established quickly.

To determine the critical gain k_c we expand the square wave relay output in a
Fourier series. Notice in the figure that the process output is practically sinusoidal
because the process attenuates higher harmonics. It is then sufficient to consider
only the first harmonic component of the input. Letting d be the relay amplitude,
this component has amplitude $4d/\pi$. If a is the amplitude of the process output,
the process gain at the critical frequency $\omega_c = 2\pi/T_c$ is $|P(i\omega_c)| = \pi a/(4d)$ and the
critical gain is $k_c = 4d/(\pi a)$. Having obtained the critical gain k_c and the critical
period T_c, the controller parameters can then be determined using the Ziegler–
Nichols rules. Improved tuning can be obtained by fitting a model to the data
obtained from the relay experiment.

The relay experiment can be automated. Since the amplitude of the oscillation
is proportional to the relay output, it is easy to control it by adjusting the relay
output. *Automatic tuning* based on relay feedback is used in many commercial PID
controllers. Tuning is accomplished simply by pushing a button that activates relay
feedback. The relay amplitude is automatically adjusted to keep the oscillations
sufficiently small, and the relay feedback is replaced by a PID controller when the
tuning is finished. The main advantage of relay tuning is that a short experiment

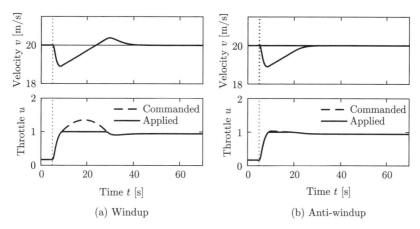

Figure 11.10: Simulation of PI cruise control with windup (a) and anti-windup (b). The figure shows the speed v and the throttle u for a car that encounters a slope that is so steep that the throttle saturates. The controller output is a dashed line. The controller parameters are $k_p = 0.5$, $k_i = 0.1$ and $k_{aw} = 2.0$. The anti-windup compensator eliminates the overshoot by preventing the error from building up in the integral term of the controller.

for identification of process dynamics is generated automatically. The original relay autotuner can be improved significantly by using an asymmetric relay, which admits determination of more parameters [41].

11.4   INTEGRAL WINDUP

Many aspects of a control system can be understood from linear models. However, there are some nonlinear phenomena that must be taken into account. These are typically limitations in the actuators: a motor has limited speed, a valve cannot be more than fully opened or fully closed, etc. For a system that operates over a wide range of conditions, it may happen that the control variable reaches the actuator limits. When this happens, the feedback loop is broken and the system runs in open loop because the actuator remains at its limit independently of the process output as long as the actuator remains saturated. The integral term will also build up since the error is typically nonzero. The integral term and the controller output may then become very large. The control signal will then remain saturated even when the error changes, and it may take a long time before the integrator and the controller output come inside the saturation range. The consequence is that there are large transients. This situation is referred to as *integrator windup*, illustrated in the following example.

Example 11.5 Cruise control
The windup effect is illustrated in Figure 11.10a, which shows what happens when a car encounters a hill that is so steep ($6°$) that the throttle saturates when the cruise controller attempts to maintain speed. When encountering the slope at time $t = 5$, the velocity decreases and the throttle increases to generate more torque. However, the torque required is so large that the throttle saturates. The error decreases slowly

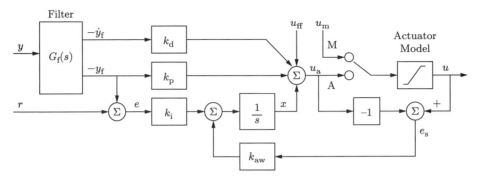

Figure 11.11: PID controller with filtering, anti-windup, and manual control. The controller has filtering of the measured signal, an input u_{ff} for the feedforward signal, and another input u_m for direct control of the output. The switch is in position A for normal operation; if it is set to M the control variable is manipulated directly. The input to the integrator ($1/s$) has a "reset" term that avoids integrator windup in addition to the normal P, I, and D terms. Notice that the reference r only enters in the integral term.

because the torque generated by the engine is just a little larger than the torque required to compensate for gravity. The error is large and the integral continues to build up until the error reaches zero at time 25, but the controller output is still larger than the saturation limit and the actuator remains saturated. The integral term starts to decrease and the velocity settles to the desired value at time $t = 40$. Also notice the large overshoot. ∇

Avoiding Windup

Windup can occur in any controller with integral action. There are many methods to avoid windup. One method for PID control is illustrated in Figure 11.11: the system has an extra feedback path that is generated from a mathematical model of the saturating actuator. The signal e_s is the difference between the outputs of the controller u_a and the actuator model u. It is fed to the input of the integrator through the gain k_{aw}. The signal e_s is zero when there is no saturation and the extra feedback loop has no effect on the system. When the actuator saturates, the signal e_s is fed back to the integrator in such a way that e_s goes toward zero. This implies that controller output is kept close to the saturation limit. The controller output will then change as soon as the error changes sign and integral windup is avoided.

The rate at which the controller output is reset is governed by the feedback gain k_{aw}; a large value of k_{aw} gives a short reset time. The parameter k_{aw} cannot be too large because measurement noise can then cause an undesirable reset. A reasonable choice is to choose k_{aw} as a multiple of the integral gain k_i.

The controller also has an input u_{ff} for feedforward control. By entering the feedforward signal as shown in Figure 11.11, the basic anti-windup scheme also deals with saturation caused by the feedforward signal.

We illustrate how integral windup can be avoided by investigating the cruise control system.

Example 11.6 Cruise control with anti-windup

Figure 11.10b shows what happens when a controller with anti-windup is applied to the system simulated in Figure 11.10a. Because of the feedback from the actuator model, the output of the integrator is quickly reset to a value such that the controller output is at the saturation limit. The behavior is drastically different from that in Figure 11.10a and the large overshoot is avoided. The tracking gain used in the simulation is $k_{aw} = 2$ which is an order of magnitude larger than the integral gain $k_i = 0.2$.   ▽

To explore if windup protection improves stability, we can redraw the block diagram so that the nonlinearity is isolated. The closed loop system then consists of a linear block and a static nonlinearity. With an ideal saturation, the nonlinearity is a sector-bounded nonlinearity modeled by equation (10.17) with $k_{low} = 0$ and $k_{high} = 1$, and the linear part has the transfer function

$$H(s) = \frac{sP(s)C(s) - k_{aw}}{s + k_{aw}} \qquad (11.11)$$

(Exercise 11.12). We can use the circle criterion in Section 10.5 to check stability of the closed loop system. We first observe that the special form of the nonlinearity implies that the circle reduces to the line Re $s = -1$. Applying the circle criterion, we find that the system with windup protection is stable if the Nyquist curve of the transfer function $H(s)$ is to the right of the line Re $s = -1$. If we use describing functions we find that oscillations may occur if the Nyquist curve $H(i\omega)$ intersects the negative real axis to the left of the critical point -1.

Manual Control and Tracking

Automatic control is often combined with manual control, where the operation modes are selected by a switch as illustrated in Figure 11.11. The switch is normally in the position A (automatic). Manual control is selected by moving the switch to position M (manual) and the control variable is then manipulated directly, often by buttons for increasing and decreasing the control signal. For example, in a cruise control system such as that shown in Figure 1.16a, the control signal increases at constant rate when pushing the increase speed (accel) button and it decreases at constant rate when the decrease speed (decel) button is pushed. In Figure 11.11 the manipulated variable is denoted by u_m.

Care has to be taken to avoid transients when switching modes. This can be accomplished by the arrangement shown in Figure 11.11. When the controller is in manual mode the feedback through the gain k_{aw} adjusts the input to the integrator so that the controller output u_a tracks the manual input u_m, resulting in no transient when switching to automatic control.

To see how the controller in Figure 11.11 is implemented, let the integrator output be z. The controller is then described by

$$\frac{dx}{dt} = k_i(r - y_f) + k_{aw}(u - u_a), \quad u_a = z - k_p y_f - k_d \dot{y}_f, \quad u = \begin{cases} F(u_a) & \text{automatic,} \\ F(u_m) & \text{manual,} \end{cases}$$

where $F(u)$ is the function that represents the actuator model. The parameter k_{aw} is typically larger than k_i and it then follows that the controller output u tracks u_m in manual mode (tracking would be ideal if the term $k_i(r - y_f)$ is zero).

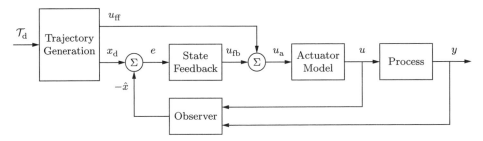

Figure 11.12: Anti-windup for a general controller architecture. Compare with the corresponding controller without anti-windup in Figure 8.11.

Anti-Windup for General Controllers

Anti-windup can also be extended to general control architectures such as the state space-based designs studied in Chapters 7 and 8. For the case of an output feedback controller with integral action via state augmentation (see Example 8.9), we modify the anti-windup compensation to adjust the entire controller state instead of just the integrator state. The approach is particularly easy to understand for controllers based on state feedback and an observer, like the one shown in Figure 8.11. Without modification, when a saturation occurs then the wrong information is sent to the observer (the commanded input instead of the saturated input). To address this, we simply introduce a model for the saturating actuator and feed its output to the observer, as illustrated in Figure 11.12.

To investigate the stability of the controller with anti-windup, we observe that if the observer model is designed so that the process actuator never saturates, the block diagram of the closed loop system can be redrawn so that it consists of a nonlinear static block representing the actuator model $F(x)$ and a linear block representing the observer and the process. We can again make use of the circle criterion described in Section 10.5 to provide conditions for stability. The linear block has the transfer function

$$H(s) = K(sI - A + LC)^{-1}(B + LC[sI - A]^{-1}B), \qquad (11.12)$$

where A, B, and C are the matrices of the state space model, K is the feedback gain matrix, and L is the gain matrix of the Kalman filter. With a simple saturating actuator, the nonlinearity is sector-bounded with $k_{\text{low}} = 0$ and $k_{\text{high}} = 1$ in equation (10.17). It then follows from the circle criterion that the closed loop is stable if the Nyquist plot of $L(i\omega)$ is to the right of the line Re $z = -1/k_{\text{high}} = -1$, and the winding number condition is satisfied.

Facilities for manual control and tracking with observers and state augmentation can be done in the same way as for the PID controller in Figure 11.11.

11.5   IMPLEMENTATION

There are many practical issues that have to be considered when implementing PID controllers. They have been developed over time based on practical experience. In this section we consider some of the most common. Similar considerations also apply to other types of controllers.

Filtering the Derivative

A drawback with derivative action is that an ideal derivative has high gain for high-frequency signals. This means that high-frequency measurement noise will generate large variations in the control signal. The effect of measurement noise may be reduced by replacing the term $k_d s$ by $k_d s/(1 + sT_f)$, which can be interpreted as an ideal derivative of a low-pass filtered signal. The time constant of the filter is typically chosen as $T_f = (k_d/k_p)/N = T_d/N$, with N in the range 5–20. Filtering is obtained automatically if the derivative is implemented by taking the difference between the signal and its filtered version as shown in Figure 11.3b; see also equation (11.5). Note that in the implementation in Figure 11.3b, the filter time constant T_f is equal to the derivative time constant T_d ($N = 1$).

Instead of filtering just the derivative, it is also possible to use an ideal controller and filter the measured signal. Choosing a second-order filter, the transfer function of the controller with the filter becomes

$$C(s) = k_p \left(1 + \frac{1}{sT_i} + sT_d\right) \frac{1}{1 + sT_f + (sT_f)^2/2}. \tag{11.13}$$

For the system in Figure 11.11, filtering is done in the box marked $G_f(s)$, which has the dynamics

$$\frac{d}{dt}\begin{pmatrix} x_1 \\ x_2 \end{pmatrix} = \begin{pmatrix} 0 & 1 \\ -2T_f^{-2} & -2T_f^{-1} \end{pmatrix} \begin{pmatrix} x_1 \\ x_2 \end{pmatrix} + \begin{pmatrix} 0 \\ 2T_f^{-2} \end{pmatrix} y. \tag{11.14}$$

The states are $x_1 = y_f$ and $x_2 = dy_f/dt$. The filter thus gives filtered versions of the measured signal and its derivative. The second-order filter also provides good high-frequency roll-off, which improves robustness.

Setpoint Weighting

Figure 11.1 shows two configurations of a PID controller. The system in Figure 11.1a has a controller with *error feedback* where proportional, integral, and derivative action acts on the error. In the simulation of PID controllers in Figure 11.2c there is a large initial peak in the control signal, which is caused by the derivative of the reference signal. The peak can be avoided by using the controller in Figure 11.1b, where proportional and derivative action acts only on the process output. An intermediate form is given by

$$u = k_p\big(\beta r - y\big) + k_i \int_0^t \big(r(\tau) - y(\tau)\big)\, d\tau + k_d\left(\gamma \frac{dr}{dt} - \frac{dy}{dt}\right), \tag{11.15}$$

where the proportional and derivative actions act on fractions β and γ of the reference. Integral action has to act on the error to make sure that the error goes to zero in steady state. The closed loop systems obtained for different values of β and γ respond to load disturbances and measurement noise in the same way. The response to reference signals is different because it depends on the values of β and γ, which are called *reference weights* or *setpoint weights*. Setpoint weighting is a simple way to obtain two degree-of-freedom action in a PID controller. A controller with $\beta = \gamma = 0$ is sometimes called an *I-PD controller*, as seen Figure 11.1b. We illustrate the effect of setpoint weighting by an example.

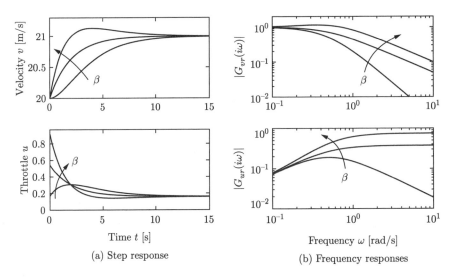

Figure 11.13: Step and frequency responses for PI cruise control with setpoint weighting. Step responses are shown in (a) and the gain curves of the frequency responses in (b). The controller gains are $k_\mathrm{p} = 0.74$ and $k_\mathrm{i} = 0.19$. The setpoint weights are $\beta = 0$, 0.5, and 1, and $\gamma = 0$.

Example 11.7 Cruise control with setpoint weighting

Consider the PI controller for the cruise control system derived in Example 11.3. Figure 11.13 shows the effect of setpoint weighting on the response of the system to a reference signal. With $\beta = 1$ (error feedback) there is an overshoot in velocity and the control signal (throttle) is initially close to the saturation limit. There is no overshoot with $\beta = 0$ and the control signal is much smaller, clearly a much better drive comfort. The frequency responses gives another view of the same effect. The parameter β is typically in the range 0–1, and γ is normally zero to avoid large transients in the control signal when the reference is changed. ∇

The controller given by equation (11.15) is a special case of the general controller structure having two degrees of freedom, which was discussed in Section 8.5.

Implementation Based on Operational Amplifiers

PID controllers have been implemented in different technologies. Figure 11.14 shows how PI and PID controllers can be implemented by feedback around operational amplifiers.

To show that the circuit in Figure 11.14b is a PID controller we will use the approximate relation given by equation (4.14), which is valid when resistances R_i are replaced by impedances Z_i (Exercise 10.1). This gives the transfer function $-Z_2/Z_1$ for the closed loop op amp circuit, noting that the gain of the operational amplifier is negative. For the PI control in Figure 11.14a the impedances are

$$Z_1 = R_1, \quad Z_2 = R_2 + \frac{1}{sC_2} = \frac{1 + R_2C_2s}{sC_2}, \quad \frac{Z_2}{Z_1} = \frac{1 + R_2C_2s}{sR_1C_2} = \frac{R_2}{R_1} + \frac{1}{R_1C_2s},$$

which shows that the circuit is an implementation of a PI controller with gains $k_\mathrm{p} = R_2/R_1$ and $k_\mathrm{i} = 1/(R_1C_2)$.

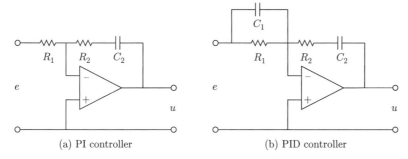

(a) PI controller   (b) PID controller

Figure 11.14: Schematic diagrams for PI and PID controllers using op amps. The circuit in (a) uses a capacitor in the feedback path to store the integral of the error. The circuit in (b) adds a filter on the input to provide derivative action.

A similar calculation for the PID controller in Figure 11.14b gives

$$Z_1(s) = \frac{R_1}{1 + R_1 C_1 s}, \quad Z_2(s) = R_2 + \frac{1}{C_2 s}, \quad \frac{Z_2}{Z_1} = \frac{(1 + R_1 C_1 s)(1 + R_2 C_2 s)}{R_1 C_2 s},$$

which shows that the circuit is an implementation of a PID controller with the parameters

$$k_\mathrm{p} = \frac{R_1 C_1 + R_2 C_2}{R_1 C_2}, \qquad T_\mathrm{i} = R_1 C_1 + R_2 C_2, \qquad T_\mathrm{d} = \frac{R_1 R_2 C_1 C_2}{R_1 C_1 + R_2 C_2}.$$

Computer Implementation

In this section we briefly describe how a PID controller may be implemented using a computer. The computer typically operates periodically, with signals from the sensors sampled and converted to digital form by the A/D converter, and the control signal computed and then converted to analog form for the actuators. The sequence of operation is as follows:

1. Wait for clock interrupt
2. Read input from sensor
3. Compute control output
4. Send output to the actuator
5. Update controller state
6. Repeat

Notice that an output is sent to the actuators as soon as it is available. The time delay is minimized by making the calculations in step 3 as short as possible and performing all updates after the output is commanded. This simple way of reducing the latency is, unfortunately, seldom used in commercial systems.

As an illustration we consider the PID controller in Figure 11.11, which has a filtered derivative, setpoint weighting, and protection against integral windup (anti-windup). The controller is a continuous-time dynamical system. To implement it using a computer, the continuous-time system has to be approximated by a discrete-time system.

In Figure 11.11, the signal u_a is the sum of the proportional, integral, and derivative terms, and the controller output is $u = \mathrm{sat}(u_\mathrm{a})$, where sat is the saturation function that models the actuator. The proportional term $P = k_\mathrm{p}(\beta r - y)$ is implemented simply by replacing the continuous variables with their sampled versions. Hence

$$P(t_k) = k_\mathrm{p}\big(\beta r(t_k) - y(t_k)\big), \tag{11.16}$$

where $\{t_k\}$ denotes the sampling instants, i.e., the times when the computer reads its input. We let h represent the sampling time, so that $t_{k+1} = t_k + h$. The integral term is obtained by approximating the integral with a sum,

$$I(t_{k+1}) = I(t_k) + k_\mathrm{i} h\, e(t_k) + \frac{h}{T_\mathrm{aw}}\big(\mathrm{sat}(u_\mathrm{a}) - u_\mathrm{a}\big), \tag{11.17}$$

where $T_\mathrm{aw} = h/k_\mathrm{aw}$ represents the anti-windup term. The filtered derivative term D is given by the differential equation

$$T_\mathrm{f}\frac{dD}{dt} + D = -k_\mathrm{d}\dot{y}.$$

Approximating the derivative with a backward difference gives

$$T_\mathrm{f}\frac{D(t_k) - D(t_{k-1})}{h} + D(t_k) = -k_\mathrm{d}\frac{y(t_k) - y(t_{k-1})}{h},$$

which can be rewritten as

$$D(t_k) = \frac{T_\mathrm{f}}{T_\mathrm{f}+h}\, D(t_{k-1}) - \frac{k_\mathrm{d}}{T_\mathrm{f}+h}\big(y(t_k) - y(t_{k-1})\big). \tag{11.18}$$

The advantage of using a backward difference is that the parameter $T_\mathrm{f}/(T_\mathrm{f}+h)$ is nonnegative and less than 1 for all $h > 0$, which guarantees that the difference equation is stable. Reorganizing equations (11.16)–(11.18), the PID controller can be described by the following pseudocode:

```
% Precompute controller coefficients
bi = ki*h
ad = Tf/(Tf+h)
bd = kd/(Tf+h)
br = h/Taw

% Initalize variables
I = 0, yold = adin(ch2)

% Control algorithm - main loop
while (running) {
   r = adin(ch1)                % read setpoint from ch1
   y = adin(ch2)                % read process variable from ch2
   P = kp*(b*r - y)             % compute proportional part
   D = ad*D - bd*(y-yold)       % compute derivative part
   ua = P + I + D               % compute temporary output
   u = sat(ua, ulow, uhigh)     % simulate actuator saturation
   daout(ch1)                   % set analog output ch1
   I = I + bi*(r-y) + br*(u-ua) % update integral state
   yold = y                     % update derivative state
   sleep(h)                     % wait until next update interval
}
```

Precomputation of the coefficients bi, ad, bd, and br saves computer time in the main loop. These calculations have to be done only when controller parameters are changed. The main loop is executed once every sampling period. The program has three states: yold, I, and D. One state variable can be eliminated at the cost of less readable code. The latency between reading the analog input and setting the analog output consists of four multiplications, four additions, and evaluation of the sat function. All computations can be done using fixed-point calculations if necessary and implemented on a programmable logical controller (PLC). Notice that the code computes the filtered derivative of the process output and that it has setpoint weighting and anti-windup protection. Note also that in this code we apply the actuator saturation inside the controller, rather than measuring the actuator output as in Figure 11.11.

11.6 FURTHER READING

The history of PID control is very rich and stretches back to the early uses of feedback. Good presentations are given by Bennett [34, 35] and Mindel [183]. Industrial perspectives on PID control are given in [44], [222], and [258], which all mention that a significant fraction of PID controllers are poorly tuned. PID algorithms have been used in many fields; an unconventional application is to explain popular monetary policy rules [115]. The Ziegler–Nichols rules for tuning PID controllers, first presented in 1942 [266], were developed based on extensive experiments with pneumatic simulators and Vannevar Bush's differential analyzer at MIT. An interesting view of the development of the Ziegler–Nichols rules is given in an interview with Ziegler [49]. The book [194] lists more than 1730 tuning rules. A detailed discussion of methods for avoiding windup is given in [261], and a comprehensive treatment of PID control is given in Åström and Hägglund [19]. Advanced relay autotuners are presented in Berner *et al.* [42].

EXERCISES

11.1 (Ideal PID controllers) Consider the systems represented by the block diagrams in Figure 11.1. Assume that the process has the transfer function $P(s) = b/(s+a)$ and show that the transfer functions from r to y are

a) $G_{yr}(s) = \dfrac{bk_d s^2 + bk_p s + bk_i}{(1+bk_d)s^2 + (a+bk_p)s + bk_i}$,

b) $G_{yr}(s) = \dfrac{bk_i}{(1+bk_d)s^2 + (a+bk_p)s + bk_i}$.

Pick some parameters and compare the step responses of the systems.

11.2 Consider a second-order process with the transfer function

$$P(s) = \frac{b}{s^2 + a_1 s + a_2}.$$

The closed loop system with a PI controller is a third-order system. Show that it is possible to position the closed loop poles as long as the sum of the poles is $-a_1$. Give equations for the parameters that give the closed loop characteristic polynomial

$$(s+\alpha_c)(s^2 + 2\zeta_c\omega_c s + \omega_c^2).$$

11.3 Consider a system with the transfer function $P(s) = (s+1)^{-2}$. Find an integral controller that gives a closed loop pole at $s = -a$ and determine the value of a that maximizes the integral gain. Determine the other poles of the system and judge if the pole can be considered dominant. Compare with the value of the integral gain given by equation (11.6).

11.4 (Tuning rules) Apply the Ziegler–Nichols and the modified tuning rules to design PI controllers for systems with the transfer functions

$$P_1 = \frac{e^{-s}}{s}, \qquad P_2 = \frac{e^{-s}}{s+1}, \qquad P_3 = e^{-s}.$$

Compute the stability margins and explore any patterns.

11.5 (Ziegler–Nichols tuning) Consider a system with transfer function $P(s) = e^{-s}/s$. Determine the parameters of P, PI, and PID controllers using Ziegler–Nichols step and frequency response methods. Compare the parameter values obtained by the different rules and discuss the results.

11.6 (Vehicle steering) Design a proportional-integral controller for the vehicle steering system that gives the closed loop characteristic polynomial

$$s^3 + 2\omega_c s^2 + 2\omega_c^2 s + \omega_c^3.$$

11.7 (Average residence time with PID control) The average residence time is a measure of the response time of the system. For a stable system with impulse response $h(t)$ and transfer function $P(s)$ it can be defined as

$$T_{ar} = \int_0^\infty t h(t)\, dt = -\frac{P'(0)}{P(0)}.$$

Consider a stable system with $P(0) \neq 0$ and a PID controller having integral gain $k_i = k_p/T_i$. Show that the average residence time of the closed loop system is given by $T_{ar} = T_i/(P(0)k_p)$.

11.8 (Web server control) Web servers can be controlled using a method known as *dynamic voltage frequency scaling* in which the processor speed is regulated by changing its supply voltage. A typical control goal is to maintain a given service rate, which is approximately equal to maintaining a specified queue length. The queue length x can be modeled by equation (3.32),

$$\frac{dx}{dt} = \lambda - \mu,$$

where λ is the arrival rate and μ is the service rate, which is manipulated by changing the processor voltage. A PI controller for keeping queue length close to x_r is given by

$$\mu = k_p(x - \beta x_r) + k_i \int_0^t (x - x_r)\, dt.$$

Choose the controller parameters k_p and k_i so that the closed loop system has the characteristic polynomial $s^2 + 1.6s + 1$, then adjust the setpoint weight β so that the response to a step in the reference signal has 2% overshoot.

11.9 (Motor drive) Consider the model of the motor drive in Exercise 3.7 with the parameter values given in Exercise 7.11. Develop an approximate second-order model of the system and use it to design an ideal PD controller that gives a closed loop system with eigenvalues $-\zeta\omega_0 \pm i\omega_0\sqrt{1-\zeta^2}$. Add low-pass filtering as shown in equation (11.13) and explore how large ω_0 can be made while maintaining a good stability margin. Simulate the closed loop system with the chosen controller and compare the results with the controller based on state feedback in Exercise 7.11.

11.10 (Windup and anti-windup) Consider a PI controller of the form $C(s) = 1 + 1/s$ for a process with input that saturates when $|u| > 1$, and whose linear dynamics are given by the transfer function $P(s) = 1/s$. Simulate the response of the system to step changes in the reference signal of magnitude 1, 2, and 10. Repeat the simulation when the windup protection scheme in Figure 11.11 is used.

11.11 (Windup protection by conditional integration) Many methods have been proposed to avoid integrator windup. One method called *conditional integration* is to update the integral only when the error is sufficiently small. To illustrate this method we consider a system with PI control described by

$$\frac{dx_1}{dt} = u, \qquad u = \mathrm{sat}_{u_0}(k_\mathrm{p}e + k_\mathrm{i}x_2), \qquad \frac{dx_2}{dt} = \begin{cases} e & \text{if } |e| < e_0, \\ 0 & \text{if } |e| \geq e_0, \end{cases}$$

where $e = r - x$. Plot the phase portrait of the system for the parameter values $k_\mathrm{p} = 1$, $k_\mathrm{i} = 1$, $u_0 = 1$, and $e_0 = 1$ and discuss the properties of the system. The example illustrates the difficulties of introducing *ad hoc* nonlinearities without careful analysis.

11.12 (Windup stability) Consider a closed loop system with controller transfer function $C(s)$ and process transfer function $P(s)$. Let the controller have windup protection with the tracking constant k_aw. Assume that the actuator model in the anti-windup scheme is chosen so that the process never saturates.

a) Use block diagram transformations to show that the closed loop system with anti-windup can be represented as a connection of a linear block with transfer function (11.11) and a nonlinear block representing the actuator model.

b) Show that the closed loop system is stable if the Nyquist plot of the transfer function (11.11) has the property $\mathrm{Re}\,H(i\omega) > -1$.

c) Assume that $P(s) = k_\mathrm{v}/s$ and $C(s) = k_\mathrm{p} + k_\mathrm{i}/s$. Show that the system with windup protection is stable if $k_\mathrm{aw} > k_\mathrm{i}/k_\mathrm{p}$.

d) Use describing function analysis to show that without the anti-windup protection, the system may not be stable and estimate the amplitude and frequency of the resulting oscillation.

e) Build a simple simulation that verifies the results from part (d).

11.13 Consider the system in Exercise 11.9 and investigate what happens if the second-order filtering of the derivative is replaced by a first-order filter.

Chapter Twelve

Frequency Domain Design

> Sensitivity improvements in one frequency range must be paid for with sensitivity deteriorations in another frequency range, and the price is higher if the plant is open loop unstable. This applies to every controller, no matter how it was designed.
>
> —Gunter Stein in the inaugural IEEE Bode Lecture, 1989 [229].

In this chapter we continue to explore the use of frequency domain techniques with a focus on the design of feedback systems. We begin with a more thorough description of the performance specifications for control systems and then introduce the concept of "loop shaping" as a mechanism for designing controllers in the frequency domain. Additional techniques discussed in this chapter include feedforward compensation, the root locus method, and nested controller design.

12.1 SENSITIVITY FUNCTIONS

In the previous chapter, we used proportional-integral-derivative (PID) feedback as a mechanism for designing a feedback controller for a given process. In this chapter we will expand our approach to include a richer repertoire of controllers and tools for shaping the frequency response of the closed loop system.

One of the key ideas in this chapter is that we can design the behavior of the closed loop system by focusing on the open loop transfer function. This same approach was used in studying stability using the Nyquist criterion: we plotted the Nyquist plot for the *open* loop transfer function to determine the stability of the *closed* loop system. From a design perspective, the use of loop analysis tools is very powerful: since the loop transfer function is $L = PC$, if we can specify the desired performance in terms of properties of L, we can directly see the impact of changes in the controller C. This is much easier, for example, than trying to reason directly about the tracking response of the closed loop system, whose transfer function is given by $G_{yr} = PC/(1 + PC)$.

We will start by investigating some key properties of a closed loop control system. A block diagram of a basic two degree-of-freedom control system is shown in Figure 12.1. The system loop is composed of two components: the process and the controller. The two degree-of-freedom controller itself has two blocks: the feedback block C and the feedforward block F. There are two disturbances acting on the process, the load disturbance v and the measurement noise w. The load disturbance represents disturbances that drive the process away from its desired behavior, while the measurement noise represents disturbances that corrupt information about the

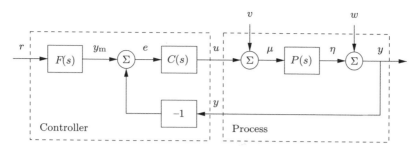

Figure 12.1: Block diagram of a control system with two degrees of freedom. The controller has a feedback block C and a feedforward block F. The external signals are the reference signal r, the load disturbance v, and the measurement noise w. The measured output is y, and the control signal is u.

process given by the sensors. For example, in a cruise control system the major load disturbances are changes in the slope of the road, and measurement noise is caused by the electronics that convert pulses measured on a rotating shaft to a velocity signal. The load disturbances typically have low frequencies, lower than the controller bandwidth, while measurement noise typically has higher frequencies. It is assumed that load disturbances enter at the process input and that the measurement noise acts at the process output. This is a simplification since disturbances may enter the process in many different ways and there may be dynamics in the sensors. These assumptions allow us to streamline the presentation without significant loss of generality.

The process output η is the variable that we want to control, and our ultimate goal is to make η track a reference signal r. To shape the response to reference signals, it is common to use a feedforward block to generate a desired (or model) reference signal y_m that represents the actual signal we attempt to track. Control is based on the difference between the model reference y_m and the measured signal y, where the measurements are corrupted by measurement noise w. The process is influenced by the controller via the control variable u. The process is thus a system with three inputs (the control variable u, the load disturbance v, and the measurement noise w) and one output (the measured signal y). The controller is a system with two inputs (the measured signal y and the reference signal r) and one output (the control signal u). Note that the control signal u is an input to the process and the output of the controller, and that the measured signal y is the output of the process and an input to the controller.

Since the control system in Figure 12.1 is composed of linear elements, the relations between the signals in the diagram can be expressed in terms of the transfer functions. The overall system has three external inputs: the reference r, the load disturbance v, and the measurement noise w. Any of the remaining signals can be relevant for design, but the most common ones are the error e, the control input u, and the output y. In addition, the process input and output, μ and η, are also useful. Table 12.1 summarizes the transfer functions between the external inputs (rows) and remaining signals (columns).

Although there are 15 entries in the table, many transfer functions appear more than once. For most control designs we focus on the following subset, which we call the *Gang of Six*:

Table 12.1: Transfer functions relating the signals of the control system in Figure 12.1. The external inputs are the reference signal r, load disturbance v, and measurement noise w, represented by each row. The columns represent the measured signal y, control input u, error e, process input μ, and process output η that are most relevant for system performance.

y	u	e	μ	η	
$\dfrac{PCF}{1+PC}$	$\dfrac{CF}{1+PC}$	$\dfrac{F}{1+PC}$	$\dfrac{CF}{1+PC}$	$\dfrac{PCF}{1+PC}$	r
$\dfrac{P}{1+PC}$	$\dfrac{-PC}{1+PC}$	$\dfrac{-P}{1+PC}$	$\dfrac{1}{1+PC}$	$\dfrac{P}{1+PC}$	v
$\dfrac{1}{1+PC}$	$\dfrac{-C}{1+PC}$	$\dfrac{-1}{1+PC}$	$\dfrac{-C}{1+PC}$	$\dfrac{-PC}{1+PC}$	w

$$
\begin{aligned}
G_{yr} &= \frac{PCF}{1+PC}, & -G_{uv} &= \frac{PC}{1+PC}, & G_{yv} &= \frac{P}{1+PC}, \\
G_{ur} &= \frac{CF}{1+PC}, & -G_{uw} &= \frac{C}{1+PC}, & G_{yw} &= \frac{1}{1+PC}.
\end{aligned}
\tag{12.1}
$$

The transfer functions in the first column of equation (12.1) give the responses of the process output y and the control signal u to the reference signal r. The second column gives the responses of the control variable u to the load disturbance v and the measurement noise w, and the final column gives the responses of the measured signal y to those two inputs. (Note that the sign convention in equation (12.1) is chosen for later convenience and does not affect the magnitude of the Gang of Six transfer functions.)

The response of the system to load disturbances and measurement noise is of particular importance and these transfer functions are referred to as *sensitivity functions*. They represent the sensitivity of the system to the various inputs, and they have specific names:

$$
\begin{array}{llll}
S = \dfrac{1}{1+PC} & \text{sensitivity function} & PS = \dfrac{P}{1+PC} & \text{load (or input) sensitivity function} \\[2em]
T = \dfrac{PC}{1+PC} & \text{complementary sensitivity function} & CS = \dfrac{C}{1+PC} & \text{noise (or output) sensitivity function}
\end{array}
\tag{12.2}
$$

Because these transfer functions are particularly important in feedback control design, they are called the *Gang of Four*, and they have many interesting properties that will be discussed in detail in the rest of the chapter. Good insight into these properties is essential in understanding the performance of feedback systems for the purposes of both analysis and design.

While the Gang of Four capture the response to disturbances, we are also interested in the response of the system to the reference signal r. The remaining two elements in the full Gang of Six capture the relationship between the reference signal and the measured output y plus the control input u:

$$G_{yr} = \frac{PCF}{1+PC}, \qquad G_{ur} = \frac{CF}{1+PC}.$$

We see that F can be used to design these responses and provides a second degree of freedom in addition to the feedback controller C. In practice, it is common to first design the feedback controller C using the Gang of Four to provide good response with respect to load disturbances and measurement noise, and then use F and the remaining transfer functions as part of the full Gang of Six to obtain good reference tracking.

In addition to the Gang of Six, other signal that can be important is the error between the reference r and the process output η (prior to the addition of measurement noise), which satisfies

$$\epsilon = r - \eta = \left(1 - \frac{PCF}{1+PC}\right)r - \frac{P}{1+PC}v - \frac{PC}{1+PC}w$$
$$= (1 - TF)r - PSv - Tw.$$

The signal ϵ is not actually present in our diagram, but is the true error that represents the tracking deviation. We see that it consists of a particular combination of transfer functions chosen from the Gang of Six.

The special case of $F = 1$ is called a system with (pure) *error feedback* because all control actions are based on feedback from the error. In this case the transfer functions given by equations (12.1) and (12.2) are the same and the system is completely characterized by the Gang of Four. In addition, the true tracking error becomes

$$\epsilon = Sr - PSv - Tw.$$

Notice that we have less freedom in design of a system with error feedback because the feedback controller C must now deal with both disturbance attenuation, robustness, and reference signal tracking.

The transfer functions in equation (12.2) have many interesting properties. For example, it follows from equation (12.2) that $S + T = 1$, which explains why T is called the complementary sensitivity function. The loop transfer function PC will typically go to zero for large s, which implies that T goes to zero and S goes to one as s goes to infinity. Thus, it will not be possible to track very high-frequency reference signals ($|G_{yr}| = |FT| \to 0$) and any high-frequency noise will propagate unfiltered to the error ($|G_{ew}| = |S| \to 1$). For controllers with integral action and processes with non-vanishing zero frequency gain, the loop transfer function PC goes to infinity for small s, which implies that S goes to zero and T goes to one as s goes to zero. Low-frequency signals are thus tracked well ($|G_{yr}| = |FT| \to 0$), and low-frequency disturbances can be completely attenuated ($|G_{ev}| = |PS| \to 0$). Many more properties of the sensitivity functions will be discussed in detail later in this chapter and in Chapters 13 and 14. Good insight into these properties is essential in understanding the performance of feedback systems for the purposes of both analysis and design. The transfer functions are also used to formulate specifications on control systems.

In Chapter 10 we focused on the loop transfer function, and we found that its properties gave useful insights into the properties of a system. The loop transfer function does not, however, always give a complete characterization of the closed loop system. In particular, it can happen that there are pole/zero cancellations in the product of P and C such that $1 + PC$ has no unstable poles, but one of the other Gang of Four transfer functions might be unstable. The following example illustrates this difficulty.

Example 12.1 The loop transfer function gives only limited insight
Consider a process with the transfer function $P(s) = 1/(s - a)$ controlled by a PI controller with error feedback having the transfer function $C(s) = k(s - a)/s$. The loop transfer function is $L = k/s$, and the sensitivity functions are

$$S = \frac{1}{1 + PC} = \frac{s}{s + k}, \qquad PS = \frac{P}{1 + PC} = \frac{s}{(s - a)(s + k)},$$

$$CS = \frac{C}{1 + PC} = \frac{k(s - a)}{s + k}, \qquad T = \frac{PC}{1 + PC} = \frac{k}{s + k}.$$

Notice that the factor $s - a$ is canceled when computing the loop transfer function and that this factor also does not appear in the sensitivity functions S and T. However, cancellation of the factor is very serious if $a > 0$ since the transfer function PS relating load disturbances to process output is then unstable. A small disturbance v then leads to an unbounded output, which is clearly not desirable. ∇

If all four of the transfer functions in equation (12.2) are stable we say that the feedback system is *internally stable*. In addition, if there is a feedforward controller F then it should also be stable in order for the full system to be internally stable. For more general systems, which may contain additional transfer functions and feedback loops, the system is internally stable if all possible input/output transfer functions are stable. For simplicity we will often say that a closed loop system is stable when we mean that it is internally stable.

As mentioned previously, the system in Figure 12.1 represents a special case because it is assumed that the load disturbance enters at the process input and that the measured output is the sum of the process variable and the measurement noise. Disturbances can enter in many different ways, and the sensors may have dynamics. A more abstract way to capture the general case is shown in Figure 12.2, which has only two blocks representing the process ($\mathcal{P}$) and the controller ($\mathcal{C}$). The process has two inputs, the control signal u and a vector of disturbances χ, and three outputs, the measured signal y, the reference signal r, and a vector of signals ξ that is used to specify performance. The system in Figure 12.1 can be captured by choosing $\chi = (r, v, w)$ and $\xi = (e, \mu, \eta, \epsilon)$. The process transfer function $\mathcal{P}$ describes the effect of χ and u on ξ, y, and r, and the controller transfer function $\mathcal{C}$ describes how u is related to y and r (see Exercise 12.2). Restricting the signal ξ to contain the errors e and ϵ, the control problem can be formulated as finding a controller $\mathcal{C}$ so that the gain of the transfer function from the disturbance $\chi = (r, v, w)$ to the generalized control error $\xi = (e, \epsilon)$ is as small as possible (discussed further in Section 13.4).

Processes with multiple inputs and outputs can be handled by regarding u and y as vectors. Representations at these higher levels of abstraction are useful for the development of theory because they make it possible to focus on fundamentals and

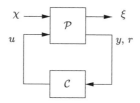

Figure 12.2: A more general representation of a feedback system. The process input u represents the control signal, which can be manipulated, and the process input χ represents the other signals that influence the process. The process output consists of the measured variable(s) y, the reference signal r, and the signal vector ξ representing the other signals of interest in the control design.

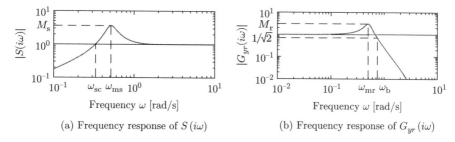

(a) Frequency response of $S(i\omega)$ (b) Frequency response of $G_{yr}(i\omega)$

Figure 12.3: Illustration of specifications in frequency domain. (a) Gain curve of sensitivity function; the maximum sensitivity M_s is a robustness measure. (b) Gain curve of the transfer function G_{yr} with peak value M_r, peak frequency ω_{mr}, and bandwidth ω_b.

to solve general problems with a wide range of applications. However, care must be exercised to maintain the coupling to the real-world control problems we intend to solve and we must keep in mind that matrix multiplication is not commutative.

12.2 PERFORMANCE SPECIFICATIONS

A key element of the control design process is how we specify the desired performance of a system. Specifications capture robustness to process variations as well performance in terms of the ability to follow reference signals and attenuate load disturbances without injecting too much measurement noise. The specifications are expressed in terms of transfer functions such as the Gang of Six and the loop transfer function, and are often represented by features of the transfer functions or their time and frequency responses.

Robustness to process variations was discussed extensively in Section 10.3, where we introduced gain margin g_m, phase margin φ_m, and stability margin s_m, as shown in Figure 10.11. The largest value of the sensitivity function $M_s = 1/s_m$ is another robustness measure, as illustrated in Figure 12.3a.

To provide specifications it is desirable to capture the characteristic properties of a system with a few parameters. Features of step responses that we have already seen are overshoot, rise time, and settling time, as shown in Figure 6.9.

Common features of frequency responses include peak value(s), peak frequency, gain crossover frequency, and bandwidth. Other features of the frequency response include the maximum value of sensitivity function M_s (occurring at frequency ω_{ms}) and the maximum value of the complementary sensitivity function M_t (occurring at frequency ω_{mt}). The *sensitivity crossover frequency* ω_{sc} is defined as the frequency where the magnitude of the sensitivity function $S(j\omega)$ is 1. The various crossover frequencies and the bandwidth are only well defined if the curves are monotone; if this is not the case the lowest such frequency is typically used.

There are interesting relationships between specifications in the time and frequency domains. Roughly speaking, the behavior of time responses for short times is related to the behavior of frequency responses at high frequencies, and vice versa. The precise relations are given by the Laplace transform. There are also useful relationships between features in the time and frequency domain; typical examples are given in Tables 7.1 and 7.2 in Section 7.3.

In the remainder of this section we consider the different types of responses that are commonly used in control design and describe the types of specifications that are relevant for each.

Response to Reference Signals

Consider the basic feedback loop in Figure 12.1. The responses of the output y and the control signal u to the reference r are described by the transfer functions $G_{yr} = PCF/(1 + PC)$ and $G_{ur} = CF/(1 + PC)$ ($F = 1$ for systems with pure error feedback). Specifications can be expressed in terms of features of the transfer function G_{yr}, such as the peak (or resonant) value M_r, the peak frequency ω_{mr}, and the bandwidth ω_b, as shown in Figure 12.3b.

In the special case where $F = 1$, the transfer function G_{yr} is equal to the complementary sensitivity function T. However, in many cases it is useful to retain the ability to shape the input/output response by using $F \neq 1$. This distinction is captured in the use of the full Gang of Six rather than just the Gang of Four.

The transfer function G_{yr} typically has unit zero frequency gain because we want to design the system so that the response to a step input has zero steady-state error. The behavior of the transfer function at low frequencies determines the tracking error for slow reference signals. We can capture this analytically by making the following series expansion of the transfer function from reference r to output e for small s:

$$G_{er}(s) \approx e_1 s + e_2 s^2 + \cdots,$$

where the coefficients e_k are called *error coefficients*. If the reference signal is $r(t)$, the tracking error is then

$$e(t) = r(t) - y(t) = G_{er}r = e_1 \frac{dr}{dt} + e_2 \frac{d^2r}{d^2t} + \cdots.$$

It follows that a ramp input $r(t) = v_0 t$ gives a steady-state tracking error $v_0 e_1$, and we can conclude that the steady-state tracking error is zero if $e_1 = 0$. A system with $e_1 = 0$ has the steady-state error $e(t) = 2ae_2$ for the input $r(t) = a_0 t^2$. The equation also supports the insight that the behavior at low frequencies (small s) corresponds to the behavior at large times, a consequence of the final value theorem (discussed briefly at the end of Section 9.3).

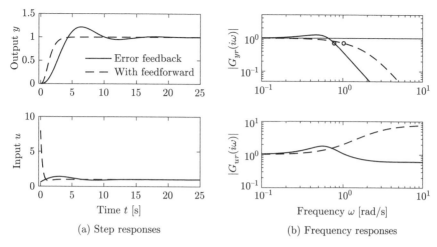

(a) Step responses   (b) Frequency responses

Figure 12.4: Reference signal responses for Example 12.2. The responses in process output y and control signal u to a unit step in the reference signal r are shown in (a), and the gain curves of G_{yr} and G_{ur} are shown in (b). Results for PI control with error feedback are shown by solid lines, and the dashed lines show results for a controller with a feedforward compensator. The bandwidth of the closed loop systems is marked in the plot of G_{yr} with open circles ($\circ$).

It has long been a practice to focus on the output when we give specifications. However, it is useful to also consider the response of the control signal because this allows us to judge the magnitude and rate of the control signal required to obtain the output response. This is illustrated in the following example.

Example 12.2 Reference signal tracking for a third-order system
Consider a process with the transfer function $P(s) = (s+1)^{-3}$ and a PI controller with error feedback having the gains $k_p = 0.6$ and $k_i = 0.5$. The responses are illustrated in Figure 12.4. The solid lines show results for a proportional-integral (PI) controller with error feedback. The dashed lines show results for a controller with feedforward controller

$$F = \frac{G_{yr}(1+PC)}{PC} = \frac{2s^4 + 6s^3 + 6s^2 + 3.2s + 1}{0.15s^4 + 1.025s^3 + 2.55s^2 + 2.7s + 1},$$

designed to give the closed loop transfer function $G_{yr} = (0.5s+1)^{-3}$. Looking at the time responses, we find that the controller with feedforward gives a faster response with no overshoot. However, much larger control signals are required to obtain the fast response. The initial value of the control signal for the controller with feedforward is 13.3, compared to 0.6 for the regular PI controller. The controller with feedforward has a larger bandwidth (marked with $\circ$) and no resonant peak. The transfer function G_{ur} also has higher gain at high frequencies. ∇

We can get some insight into the relationship between time and frequency responses from Figure 12.4. The figures in the top row show the unit step response and the frequency response for the transfer function G_{yr}, and the lower plots show the same quantities for G_{ur}. The dashed time and frequency responses have no peaks while the solid responses have peaks. The peaks are related in the sense that a large overshoot in the time response corresponds to a large resonant peak in the frequency response. The time responses in the bottom plot of Figure 12.4 have the initial values 8 (dashed) and 6 (solid), and the frequency responses have the same final values. In general, it can be shown using the Laplace transform (or appropriate exponential responses) and the initial and final value theorems that for a unit reference signal $r(t)$ we have that $u(t) \to G_{ur}(0)$ as $t \to \infty$ and if $x(0) = 0$ then $u(0) = G_{ur}(\infty)$.

The dashed time response is faster than the solid time response and the dashed frequency response has larger bandwidth than the solid frequency response. The product of the rise time of the unit step response and the bandwidth of a transfer function (the *rise time-bandwidth product*) is a dimension-free variable that is a useful characteristic. The time responses in Figure 12.4 have rise times of $T_r = 1.7$ (dashed) and 3.0 (solid), and the corresponding bandwidths are $\omega_b = 1.9$ (dashed) and 0.8 (solid), which gives the products $T_r \omega_b = 3.2$ (dashed) and 2.4 (solid). A similar observation can be made from Tables 7.1 and 7.2 in Section 7.3, which gives $T_r \omega_b \approx 2.7$–2.8. It thus appears that the product of the rise time of the step response and the bandwidth of the frequency response is approximately constant ($T_r \omega_b \approx 3$). It can be shown that the rise time-bandwidth product increases if the frequency response has a faster roll-off (see Exercise 12.5, which uses a slightly different definition of bandwidth).

Response to Load Disturbances and Measurement Noise

A simple criterion for disturbance attenuation is to compare the output of the closed loop system in Figure 12.1 with the output of the corresponding open loop system, which can be obtained by setting $C = 0$ in the figure. With identical disturbances for the open and closed loop systems, the output of the closed loop system can be obtained simply by sending the open loop output through a system with the transfer function S (Exercise 12.6). The sensitivity S function thus directly shows how feedback influences the response of the output to disturbances both in the form of load disturbances and measurement noise. Disturbances with frequencies such that $|S(i\omega)| < 1$ are attenuated, but disturbances with frequencies such that $|S(i\omega)| > 1$ are amplified by feedback. The *sensitivity crossover frequency* ω_{sc} is the (lowest) frequency where $|S(i\omega)| = 1$, as shown in Figure 12.5a.

Since the sensitivity function is related to the loop transfer function by $S = 1/(1 + L)$, disturbance attenuation can be visualized graphically by the Nyquist plot of the loop transfer function, as shown in Figure 12.5b. The complex number $1 + L(i\omega)$, which is the inverse of the sensitivity function $S(i\omega)$, can be represented as the vector from the point -1 to the point $L(i\omega)$ on the Nyquist curve. The sensitivity is thus less than 1 for all points outside a circle with radius 1 and center at -1. Disturbances with frequencies in this range are attenuated by the feedback, while disturbances with frequencies corresponding to points inside the circle are amplified.

The maximum sensitivity M_s, which occurs at the frequency ω_{ms}, is a measure of the largest amplification of the disturbances. The sensitivity crossover frequency ω_{sc} and the maximum sensitivity M_s are two parameters that give a gross characterization of load disturbance attenuation. For systems where the phase margin is $\varphi_m = 60°$, it can be shown that the sensitivity crossover frequency ω_{sc} is equal to the gain crossover frequency ω_{gc} and the complementary sensitivity function

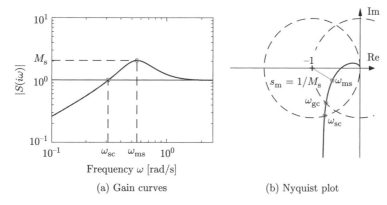

(a) Gain curves   (b) Nyquist plot

Figure 12.5: Illustration of sensitivity to disturbances. The gain curves of the sensitivity function S and the loop transfer function L are shown in (a). The Nyquist plot of the loop transfer function L is shown in (b). Disturbances with frequencies less than the sensitivity crossover frequency, to the left of ω_{sc} in (a) and inside the dashed circle in (b), are attenuated by feedback. Disturbances with frequencies higher than ω_{sc} are amplified. The largest amplification occurs for the frequency ω_{ms}, where the sensitivity has its largest value M_s, the point where the Nyquist curve is closest to the critical point -1 in (b).

crossover frequency ω_{tc}. Notice that the maximum magnitude of $1/(1 + L(i\omega))$ corresponds to the minimum of $|1 + L(i\omega)|$, which is the stability margin s_m defined in Section 10.3, so that $M_s = 1/s_m$. The maximum sensitivity is therefore also a robustness measure.

The transfer function G_{yv} from load disturbance v to process output y for the system in Figure 12.1 is

$$G_{yv} = \frac{P}{1 + PC} = PS = \frac{T}{C}. \tag{12.3}$$

Load disturbances typically have low frequencies. For small s (low frequencies) we have $T \approx 1$ which gives $G_{yv} \approx 1/C$. For processes with $P(0) \neq 0$ and controllers with integral action we have $C(s) \approx k_i/s$ for small s and $G_{yv} \approx s/k_i$. A controller with integral action thus attenuates disturbances with low frequencies effectively, and the integral gain k_i is a measure of disturbance attenuation. For high frequencies we have $S \approx 1$ which implies that $G_{yv} \approx P$ for large s.

Measurement noise, which typically has high frequencies, generates rapid variations in the control variable that are detrimental because they cause wear in the actuators and can even saturate an actuator. It is thus important to keep variations in the control signal due to measurement noise at reasonable levels—a typical requirement is that the variations are only a fraction of the allowable range of the control signal. The effects of measurement noise are captured by the transfer function from the measurement noise to the control signal,

$$-G_{uw} = \frac{C}{1 + PC} = \frac{T}{P} = CS. \tag{12.4}$$

Under the assumption that $S \approx 1$ for large s (high frequencies, which is appropriate for measurement noise), we have $-G_{uw} \approx C$. The formula clearly shows it is useful

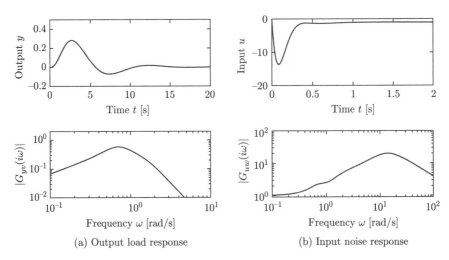

Figure 12.6: Closed loop disturbance responses for Example 12.3. The closed loop unit step response and frequency response for the transfer function G_{yv} from load disturbance v to process output y are shown in (a) and the corresponding responses for the transfer function G_{uw} from measurement noise w to the control signal u are shown in (b).

to filter the derivative so that the transfer function $C(s)$ goes to zero for large s (high-frequency roll-off).

Example 12.3 Disturbance attenuation for a third-order system
Consider a process with the transfer function $P(s) = (s+1)^{-3}$ and a proportional-integral-derivative (PID) controller with gains $k_p = 2$, $k_i = 1.5$, and $k_d = 2.0$. We augment the controller with a second-order noise filter with damping ratio $1/\sqrt{2}$ and $T_f = 0.1$. The controller transfer function then becomes

$$C(s) = \frac{k_d s^2 + k_p s + k_i}{s(s^2 T_f^2/2 + s T_f + 1)}. \tag{12.5}$$

The closed loop system responses are illustrated in Figure 12.6.

The closed loop response of the output y to a unit step in the load disturbance v in the upper part of Figure 12.6a has a peak of 0.28 at time $t = 2.73$ s. The frequency response in Figure 12.6a shows that the gain has a maximum of 0.58 at $\omega = 0.7$ rad/s.

The closed loop response of the control signal u to a step in measurement noise w is shown in Figure 12.6b. The high-frequency roll-off of the transfer function $G_{uw}(i\omega)$ is due to filtering; without it the gain curve in Figure 12.6b would continue to rise after 20 rad/s. The step response has a valley of -14 at $t = 0.08$ s. The frequency response has a peak of 20 at $\omega = 14$ rad/s. Notice that the peak occurs at a frequency far above the peak of the response to load disturbances and far above the gain crossover frequency $\omega_{gc} = 0.78$ rad/s. An approximation derived in Exercise 12.7 gives $\max |CS(i\omega)| \approx k_d/T_f = 20$ for $\omega = \sqrt{2}/T_d = 14.1$ rad/s. ∇

Figure 12.6 also gives insight into the relationship between the time and frequency responses. The frequency response of the transfer functions G_{yv} and G_{uw} have band-pass characteristics and their gains go to zero for high and low frequencies.

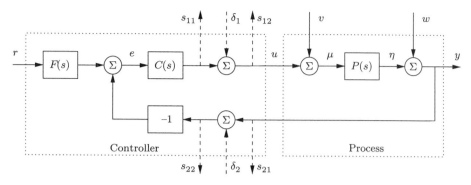

Figure 12.7: Specifications can be tested by injecting signals at test points δ_k and measuring responses at s_{ij}. Compare with Figure 12.1.

A consequence is that the corresponding step responses are zero both for small and large times. The frequency response G_{yv} in Figure 12.6a has a peak of 0.6 for $\omega_p = 0.7$ and the time response has a peak of 0.3 for $t_p = 2.7$, hence $\omega_p t_p = 1.9$. Figure 12.6b shows that the low-frequency gain of the transfer function G_{uw} and steady-state time response are both 1, and the time response starts at zero because the frequency response goes to zero at high frequencies. The frequency response has a peak of 20 for $\omega_p = 14$ and the time response has a peak of 14 for $t_p = 0.08$, hence $\omega_p t_p = 1.1$. These observations support the simple rules for transfer functions with a band-pass character: the product of the peak time of the step response and the resonant peak of the frequency response is in the range of 1 to 2 (Exercise 12.8).

Measuring Specifications

Many specifications are expressed in terms of properties of the transfer functions in the Gang of Six and they can easily be checked simply by computing the transfer functions numerically. To test a real system it is necessary to provide the controller with test points for injecting and measuring signals. Some possible test points are shown in Figure 12.7. As an example, the transfer function G_{yv}, which characterizes response of process output to load a disturbance, can be found by injecting a signal at δ_1 and measuring the output s_{21}. A frequency analyzer that measures the transfer function directly is very convenient for such a test. By measuring the transfer functions we can ensure that robustness and performance are maintained during the design phase and operation of a system.

12.3 FEEDBACK DESIGN VIA LOOP SHAPING

One advantage of the Nyquist stability theorem is that it is based on the loop transfer function $L = PC$, which is the product of the transfer functions of the process and the controller. It is thus easy to see how the controller influences the loop transfer function. For example, to make an unstable system stable we simply have to bend the Nyquist curve away from the critical point. This simple idea is the basis of several different design methods collectively called *loop shaping*. These methods are based on choosing a compensator that gives a loop transfer function

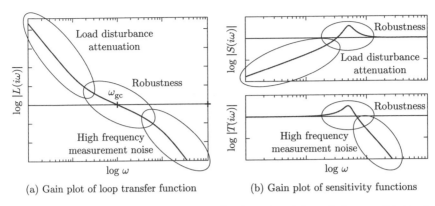

(a) Gain plot of loop transfer function   (b) Gain plot of sensitivity functions

Figure 12.8: Gain plots of the loop transfer function (a) and the sensitivity functions (b) for typical loop transfer functions. The plot on the left shows the gain curve and the plots on the right show the sensitivity function and complementary sensitivity function. The crossover frequency ω_{gc} determines the attenuation of load disturbances, bandwidth, and response time of the closed loop system. The slope n_{gc} of the gain curve of $L(s)$ at the gain crossover frequency ω_{gc} determines the robustness of the closed loop systems (equation (12.6)). At low frequencies, a large magnitude of L provides good load disturbance rejection and reference tracking, while at high frequencies a small loop gain avoids injecting too much measurement noise.

with a desired shape. One possibility is to determine a loop transfer function that gives a closed loop system with the desired properties and to compute the controller as $C = L/P$. This approach may lead to controllers of high order and there are limits if the process transfer function has poles and zeros in the right half-plane, as discussed briefly in Section 12.4 and in more detail in Section 14.3. Another possibility is to start with the process transfer function, change its gain to obtain the desired bandwidth, and then add poles and zeros until the desired shape is obtained. In this section we will explore different loop-shaping methods for control law design.

Design Considerations

We will first discuss a suitable shape for the loop transfer function that gives good performance and good stability margins. Figure 12.8 shows a typical loop transfer function. Good performance requires that the loop transfer function is large for frequencies where we desire good tracking of reference signals and good attenuation of low-frequency load disturbances. Since $S = 1/(1 + L)$, it follows that for frequencies where $|L| > 100$ disturbances will be attenuated by approximately a factor of 100 or more and the tracking error is less than 1%. The transfer function from measurement noise to control action is $CS = C/(1 + L)$. To avoid injecting too much measurement noise, which can create undesirable control actions, the controller transfer function should have low gain at high frequencies, a property called *high-frequency roll-off*. The loop transfer function should thus have roughly the shape shown in Figure 12.8. It has unit gain at the gain crossover frequency ($|L(i\omega_{gc})| = 1$), large gain for lower frequencies, and small gain for higher frequencies.

Robustness is determined by the shape of the loop transfer function around the crossover frequency. Good robustness requires good stability margins, which imposes requirements on the loop transfer function around the gain crossover frequency ω_{gc}. It would be desirable to transition from high loop gain $|L(i\omega))|$ at low frequencies to low loop gain as quickly as possible, but robustness requirements expressed via Bode's relations (Section 10.4) impose restrictions on how fast the gain can decrease. Equation (10.9) implies that the slope of the gain curve at ω_{gc} cannot be too steep. If the gain curve has a constant slope around ω_{gc}, we have the following relationship between slope n_{gc} and phase margin φ_m (in degrees):

$$n_{gc} \approx -2 + \frac{\varphi_m}{90}, \tag{12.6}$$

for a minimum-phase system. A steeper slope thus gives a smaller phase margin. The equation is a reasonable approximation when the gain curve does not deviate too much from a straight line. It follows from equation (12.6) that the phase margins $30°$, $45°$, and $60°$ correspond to the slopes $-5/3$, $-3/2$, and $-4/3$, with a steeper slope giving smaller phase margin. Time delays and poles and zeros in the right half-plane impose further restrictions as will be discussed in Chapter 14.

Loop shaping is a trial-and-error procedure. We typically start with a Bode plot of the process transfer function. Choosing the gain crossover frequency ω_{gc} is a major design decision and is a compromise between attenuation of load disturbances and injection of measurement noise. Notice that the gain crossover frequency and the sensitivity crossover frequencies are the same if the phase margin is $\varphi_m = 60°$, while for smaller phase margins we have $\omega_{gc} < \omega_{sc}$. Having determined the gain crossover frequency we then attempt to shape the loop transfer function by changing the controller gain and adding poles and zeros to the controller transfer function. As we shall see, the controller gain at low frequencies can be increased by so-called "lag compensation," and the behavior around the crossover frequency can be changed by so-called "lead compensation." Different performance specifications are evaluated for each controller as we attempt to balance many different requirements by adjusting controller parameters and complexity.

Loop shaping is straightforward to apply to single-input, single-output systems. It can also be applied to systems with one input and many outputs by closing the loops one at a time. The only limitation for minimum phase systems is that large phase leads and high controller gains may be required to obtain closed loop systems with a fast response. Many specific procedures are available: they all require experience, but they also give good insight into the conflicting specifications. There are fundamental limits to what can be achieved for systems that are not minimum phase; they will be discussed in Section 14.3.

Lead and Lag Compensation

A simple way to do loop shaping is to start with the transfer function of the process and add simple compensators with transfer function

$$C(s) = k\frac{s+a}{s+b}, \quad a > 0, \ b > 0. \tag{12.7}$$

The compensator is called a *lead compensator* if $a < b$, and a *lag compensator* if $a > b$. The PI controller is a special case of a lag compensator with $b = 0$. A lead

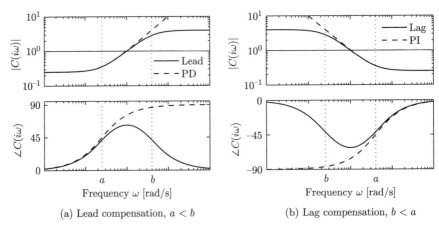

Figure 12.9: Frequency response for lead and lag compensators $C(s) = k(s+a)/(s+b)$. Lead compensation (a) occurs when $a < b$ and provides phase lead between $\omega = a$ and $\omega = b$. Lag compensation (b) corresponds to $a > b$ and provides low-frequency gain. PI control is a special case of lag compensation and PD control is a special case of lead compensation. PI/PD frequency responses are shown by dashed curves. The parameters are $a = 0.25$, $b = 4$, $k = 16$ in (a) and $a = 4$, $b = 0.25$, $k = 1$ in (b).

compensator is essentially the same as a PD controller with filtering. As described in Section 11.5, we often use a filter for the derivative action of a PID controller to limit the high-frequency gain. This same effect is present in a lead compensator through the pole at $s = b$. Equation (12.7) is a first-order compensator and can provide up to 90° of phase lead. Larger phase lead can be obtained by using a higher-order lead compensator (Exercise 12.17):

$$C(s) = k\frac{(s+a)^n}{(s+b)^n}, \qquad a < b.$$

Bode plots of lead and lag compensators are shown in Figure 12.9.

Lag compensation, which increases the gain at low frequencies, is typically used to improve tracking performance and disturbance attenuation at low frequencies. Lead compensation is typically used to improve phase margin. If we set $a < b$ in equation (12.7), we add phase lead in the frequency range between the pole/zero pair (and extending approximately 10× in frequency in each direction). By appropriately choosing the location of this phase lead, we can provide additional phase margin at the gain crossover frequency.

Lead compensation is associated with an increase of the high-frequency gain. Let $G(s)$ be a transfer function with $G(0) > 0$, with no poles and zeros in the right half plane, and assume that $\lim_{s \to \infty} G(s) = G(\infty) > 0$. Then

$$\log \frac{G(\infty)}{G(0)} = \frac{2}{\pi} \int_0^\infty \arg G(i\omega)\, d\log \omega = \frac{2}{\pi} \int_{-\infty}^\infty \arg G(ie^u)\, du. \qquad (12.8)$$

This formula, which we call *Bode's phase area formula*, implies that the logarithm of the gain ratio $G(\infty)/G(0)$ for a transfer function is proportional to the area of the phase curve in the Bode plot. The equation was derived by Bode [51, page 286]

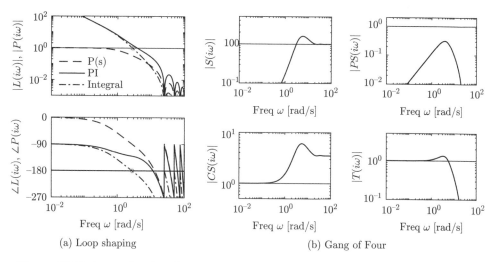

(a) Loop shaping   (b) Gang of Four

Figure 12.10: Loop-shaping design of a controller for an atomic force microscope in tapping mode. (a) Bode plots of the process (dashed), the loop transfer function for an integral controller with critical gain (dash-dotted), and a PI controller (solid) adjusted to give reasonable robustness. (b) Gain curves for the Gang of Four for the system.

using the theory of complex variables. Lead compensation thus requires high gain at high frequencies and increases the sensitivity to measurement noise.

Lead and lag compensators can also be combined to form a lead-lag compensator (Exercise 12.11). Compensators that are tailored to specific disturbances can be also designed, as shown in Exercise 12.12. The following examples illustrate the use of lag compensation (via PI control) and lead compensation (to increase phase margin).

Example 12.4 Atomic force microscope in tapping mode
A simple model of the dynamics of the vertical motion of an atomic force microscope in tapping mode was given in Exercise 10.2. The transfer function for the system dynamics is

$$P(s) = \frac{a(1 - e^{-s\tau})}{s\tau(s + a)},$$

and the parameters $a = \zeta\omega_0$, $\tau = 2\pi n/\omega_0$ are explained in Example 11.2. The gain has been normalized to 1. A Bode plot of this transfer function for the parameters $a = 1$ and $\tau = 0.25$ is shown using dashed curves in Figure 12.10a. To improve the attenuation of load disturbances we increase the low-frequency gain by introducing an integral controller. The loop transfer function then becomes $L = k_i P(s)/s$, and we start by adjusting the gain k_i so that the closed loop system is marginally stable, giving $k_i = 8.3$. The Bode plot is shown by the dash-dotted line in Figure 12.10a, where the critical point is indicated by ∘. Notice the increase of the gain at low frequencies. To obtain a reasonable phase margin we introduce proportional action and we increase the proportional gain k_p gradually until reasonable values of the sensitivities are obtained. The value $k_p = 3.5$ gives maximum sensitivity $M_s = 1.6$ and maximum complementary sensitivity $M_t = 1.3$. The loop transfer function is shown in solid lines in Figure 12.10a. Notice the significant increase of the phase margin compared with the purely integral controller (dash-dotted line).

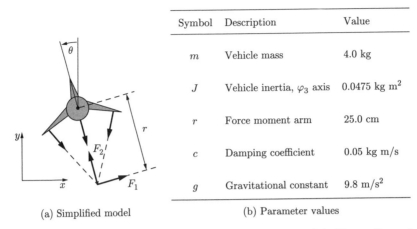

Symbol	Description	Value
m	Vehicle mass	4.0 kg
J	Vehicle inertia, φ_3 axis	0.0475 kg m^2
r	Force moment arm	25.0 cm
c	Damping coefficient	0.05 kg m/s
g	Gravitational constant	9.8 m/s^2

(a) Simplified model   (b) Parameter values

Figure 12.11: Roll control of a vectored thrust aircraft. (a) The roll angle θ is controlled by applying maneuvering thrusters, resulting in a moment generated by F_1. (b) The table lists the parameter values for a laboratory version of the system.

To evaluate the design we also compute the gain curves of the transfer functions in the Gang of Four. They are shown in Figure 12.10b. The peaks of the sensitivity curves are reasonable, and the plot of PS shows that the largest value of PS is 0.3, which implies that the load disturbances are well attenuated. The plot of CS shows that the largest noise gain $|C(i\omega)S(i\omega)|$ is 6. The controller has a gain $k_p = 3.5$ at high frequencies, and hence we may consider adding high-frequency roll-off to make CS smaller at high frequencies. ∇

Example 12.5 Roll control for a vectored thrust aircraft

Consider the control of the roll of a vectored thrust aircraft such as the one illustrated in Figure 12.11. Following Exercise 9.11, we model the system with a second-order transfer function of the form

$$P(s) = \frac{r}{Js^2},$$

with the parameters given in Figure 12.11b. We take as our performance specification that we would like less than 1% error in steady state and less than 10% tracking error up to 10 rad/s.

The open loop transfer function from F_1 to θ is shown in Figure 12.12a. To achieve our performance specification, we would like to have a gain of at least 10 at a frequency of 10 rad/s, requiring the gain crossover frequency to be at a higher frequency. We see from the loop shape that in order to achieve the desired performance we cannot simply increase the gain since this would give a very low phase margin. Instead, we must increase the phase at the desired crossover frequency.

To accomplish this, we use a lead compensator (12.7) with $a = 2$, $b = 50$, and $k = 200$. We then set the gain of the system to provide a large loop gain up to the desired bandwidth, as shown in Figure 12.12b. We see that this system has a gain of greater than 10 at all frequencies up to 10 rad/s and that it has more than 60° of phase margin. ∇

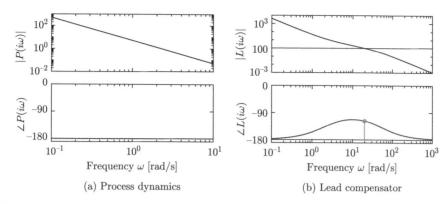

(a) Process dynamics (b) Lead compensator

Figure 12.12: Control design for a vectored thrust aircraft using lead compensation. The Bode plot for the open loop process P is shown in (a) and (b) shows the Bode plot for the loop transfer function $L = PC$, where C is the lead given by equation (12.7) with $a = 2$, $b = 50$, and $k = 200$. Note the phase lead in the crossover region near $\omega = 20$ rad/s.

12.4 FEEDFORWARD DESIGN

Feedforward is a simple and powerful technique that complements feedback. It can be used both to improve the response to reference signals and to reduce the effect of measurable disturbances. Design of feedforward for controllers based on state feedback and observers was developed in Section 8.5 (Figure 8.11). Section 11.5 presented setpoint weighting as simple form of feedforward for PID controllers (equation (11.15)). In this section we will use transfer functions to develop more advanced methods for feedforward design.

Combining Feedforward and Feedback

Figure 12.13 shows a block diagram of a system with feedback and feedforward control. The process dynamics are separated into two blocks $P_1(s)$ and $P_2(s)$, where the measured disturbance v enters at the input of the block P_2, and we define $P(s) = P_1(s)P_2(s)$. The transfer function F_m represents the desired (model) response to reference signals. There are two feedforward blocks with transfer functions F_r and F_v to deal with the reference signal r and the measured disturbances v.

A major advantage of controllers with two degrees of freedom that combine feedback and feedforward is that the control design problem can be split in two parts. The feedback transfer function C can be designed to give good robustness and effective disturbance attenuation, and the feedforward transfer functions F_r and F_v can be designed independently to give the desired responses to reference signals and to reduce effects of measured disturbances.

We will first explore the response to reference signals. The transfer function $G_{yr}(s)$ from reference input r to process output y in Figure 12.13 is

$$G_{yr}(s) = \frac{P(CF_m + F_r)}{1 + PC} = TF_m + SPF_r = F_m + S(PF_r - F_m) \qquad (12.9)$$

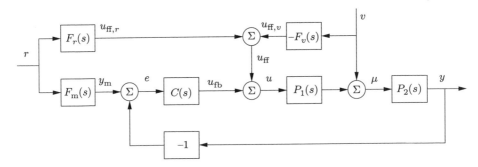

Figure 12.13: Block diagram of a system with feedforward compensation for improved response to reference signals and measured disturbances (2 degree-of-freedom system). Three feedforward elements are present: $F_m(s)$ sets the desired output value, $F_r(s)$ generates the feedforward command $u_{ff,r}$ to improve reference signal response and $F_v(s)$ generates the feedforward signal $u_{ff,v}$ that reduces the effect of the measured disturbance v.

where S is the sensitivity function and T the complementary sensitivity function (equation (12.2)) and we use the fact that $T = 1 - S$. We can make G_{yr} close to the desired transfer function F_m in two different ways: by choosing the feedforward transfer function F_r so that $PF_r - F_m$ is small, or by choosing the feedback transfer function C so that the sensitivity $S = 1/(1 + PC)$ is small. Perfect feedforward compensation is obtained by choosing

$$F_r = \frac{F_m}{P_1 P_2} = \frac{F_m}{P}, \tag{12.10}$$

which gives $G_{yr} = F_m$. Notice that the feedforward compensator F_r contains an inverse model of the process dynamics.

Next we will consider attenuation of disturbances that can be measured. The transfer function from load disturbance v to process output y is given by

$$G_{yv} = \frac{P_2(1 - P_1 F_v)}{1 + PC} = P_2 S(1 - P_1 F_v). \tag{12.11}$$

The transfer function G_{yv} can be made small in two different ways: by choosing the feedforward transfer function F_v so that $1 - P_1 F_v$ is small, or by choosing the feedback transfer function C so that the sensitivity $S = 1/(1 + PC)$ is small. Perfect compensation is obtained by choosing

$$F_v = \frac{1}{P_1}. \tag{12.12}$$

Design of feedforward to improve responses to reference signals and disturbances using transfer functions is thus a simple task, but it requires inversion of process models. We illustrate with an example.

Example 12.6 Vehicle steering

A linearized model for vehicle steering was given in Example 7.4. The normalized transfer function from steering angle δ to lateral deviation y is $P(s) = (\gamma s + 1)/s^2$. For a lane transfer system we would like to have a nice response without overshoot,

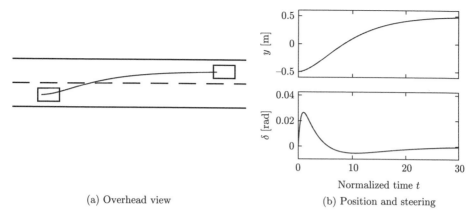

(a) Overhead view   (b) Position and steering

Figure 12.14: Feedforward control for vehicle steering. The plot in (a) shows the trajectory generated by the controller for changing lanes. The plots in (b) show the lateral deviation y (top) and the steering angle δ (bottom) for a smooth lane change control using feedforward (based on the linearized model).

and we therefore choose the desired response as $F_m(s) = \omega_c^2/(s+\omega_c)^2$, where the response speed or aggressiveness of the steering is governed by the parameter ω_c. Equation (12.10) gives

$$F_r = \frac{F_m}{P} = \frac{\omega_c^2 s^2}{(\gamma s+1)(s+\omega_c)^2},$$

which is a stable transfer function as long as $\gamma > 0$. Figure 12.14 shows the responses of the system for $\omega_c = 0.2$.

The figure shows that a lane change is accomplished in about 20 vehicle lengths with smooth steering angles. The largest steering angle is slightly larger than 0.2 rad (12°). Using the scaled variables, the curve showing lateral deviations (y as a function of t) can also be interpreted as the vehicle path (y as a function of x) with the vehicle length as the length unit.   ▽

Difficulties with Feedforward

The ideal feedforward compensators for Figure 12.13 are given by

$$F_r = \frac{F_m}{P_1 P_2}, \qquad F_v = \frac{1}{P_1}. \tag{12.13}$$

Both transfer functions require inversion of process transfer functions and there can be problems with inversion if the process transfer function has time delays, right half-plane zeros, or high pole excess. Inversion of time delays requires prediction, which cannot be done perfectly except in the situation when the command signal is known in advance. If the process transfer function has zeros in the right half-plane, the inverse process transfer function is unstable and approximate inverses may have to be used. Finally, if the pole excess of the process transfer function is greater than zero, then the inverse requires differentiation. In this case the reference signal must then be sufficiently smooth and there may also be problems with noise.

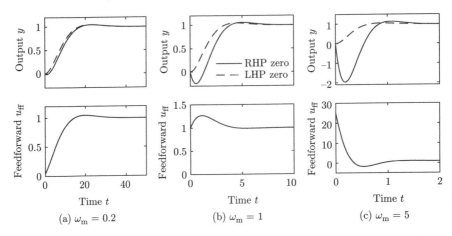

(a) $\omega_m = 0.2$ (b) $\omega_m = 1$ (c) $\omega_m = 5$

Figure 12.15: Feedforward control for a process with a right half-plane zero (Example 12.7). Outputs y (upper plots) and feedforward signals u_{ff} (lower plots) for a unit step command signal. The design parameter has the values $\omega_m = 0.2$, 1, and 0.5 for a unit step command in the reference signal. The dashed curve shows the response that could be achieved if the process did not have the right half-plane zero.

There is some extra freedom when finding the transfer function F_r because it also contains the transfer function F_m, which specifies the ideal behavior. A stable feedforward transfer function can be obtained if F_m has the same time delays and right half-plane zeros as the process. We illustrate with an example.

Example 12.7 Feedforward for a process with a right half-plane zero
Let the process and the desired response have the transfer functions

$$P(s) = \frac{1-s}{(s+1)^2}, \qquad F_m(s) = \frac{\omega_m^2(1-s)}{s^2 + 2\zeta_c\omega_m s + \omega_m^2}.$$

Since the process has a right half-plane zero at $s = 1$, the desired transfer function $F_m(s)$ must have the same zeros to avoid having an unstable feedforward transfer function F_r. Equation (12.10) gives the feedforward transfer function:

$$F_r(s) = \frac{\omega_m^2(s+1)^2}{s^2 + 2\zeta_c\omega_m s + \omega_m^2}. \tag{12.14}$$

Figure 12.15 shows the outputs y and the feedforward signals u_{ff} for different values of ω_m. The response to the reference signal goes in the wrong direction initially because of the right half-plane zero at $s = 1$. This effect, called *inverse response*, is barely noticeable if the response is slow ($\omega_m = 1$) but it increases with increasing response speed. For $\omega_m = 5$ the undershoot is more than 200%. The large undershoot is an indication that a right half-plane zero limits the achievable bandwidth, as will be discussed in depth in Chapter 14. A reasonable choice of ω_m is in the range 0.2 to 0.5. Notice that the same feedforward transfer function (12.14) is obtained if the process and the desired model have the transfer functions

$$P(s) = \frac{1}{(s+1)^2}, \qquad F_{\mathrm{m}}(s) = \frac{\omega_{\mathrm{m}}^2}{s^2 + 2\zeta_{\mathrm{c}}\omega_{\mathrm{m}}s + \omega_{\mathrm{m}}^2}.$$

The corresponding responses are shown as dashed lines in Figure 12.15. When there is no right half-plane zero it is thus possible to obtain well-behaved, fast responses.

The control signals for different values of ω_{m} differ significantly, as shown in the bottom row of plots in Figure 12.15. Since $r = 1$ and the zero frequency gain of the feedforward transfer function is $F_{\mathrm{r}}(0) = 1$, the control signal goes to 1 as time goes to infinity in all cases. The feedforward transfer function also has constant gain $F_{\mathrm{r}}(\infty) = \omega_{\mathrm{m}}^2$ for high frequencies, which means that gain for high-frequency signals is ω_{m}^2 and this can be undesirable if ω_{m} is large. The initial response to a unit step signal is then $u_{\mathrm{ff}}(0) = F_{\mathrm{r}}(\infty) = \omega_{\mathrm{m}}^2$ (using the initial value theorem). For $\omega_{\mathrm{m}} = 0.2$ the control signal grows from 0.04 to the final value 1 with a small overshoot. For $\omega_{\mathrm{m}} = 1$ the control signal starts from 1, has an overshoot, and then settles on the final value 1. For $\omega_{\mathrm{m}} = 5$ the control signal starts at 25 and decays towards the final value 1 with an undershoot. $\qquad\qquad\qquad\qquad\qquad\qquad \nabla$

Approximate Inverses

Processes with right half-plane zeros do not have stable inverses. To design feedforward compensators for such systems we need to use approximate inverses that are stable. The following theorem, which is presented without proof, provides a means of constructing such approximate inverses.

Theorem 12.1 (Approximate inverse). *Let the rational transfer function $G(s)$ have all its poles in the left half-plane and no zeros on the imaginary axis. Factor the transfer function as $G(s) = G^+(s)G^-(s)$, where $G^+(s)$ has all its zeros in the left half-plane and $G^-(s)$ has all its zeros in the right half-plane. An approximate stable inverse of $G(s)$ that minimizes the mean square error for a step input is*

$$G^\dagger(s) = \frac{1}{G^+(s)G^-(-s)}. \tag{12.15}$$

We illustrate the theorem with an example.

Example 12.8 Approximate inverse for a system with a right half-plane zero

Let the transfer functions of the process and the reference model (desired response) be

$$P(s) = \frac{1-s}{(s+1)^2}, \qquad F_{\mathrm{m}}(s) = \frac{\omega_m^2}{s^2 + 2\zeta_{\mathrm{c}}\omega_m s + \omega_m^2}.$$

Note that in comparison to Example 12.7, we do not include the right half-plane zero in F_{m}. The process transfer function can be factored as

$$P^-(s) = 1 - s, \quad P^+(s) = \frac{1}{(s+1)^2}.$$

Theorem 12.1 then gives the following approximate inverse:

$$P^\dagger(s) = \frac{1}{P^+(s)P^-(-s)} = \frac{(s+1)^2}{1+s} = s+1.$$

The feedforward transfer function is then

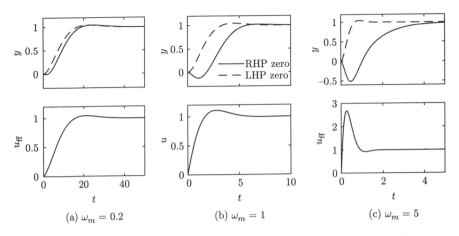

(a) $\omega_m = 0.2$ (b) $\omega_m = 1$ (c) $\omega_m = 5$

Figure 12.16: Feedforward design based on an approximate inverse. Outputs y (upper plots) and feedforward signals u_{ff} (lower plots) for a unit step reference signal. The design parameter has the values $\omega_m = 0.2$, 1, and 0.5 for a unit step command in the reference signal. The dashed curves show the responses for a process without the right half-plane zero.

$$F_r(s) = F_{\mathrm{m}}(s)P^{\dagger}(s) = \frac{\omega_m^2(s+1)}{s^2 + 2\zeta_{\mathrm{c}}\omega_m s + \omega_m^2},$$

which is similar to equation (12.14) but no longer relies on cancellation of the right half-plane zero to obtain a stable feedforward transfer function. The transfer function from reference r to output y is then

$$G_{yr}(s) = P(s)F_r(s) = \frac{1-s}{(s^2 + 2\zeta_{\mathrm{c}}\omega_m s + \omega_m^2)(s+1)}.$$

Figure 12.16 shows the step responses for different values of ω_m.

Comparing Figures 12.15 and 12.16 we find that there are small differences for $\omega_m = 0.2$, but large differences for $\omega_m = 5$. Notice in particular the shapes of the feedforward signals u_{ff}. The design based on the approximate inverse has smaller undershoot but the time responses have somewhat longer settling times. ∇

In summary, we see that feedforward can be used to improve the response to reference signals and to reduce the effects of load disturbances that can be measured. There are limits if the process has time delays, right half-plane zeros, or high pole excess. The zeros depend on the sensors and we can change them by moving or adding sensors. In addition, we will see in Chapter 13 that feedforward controllers can be sensitive to model uncertainty (Section 13.3 and Exercise 13.5), and hence feedforward control is usually combined with feedback control to obtain robust performance.

12.5 THE ROOT LOCUS METHOD

In design methods such as eigenvalue assignment, discussed in Sections 7.2 and 8.3, we designed controllers that give desired closed loop poles. The controllers were sufficiently complex so that all closed loop poles could be specified. The complexity of the controller is thus directly related to the complexity of the process. In practice we may have to use a simple controller for a complex process, and it is then not possible to find a controller that gives all closed poles their desired values. It is interesting to explore what can be done with a controller having restricted complexity as was the case for PID control in Chapter 11 and loop shaping in Section 12.3. The simplest case with only one selectable controller parameter can be investigated with the *root locus method*. The *root locus* is a graph of the roots of the characteristic polynomial as a function of a parameter, and the method gives insight into the effects of the controller parameter. It is straightforward to obtain the root locus by finding the roots of the closed loop characteristic polynomial for different values of the parameter. There are also good computer tools for generating the root locus. Of greater interest is the fact that the general shape of the root locus can be obtained with very little effort, and that it often gives considerable insight.

To illustrate the root locus method we consider a process with the transfer function

$$P(s) = \frac{b(s)}{a(s)} = \frac{b_0 s^m + b_1 s^{m-1} + \cdots b_m}{s^n + a_1 s^{n-1} + \cdots a_n} = b_0 \frac{(s - z_1)(s - z_2) \ldots (s - z_m)}{(s - p_1)(s - p_2) \cdots (s - p_n)}.$$

The polynomial $a(s)$ has degree n and the polynomial $b(s)$ has degree m. We assume that pole excess $n_{\text{pe}} = n - m$ is positive or zero. The controller is assumed to be a proportional controller with the transfer function $C(s) = k$. We will explore the poles of the closed loop system when the gain k of the proportional controller ranges from 0 to ∞.

The closed loop characteristic polynomial is

$$a_{\text{cl}}(s) = a(s) + kb(s) \tag{12.16}$$

and the closed loop poles are the roots of $a_{\text{cl}}(s)$. The root locus is a graph of the roots of $a_{\text{cl}}(s)$ as the gain k is varied from 0 to ∞. Since the polynomial $a_{\text{cl}}(s)$ has degree n, the plot will have n branches.

When the gain k is zero we have $a_{\text{cl}}(s) = a(s)$ and the closed loop poles are equal to the open loop poles. When there are open loop poles at $s = p_l$ with multiplicity m, the characteristic equation can be written as

$$(s - p_l)^m \tilde{a}(s) + kb(s) \approx (s - p_l)^m \tilde{a}(p_l) + kb(p_l) = 0,$$

where $\tilde{a}(s)$ represents the polynomial $a(s)$ with the poles at $s = p_l$ factored out. For small values of k the roots of this equation are given by $s = p_l + \sqrt[m]{-kb(p_l)/\tilde{a}(p_l)}$. The root locus thus has a star pattern with m branches emanating from the open loop pole $s = p_l$. The angle between two neighboring branches is $2\pi/m$.

To explore what happens for large gain we approximate the characteristic polynomial (12.16) for large s and k, which gives

$$a_{\text{cl}}(s) = b(s) \left(\frac{a(s)}{b(s)} + k \right) \approx b(s) \left(\frac{s^{n_{\text{pe}}}}{b_0} + k \right). \tag{12.17}$$

For large k the closed loop poles are approximately the roots of $b(s)$ and $\sqrt[n_{\text{pe}}]{-b_0 k}$. A better approximation of the roots of equation (12.17) is

$$s = s_0 + \sqrt[n_{\text{pe}}]{-kb_0}, \qquad s_0 = \frac{1}{n_{\text{pe}}} \left(\sum_{k=1}^{n} p_k - \sum_{k=1}^{m} z_k \right) \tag{12.18}$$

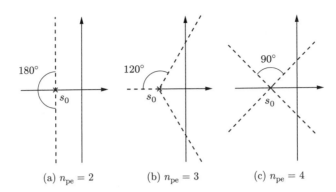

(a) $n_{\mathrm{pe}} = 2$ (b) $n_{\mathrm{pe}} = 3$ (c) $n_{\mathrm{pe}} = 4$

Figure 12.17: Asymptotes of the root locus for systems with pole excess $n_{\mathrm{pe}} = 2$, 3, and 4. There are n_{pe} asymptotes that radiate from the point $s = s_0$ given by equation (12.18), and the angles between the asymptotes are $360°/n_{\mathrm{pe}}$.

(Exercise 12.15). The asymptotes are thus n_{pe} lines that radiate from $s = s_0$, the center of mass of poles and zeros. When $b_0 k > 0$ the lines have the angles $(\pi + 2l\pi)/n_{\mathrm{pe}}$, $l = 1, \ldots, n_{\mathrm{pe}}$ with respect to the real line. Figure 12.17 shows the asymptotes of the root locus for large gain for different values of the pole excess n_{pe}.

Summarizing, we find that the root locus plot with the loop gain as the varied parameter has n branches that start at the open loop poles and end either at the open loop zeros or at infinity. The branches that end at infinity have star-patterned asymptotes given by equation (12.18). An immediate consequence is that open loop systems with right half-plane zeros or a pole excess larger than 2 will always be unstable for sufficiently large gains.

There are simple rules for sketching the root locus. We describe here a few of them. As discussed already, the root locus has a (locally) symmetric star pattern at points where there are multiple roots; the number of branches depends on the multiplicity of the roots. For systems with $kb_0 > 0$ the root locus has segments on the real line where there are odd numbers of real poles and zeros to the right of the segment (Exercise 12.16). It is also straightforward to find directions where a branch of the root locus leaves a pole, as discussed in Exercise 12.19.

Figure 12.18 shows root loci for systems with $k > 0$ and the transfer functions

$$P_{\mathrm{a}}(s) = k\,\frac{s+1}{s^2}, \qquad P_{\mathrm{b}}(s) = k\,\frac{s+1}{s(s+2)(s^2+2s+4)},$$

$$P_{\mathrm{c}}(s) = k\,\frac{s+1}{s(s^2+1)}, \qquad P_{\mathrm{d}}(s) = k\,\frac{s^2+2s+2}{s(s^2+1)}. \tag{12.19}$$

The locus of $P_{\mathrm{a}}(s)$ in Figure 12.18a starts with two roots at the origin and the pattern locally has the star configuration with $m = 2$. As the gain increases the locus bends because of the attraction of the zero. In this particular case the locus is actually a circle around the zero $s = -1$. Two roots meet at the real axis and depart forming a star pattern. One root goes towards the zero and the other one goes to infinity along the negative real axis as the gain k increases. The root locus thus has the segment $(-\infty, -1]$ on the real axis. The locus in Figure 12.18b starts at the open loop poles $s = -2$, 0, and $-1 \pm i\sqrt{3}$. The pole excess is $n_{\mathrm{pe}} = 3$ and the asymptotes that originate from $s_0 = -1$ have the corresponding pattern. The locus in Figure 12.18c has vertical asymptotes since $n_{\mathrm{pe}} = 2$ (see Figure 12.17). The

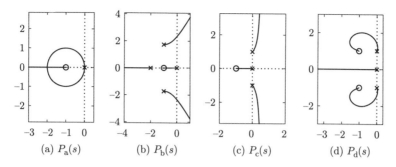

Figure 12.18: Examples of root loci for processes with the transfer functions $P_a(s)$, $P_b(s)$, $P_c(s)$, and $P_d(s)$ given by equation (12.19).

asymptotes originate from $s_0 = 0.5$. The root locus has the segment $[-1, 0]$ on the real line. The locus in Figure 12.18d has three branches: one is the segment $(-\infty, 0]$ on the real line and the other two originate on the complex open loop poles and end at the open loop zeros.

The root locus can also be used for design. Consider for example the system in Figure 12.18c, which can represent PI control of a system with the transfer functions

$$P(s) = \frac{1}{s^2 + 1}, \qquad C(s) = k \frac{s + 2}{s}.$$

The root locus in Figure 12.18c shows that the system is unstable for all values of the controller gain and we can immediately conclude that the process cannot be stabilized with a PI controller. To obtain a stable closed loop system we can attempt to choose a PID controller with zeros to the left of the undamped poles, for example

$$C(s) = k \frac{s^2 + 2s + 2}{s}.$$

The root locus obtained with this controller is shown in Figure 12.18d. We see that this system is stable for $k > 0$ and we can choose k to place the poles in reasonable locations.

We have illustrated the root locus with a closed loop system with a proportional controller where the parameter is the gain. The root locus can also be used to find the effects of other parameters, as was illustrated in Example 5.17.

12.6 DESIGN EXAMPLE

In this final section we present a detailed example that illustrates some of the design techniques described in this chapter.

Example 12.9 Lateral control of a vectored thrust aircraft
The problem of controlling the motion of a vertical takeoff and landing (VTOL) aircraft was introduced in Example 3.12 and in Example 12.5, where we designed a controller for the roll dynamics. We now wish to control the position of the aircraft, a problem that requires stabilization of the attitude.

To control the lateral dynamics of the vectored thrust aircraft, we make use of an "inner/outer" loop design methodology, as illustrated in Figure 12.19. This diagram shows the process dynamics and controller divided into two components: an *inner*

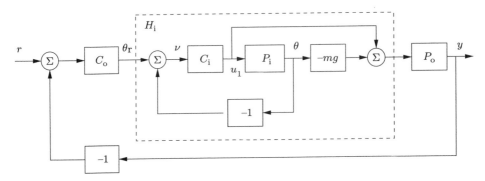

Figure 12.19: Inner/outer control design for a vectored thrust aircraft. The inner loop H_i controls the roll angle of the aircraft using the vectored thrust. The outer loop controller C_o commands the roll angle to regulate the lateral position. The process dynamics are decomposed into inner loop (P_i) and outer loop (P_o) dynamics, which combine to form the full dynamics for the aircraft.

loop consisting of the roll dynamics and controller and an *outer loop* consisting of the lateral position dynamics and controller. This decomposition follows the block diagram representation of the dynamics given in Exercise 9.11.

The approach that we take is to design a controller C_i for the inner loop so that the resulting closed loop system H_i assures that the roll angle θ follows its reference θ_r quickly and accurately. We then design a controller for the lateral position y that uses the approximation that we can directly control the roll angle as an input θ to the dynamics controlling the position. Under the assumption that the dynamics of the roll controller are fast relative to the desired bandwidth of the lateral position control, we can then combine the inner and outer loop controllers to get a single controller for the entire system. As a performance specification for the entire system, we would like to have zero steady-state error in the lateral position, a bandwidth of approximately 1 rad/s, and a phase margin of 45°.

For the inner loop, we choose our design specification to provide the outer loop with accurate and fast control of the roll. The inner loop dynamics are given by

$$P_i(s) = H_{\theta u_1}(s) = \frac{r}{Js^2}.$$

We choose the desired bandwidth to be 10 rad/s (10 times that of the outer loop) and the low-frequency error to be no more than 5%. This specification is satisfied using the lead compensator of Example 12.5 designed previously, so we choose

$$C_i(s) = k\frac{s+a}{s+b}, \qquad a = 2, \quad b = 50, \quad k = 200.$$

The closed loop dynamics for the system satisfy

$$H_i = \frac{C_i}{1 + C_i P_i} - mg\frac{C_i P_i}{1 + C_i P_i} = \frac{C_i(1 - mgP_i)}{1 + C_i P_i}.$$

A plot of the magnitude of this transfer function is shown in Figure 12.20b, and we see that $H_i \approx -mg = -39.2$ is a good approximation up to 10 rad/s.

To design the outer loop controller, we assume the inner loop roll control is perfect, so that we can take θ_r as the input to our lateral dynamics. Following the

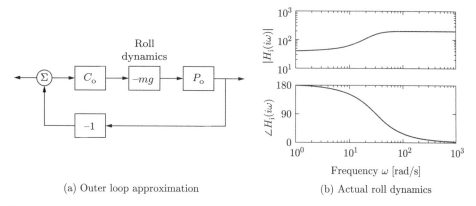

(a) Outer loop approximation (b) Actual roll dynamics

Figure 12.20: Outer loop control design for a vectored thrust aircraft. (a) The outer loop approximates the roll dynamics as a state gain $-mg$. (b) The Bode plot for the roll dynamics, indicating that this approximation is accurate up to approximately 10 rad/s.

diagram shown in Exercise 9.11, the outer loop dynamics can be written as

$$P(s) = H_i(0)P_o(s) = \frac{H_i(0)}{ms^2 + cs},$$

where we replace $H_i(s)$ with $H_i(0)$ to reflect our approximation that the inner loop will eventually track our commanded input. Of course, this approximation may not be valid, and so we must verify this when we complete our design.

Our control goal is now to design a controller that gives zero steady-state error in y for a step input and has a bandwidth of 1 rad/s. The outer loop process dynamics are given by a double integrator, and we can again use a simple lead compensator to satisfy the specifications. We also choose the design such that the loop transfer function for the outer loop has $|L_o| < 0.1$ for $\omega > 10$ rad/s, so that the H_i high-frequency dynamics can be neglected. We choose the controller to be of the form

$$C_o(s) = -k_o \frac{s + a_o}{s + b_o},$$

with the negative sign to cancel the negative sign in the process dynamics. To find the location of the poles, we note that the phase lead flattens out at approximately $b_o/10$. We desire phase lead at crossover, and we desire the crossover at $\omega_{gc} = 1$ rad/s, so this gives $b_o = 10$. To ensure that we have adequate phase lead, we must choose a_o such that $b_o/10 < 10a_o < b_o$, which implies that a_o should be between 0.1 and 1. We choose $a_o = 0.3$. Finally, we need to set the gain of the system such that at the desired crossover frequency the loop gain has magnitude 1 or more. A simple calculation shows that $k_o = 2$ satisfies this objective. Thus, the final outer loop controller becomes

$$C_o(s) = -2\frac{s + 0.3}{s + 10}.$$

Finally, we can combine the inner and outer loop controllers and verify that the system has the desired closed loop performance. The Bode and Nyquist plots

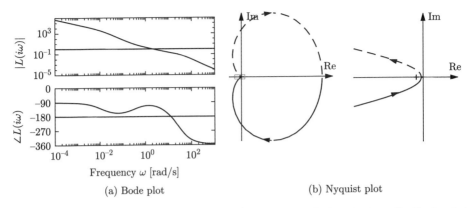

Figure 12.21: Inner/outer loop controller for a vectored thrust aircraft. Bode plot (a) and Nyquist plot (b) for the loop transfer function cut at θ_r, for the complete system. The system has a phase margin of 68° and a gain margin of 6.2.

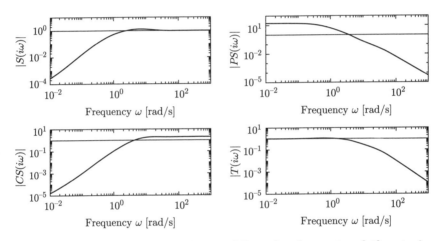

Figure 12.22: Gain curves for the Gang of Four for the vectored thrust aircraft system.

corresponding to Figure 12.19 with inner and outer loop controllers are shown in Figure 12.21, and we see that the specifications are satisfied. In addition, we show the gain curves of the Gang of Four in Figure 12.22, and we see that the transfer functions between all inputs and outputs are reasonable. The sensitivity to load disturbances PS is large at low frequency because the controller does not have integral action.

The approach of splitting the dynamics into an inner and an outer loop is common in many control applications and can lead to simpler designs for complex systems. Indeed, for the aircraft dynamics studied in this example, it is very challenging to directly design a controller from the lateral position y to the input u_1. The use of the additional measurement of θ greatly simplifies the design because it can be broken up into simpler pieces. ∇

12.7 FURTHER READING

Loop shaping design emerged at Bell Labs in connection with the development of Black's [45] electronic amplifier with negative feedback. Nyquist [192] derived his stability criterion to understand and avoid instabilities or "singing," as it was called at the time. Bode [50] used the theory of complex variables to establish important fundamental results such as the relation between amplitude and phase for a minimum phase system, the ideal loop transfer functions, and the phase area formula. His results are nicely summarized in the book [51]. Design by loop shaping became a key element in the early development of control, and many design methods were developed; see James, Nichols, and Phillips [130], Chestnut and Mayer [62], Truxal [241], and Thaler [238]. Loop shaping is also treated in standard textbooks such as Franklin, Powell, and Emami-Naeini [93], Dorf and Bishop [73], Kuo and Golnaraghi [157], and Ogata [195]. Horowitz [121] developed the notion of systems with two degrees of freedom. Much of the early work was based on the loop transfer function; the importance of the sensitivity functions appeared in connection with developments in the 1980s that resulted in H_∞ design methods. A compact presentation is given in the texts by Doyle, Francis, and Tannenbaum [76] and Zhou, Doyle, and Glover [265]. Loop shaping was integrated with robust control theory in McFarlane and Glover [181] and Vinnicombe [248]. Comprehensive treatments of control system design are given in Maciejowski [172] and Goodwin, Graebe, and Salgado [107]. There are fundamental limits to what can be achieved given by nonlinearities of the process and the poles and zeros. These will be discussed in Chapter 14.

EXERCISES

12.1 Consider the system in Example 12.1, where the process and controller transfer functions are given by

$$P(s) = 1/(s-a), \qquad C(s) = k(s-a)/s.$$

Choose the parameter $a = -1$ and compute the time (step) and frequency responses for all the transfer functions in the Gang of Four for controllers with $k = 0.2$ and $k = 5$.

12.2 (Equivalence of Figures 12.1 and 12.2) Consider the system in Figure 12.1 and let the outputs of interest be $\xi = (\mu, \eta)$ and the major disturbances be $\chi = (w, v)$. Show that the system can be represented by Figure 12.2 and give the matrix transfer functions $\mathcal{P}$ and $\mathcal{C}$. Verify that the elements of the closed loop transfer function $H_{\xi\chi}$ are the Gang of Four.

12.3 (Equivalence of controllers with two degrees of freedom) Show that the systems in Figures 12.1 and 12.13 give the same responses to command signals if $F_{\mathrm{m}}C + F_u = CF$.

12.4 (Web server control) Feedback and feedforward are increasingly used for complex computer systems such as web servers. Control of a single server is an example. A model for a virtual server is given by equation (3.32),

$$\frac{dx}{dt} = \lambda - \mu,$$

where x is the queue length, λ is the arrival rate, and μ is the server rate. The objective of control is to maintain a given queue length. The service rate μ can be changed by dynamic voltage and frequency scaling (DVFS). Determine a PI controller that gives a closed loop system with the characteristic polynomial $s^2 + 4s + 4$. Use feedforward in the form of setpoint weighting to reduce the overshoot for step changes in reference signals; simulate the closed loop system to determine the setpoint weighting.

 12.5 (Rise time-bandwidth product) Consider a stable system with the transfer function $G(s)$ whose frequency response is an ideal low-pass filter with $|G(i\omega)| = 1$ for $\omega \leq \omega_b$ and $|G(i\omega)| = 0$ for $\omega > \omega_b$ and which has low-pass character. Define the rise time T_r as the inverse of the largest slope of the unit step response and the bandwidth as $\widetilde{\omega}_b = \int_0^\infty |G(i\omega)|/G(0)\, d\omega$. Show that with this definition of the bandwidth the *rise time-bandwidth product* satisfies $T_r \widetilde{\omega}_b \geq \pi$.

12.6 (Disturbance attenuation) Consider the feedback system shown in Figure 12.1. Assume that the reference signal is constant. Let y_{ol} be the measured output when there is no feedback and y_{cl} be the output with feedback. Show that $y_{cl} = S(s)y_{ol}$, where y_{cl} and y_{ol} are exponential signals and S is the sensitivity function.

12.7 (Approximate expression for noise sensitivity) Show that the effect of high-frequency measurement noise on the control signal for the system in Example 12.3 can be approximated by

$$CS \approx C = \frac{k_d s}{(sT_f)^2/2 + sT_f + 1},$$

and that the largest value of $|CS(i\omega)|$ is k_d/T_f which occurs for $\omega = \sqrt{2}/T_f$.

12.8 (Peak frequency-peak time product) Consider the transfer function for a second-order system

$$G(s) = \frac{\omega_0 s}{s^2 + 2\zeta\omega_0 s + \omega_0^2},$$

which has the unit step response

$$y(t) = \frac{1}{\sqrt{1-\zeta^2}} e^{-\zeta\omega_0 t} \sin \omega_0 t \sqrt{1-\zeta^2}.$$

Let $M_r = \max_\omega |G(i\omega)|$ be the largest gain of $G(s)$, which is assumed to occur at ω_{mr}, and let $y_p = \max_t y(t)$ be the largest value of $y(t)$, which is assumed to occur at t_p. Show that

$$t_p \omega_{mr} = \frac{\arccos \zeta}{\sqrt{1-\zeta^2}}, \qquad \frac{y_p}{M_r} = 2\zeta e^{-\zeta\varphi},$$

and evaluate the right-hand sides of the above equations for $\zeta = 0.5, 0.707$, and 1.0.

12.9 (Disturbance reduction through feedback) Consider a problem in which an output variable has been measured to estimate the potential for disturbance attenuation by feedback. Suppose an analysis shows that it is possible to design a closed loop system with the sensitivity function

$$S(s) = \frac{s}{s^2 + s + 1}.$$

Estimate the possible disturbance reduction when the measured disturbance response is

$$y(t) = 5 \sin (0.1\, t) + 3 \sin (0.17\, t) + 0.5 \cos (0.9\, t) + 0.1\, t.$$

12.10 (Bode's formula) Consider the lead compensator

$$G(s) = 16 \frac{s + 0.25}{s + 4}.$$

Verify Bode's phase area formula (12.8) and show that $G(\infty) = 16G(0)$ by numerical integration.

12.11 (Lead-lag compensation) Lead and lag compensators can be combined into a lead-lag compensator that has the transfer function

$$C(s) = k \frac{(s + a_1)(s + a_2)}{(s + b_1)(s + b_2)}.$$

Show that the controller reduces to a PID controller with special choice of parameters and give the relations between the parameters.

12.12 (Attenuation of low-frequency sinusoidal disturbances) Integral action eliminates constant disturbances and reduces low-frequency disturbances because the controller gain is infinite at zero frequency. A similar idea can be used to reduce the effects of sinusoidal disturbances of known frequency ω_0 by using the controller

$$C(s) = k_\mathrm{p} + \frac{k_s s}{s^2 + 2\zeta\omega_0 s + \omega_0^2}.$$

This controller has the gain $C_s(i\omega_0) = k_\mathrm{p} + k_s/(2\zeta)$ for the frequency ω_0, which can be large by choosing a small value of ζ. Assume that the process has the transfer function $P(s) = 1/s$. Determine the Bode plot of the loop transfer function and simulate the system. Compare the results with PI control.

12.13 (Performance specifications and transfer functions) Find the transfer function of a second-order system that satisfies the following closed loop specifications: zero steady-state error, 2% settling time less than 2 s, rise time less than 0.8 s, and overshoot less than 3%.

12.14 Consider the spring–mass system given by equation (3.16), which has the transfer function

$$P(s) = \frac{1}{ms^2 + cs + k}.$$

Design a feedforward compensator that gives a response with critical damping ($\zeta = 1$).

12.15 (Asymptotes of root locus) Consider proportional control of a system with the transfer function

$$P(s) = \frac{b(s)}{a(s)} = \frac{b_0 s^m + b_1 s^{m-1} + \cdots b_m}{s^n + a_1 s^{n-1} + \cdots a_n} = b_0 \frac{(s - z_1)(s - z_2) \ldots (s - z_m)}{(s - p_1)(s - p_2) \cdots (s - p_n)}.$$

Show that the root locus has asymptotes that are straight lines that emerge from the point

$$s_0 = \frac{1}{n_e}\left(\sum_{k=1}^{n} p_k - \sum_{k=1}^{m} z_k\right),$$

where $n_e = n - m$ is the pole excess of the transfer function.

12.16 (Real line segments of root locus) Consider proportional control of a process with a rational transfer function. Assuming that $b_0 k > 0$, show that the root locus has segments on the real line where there are an odd number of real poles and zeros to the right of the segment.

12.17 Consider a lead compensator with the transfer function

$$C_n(s) = \left(\frac{s\sqrt[n]{k}+a}{s+a}\right)^n,$$

which has zero frequency gain $C(0) = 1$ and high-frequency gain $C(\infty) = k$. Show that the gain required to provide a given phase lead φ is

$$k = \left(1 + 2\tan^2(\varphi/n) + 2\tan(\varphi/n)\sqrt{1+\tan^2(\varphi/n)}\right)^n,$$

and that $\lim_{n\to\infty} k = e^{2\varphi}$.

12.18 (Phase margin formulas) Show that the relationship between the phase margin and the values of the sensitivity functions at gain crossover is given by

$$|S(i\omega_{gc})| = |T(i\omega_{gc})| = \frac{1}{2\sin(\varphi_m/2)}.$$

12.19 (Initial direction of root locus) Consider proportional control of a system with the transfer function

$$P(s) = \frac{b(s)}{a(s)} = \frac{b_0 s^m + b_1 s^{m-1} + \cdots b_m}{s^n + a_1 s^{n-1} + \cdots a_n} = b_0 \frac{(s-z_1)(s-z_2)\ldots(s-z_m)}{(s-p_1)(s-p_2)\cdots(s-p_n)}.$$

Let p_j be an isolated pole and assume that $k b_0 > 0$. Show that the root locus starting at p_j has the initial direction.

$$\angle(s-p_j) = \pi + \Sigma_{k=1}^{m}\angle(p_j - s_k) - \Sigma_{k\neq j}\angle(p_j - p_k).$$

Give a geometric interpretation of the result.

Chapter Thirteen

Robust Performance

> However, by building an amplifier whose gain is deliberately made, say 40 decibels higher than necessary (10000 fold excess on energy basis), and then feeding the output back on the input in such a way as to throw away the excess gain, it has been found possible to effect extraordinary improvement in constancy of amplification and freedom from non-linearity.
>
> —Harold S. Black, "Stabilized Feedback Amplifiers," 1934 [45].

This chapter focuses on the analysis of robustness of feedback systems, a vast topic for which we provide only an introduction to some of the key concepts. We consider the stability and performance of systems whose process dynamics are uncertain. We make use of generalizations of Nyquist's stability criterion as a mechanism to characterize *robust* stability and performance. To do this we develop ways to describe uncertainty, both in the form of parameter variations and in the form of neglected dynamics. We also briefly mention some methods for designing controllers to achieve robust performance.

13.1 MODELING UNCERTAINTY

Harold Black's quote illustrates that one of the key uses of feedback is to provide robustness to uncertainty ("constancy of amplification"). It is one of the most useful properties of feedback and is what makes it possible to design feedback systems based on strongly simplified models. In this section we explore different types of uncertainty in our knowledge of the dynamics of the system, including the important problem of determining when two systems are similar from a controls perspective.

Parametric Uncertainty

One form of uncertainty in dynamical systems is *parametric uncertainty* in which the parameters describing the system are not precisely known. A typical example is the variation of the mass of a car, which changes with the number of passengers and the weight of the baggage. When linearizing a nonlinear system, the parameters of the linearized model also depend on the operating conditions. It is straightforward to investigate the effects of parametric uncertainty simply by evaluating the performance criteria for a range of parameters. Such a calculation reveals the consequences of parameter variations. We illustrate by an example.

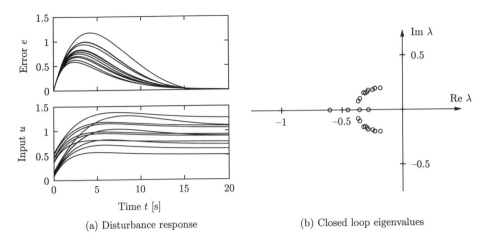

(a) Disturbance response   (b) Closed loop eigenvalues

Figure 13.1: Responses of the cruise control system to a slope increase of 4° (a) and the eigenvalues of the closed loop system (b). Model parameters are swept over a wide range. The closed loop system is of second order.

Example 13.1 Cruise control

The cruise control problem is described in Section 4.1, and a PI controller was designed in Example 11.3. To investigate the effect of parameter variations, we will choose a controller designed for a nominal operating condition corresponding to mass $m = 1600$ kg, fourth gear ($\alpha = 12$), and speed $v_e = 20$ m/s; the controller gains are $k_p = 0.5$ and $k_i = 0.1$. Figure 13.1a shows the velocity error e and the throttle u when encountering a hill with a 4° slope with masses in the range $1600 < m < 2000$ kg, gear ratios 3–5 ($\alpha = 10$, 12, and 16), and velocity $10 \leq v \leq 40$ m/s. The simulations were done using models that were linearized around the different operating conditions. The figure shows that there are variations in the response but that they are all quite reasonable. The largest velocity error is in the range of 0.5–1.2 m/s, and the settling time is about 15 s. The control signal is larger than 1 in some cases, which implies that the throttle is fully open. (A full nonlinear simulation using a controller with windup protection is required if we want to explore these cases in more detail.) The closed loop system has two eigenvalues, shown in Figure 13.1b for the different operating conditions. We see that the closed loop system is well damped in all cases.   ∇

This example indicates that at least as far as parametric variations are concerned, a design based on a simple nominal model will give satisfactory control. The example also indicates that a controller with fixed parameters can be used in all cases. Notice that we have not considered operating conditions in low gear and at low speed, but cruise controllers are not typically used in these cases.

Unmodeled Dynamics

It is generally easy to investigate the effects of parametric variations. However, there are other uncertainties that also are important, as discussed at the end of Section 3.1. The simple model of the cruise control system captures only the dynamics of the forward motion of the vehicle and the torque characteristics of the engine and transmission. It does not, for example, include a detailed model of the engine

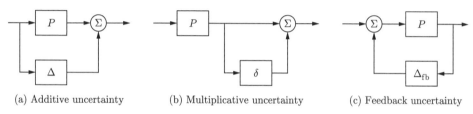

(a) Additive uncertainty   (b) Multiplicative uncertainty   (c) Feedback uncertainty

Figure 13.2: Unmodeled dynamics in linear systems. Uncertainty can be represented using additive perturbations (a), multiplicative perturbations (b), or feedback perturbations (c). The nominal system is P, and Δ, δ, and Δ_{fb} represent unmodeled dynamics.

dynamics (whose combustion processes are extremely complex) or the slight delays that can occur in modern electronically-controlled engines (as a result of the processing time of the embedded computers). These neglected mechanisms are called *unmodeled dynamics*.

One way to account for unmodeled dynamics is by developing a more complex model that includes additional details that are deemed important for control design. Such models are commonly used for controller development, but substantial effort is required to generate them. In addition, these models are themselves likely to be uncertain, since the parameter values may vary over time or between units. Performing parametric analysis on complex models can be very time-consuming, especially if the parameter space is large.

An alternative is to investigate whether the closed loop system can be made insensitive to generic forms of unmodeled dynamics. The basic idea is to augment the nominal model with a bounded input/output transfer function that captures the gross features of the unmodeled dynamics. For example, in the cruise control example the model of the engine can be a static model that provides the torque instantaneously and the augmented model can include a time delay with an unknown but bounded value. Describing unmodeled dynamics with transfer functions permits us to handle infinite-dimensional systems like time delays.

Figure 13.2 illustrates some ways in which unmodeled dynamics can be captured. The transfer functions Δ, δ, Δ_{fb} are taken as bounded input/output operators that represent the unmodeled dynamics. For example, in Figure 13.2a we assume that the transfer function of the process is $\tilde{P}(s) = P(s) + \Delta(s)$, where $P(s)$ is the nominal simplified transfer function and $\Delta(s)$ is a transfer function that represents the unmodeled dynamics in terms of *additive uncertainty*. If we can show that the closed loop system is stable for all $\Delta(s)$ satisfying a given bound (e.g., $|\Delta(s)| < \epsilon$), then the system is said to be *robustly stable*.

Different representations are possible in addition to additive uncertainty. Figure 13.2b shows a representation for *multiplicative uncertainty* and Figure 13.2c represents *feedback uncertainty*. The specific form that is used depends on what provides the best representation of the unmodeled dynamics. The different types of uncertainty can also be related to each other:

$$\delta = \frac{\Delta}{P}, \qquad \Delta_{\text{fb}} = \frac{\Delta}{P(P + \Delta)} = \frac{\delta}{P(1 + \delta)}.$$

We will return to these representations in the next section, where we develop conditions for robust stability in the presence of unmodeled dynamics.

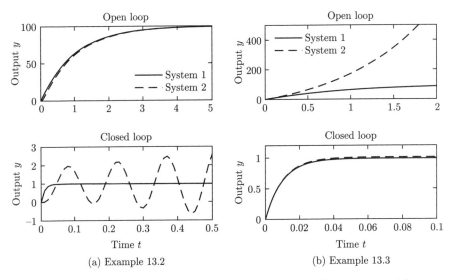

Figure 13.3: Determining when two systems are close. The plots in (a) show a situation when the open loop responses are almost identical, but the closed loop responses are very different. The processes are given by equation (13.1) with $k = 100$ and $T = 0.025$. The plots in (b) show the opposite situation: the systems are different in open loop but similar in closed loop. The processes are given by equation (13.3) with $k = 100$.

When Are Two Systems Similar?

A fundamental issue in describing robustness is to determine when two systems are close. Given such a characterization, we can then attempt to describe robustness according to how close the actual system must be to the model in order to still achieve the desired levels of performance. This seemingly innocent problem is not as simple as it may appear. A naive approach is to say that two systems are close if their open loop responses are close. Even if this appears natural, there are complications, as illustrated by the following examples.

Example 13.2 Systems similar in open loop but different in closed loop
The systems with the transfer functions

$$P_1(s) = \frac{k}{s+1}, \qquad P_2(s) = \frac{k}{(s+1)(sT+1)^2} \tag{13.1}$$

have very similar open loop step responses for small values of T, as illustrated in the upper plot in Figure 13.3a, which corresponds to $T = 0.025$ and $k = 100$.

    The differences between the open loop step responses are barely noticeable in the figure. Closing a feedback loop with unit gain ($C = 1$) around the systems gives closed loop systems with the transfer functions

$$T_1(s) = \frac{k}{s+1+k}, \qquad T_2(s) = \frac{k}{s^3T^2 + (T^2+2T)s^2T + (1+2T)s+1+k}. \tag{13.2}$$

We find that T_1 is stable for $k > -1$ and T_2 is stable for $-1 < k < 2T+4+2/T$. With the numerical values $k = 100$ and $T = 0.025$ the transfer function T_1 is stable and T_2 is unstable, which is clearly seen in the closed loop step responses in the lower plot in Figure 13.3a. ▽

Example 13.3 Systems different in open loop but similar in closed loop
Consider the systems

$$P_1(s) = \frac{k}{s+1}, \qquad P_2(s) = \frac{k}{s-1}. \tag{13.3}$$

The open loop responses are different because P_1 is stable and P_2 is unstable, as shown in the upper plot in Figure 13.3b. Closing a feedback loop with unit gain $(C = 1)$ around the systems, we find that the closed loop transfer functions are

$$T_1(s) = \frac{k}{s+k+1}, \qquad T_2(s) = \frac{k}{s+k-1}, \tag{13.4}$$

which are very close for large k, as shown in the lower plot in Figure 13.3b. ▽

The examples we have just discussed indicate that comparing time responses may not be a good way to compare systems. We will next compare frequency responses.

Example 13.4 Comparison of systems via frequency responses
Consider the systems

$$P_1(s) = \frac{2}{(1+5s)^3(1-0.05s)}, \qquad P_2(s) = \frac{2}{(1+5s)^3(1+0.05s)}. \tag{13.5}$$

Bode and Nyquist plots of these transfer functions are shown in Figure 13.4. The figure shows that both systems have very similar Bode and Nyquist plots. In spite of this, the closed loop systems obtained with unit feedback are very different. Neither system has any zeros, but P_1 has two poles in the left half-plane and one pole in the right half-plane while P_2 has all its poles in the left half-plane. Both $1 + P_1$ and $1 + P_2$ have winding number $n_w = 0$. Since P_1 has a pole in the right half-plane it follows from the Nyquist criterion (Theorem 10.3) that the characteristic polynomial of the closed loop system obtained with unit feedback has one zero in the right half-plane ($f = 1 + P_1$, $n_{z,D} = n_{w,\Gamma} + n_{p,D}$ in the principle of variation of the argument, Theorem 10.2). Thus the closed loop system using P_1 is unstable while the closed loop system using P_2 is stable. ▽

The important lesson to learn from this example is that two systems may not be close from the point of view of feedback even if their open loop frequency responses are similar. It is also necessary that both systems satisfy the winding number condition.

The Vinnicombe Metric

Examples 13.2 and 13.3 show that comparing open loop time responses is not a good way to judge closed loop behavior. Example 13.4 shows that it is necessary to have a winding number condition if frequency responses are compared. We will now introduce the *Vinnicombe metric*, which is the proper way to compare open loop systems in a way that reflects their closed loop behavior. The metric is closely related to the Nyquist plot; more information is available in [247, 248].

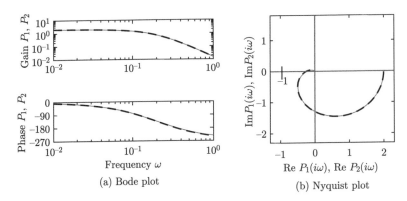

Figure 13.4: Comparison of frequency response of $P_1(s)$ (solid) and $P_2(s)$ (dashed). (a) Bode plot and (b) Nyquist plot.

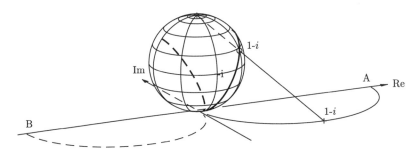

Figure 13.5: Geometric interpretation of the chordal metric $d(P_1, P_2)$ on a Nyquist plot with a Riemann sphere. At each frequency, the points on the Nyquist curve for P_1 (solid, starting at A) and P_2 (dashed, starting at B) are projected onto the sphere of diameter 1 positioned at the origin of the complex plane. The projection of the point $1 - i$ is shown in the figure. The distance between the two systems is defined as the maximum distance between the projections of $P_1(i\omega)$ and $P_2(i\omega)$ over all frequencies ω. The figure is plotted for the transfer functions $P_1(s) = 2/(s + 1)$ and $P_2(s) = 2/(s - 1)$. (Diagram courtesy G. Vinnicombe.)

We start by introducing the *chordal metric*, which is a function $\mathbb{C} \times \mathbb{C} \to [0\ 1]$ that maps two complex numbers to a real variable in the range $0 \le x \le 1$. Applied to the transfer functions $P_1(s)$ and $P_2(s)$ the chordal metric is defined as

$$d_{P_1 P_2}(\omega) := \frac{|P_1(i\omega) - P_2(i\omega)|}{\sqrt{1 + |P_1(i\omega)|^2}\sqrt{1 + |P_2(i\omega)|^2}}. \tag{13.6}$$

The chordal metric $d_{P_1 P_2}$ has a nice geometric interpretation, illustrated in Figure 13.5. The points $P_1(i\omega)$ and $P_2(i\omega)$ are projected onto a sphere with diameter 1 positioned at the origin of the complex plane (the Riemann sphere). The projection is the intersection of the sphere with a straight line from the point to the north pole of the sphere (inverse stereographic projection). The chordal distance is then the Euclidean distance between the two points on the sphere.

To define a metric between two transfer functions, Vinnicombe introduced the following set $\mathcal{C}$ of rational transfer functions P_1 and P_2:

$$\mathcal{C} = \Big\{ P_1, P_2 : 1 + P_1(i\omega)P_2(-i\omega) \neq 0 \; \forall \, \omega,$$

$$n_{w,\Gamma}(1 + P_1(s)P_2(-s)) + n_{p,rhp}(P_1(s)) - n_{p,rhp}(P_2(-s)) = 0 \Big\}, \tag{13.7}$$

where $n_{w,\Gamma}(f)$ is the winding number for the function $f(s)$ around the Nyquist contour Γ and $n_{p,rhp}(f)$ is the number of poles of the $f(s)$ in the open right half-plane. (Compare with the corresponding conditions in Nyquist's criterion in Theorem 10.3.) The metric is then defined as follows.

Definition 13.1 (The ν-gap metric). Let $P_1(s)$ and $P_2(s)$ be rational transfer functions. The ν-gap metric is

$$\delta_\nu(P_1, P_2) = \begin{cases} \sup_\omega d_{P_1 P_2}(\omega), & \text{if } (P_1, P_2) \in \mathcal{C}, \\ 1, & \text{otherwise}, \end{cases} \tag{13.8}$$

where $d_{P_1 P_2}(\omega)$ is given by equation (13.6).

We will also call this metric the *Vinnicombe metric* after its inventor. Vinnicombe showed that $\delta_\nu(P_1, P_2)$ is indeed a metric. He extended it to multivariable and infinite-dimensional systems, and he gave strong robustness results that will be discussed later. There is a MATLAB command `gapmetric` for computing the Vinnicombe metric.

Vinnicombe gave several interpretations of the winding number condition that determines if (P_1, P_2) belong to $\mathcal{C}$. He showed that the condition implies that the closed loop system obtained when $P_1(s)$ is connected in a feedback loop with $P_1(-s)$ has the same number of right half-plane poles as when $P_1(s)$ is connected with $P_2(-s)$. A necessary condition is that the rational functions $1 + P_1(s)P_1(-s)$ and $1 + P_1(s)P_2(-s)$ have the same number of zeros in the right half-plane. This condition can be interpreted as a continuity condition: the transfer function P can be continuously perturbed from P_1 to P_2 in such a way that there is no intermediate transfer function P where $d_{P_1 P}(\omega) = 1$.

We illustrate the Vinnicombe metric by computing it for the systems in Examples 13.2 and 13.3.

Example 13.5 Vinnicombe metric for Example 13.2

The transfer functions P_1 and P_2 for the systems in Example 13.2 are given by equation (13.1). We have

$$f(s) = 1 + P_1(s)P_2(-s) = 1 + \frac{k^2}{(1 - s^2)(1 - sT)^2}, \quad k = 100.$$

The graph of $f(i\omega)$ for $-\infty \leq \omega \leq \infty$ is a closed contour in the right half-plane that does not encircle the origin (see Figure 13.6a and 13.6b for an enlargement of the region around the origin), hence $n_{w,\Gamma}(1 + P_1(s)P_2(-s)) = 0$. In addition, the transfer functions P_1 and P_2 have no poles in the right half-plane and we can conclude that $(P_1, P_2) \in \mathcal{C}$ (equation (13.7)). An alternative to verify the winding number condition is to compute the number of right half-plane zeros of the transfer functions $1 + P_1(s)P_1(-s)$ and $1 + P_1(s)P_2(-s)$. A direct computation shows that both transfer functions have one zero in the open right half-plane. It follows from

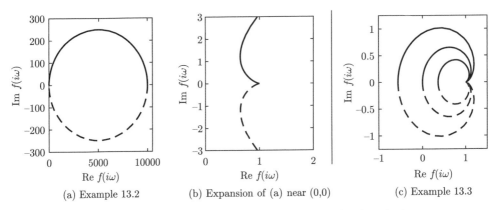

(a) Example 13.2   (b) Expansion of (a) near (0,0)   (c) Example 13.3

Figure 13.6: Graphs of the function $f(i\omega) = 1 + P_1(i\omega)P_2(-i\omega)$ for $-\infty \leq \omega \leq \infty$. The plots for Example 13.2 with $P_1(s) = 100/(s+1)$ and $P_2(s) = 100/((s+1)(0.025s+1)^2$ are shown in (a), with an enlargement in the area close to the origin in (b). The plots for Example 13.3 with $P_1(s) = k/(s+1)$ and $P_2(s) = k/(-s+1)$ are shown in (c), with gains $k = 1.25$ (outer), $k = 1$, and $k = 0.8$ (inner). Values for positive ω are shown as solid lines and negative values are shown as dashed lines.

equation (13.8) that the Vinnicombe metric is $\delta_\nu(P_1, P_2) = 0.89$, which is large since 1.0 is as big as it can get, confirming that P_1 and P_2 are quite different. ∇

Example 13.6 Vinnicombe metric for Example 13.3

The transfer functions P_1 and P_2 for the systems in Example 13.3 are given by equation (13.3). We have

$$1 + P_1(i\omega)P_2(-i\omega) = 1 - \frac{k^2}{(1+i\omega)^2} = 1 - \frac{k^2(1-\omega^2)}{(1+\omega^2)^2} + \frac{2k^2 i\omega}{(1+\omega^2)^2}.$$

The imaginary part of the function $1 + P_1(i\omega)P_2(-i\omega)$ is zero for $\omega = 0$ and $\omega = \infty$ and the corresponding values of the real part are $1 - k^2$ and 1. The function is thus zero only for $\omega = 0$ and $k = 1$. Furthermore

$$f(s) = 1 + P_1(s)P_2(-s) = 1 - \frac{k^2}{(s+1)^2} = \frac{s^2 + 2s + 1 - k^2}{(s+1)^2}.$$

The function $f(s)$ has a zero in the open right half-plane if $k > 1$. The winding number of $1 + P_1(s)P_2(-s)$ is 0 if $k \leq 1$ and 1 if $k > 1$, as seen in Figure 13.6c. Since P_1 has no poles in the right half-plane and P_2 has one pole in the right half-plane, equation (13.8) implies that $\delta_\nu(P_1, P_2) = 1$ if $k \leq 1$.

We have thus found that $(P_1, P_2) \in \mathcal{C}$ if $k > 1$, and equation (13.6) implies that

$$d_{P_1 P_2}(\omega) = \frac{2k}{1 + k^2 + \omega^2}.$$

The largest value occurs for $\omega = 0$, and the Vinnicombe metric, equation (13.8), becomes

$$\delta_\nu(P_1, P_2) = \begin{cases} 1 & \text{if } k \leq 1, \\ \dfrac{2k}{1 + k^2} & \text{if } k > 1. \end{cases}$$

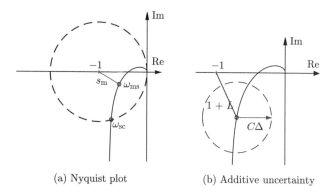

(a) Nyquist plot   (b) Additive uncertainty

Figure 13.7: Illustrations of robust stability in Nyquist plots. The plot (a) shows the stability margin $s_m = 1/M_s$. The plot (b) shows the Nyquist curve and the circle shows uncertainty due to stable additive process variations Δ.

With $k = 100$ we get $\delta_\nu(P_1, P_2) = 0.02$, indicating that the closed loop transfer functions are very close, as illustrated in Figure 13.3b. ∇

13.2   STABILITY IN THE PRESENCE OF UNCERTAINTY

Having discussed how to describe uncertainty and the similarity between two systems, we now consider the problem of robust stability: when can we show that the stability of a system is robust with respect to process variations? This is an important question since the potential for instability is one of the main drawbacks of feedback. Hence we want to ensure that even if we have small inaccuracies in our model, we can still guarantee stability and performance of the closed loop system.

Robust Stability Using Nyquist's Criterion

The Nyquist criterion provides a powerful and elegant way to study the effects of uncertainty for linear systems. A simple criterion for a stable system is that the Nyquist curve be sufficiently far from the critical point -1. Recall that the shortest distance from the Nyquist curve to the critical point is $s_m = 1/M_s$, where M_s is the maximum of the sensitivity function and s_m is the stability margin introduced in Section 10.3. The maximum sensitivity M_s or the stability margin s_m is thus a good robustness measure, as illustrated in Figure 13.7a.

We will now derive explicit conditions on the controller C such that stability is guaranteed for process perturbations where $|\Delta|$ is less than a given bound. Consider a stable feedback system with a process P and a controller C. If the process is changed from P to $P + \Delta$, the loop transfer function changes from PC to $PC + C\Delta$, as illustrated in Figure 13.7b. The additive perturbation Δ must be a stable transfer function to satisfy the winding number condition in the Nyquist criterion. If we have a bound on the size of Δ (represented by the dashed circle in the figure), then the system remains stable as long as the perturbed loop transfer function $|1 + (P + \Delta)C|$ never reaches the critical point -1, since the number of encirclements of -1 remains unchanged.

We will now compute an analytical bound on the allowable process disturbances. The distance from the critical point -1 to the loop transfer function $L = PC$ is $|1 + L|$. This means that the perturbed Nyquist curve will not reach the critical point -1 provided that $|C\Delta| < |1 + L|$, which is guaranteed if

$$|\Delta| < \left|\frac{1+PC}{C}\right| = \left|\frac{1+L}{C}\right| \qquad \text{or} \qquad |\delta| < \frac{1}{|T|}, \qquad \text{where} \quad \delta := \frac{\Delta}{P}. \qquad (13.9)$$

This condition must be valid for all points on the Nyquist curve, i.e., pointwise for all frequencies. The condition for robust stability can thus be written as

$$|\delta(i\omega)| = \left|\frac{\Delta(i\omega)}{P(i\omega)}\right| < \left|\frac{1+L(i\omega)}{L(i\omega)}\right| = \frac{1}{|T(i\omega)|} \qquad \text{for all } \omega \geq 0. \qquad (13.10)$$

Notice that the condition is conservative in the sense that the critical perturbation is in the direction toward the critical point -1. Larger perturbations can be permitted in the other directions.

Robustness is normally defined as the margin to maintain stability. It is easy to modify the criterion and obtain a robustness condition that guarantees a specified stability margin (Exercise 13.6).

The condition in equation (13.10) allows us to reason about uncertainty without exact knowledge of the process perturbations. Namely, we can verify stability for *any* uncertainty Δ that satisfies the given bound. From an analysis perspective, this gives us a measure of the robustness for a given design. Conversely, if we require robustness of a given level, we can attempt to choose our controller C such that the desired level of robustness is available (by asking that T be small) in the appropriate frequency bands.

Equation (13.10) is one of the reasons why feedback systems work so well in practice. The mathematical models used to design control systems are often simplified, and the properties of a process may change during operation. Equation (13.10) implies that the closed loop system will at least be stable for substantial variations in the process dynamics.

It follows from equation (13.10) that the variations can be large for those frequencies where T is small and that smaller variations are allowed for frequencies where T is large. A conservative estimate of permissible process variations that will not cause instability is given by

$$|\delta(i\omega)| = \left|\frac{\Delta(i\omega)}{P(i\omega)}\right| < \frac{1}{|T(i\omega)|} \leq \frac{1}{M_t}, \qquad (13.11)$$

where M_t is the largest value of the complementary sensitivity

$$M_t = \sup_{\omega} |T(i\omega)| = \left\|\frac{PC}{1+PC}\right\|_{\infty}. \qquad (13.12)$$

Reasonable values of M_t are in the range of 1.2 to 2. It is shown in Exercise 13.7 that if $M_t = 2$ then pure gain variations of 50% or pure phase variations of 30° are permitted without making the closed loop system unstable.

Example 13.7 Cruise control
Consider the cruise control system discussed in Section 4.1. Using the parameters from Example 6.11, the model of the car in fourth gear at speed 20 m/s is

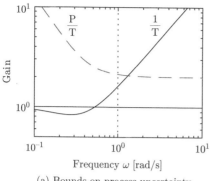
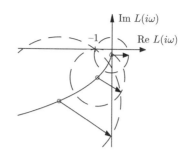

(a) Bounds on process uncertainty   (b) Nyquist plot representation of bounds

Figure 13.8: Robustness for a cruise controller. (a) The maximum relative error $1/|T|$ (solid) and the absolute error $|P|/|T|$ (dashed) for the process uncertainty Δ. (b) The Nyquist plot of the loop transfer function L (zoomed in to the region around the critical point) is shown as a solid line. The dashed circles show allowable perturbations in the process dynamics, $|C\Delta| = |CP|/|T|$, at the frequencies $\omega = 0.2$, 0.4, and 2, which are marked with small circles.

$$P(s) = \frac{1.32}{s + 0.01},$$

and the controller is a PI controller with gains $k_{\mathrm{p}} = 0.5$ and $k_{\mathrm{i}} = 0.1$. Figure 13.8 plots the allowable size of the process uncertainty using the bound in equation (13.10).

At low frequencies $T \to 1$ and so the perturbations can be as large as the original process ($|\delta| = |\Delta/P| < 1$). The complementary sensitivity has its maximum $M_{\mathrm{t}} = 1.17$ at $\omega_{\mathrm{mt}} = 0.26$, and hence this gives the lowest allowable process uncertainty, with $|\delta| < 0.86$ or $|\Delta| < 4.36$. Finally, at high frequencies, $T \to 0$ and hence the relative error can get very large. For example, at $\omega = 5$ rad/s we have $|T(i\omega)| = 0.264$, which means that the stability requirement is $|\delta| < 3.8$. The analysis clearly indicates that the system has good robustness and that the high-frequency properties of the transmission system are not important for the design of the cruise controller.

Another illustration of the robustness of the system is given in Figure 13.8b, which shows the Nyquist curve of the loop transfer function L along with the allowable perturbations. We see that the system can tolerate large amounts of uncertainty and still maintain stability of the closed loop. ∇

The situation illustrated in the previous example is typical of many processes: moderately small uncertainties are required only around the gain crossover frequencies, but large uncertainties can be permitted at higher and lower frequencies. A consequence of this is that a simple model that describes the process dynamics well around the crossover frequency is often sufficient for design. Systems with many resonant peaks are an exception to this rule because the process transfer function for such systems may also have large gains for higher frequencies, as shown for instance in Example 10.9.

The robustness condition given by equation (13.10) can be given another interpretation by using the small gain theorem (Theorem 10.4). To apply the theorem we start with block diagrams of a closed loop system with a perturbed process and make a sequence of transformations of the block diagram that isolate the block

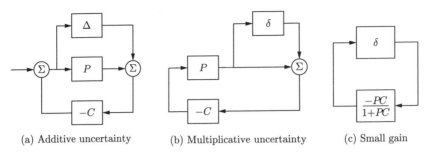

(a) Additive uncertainty   (b) Multiplicative uncertainty   (c) Small gain

Figure 13.9: Illustration of robustness to process perturbations. A system with additive uncertainty (a) can be manipulated via block diagram algebra to one with multiplicative uncertainty $\delta = \Delta/P$ (b). Additional manipulations isolate the uncertainty in a manner that allows application of the small gain theorem (c).

Table 13.1: Conditions for robust stability for different types of uncertainty.

Process	Uncertainty Type	Robust Stability
$P + \Delta$	Additive	$\|CS\Delta\|_\infty < 1$
$P(1 + \delta)$	Multiplicative	$\|T\delta\|_\infty < 1$
$P/(1 + \Delta_{\mathrm{fb}} \cdot P)$	Feedback	$\|PS\Delta_{\mathrm{fb}}\|_\infty < 1$

representing the uncertainty, as shown in Figure 13.9. The result is the two-block interconnection shown in Figure 13.9c, which has the loop transfer function

$$L = \frac{PC}{1+PC} \frac{\Delta}{P} = T\delta.$$

Equation (13.10) implies that the largest loop gain is less than 1 and hence the system is stable via the small gain theorem.

The small gain theorem can be used to check robust stability for uncertainty in a variety of other situations. Table 13.1 summarizes a few of the common cases; the proofs (all via the small gain theorem) are left as exercises.

The circle criterion can also be used to understand robustness to nonlinear gain variations, as illustrated by the following example.

Example 13.8 Robustness for sector-bounded nonlinearities
Consider a system with a nonlinear gain $F(x)$ that can be isolated through appropriate manipulation of the block diagram, resulting in a system that is a feedback composition of the nonlinear block $F(x)$ and a linear part with the transfer function $H(s)$. If the nonlinearity is sector bounded,

$$k_{\mathrm{low}} \; x < F(x) < k_{\mathrm{high}} \; x,$$

and the nominal system has been designed to have maximum sensitivities M_{s} and M_{t}, we can use the circle criterion to verify stability of the closed loop system. In particular, the system can be shown to be stable for sector-bounded nonlinearities with

$$k_{\text{low}} = \frac{M_s}{M_s + 1} \quad \text{or} \quad \frac{M_t - 1}{M_t}, \qquad k_{\text{high}} = \frac{M_s}{M_s - 1} \quad \text{or} \quad \frac{M_t + 1}{M_t}.$$

With $M_s = M_t = 1.4$ we can thus permit gain variations from 0.3 to 3.5, and for a design with $M_s = M_t = 2$ we can allow gain variations of 0.5 to 2 without the system becoming unstable. $\triangledown$

The following example illustrates that it is possible to design systems that are robust to parameter variations.

Example 13.9 Bode's ideal loop transfer function

A major problem in the design of electronic amplifiers is to obtain a closed loop system that is insensitive to changes in the gain of the electronic components. Bode found that the loop transfer function

$$L(s) = ks^{-n}, \qquad 1 \le n \le 5/3 \tag{13.13}$$

had very useful robustness properties. The gain curve of the Bode plot is a straight line with slope $-n$ and the phase is constant $\arg L(i\omega) = -n\pi/2$. The phase margin is thus $\varphi_m = 90(2 - n)°$ for all values of the gain k and the stability margin is $s_m = \sin \pi(1 - n/2)$. Bode called the transfer function the "ideal cut-off characteristic" [52, pp. 454–458]; we will call it *Bode's ideal loop transfer function* in honor of Bode. The transfer function cannot be realized with lumped physical components unless n is an integer, but it can be approximated over a given frequency range with a proper rational function for any n (Exercise 13.8). An operational amplifier circuit that has the approximate transfer function $G(s) = k/(s + a)$ is a realization of Bode's ideal transfer function with $n = 1$, as described in Example 9.2. Designers of operational amplifiers go to great efforts to make the approximation valid over a wide frequency range. $\triangledown$

Youla Parameterization

Since stability is such an essential property, it is useful to characterize all controllers that stabilize a given process. Such a representation, which is called a *Youla parameterization*, is also very useful when solving design problems because it makes it possible to search over all stabilizing controllers without the need to test stability explicitly.

We will first derive Youla's parameterization for a stable process with a rational transfer function P. A system with a given complementary sensitivity function T can be obtained by feedforward control with the stable transfer function Q where $T = PQ$. Assume that we want to implement the transfer function T by feedback with the controller C. Since $T = PC/(1 + PC) = PQ$, the controller transfer function and its input-output relation are

$$C = \frac{Q}{1 - PQ}, \qquad u = Q(r - y + Py). \tag{13.14}$$

A straightforward calculation gives the transfer functions for the Gang of Four as

$$S = 1 - PQ, \qquad PS = P(1 - PQ), \qquad CS = Q, \qquad T = PQ.$$

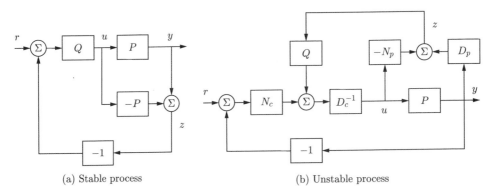

(a) Stable process   (b) Unstable process

Figure 13.10: Block diagrams of Youla parameterizations for a stable process (a) and an unstable process (b). Notice that the signal z is zero in steady state in both cases.

These transfer functions are all stable if P and Q are stable and the controller given by equation (13.14) is thus stabilizing. Indeed, it can be shown that all stabilizing controllers for a stable process are in the form given by equation (13.14) for some choice of Q.

The closed loop system with the controller (13.14) can be represented by the block diagram in Figure 13.10a. Notice that the signal z is always zero in steady state, because it is a subtraction of identical signals. Using block diagram algebra we find from the figure that the transfer function of the closed loop system is PQ. The fact that there are two blocks with transfer function P in parallel in the block diagram implies that there are modes, corresponding to the poles of P, that are not reachable and observable. These modes are stable because we assumed that P was stable.

The scheme in Figure 13.10a cannot be used when the process is unstable but we can make a similar construct. Consider a closed loop system where the process is a rational transfer function $P = n_\mathrm{p}/d_\mathrm{p}$, where d_p and n_p are polynomials with no common factors. Assume that the controller $C = n_\mathrm{c}/d_\mathrm{c}$, where d_c and n_c are polynomials without common factors, stabilizes the system in the sense that all sensitivity functions are stable. By introducing stable polynomials f_p and f_c we obtain

$$P = \frac{n_\mathrm{p}}{d_\mathrm{p}} = \frac{N_\mathrm{p}}{D_\mathrm{p}}, \qquad C = \frac{n_\mathrm{c}}{d_\mathrm{c}} = \frac{N_\mathrm{c}}{D_\mathrm{c}}, \tag{13.15}$$

where $N_\mathrm{p} = d_\mathrm{p}/f_\mathrm{p}$, $D_\mathrm{p} = n_\mathrm{p}/f_\mathrm{p}$, $N_\mathrm{c} = n_\mathrm{c}/f_\mathrm{c}$, and $D_\mathrm{c} = d_\mathrm{c}/f_\mathrm{c}$ are rational functions with no zeros in the right half-plane (stable rational functions). The sensitivity functions are

$$S = \frac{1}{1 + PC} = \frac{D_\mathrm{p} D_\mathrm{c}}{D_\mathrm{p} D_\mathrm{c} + N_\mathrm{p} N_\mathrm{c}}, \qquad PS = \frac{P}{1 + PC} = \frac{N_\mathrm{p} D_\mathrm{c}}{D_\mathrm{p} D_\mathrm{c} + N_\mathrm{p} N_\mathrm{c}},$$

$$CS = \frac{C}{1 + PC} = \frac{D_\mathrm{p} N_\mathrm{c}}{D_\mathrm{p} D_\mathrm{c} + N_\mathrm{p} N_\mathrm{c}}, \qquad T = \frac{PC}{1 + PC} = \frac{N_\mathrm{p} N_\mathrm{c}}{D_\mathrm{p} D_\mathrm{c} + N_\mathrm{p} N_\mathrm{c}}.$$

The controller C is stabilizing if and only if the rational function $D_\mathrm{p} D_\mathrm{c} + N_\mathrm{p} N_\mathrm{c}$ does not have any zeros in the right half-plane. Letting Q be a stable rational function, we observe that the closed loop poles do not change if the controller C is changed

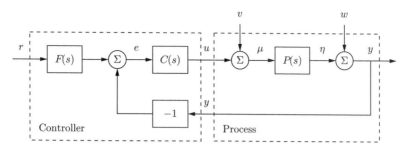

Figure 13.11: Block diagram of a basic feedback loop. The external signals are the reference signal r, the load disturbance v, and the measurement noise w. The process output is y, and the control signal is u. The process P may include unmodeled dynamics, such as additive perturbations.

by adding $N_\mathrm{p}Q$ to D_c and subtracting $D_\mathrm{p}Q$ from N_c, resulting in the controller

$$\overline{C} = \frac{N_\mathrm{c} - D_\mathrm{p}Q}{D_\mathrm{c} + N_\mathrm{p}Q}, \qquad D_\mathrm{c}u = N_\mathrm{c}(r - y) + Q(D_\mathrm{p}y - N_\mathrm{p}u). \tag{13.16}$$

A block diagram of the closed loop system is shown in Figure 13.10b.

Figure 13.10b and 13.10a share the same basic structure, despite their difference in appearance. In both cases we form a signal z that is zero in steady state and feed it back into the system via the stable transfer function Q. The sensitivity functions of the closed loop system are

$$S = \frac{1}{1 + P\overline{C}} = \frac{D_\mathrm{p}(D_\mathrm{c} + N_\mathrm{p}Q)}{D_\mathrm{p}D_\mathrm{c} + N_\mathrm{p}N_\mathrm{c}}, \qquad PS = \frac{P}{1 + P\overline{C}} = \frac{N_\mathrm{p}(D_\mathrm{c} + N_\mathrm{p}Q)}{D_\mathrm{p}D_\mathrm{c} + N_\mathrm{p}N_\mathrm{c}},$$

$$\overline{C}S = \frac{\overline{C}}{1 + P\overline{C}} = \frac{D_\mathrm{p}(N_\mathrm{c} - D_\mathrm{p}Q)}{D_\mathrm{p}D_\mathrm{c} + N_\mathrm{p}N_\mathrm{c}}, \qquad T = \frac{P\overline{C}}{1 + P\overline{C}} = \frac{N_\mathrm{p}(N_\mathrm{c} - D_\mathrm{p}Q)}{D_\mathrm{p}D_\mathrm{c} + N_\mathrm{p}N_\mathrm{c}}. \tag{13.17}$$

These transfer functions are all stable and equation (13.16) is therefore a parameterization of controllers that stabilize the process P. Conversely it can be shown that all stabilizing controllers can be represented by the controller (13.16); see [246, Section 3.1]. The controller $\overline{C}$ is a called a *Youla parameterization* of the controller C.

The Youla parameterization is very useful for controller design because it characterizes all controllers that stabilize a given process. The fact that the transfer function Q appears affinely in the expressions for the Gang of Four in equation (13.17) is very useful if we want to use optimization techniques to find a transfer function Q that yields desired closed loop properties.

13.3 PERFORMANCE IN THE PRESENCE OF UNCERTAINTY

So far we have investigated the risk for instability and robustness to process uncertainty. We will now explore how responses to load disturbances, measurement noise, and reference signals are influenced by process variations. To do this we will analyze the system in Figure 13.11, which is identical to the basic feedback loop analyzed in Chapter 12.

Disturbance Attenuation

The sensitivity function S gives a rough characterization of the effect of feedback on disturbances, as was discussed in Section 12.2. A more detailed characterization is given by the transfer function from load disturbances to process output:

$$G_{yv} = \frac{P}{1 + PC} = PS. \tag{13.18}$$

Load disturbances typically have low frequencies, and it is therefore important that the transfer function G_{yv} is small for low frequencies. For processes P with constant low-frequency gain and a controller with integral action it follows from equation (13.18) that $G_{yv} \approx s/k_i$. The integral gain k_i is thus a simple measure of the attenuation of low-frequency load disturbances.

To find out how the transfer function G_{yv} is influenced by small variations in the process transfer function we differentiate equation (13.18) with respect to P, yielding

$$\frac{dG_{yv}}{dP} = \frac{1}{(1 + PC)^2} = \frac{SP}{P(1 + PC)} = S\frac{G_{yv}}{P},$$

and it follows that

$$\frac{dG_{yv}}{G_{yv}} = S\frac{dP}{P}, \tag{13.19}$$

where we write dG and dP as a reminder that this expression holds for small variations.

In this form, we see that the relative error in the transfer function G_{yu} is determined by the relative error in the process transfer function, scaled by the sensitivity function S. The response to load disturbances is thus insensitive to process variations for frequencies where $|S(i\omega)|$ is small.

A drawback with feedback is that the controller feeds measurement noise into the system. It is thus also important that the control actions generated by measurement noise are not too large. It follows from Figure 13.11 that the transfer function G_{uw} from measurement noise to controller output is given by

$$G_{uw} = -\frac{C}{1 + PC} = -\frac{T}{P}. \tag{13.20}$$

Since measurement noise typically has high frequencies, the transfer function G_{uw} should not be too large for high frequencies. The loop transfer function PC is typically small for high frequencies, which implies that $G_{uw} \approx C$ for large s. To avoid injecting too much measurement noise the high-frequency gain of the controller transfer function $C(s)$ should thus be small. This property is called *high-frequency roll-off*. Low-pass filtering of the measured signal is a simple way to achieve this property, and this is common practice in PID control; see Section 11.5.

To determine how the transfer function G_{uw} is influenced by small variations in the process transfer function, we differentiate equation (13.20) with respect to P:

$$\frac{dG_{uw}}{dP} = \frac{d}{dP}\left(-\frac{C}{1 + PC}\right) = \frac{C}{(1 + PC)^2}C = -T\frac{G_{uw}}{P}.$$

Rearranging the terms gives

$$\frac{dG_{uw}}{G_{uw}} = -T\frac{dP}{P}. \tag{13.21}$$

If PC is small for high frequencies the complementary sensitivity function is also small, and we find that process uncertainty has little influence on the transfer function G_{uw} for those frequencies.

Response to Reference Signals

The transfer function from reference to output is given by

$$G_{yr} = \frac{PCF}{1+PC} = TF, \tag{13.22}$$

which contains the complementary sensitivity function. To see how variations in P affect the performance of the system, we differentiate equation (13.22) with respect to the process transfer function:

$$\frac{dG_{yr}}{dP} = \frac{CF}{1+PC} - \frac{PCFC}{(1+PC)^2} = \frac{CF}{(1+PC)^2} = S\frac{G_{yr}}{P},$$

and it follows that

$$\frac{dG_{yr}}{G_{yr}} = S\frac{dP}{P}. \tag{13.23}$$

The relative error in the closed loop transfer function thus equals the product of the sensitivity function and the relative error in the process. In particular, it follows from equation (13.23) that the relative error in the closed loop transfer function is small when the sensitivity is small. This is one of the useful properties of feedback.

As in the previous section, there are some mathematical assumptions that are required for the analysis presented here to hold. As already stated, we require that the perturbations Δ be small (as indicated by writing dP). Second, we require that the perturbations be stable, so that we do not introduce any new right half-plane poles that would require additional encirclements in the Nyquist criterion. Also, as before, this condition is conservative: it allows for any perturbation that satisfies the given bounds, while in practice the perturbations may be more restricted.

Example 13.10 Operational amplifier circuit

To illustrate the use of these tools, consider the performance of an op amp-based amplifier, as shown in Figure 13.12a. We wish to analyze the performance of the amplifier in the presence of uncertainty in the dynamic response of the op amp and changes in the loading on the output. We model the system using the block diagram in Figure 13.12b, which is based on the derivation in Exercise 10.1.

Consider first the effect of unknown dynamics for the operational amplifier. Letting the dynamics of the op amp be modeled as $v_2 = -G(s)v$, it follows from the block diagram in Figure 13.12b that the transfer function for the overall circuit is

$$G_{v_2 v_1} = -\frac{R_2}{R_1}\frac{G(s)}{G(s) + R_2/R_1 + 1}.$$

We see that if $G(s)$ is large over the desired frequency range, then the closed loop system is very close to the ideal response $\alpha := R_2/R_1$. Assuming $G(s) = b/(s+a)$,

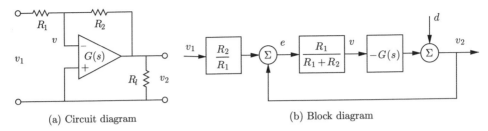

(a) Circuit diagram   (b) Block diagram

Figure 13.12: Operational amplifier with uncertain dynamics. The circuit in (a) is modeled using the transfer function $G(s)$ to capture its dynamic properties and it has a load at the output. The block diagram in (b) shows the input/output relationships. The load is represented as a disturbance d applied at the output of $G(s)$.

where $b = ak$ is the gain-bandwidth product of the amplifier (as discussed in Example 9.2), the sensitivity function and the complementary sensitivity function become

$$S = \frac{s+a}{s+a+\alpha b}, \qquad T = \frac{\alpha b}{s+a+\alpha b}.$$

The sensitivity function around the nominal values tells us how the tracking response varies as a function of process perturbations:

$$\frac{dG_{v_2 v_1}}{G_{v_2 v_1}} = S\frac{dP}{P}.$$

We see that for low frequencies, where S is small, variations in the bandwidth a or the gain-bandwidth product b will have relatively little effect on the performance of the amplifier (under the assumption that b is sufficiently large).

To model the effects of an unknown load, we consider the addition of a disturbance d at the output of the system, as shown in Figure 13.12b. This disturbance represents changes in the output voltage due to loading effects. The transfer function $G_{v_2 d} = S$ gives the response of the output to the load disturbance, and we see that if S is small, then we are able to reject such disturbances. The sensitivity of $G_{v_2 d}$ to perturbations in the process dynamics can be computed by taking the derivative of $G_{v_2 d}$ with respect to P:

$$\frac{dG_{v_2 d}}{dP} = \frac{-C}{(1+PC)^2} = -\frac{T}{P}G_{v_2 d} \qquad \Longrightarrow \qquad \frac{dG_{v_2 d}}{G_{v_2 d}} = -T\frac{dP}{P}.$$

Thus we see that the relative changes in disturbance rejection are roughly the same as the process perturbations at low frequencies (when T is approximately 1) and drop off at higher frequencies. However, it is important to remember that $G_{v_2 d}$ itself is small at low frequency, and so these variations in relative performance may not be an issue in many applications. ∇

Analysis of the sensitivity to small process perturbations can performed for many other system configurations. The analysis for the system in Figure 12.13, where the reference signal response is improved by feedforward and the load disturbance response is improved by feedforward from measured disturbances, is presented in Exercise 13.11.

13.4   DESIGN FOR ROBUST PERFORMANCE

Control design is a rich problem where many factors have to be taken into account. Typical requirements are that load disturbances should be attenuated, the controller should inject only a moderate amount of measurement noise, the output should follow variations in the command signal well, and the closed loop system should be insensitive to process variations. For the system in Figure 13.11 these requirements can be captured by specifications on the sensitivity functions S and T and the transfer functions G_{yv}, G_{uw}, G_{yr}, and G_{ur}. Notice that it is necessary to consider at least six transfer functions, as discussed in Section 12.1. The requirements are mutually conflicting, and we have to make trade-offs. The attenuation of load disturbances will be improved if the bandwidth is increased, but the noise injection will be worse. The following example is an illustration.

Example 13.11 Nanopositioning system for an atomic force microscope
A simple nanopositioner with the process transfer function

$$P(s) = \frac{\omega_0^2}{s^2 + 2\zeta\omega_0 s + \omega_0^2}$$

was explored in Example 10.9. It was shown that the system could be controlled using an integral controller. The closed loop performance was poor because the gain crossover frequency was limited to $\omega_{gc} < 2\zeta\omega_0(1 - s_m)$ to have good robustness with the integral controller. It can be shown that little improvement is obtained by using a PI controller. We will explore if better performance can be obtained with PID control. As justified in Example 14.11 in the next chapter, we trying choosing a controller zero that is near the fast stable process pole. The controller transfer function should thus be chosen as

$$C(s) = \frac{k_d s^2 + k_p s + k_i}{s} = \frac{k_i}{s} \frac{s^2 + 2\zeta\omega_0 s + \omega_0^2}{\omega_0^2}, \tag{13.24}$$

which gives $k_p = 2\zeta k_i/\omega_0$ and $k_d = k_i/\omega_0^2$. The loop transfer function becomes $L(s) = k_i/s$.

Figure 13.13 shows, in dashed lines, the gain curves for the Gang of Four for a system designed with $k_i = 0.5$. A comparison with Figure 10.14 shows that the bandwidth is increased significantly from $\omega_{gc} = 0.01$ to $\omega_{gc} = k_i = 0.5$. However, since the process pole is canceled, the system will be very sensitive to load disturbances with frequencies close to the resonant frequency, as seen by the peak in PS at $\omega/\omega_0 = 1$. The gain curve of CS has a dip or a notch at the resonant frequency ω_0, which implies that the controller gain is very low for frequencies around the resonance. The gain curve also shows that the system is very sensitive to high-frequency noise. The system will likely be unusable because the gain goes to infinity for high frequencies.

The sensitivity to high-frequency noise can be reduced by modifying the controller to be

$$C(s) = \frac{k_i}{s} \frac{s^2 + 2\zeta\omega_0 s + \omega_0^2}{\omega_0^2(1 + sT_f + (sT_f)^2/2)}, \tag{13.25}$$

which has high-frequency roll-off. Selection of the constant T_f for the filter is a compromise between attenuation of high-frequency measurement noise and robustness. A large value of T_f reduces the effects of sensor noise significantly, but it also reduces

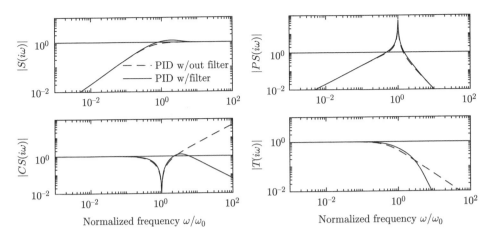

Figure 13.13: Nanopositioning system control via cancellation of the fast process pole. Gain curves for the Gang of Four for PID control with second-order filtering (13.25) are shown by solid lines, and the dashed lines show results for a PID controller without filtering (13.24).

the stability margin. Since the gain crossover frequency without filtering is k_i, a reasonable choice is $T_f = 0.2/k_i$, as shown by the solid curves in Figure 13.13. The plots of $|CS(i\omega)|$ and $|S(i\omega)|$ show that the sensitivity to high-frequency measurement noise is reduced dramatically at the cost of a marginal increase of sensitivity. Notice that the poor attenuation of disturbances with frequencies close to the resonance is not visible in the sensitivity function because of the cancellation of the resonant poles (but it can be seen in PS).

The designs thus far have the drawback that load disturbances with frequencies close to the resonance are not attenuated, since $|S(i\omega_0)|$ is close to one. We will now consider a design that actively attenuates the poorly damped modes. We start with an ideal PID controller where the design can be done analytically, and we add high-frequency roll-off. The loop transfer function obtained with this controller is

$$L(s) = \frac{\omega_0^2(k_d s^2 + k_p s + k_i)}{s(s^2 + 2\zeta\omega_0 s + \omega_0^2)}. \tag{13.26}$$

The closed loop system is of third order, and its characteristic polynomial is

$$s^3 + (k_d\omega_0^2 + 2\zeta\omega_0)s^2 + (k_p + 1)\omega_0^2 s + k_i\omega_0^2. \tag{13.27}$$

A general third-order polynomial can be parameterized as

$$s^3 + (\alpha_c + 2\zeta_c)\omega_c s^2 + (1 + 2\alpha_c\zeta_c)\omega_c^2 s + \alpha_c\omega_c^3. \tag{13.28}$$

The parameters α_c and ζ_c give the relative configuration of the poles, and the parameter ω_c gives their magnitudes, and therefore also the bandwidth of the system.

The identification of coefficients of equal powers of s with equation (13.27) gives a linear equation for the controller parameters, which has the solution

$$k_p = \frac{(1 + 2\alpha_c\zeta_c)\omega_c^2}{\omega_0^2} - 1, \qquad k_i = \frac{\alpha_c\omega_c^3}{\omega_0^2}, \qquad k_d = \frac{(\alpha_c + 2\zeta_c)\omega_c}{\omega_0^2} - \frac{2\zeta_c}{\omega_0}. \tag{13.29}$$

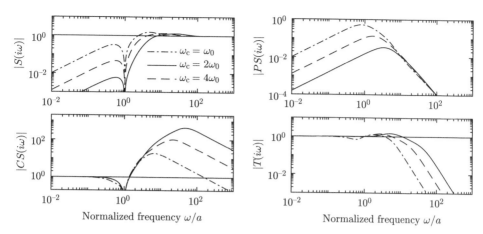

Figure 13.14: Nanopositioner control using active damping. Gain curves for the Gang of Four for PID control of the nanopositioner designed for $\omega_c = \omega_0$ (dash-dotted), $2\omega_0$ (dashed), and $4\omega_0$ (solid). The controller has high-frequency roll-off and has been designed to give active damping of the oscillatory mode. The different curves correspond to different choices of magnitudes of the poles, parameterized by ω_c in equation (13.27).

Adding high-frequency roll-off, the controller becomes

$$C(s) = \frac{k_d s^2 + k_p s + k}{s(1 + sT_f + (sT_f)^2/2)}.\tag{13.30}$$

If the PID controller is designed without the filter, the filter time constant must be significantly smaller than T_d to avoid introducing extra phase lag; a reasonable value is $T_f = T_d/10 = 0.1\,k_d/k$. If more filtering is desired it is necessary to account for the filter dynamics in the design.

Figure 13.14 shows the gain curves of the Gang of Four for designs with $\zeta_c = 0.707$, $\alpha_c = 1$, and $\omega_c = \omega_0$, $2\omega_0$, and $4\omega_0$. The figure shows that the largest values of the sensitivity functions S and T are small. The gain curve for PS shows that the load disturbances are now well attenuated over the whole frequency range, and attenuation increases with increasing ω_0. The gain curve for CS shows that large control signals are required to provide active damping. The high gain of CS for high frequencies also shows that low-noise sensors and actuators with a wide range are required. The largest gains for CS are 19, 103, and 434 for $\omega_c = \omega_0$, $2\omega_0$, and $4\omega_0$, respectively. There is clearly a trade-off between disturbance attenuation and controller gain. A comparison of Figures 13.13 and 13.14 illustrates the trade-offs between control action and disturbance attenuation for the designs with cancellation of the fast process pole and active damping.   ∇

It is highly desirable to have design methods that can guarantee robust performance. Such design methods did not appear until the late 1980s. Many of these design methods result in controllers having the same structure as the controller based on state feedback and an observer. In the remainder of this section we provide a brief review of some of the techniques as a preview for those interested in more specialized study.

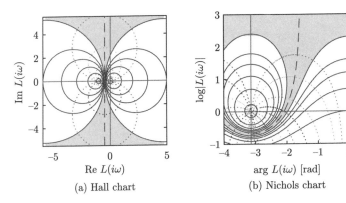

(a) Hall chart   (b) Nichols chart

Figure 13.15: Hall and Nichols charts. The Hall chart (a) is a Nyquist plot with curves for constant gain (solid) and phase (dotted) of the complementary sensitivity function T. The Nichols chart (b) is the conformal map of the Hall chart under the transformation $N = \log L$ (with the scale flipped). The dashed curve is the line where $|T(i\omega)| = 1$, and the shaded region corresponds to loop transfer functions whose complementary sensitivity changes by no more than $\pm 10\%$.

Quantitative Feedback Theory

Quantitative feedback theory (QFT) is a graphical design method for robust loop shaping that was developed by I. M. Horowitz [123]. The idea is to first determine a controller that gives a complementary sensitivity that is robust to process variations and then to shape the response to reference signals by feedforward. The idea is illustrated in Figure 13.15a, which shows the level curves of the gain $|T(i\omega)|$ of the complementary sensitivity function on a Nyquist plot (this type of Nyquist plot is also called a *Hall chart*). The complementary sensitivity function has unit gain on the line $\text{Re } L(i\omega) = -0.5$. In the neighborhood of this line, significant variations in process dynamics only give moderate changes in the complementary transfer function. The shaded part of the figure corresponds to the region $0.9 < |T(i\omega)| < 1.1$. To use the design method, we represent the uncertainty for each frequency by a region and attempt to shape the loop transfer function so that the variation in T is as small as possible. The design is often performed using the Nichols chart shown in Figure 13.15b.

Linear Quadratic Control

One way to make the trade-off between the attenuation of load disturbances and the injection of measurement noise is to design a controller that minimizes the cost function

$$J = \int_0^\infty \left(y^2(t) + \rho u^2(t) \right) dt,$$

where ρ is a weighting parameter as discussed in Section 8.4. This cost function gives a compromise between load disturbance attenuation and disturbance injection because it balances control actions against deviations in the output. If all state variables are measured, the controller is a state feedback $u = -Kx$ as described in Section 7.5. It has been shown that this controller is very robust: it has a phase margin of at least 60° and an infinite gain margin. This controller is called a *linear*

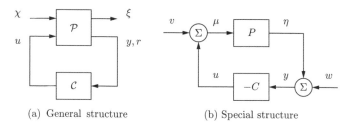

(a) General structure   (b) Special structure

Figure 13.16: H_∞ robust control formulation. (a) General representation of a control problem used in robust control. The input u represents the control signal, the input χ represents the external influences on the system, the output ξ is the generalized error, and the output y is the measured signal. (b) Special case of the basic feedback loop in Figure 13.11 where the reference signal is zero.

quadratic regulator or *LQR controller* because the process model is linear and the criterion is quadratic.

When all state variables are not measured, the state can be reconstructed using an observer, as discussed in Section 8.3. It is also possible to introduce process disturbances and measurement noise explicitly in the model and to reconstruct the states using a Kalman filter, as discussed briefly in Section 8.4. The Kalman filter has the same structure as the observer designed by eigenvalue assignment in Section 8.3, but the observer gains L are now obtained by solving an optimization problem.

The control law obtained by combining linear quadratic control with a Kalman filter is called *linear quadratic Gaussian control* or *LQG control*. The Kalman filter is optimal when the models for load disturbances and measurement noise are Gaussian. There are efficient programs to compute these feedback and observer gains. The basic task is to solve algebraic Riccati equations. For numerical calculations we can use the MATLAB commands `care` for continuous time systems and `dare` for discrete time systems. The are also MATLAB commands `lqg`, `lqi`, and `kalman` that perform the complete design.

It is interesting that the solution to the optimization problem leads to a controller having the structure of a state feedback and an observer. The state feedback gains depend on the parameter ρ, and the filter gains depend on the parameters in the model that characterize process noise and measurement noise (see Section 8.4).

The nice robustness properties of state feedback are unfortunately lost when the observer is added [75]. There are parameters that give closed loop systems with poor robustness, and hence there is a fundamental difference between directly measuring the states of a system and reconstructing the states using an observer.

H_∞ Control

An elegant method for robust control design is called H_∞ control because it can be formulated as minimization of the H_∞ norm of a matrix of transfer functions, defined in equation (10.15). The basic ideas are simple, but the details are complicated and we will therefore just give the flavor of the results. A key idea is illustrated in Figure 13.16a, where the closed loop system is represented by two blocks, the process $\mathcal{P}$ and the controller $\mathcal{C}$ as discussed in Section 12.1. The process $\mathcal{P}$ has two inputs, the control signal u, which can be manipulated by the controller, and the generalized disturbance χ, which represents all external influences, e.g., command signals, load disturbances, and measurement noise. The process has two outputs, the generalized error ξ, which is a vector of error signals representing the deviation of signals from their desired values, and the measured signal y, which can be used by the controller to compute u. For a linear system and a linear controller the closed loop system can be represented by a linear system

$$\xi = \mathcal{G}(P(s), C(s))\chi, \tag{13.31}$$

which tells how the generalized error ξ depends on the generalized disturbances χ. The control design problem is to find a controller C such that the gain of the transfer function $\mathcal{G}$ is small even when the process has uncertainties. There are many different ways to specify uncertainty and gain, giving rise to different designs depending on the chosen norms.

To illustrate the ideas we will consider a regulation problem for a system where the reference signal is assumed to be zero and the external signals are the load disturbance v and the measurement noise w, as shown in Figure 13.16b. The generalized error is defined as $\xi = (\mu, \eta)$, where $\mu = v - u$ is the part of the load disturbance that is not compensated by the controller and η is the process output. The generalized input is $\chi = (v, -w)$ (the negative sign of w is not essential but is chosen to obtain somewhat nicer equations). The closed loop system is thus modeled by

$$\xi = \begin{pmatrix} \mu \\ \eta \end{pmatrix} = \begin{pmatrix} \dfrac{1}{1+PC} & \dfrac{C}{1+PC} \\ \dfrac{P}{1+PC} & \dfrac{PC}{1+PC} \end{pmatrix} \begin{pmatrix} v \\ -w \end{pmatrix} =: \mathcal{G}(P, C)\chi, \tag{13.32}$$

which is a special case of equation (13.31). If C is stabilizing we have

$$\|\mathcal{G}(P, C))\|_\infty = \sup_\omega \bar{\sigma}(\mathcal{G}) = \sup_\omega \frac{\sqrt{(1+|P(i\omega)|^2)(1+|C(i\omega)|^2)}}{|1 + P(i\omega)C(i\omega)|}, \tag{13.33}$$

where $\bar{\sigma}$ is the largest singular value. Notice that the elements of $\mathcal{G}$ are the Gang of Four. The diagonal elements of $\mathcal{G}$ are the sensitivity functions $S = 1/(1 + PC)$ and $T = PC/(1 + PC)$, which capture robustness. The off-diagonal elements $P/(1 + PC) = G_{yv}$ and $C/(1 + PC) = -G_{uw}$ represent the responses of the output to load disturbances and of the control signal to measurement noise, and they capture performance. If we minimize $\|\mathcal{G}(P, C)\|_\infty$, we thus balance performance and robustness.

There are numerical methods for finding a stabilizing controller C that minimizes $\|\mathcal{G}(P, C)\|_\infty$, if such a controller exists. This controller has the same structure as the controller based on state feedback and an observer; see Figure 8.7 and Theorem 8.3. The controller gains are given by *algebraic Riccati* equations. They can be computed numerically by the MATLAB command `hinfsyn`.

The Generalized Stability Margin

In Section 13.2 we introduced the stability margin as $s_\mathrm{m} = \inf_\omega |1 + P(i\omega)C(i\omega)|$ for systems such that C stabilizes P. The margin can be interpreted as the shortest distance between the Nyquist plot of the loop transfer function PC and the critical point -1, as shown in Figure 13.7a. We also found that $s_m = 1/M_s$ where M_s is the maximum sensitivity. We now define the *generalized stability margin*

$$\sigma_m = \begin{cases} \inf_\omega \dfrac{|1 + P(i\omega)C(i\omega)|}{\sqrt{(1 + |P(i\omega)|^2)(1 + |C(i\omega)|^2)}} & \text{if } C \text{ stabilizes } P, \\ 0 & \text{otherwise.} \end{cases} \tag{13.34}$$

It can be shown that

$$\inf_\omega \frac{|1 + P(i\omega)C(i\omega)|}{\sqrt{(1 + |P(i\omega)|^2)(1 + |C(i\omega)|^2)}} = \inf_\omega \frac{|P(i\omega) + 1/C(i\omega)|}{\sqrt{(1 + |P(i\omega)|^2)(1 + |1/C(i\omega)|^2)}},$$

and it follows that σ_m can be interpreted as the shortest chordal distance between $P(i\omega)$ and $-1/C(i\omega)$. Furthermore equations (13.6) and (13.33) imply that ·

$$\sigma_m(P, C) = \begin{cases} \dfrac{1}{\|\mathcal{G}(P, C)\|_\infty} & \text{if } C \text{ stabilizes } P, \\ 0 & \text{otherwise.} \end{cases} \tag{13.35}$$

Using the generalized stability margin we have the following fundamental robustness theorem, which is proved in [248].

Theorem 13.1 (Vinnicombe's robustness theorem). *Consider a process with transfer function P. Assume that the controller C is designed to give the generalized stability margin σ_m. Then the controller C will stabilize all processes P_1 such that $\delta_\nu(P, P_1) < \sigma_m(P, C)$, where δ_ν is the Vinnicombe metric.*

The theorem is a generalization of equation (13.11). The generalized stability margins can be related to the classical gain and phase margins. It follows from equation (13.34) that

$$|1 + P(i\omega)C(i\omega)|^2 \geq \sigma_m^2(1 + |P(i\omega)|^2)(1 + |C(i\omega)|^2). \tag{13.36}$$

If the Nyquist curve of the loop transfer function PC intersects the negative real axis for some ω we have $P(i\omega)C(i\omega) = -k$ for some $0 < k < 1$ and equation (13.36) becomes

$$|1 - k|^2 \geq \sigma_m^2(1 + |P(i\omega)|^2 + |C(i\omega)|^2 + k^2) \geq \sigma_m^2(1 + k)^2,$$

which implies that

$$k \leq \frac{1 - \sigma_m}{1 + \sigma_m}, \qquad g_m = \frac{1}{k} \geq \frac{1 + \sigma_m}{1 - \sigma_m}. \tag{13.37}$$

If the loop transfer function intersects the unit circle so that the phase margin is φ_m we have $P(i\omega)C(i\omega) = e^{i(\pi + \varphi_m)} = -e^{i\varphi_m}$ and equation (13.36) becomes

$$|1 - e^{i\varphi_m}|^2 \geq \sigma_m^2(1 + |P(i\omega)|^2 + 1/|P(i\omega)|^2 + 1) \geq 4\sigma_m^2,$$

where the last inequality follows from the fact that $|x| + 1/|x| \geq 2$. Since $|1 - e^{i\varphi_m}| = 2\sin(\varphi_m/2)$ (think geometrically) it follows that the above inequality can be written as

$$4\sin(\varphi_m/2) \geq 4\sigma_m^2, \qquad \varphi_m \geq 2\arcsin\sigma_m \tag{13.38}$$

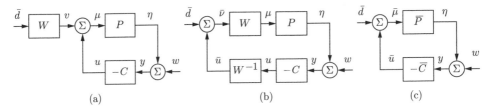

Figure 13.17: Block diagrams that illustrate frequency weighting of load disturbances. A frequency weight W is introduced on the load disturbance in (a). Block diagram transformations are used in (b) to obtain a system in standard form, which is redrawn in (c) using $\overline{P} = PW$ and $\overline{C} = W^{-1}C$.

(compare with equation 10.7). For $\sigma_\mathrm{m} = 1/3$, $1/2$, $2/3$ we have $g_\mathrm{m} \geq 2$, 3, 5 and $\varphi_\mathrm{m} \geq 39°$, $60°$, $84°$.

Disturbance Weighting

H_∞ control attempts to find a controller that minimizes the effect of external signals (χ in Figure 13.16a or ν and w in Figure 13.16b) on the generalized error ξ, in the sense that the largest singular value of the matrix $\|\mathcal{G}(P,C)\|_\infty$ is as small as possible. The solution of the problem can be changed by introducing weights W, which is illustrated in Figure 13.17a.

Figures 13.17b and 13.17c show how the problem with a weight W can be transformed into a problem of the same form as in Figure 13.17a. This allows the weighted problem to be solved using the same tools as the unweighted problem. In the transformed problem the process transfer function P is replaced by $\overline{P} = PW$ and the controller transfer function is replaced by $\overline{C} = W^{-1}C$. The relation between the transformed signals then becomes

$$\bar{\xi} = \begin{pmatrix} \bar{\mu} \\ \bar{\eta} \end{pmatrix} \begin{pmatrix} \dfrac{1}{1+\overline{P}\,\overline{C}} & \dfrac{\overline{P}}{1+\overline{P}\,\overline{C}} \\[2ex] \dfrac{\overline{C}}{1+\overline{P}\,\overline{C}} & \dfrac{\overline{P}\,\overline{C}}{1+\overline{P}\,\overline{C}} \end{pmatrix} \begin{pmatrix} \bar{v} \\ -w \end{pmatrix} = \mathcal{G}(\overline{P},\overline{C})\bar{\chi}.$$

Notice that $\overline{P}\,\overline{C} = PC$, which means that only the off diagonal block elements in the matrix $\mathcal{G}(\overline{P},\overline{C})$ are different from those in $\mathcal{G}(P,C)$. Weighting thus does not change the sensitivity and complementary sensitivity functions. The matrix element corresponding to load disturbances changes from $P/(1+PC)$ to $PW/(1+PC)$ and the matrix element corresponding to measurement noise changes from $C/(1+PC)$ to $CW^{-1}/(1+PC)$.

Having chosen the desired weight W, the solution to the weighted H_∞ problem gives the controller $\overline{C}$. Transforming back then gives the real controller $C = W\overline{C}$. Choosing proper weights allows the designer to obtain a controller that reflects the design specifications. If W is a scalar greater than one it means that we are increasing the effect of the load disturbances and reducing the effect of the measurement noise. The weighting can also be made frequency dependent. For example, choosing the weight as $W = k/s$ will automatically give a controller with integral action. Similarly a weighting that emphasizes high frequencies will give a controller with high-frequency roll-off. Frequency weighting allows the designer to modify the solution to reflect the many different design specifications, making H_∞ loop shaping a very powerful design method.

Limits of Robust Design

There is a limit to what can be achieved by robust design. In spite of the nice properties of feedback, there are situations where the process variations are so large that it is not possible to find a linear controller that gives a robust system with good performance. It is then necessary to use other types of controllers. In some cases it is possible to measure a variable that is well correlated with the process variations. Controllers for different parameter values can then be designed and the corresponding controller can be chosen based on the measured signal. This type of control design is called *gain scheduling* and it was discussed briefly in Section 8.5. The cruise controller is a typical example where the measured signal could be gear position and velocity. Gain scheduling is the common solution for high-performance aircraft where scheduling is done based on Mach number and dynamic pressure. When using gain scheduling, it is important to make sure that switches between the controllers do not create undesirable transients (often referred to as the *bumpless transfer* problem).

It is often not possible to measure variables related to the parameters, in which case *automatic tuning* and *adaptive control* can be used. In automatic tuning the process dynamics are measured by perturbing the system, and a controller is then designed automatically. Automatic tuning requires that parameters remain constant, and it has been widely applied for PID control. It is a reasonable guess that in the future many controllers will have features for automatic tuning. If parameters are changing, it is possible to use adaptive methods where process dynamics are measured online.

13.5   FURTHER READING

The topic of robust control is a large one, with many articles and textbooks devoted to the subject. Robustness was a central issue in classical control as described in the books by Bode [51], James, Nichols, and Phillips [130], and Horowitz [121]. Quantitative feedback theory (QFT) [124] can be regarded as an extension of Bode's work. The interesting properties of Bode's ideal loop transfer function were rediscovered in the late 1990s, creating an interest in fractional transfer functions [185]. It took a long time before the fundamental question of when two systems are similar was clearly formulated. The gap metric was introduced by Zames and El-Sakkary [264], and Vidyasagar introduced the graph metric a few year later [245, 246].

The ν-gap metric, which is the proper notion, is due to Vinnicombe [247, 248]. Robustness was de-emphasized in the euphoria of the development of design methods based on optimization. The strong robustness of controllers based on state feedback, shown by Anderson and Moore [9], contributed to the optimism. The poor robustness of output feedback was pointed out by Rosenbrock [210], Horowitz [122], and Doyle [75] and resulted in a renewed interest in robustness. A major step forward was the development of design methods where robustness was explicitly taken into account, such as the seminal work of Zames [263].

Robust control was originally developed using powerful results from the theory of complex variables, which gave controllers of high order. A major breakthrough was made by Doyle, Glover, Khargonekar, and Francis [77], who showed that the solution to the problem could be obtained using Riccati equations and that a controller of low order could be found. This paper led to an extensive treatment of H_∞ control, including books by Francis [92], McFarlane and Glover [181], Doyle, Francis, and Tannenbaum [76], Green and Limebeer [109], Zhou, Doyle, and Glover [265], Skogestad and Postlethwaite [223], and Vinnicombe [248]. A major advantage of the theory is that it combines much of the intuition from servomechanism theory with sound numerical algorithms based on numerical linear algebra and optimization. The results have been extended to nonlinear systems by treating the design problem as a game where the disturbances are generated by an adversary, as described in the book by Basar and Bernhard [29]. Gain scheduling and adaptation are discussed in the book by Åström and Wittenmark [23].

EXERCISES

13.1 Consider systems with the transfer functions $P_1 = 1/(s+1)$ and $P_2 = 1/(s+a)$. Show that P_1 can be changed continuously to P_2 with bounded additive and multiplicative uncertainty if $a > 0$ but not if $a < 0$. Also show that no restriction on a is required for feedback uncertainty.

13.2 Consider systems with the transfer functions $P_1 = (s+1)/(s+1)^2$ and $P_2 = (s+a)/(s+1)^2$. Show that P_1 can be changed continuously to P_2 with bounded feedback uncertainty if $a > 0$ but not if $a < 0$. Also show that no restriction on a is required for additive and multiplicative uncertainties.

13.3 (Difference in sensitivity functions) Let $T(P, C)$ be the complementary sensitivity function for a system with process P and controller C. Show that

$$T(P_1, C) - T(P_2, C) = \frac{(P_1 - P_2)C}{(1 + P_1 C)(1 + P_2 C)},$$

and compare with equation (13.6). Derive a similar formula for the sensitivity function.

13.4 (Vinnicombe metrics) Consider the transfer functions

$$P_1(s) = \frac{k}{4s+1}, \qquad P_2(s) = \frac{k}{(2s+1)^2}, \qquad P_3(s) = \frac{k}{(s+1)^4}.$$

Compute the Vinnicombe metric for all combinations of the transfer functions when $k = 1$ and $k = 2$. Discuss the results.

13.5 (Sensitivity of feedback and feedforward) Consider the system in Figure 13.11 and let G_{yr} be the transfer function relating the measured signal y to the reference r. Show that the sensitivities of G_{yr} with respect to the feedforward and feedback transfer functions F and C are given by $dG_{yr}/dF = CP/(1+PC)$ and $dG_{yr}/dC = FP/(1+PC)^2 = G_{yr}S/C$.

13.6 (Guaranteed stability margin) The inequality given by equation (13.10) guarantees that the closed loop system is stable for process uncertainties. Let $s_m^0 = 1/M_s^0$ be a specified stability margin. Show that the inequality

$$|\delta(i\omega)| < \frac{1 - s_m^0|S(i\omega)|}{|T(i\omega)|} = \frac{1 - |S(i\omega)|/M_s^0}{|T(i\omega)|}, \qquad \text{for all } \omega \geq 0,$$

guarantees that the closed loop system has a stability margin greater than s_m^0 for all perturbations (compare with equation (13.10)).

13.7 (Stability margins) Consider a feedback loop with the process and the controller having transfer functions P and C. Assume that the maximum sensitivity is $M_s = 2$. Show that the phase margin is at least $30°$ and that the closed loop system will be stable if the gain is changed by 50%.

13.8 Consider a process with the transfer function $P(s) = k/(s(s+1))$, where the gain can vary between 0.1 and 10. A controller that has a phase margin close to $\varphi_m = 45°$ for the gain variations can be obtained by finding a controller that gives the loop transfer function $L(s) = 1/(s\sqrt{s})$. Suggest how the transfer function can be implemented by approximating it by a rational function.

13.9 (Bode's ideal loop transfer function) Bode's ideal loop transfer function is given in Example 13.9. Show that the phase margin is $\varphi_m = 180° - 90°n$ and that the stability margin is $s_m = \sin \pi(1 - n/2)$. Make Bode and Nyquist plots of the transfer function for $n = 5/3$.

13.10 (Ideal delay compensator) Consider a process whose dynamics are a pure time delay with transfer function $P(s) = e^{-s}$. The ideal delay compensator is a controller with the transfer function $C(s) = 1/(1 - e^{-s})$. Show that the sensitivity functions are $T(s) = e^{-s}$ and $S(s) = 1 - e^{-s}$ and that the closed loop system will be unstable for arbitrarily small changes in the delay.

13.11 (Sensitivity of two degree-of-freedom controllers to process variations) Consider the two degree-of-freedom controller shown in Figure 12.13, which uses feedforward compensation to provide improved response to reference signals and measured disturbances. Show that the input/output transfer functions and the corresponding sensitivities to process variations for the feedforward, feedback, and combined controllers are given by

Controller	G_{yr}	$\dfrac{dG_{yr}}{G_{yr}}$	G_{yv}	$\dfrac{dG_{yv}}{dP_1}$
Feedforward ($C = 0$)	F_m	$\dfrac{dP}{P}$	0	$-\dfrac{P_2}{P_1}$
Feedback ($F_r, F_v = 0$)	TF_m	$S\dfrac{dP}{P}$	SP_2	$-S\dfrac{P_2}{P_1}$
Feedforward and Feedback	F_m	$S\dfrac{dP}{P}$	0	$S\dfrac{P_2}{P_1}$

13.12 (H_∞ control) Consider the matrix $\mathcal{G}(P, C)$ in equation (13.32). Show that it has the singular values

$$\sigma_1 = 0, \qquad \sigma_2 = \bar{\sigma} = \sup_\omega \frac{\sqrt{(1 + |P(i\omega)|^2)(1 + |C(i\omega)|^2)}}{|1 + P(i\omega)C(i\omega)|} = \|\mathcal{G}(P, C)\|_\infty.$$

Also show that $\bar{\sigma} = 1/\delta_\nu(P, -1/C)$, which implies that $1/\bar{\sigma}$ is a generalization of the closest distance of the Nyquist plot to the critical point and hence also serves as a measure of the stability margin.

13.13 (Disturbance weighting) Consider an H_∞ control problem with the disturbance weight W ($\overline{P} = PW$ and $\overline{C} = W^{-1}C$). Show that

$$\|\mathcal{G}(\overline{P}, \overline{C})\|_\infty \geq \sup_\omega \big(|S(i\omega)| + |T(i\omega)|\big).$$

Chapter Fourteen

Fundamental Limits

Many people have seen theoretical advantages in the facts that front-drive rear-steering recumbent bicycles would have simpler transmissions than rear-drive recumbents and could have the center of mass nearer to the front wheel than the rear. The U.S. Department of Transportation commissioned the construction of a safe motorcycle with this configuration. It turned out to be safe in an unexpected way: No one could ride it.

—F. R. Whitt and D. G. Wilson, *Bicycling Science*, 1997 [250].

In this chapter we discuss properties that limit performance and robustness of control systems. Non-minimum phase dynamics, due to time delays and right half-plane poles and zeros impose severe limits. There are also nonlinear behaviors that appear at large and small signal levels. Large signal limits can be caused by limited rate and power of actuators, or by constraints required to protect the process. Small signal limits can be caused by measurement noise, friction, and quantization in converters. We also discuss consequences of the limits for loop shaping, and give rules for pole placement design.

14.1 SYSTEM DESIGN CONSIDERATIONS

The initial design of a system can have a significant impact on the ability to use feedback to provide robustness and performance improvements. It is particularly important to recognize fundamental limits in the performance of feedback systems early in the design process. For example, we may expect that a system with time delays cannot admit fast control because control actions are delayed. Similarly it seems reasonable that unstable systems will require fast controllers, which will depend on the bandwidth of sensors and actuators. These limits are caused by properties of the system dynamics and can often be captured by conditions on the poles and zeros of the process.

The freedom for the control designer depends very much on the situation. The designer can be faced with a process with given sensors and actuators and his or her task is to design a suitable controller. The designer then has limited freedom. In other cases he or she may be able to choose sensors, and in yet other cases the location and characteristics of sensors, actuators, and controller are designed simultaneously. The designer then has significant freedom. The typical case is somewhere in between these extremes.

In any case, design engineers should be aware of the fundamental limits of feedback systems and be able to deal with them as early as possible in the design process. Awareness of the limits and co-design of the process and the controller are good ways to avoid potential difficulties both for system and control designers. The limits alluded to in the chapter quote are due to process dynamics and limits on actuation power and actuation rate. The dynamics limitations can be captured by time delays and poles and zeros in the right half-plane. It seems intuitively clear that a time delay in the process limits the achievable response speed. A less obvious case is that a process with a right half-plane pole/zero pair cannot be controlled robustly if the pole is close to the zero. Restriction in actuation can be captured by actuation power and actuation rates. These are all examples of fundamental limits whose potential impacts should be taken into account during initial system design.

Stabilizability and Strong Stabilizability

One of the most fundamental properties of a control system is the ability to design the dynamics of the (closed loop) system to meet a set of performance specifications. Often this can be captured by the location of poles and zeros in the relevant transfer functions, such as the Gang of Four. In Section 7.2 we found that a system must be reachable in order to find a state feedback that places closed loop eigenvalues in arbitrary positions. The corresponding condition for output feedback is that the system is both reachable and observable (Section 8.3). There are also trade-offs that are captured by the stability margin, bandwidth, peak values and locations of sensitivity functions, and many other features that we have encountered in the previous chapters.

One question of particular interest for systems whose natural dynamics are unstable is when a system can be stabilized using feedback and whether it can be stabilized using a *stable* controller (a condition known as *strong stabilizability*). The question of stabilizability is slightly different than reachability since it may turn out that there are stable eigenvalues that cannot be modified by feedback, but we can still modify all unstable eigenvalues. Strong stabilizability is important for system-level design since we may not want to implement an unstable controller unless it is necessary to do so. (Note that just having the controller be unstable does not mean that the closed loop system is unstable.)

A linear system with state feedback is always stabilizable if it is reachable. If a linear system is not reachable, it follows from Kalman's decomposition theorem (Section 8.3) that the system dynamics can be written as

$$\frac{dx}{dt} = \frac{d}{dt} \begin{pmatrix} x_{\mathrm{r}} \\ x_{\bar{\mathrm{r}}} \end{pmatrix} = \begin{pmatrix} A_{\mathrm{r}} & 0 \\ * & A_{\bar{\mathrm{r}}} \end{pmatrix} \begin{pmatrix} x_{\mathrm{r}} \\ x_{\bar{\mathrm{r}}} \end{pmatrix} + \begin{pmatrix} B_{\mathrm{r}} \\ 0 \end{pmatrix} u, \qquad (14.1)$$

where the states have been decomposed into two parts: the reachable states x_{r} and the unreachable states $x_{\bar{\mathrm{r}}}$. The dynamics in the invariant subspace represented by x_{r} are reachable and it follows that we can always find a state feedback K_{r} such that $A_{\mathrm{r}} - K_{\mathrm{r}}B_{\mathrm{r}}$ has arbitrary eigenvalues. The system (14.1) is then stabilizable if and only if the eigenvalues of $A_{\bar{\mathrm{r}}}$ are in the left half-plane. A system with state feedback is thus stabilizable if the unreachable part of the system is stable.

Reachability and stabilizability for systems with state feedback can also be stated as a rank condition. A system with dynamics and control matrices A, B having n state variables is reachable if and only if

$$\mathrm{rank} \begin{pmatrix} A - sI & B \end{pmatrix} = n \qquad (14.2)$$

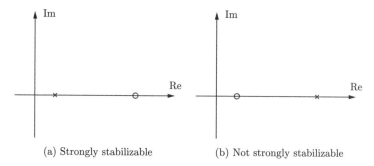

(a) Strongly stabilizable   (b) Not strongly stabilizable

Figure 14.1: Pole zero diagrams for strongly stabilizable and non-strongly stabilizable systems. The system in (a) can be stabilized with a stable controller, but stabilization of the system in (b) requires a controller with a pole in the right half-plane.

for all values of $\lambda \in \mathbb{C}$. This test is known as the Popov-Belevitch-Hautus (PBH) test. The system is *stabilizable* if the condition holds for all λ in the right half-plane $\operatorname{Re} s \geq 0$ ($\operatorname{Re} s > 0$ for *strict* stabilizability). Stabilizabilty for systems with state feedback is thus a weaker condition than reachability.

For a linear system with output feedback a controller can be constructed using an estimator and linear feedback on the state estimate, and the resulting controller has the input/output dynamics given in equation (8.16), repeated here:

$$\frac{d\hat{x}}{dt} = A\hat{x} + Bu + L(y - C\hat{x}), \qquad u = -K\hat{x}. \tag{14.3}$$

The controller poles are the eigenvalues of the matrix $A - BK - LC$, and the controller zeros are the values of s where the matrix

$$\begin{pmatrix} A - BK - LC - sI & L \\ K & 0 \end{pmatrix} \tag{14.4}$$

loses rank. If a system is stabilizable and observable, it is always possible to construct a stabilizing controller. However, the question about whether this controller is stable (corresponding to *strong* stabilizability) is more subtle. Strong stabilizability can be expressed as conditions on the transfer function, as described in the following theorem.

Theorem 14.1 (Strong stabilizability). *Consider a linear system with the rational transfer function $P(s) = n(s)/d(s)$, where the polynomials $n(s)$ and $d(s)$ do not have a factor in common. The system can be strongly stabilized if and only if all $d(z_k)$ have the same sign for all z_k such that $n(z_k) = 0$.*

This theorem is proven in Vidyasagar [246, Theorem 3.1 and Corollary 3.3] (see also Youla [260]). For a system with a single pole at p and zero at z, this result implies that a process with $p > z$ requires a controller with a pole in the right half-plane, hence an *unstable* controller. This situation is illustrated in Figure 14.1. An example is given in Exercise 14.1. The root locus method gives significant insight into these cases.

Another characterization of strong stabilizability is given in Doyle, Frances, and Tannenbaum [76, Theorem 3, Chapter 5]:

Theorem 14.2. *A linear system P is strongly stabilizable if and only if it has an even number of real poles between every pair of real zeros in $\mathrm{Re}\, s \geq 0$.*

These two results show that strong stabilizability depends on the *patterns* of poles and zeros, which are often determined in the early stages of system design. Note that this does not imply that unstable systems should always be avoided, because instability may actually have advantages. A typical example is when high maneuverability is desired, such as in high-performance aircraft.

Right Half-Plane Zeros and Time Delays

In addition to questions related to stabilizability, we will see throughout this chapter that there can be significant limitations on closed loop performance when a system has zeros in the right half-plane or time delays in the loop transfer function. A natural question to ask is whether these features can be avoided at the time of system design.

The poles of a system depend on the intrinsic dynamics of the system. They represent the modes of the system and they are given by the eigenvalues of the dynamics matrix A of the linearized model. Sensors and actuators have no effect on the poles: the only way to change poles is by feedback or by redesign of the process. However, the zeros of a system depend on how the sensors and actuators are connected to the process. Zeros can thus be changed by moving or adding sensors and actuators, which is often simpler than redesigning the process dynamics.

The following example illustrates how the location of zeros can be determined through placement of sensors.

Example 14.1 Vehicle steering

Consider the vehicle steering system introduced in Example 3.11. The linearized (but non-normalized) model of the dynamics of the system relating lateral velocity to steering angle was given in Example 10.11 and has the form

$$P(s) = \frac{av_0 s + v_0^2}{bs},$$

where v_0 is the velocity of the vehicle, a is the offset to the reference point for the vehicle position, and b is the wheelbase. We observed that the system has a right half-plane zero when the velocity of the vehicle is negative and this can lead to limits in the closed loop performance of the system, such as those described in Example 10.11.

The existence of the right half-plane zero can be removed if we choose to measure the location of the vehicle by the position of the center of the rear wheels instead of the center of mass. This gives $a = 0$ and our dynamics become

$$G(s) = \frac{v_0^2}{bs},$$

which no longer has a right half-plane zero. Choosing this output can simplify the design constraints and is easily implemented by calibrating the position sensor for the vehicle so that it returns the position of the center of the rear wheels.

We note that this choice of "sensor" is subject to calibration errors and this can lead to a zero of the process transfer function at v_0/ϵ, where ϵ represents the calibration error and the sign of the zero depends on the sign of the calibration error and the direction of travel. We will see later in the chapter that this corresponds that what we call a "fast" zero and its impact on fundamental limits is relatively minor. Thus it can be advantageous to choose the system output to be at a different point in order to simplify the feedback controller design. ▽

Another source of limitations is due to time delays, which can add significant phase lag to the loop transfer function, making it difficult to maintain sufficient phase margin. Time delays may appear in the process, in communication channels, and in computations. Time delays have effects similar to right half-plane zeros. One way to see this is to consider the Padé approximation for a time delay, which provides a unity gain, rational transfer function whose phase approximates that of a time delay. The first- and second-order Padé approximations are given by

$$G_1(s) = \frac{1 - s\tau/2}{1 + s\tau/2}, \qquad G_2(s) = \frac{1 - \tau s/2 + (\tau s)^2/12}{1 + \tau s/2 + (\tau s)^2/12}.$$

The first-order Padé approximation has a right half-plane zero at $2/\tau$ and the second-order Padé approximation has a complex conjugate pair of right half-plane zeros at $s = (3 \pm i\sqrt{3})/\tau$.

Unlike zeros, time delays cannot generally be avoided by choice of sensor or actuator location, and hence they should be avoided by proper design of the system's computing and communications architecture. Minimizing time delays whenever possible is usually a good design guideline for feedback control systems.

14.2 BODE'S INTEGRAL FORMULA

One of the most important limits in feedback control design was obtained by Bode, who showed that it was not possible to uniformly improve the performance of certain closed loop performance characteristics. Bode's result makes use of the sensitivity function S introduced in Section 12.1, which gives an overview of performance and robustness of a closed loop system. Specifically, it describes how disturbances are attenuated by feedback and allows comparison of disturbance attenuation of open and closed loop systems. We recall that disturbances with frequency ω are attenuated by feedback if $|S(i\omega)| < 1$, and disturbances with frequencies such that $|S(i\omega)| > 1$ are amplified. The maximum sensitivity $M_s = \max_\omega |S(i\omega)|$ gives the largest amplification and is also a robustness measure, since $1/M_s$ is equal to the stability margin s_m (see Figure 10.12).

A key observation is that the sensitivity function cannot be made small over a wide frequency range. There is an invariant (conserved quantity) called *Bode's integral formula* that implies that reducing the sensitivity at one frequency increases it at another, and the situation is worse if the process has right half-plane poles. Control design is thus always a compromise. The following theorem captures limits of performance under feedback.

Theorem 14.3 (Bode's integral formula). *Let $S(s)$ be the sensitivity function of an internally stable closed loop system with loop transfer function $L(s)$. Assume that the loop transfer function $L(s)$ is such that $sL(s)$ goes to zero as $s \to \infty$. Then the sensitivity function has the property*

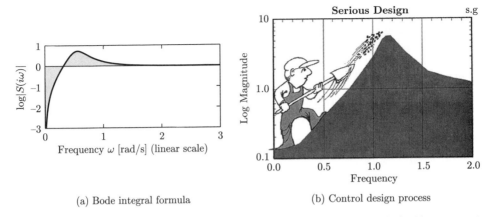

(a) Bode integral formula   (b) Control design process

Figure 14.2: Interpretation of the *waterbed effect*. The function $\log|S(i\omega)|$ is plotted versus ω using a linear scale in (a). According to Bode's integral formula (14.5), the area of $\log|S(i\omega)|$ above zero must be equal to the area below zero. Gunter Stein's interpretation of design as a trade-off of sensitivities at different frequencies is shown in (b) (from [229]).

$$\int_0^\infty \log|S(i\omega)|\,d\omega = \int_0^\infty \log\frac{1}{|1+L(i\omega)|}\,d\omega = \pi\sum p_k, \qquad (14.5)$$

where the sum is over the right half-plane poles p_k of $L(s)$.

Equation (14.5) implies that if we design a controller that decreases the effect of disturbances for some frequencies it will increase the effect for other frequencies because the integral of $\log|S(i\omega)|$ remains constant. This property is sometimes referred to as the *waterbed effect*. It also follows that systems with open loop poles in the right half-plane have larger overall sensitivity than stable systems.

Equation (14.5) can be regarded as a *conservation law*: if the loop transfer function has no poles in the right half-plane, the equation simplifies to

$$\int_0^\infty \log|S(i\omega)|\,d\omega = 0.$$

This formula can be given a nice geometric interpretation as illustrated in Figure 14.2, which shows $\log|S(i\omega)|$ as a function of ω. The area over the horizontal axis must be equal to the area under the axis when the frequency is plotted on a *linear* scale. Thus if we wish to make the sensitivity smaller up to some frequency ω_{sc}, we must balance this by increased sensitivity above ω_{sc}. Control system design can be viewed as trading the disturbance attenuation at some frequencies for disturbance amplification at other frequencies. Notice that the assumption $\lim_{s\to\infty} sL(s) = 0$ is essential. Exercise 14.2 shows that without this assumption the sensitivity can be made arbitrarily small. A modification that covers $\lim_{s\to\infty} sL(s) = k$ is given in Exercise 14.3.

An equation similar to equation (14.5) holds for the complementary sensitivity function:

$$\int_0^\infty \frac{\log|T(i\omega)|}{\omega^2}\,d\omega = \pi\sum\frac{1}{z_i}, \quad T(s) = \frac{L(s)}{1+L(s)}, \qquad (14.6)$$

(a) X-29 aircraft

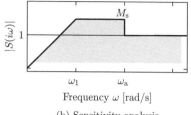

(b) Sensitivity analysis

Figure 14.3: X-29 flight control system. The aircraft makes use of forward swept wings and a set of canards on the fuselage to achieve high maneuverability (a). The desired sensitivity for the closed loop system is shown in (b). We seek to use our control authority to shape the sensitivity curve so that we have low sensitivity (good performance) up to frequency ω_1 by creating higher sensitivity up to our actuator bandwidth ω_a.

where the summation is over all right half-plane zeros of the loop transfer function $L(s) = P(s)C(s)$ (Exercise 14.4). It follows from equation (14.6) that slow right half-plane zeros are worse than fast ones, just as equation (14.5) implies that fast right half-plane poles are worse than slow ones.

Example 14.2 The X-29 aircraft

As an illustration of Bode's integral formula, we present an analysis of the control system for the X-29 aircraft (see Figure 14.3a), which has an unusual configuration of aerodynamic surfaces that is designed to enhance its maneuverability. This analysis was originally carried out by Gunter Stein in his inaugural IEEE Bode lecture "Respect the Unstable" [229].

To analyze the system, we make use of a small set of parameters that describe the key properties of the system. A typical robustness requirement in aerospace systems is that the phase margins should be at least $\varphi_m = 45°$. The X-29 has longitudinal dynamics that are similar to inverted pendulum dynamics (Exercise 9.3). It has a right half-plane pole at approximately $p = 6 \, \text{rad/s}$ and a right half-plane zero at $z = 26 \, \text{rad/s}$. The actuators that stabilize the pitch have a bandwidth of $\omega_a = 40 \, \text{rad/s}$ and the desired bandwidth of the pitch control loop is $\omega_1 = 3 \, \text{rad/s}$.

To evaluate the achievable performance, we search for a control law such that the sensitivity function is small up to the desired bandwidth and not greater than M_s beyond that frequency. Because of Bode's integral formula, we know that M_s must be greater than 1 at high frequencies to balance the small sensitivity at low frequency. We thus ask if we can find a controller that has the shape shown in Figure 14.3b with the smallest possible value of M_s. Note that the sensitivity above the frequency ω_a is 1 since we have no actuator authority above those frequencies. Thus, we desire to design a closed loop system that has low sensitivity at frequencies below ω_1 and sensitivity that is not too large between ω_1 and ω_a.

From Bode's integral formula, we know that whatever controller we choose, equation (14.5) must hold. We will assume that the sensitivity function is given by

$$|S(i\omega)| = \begin{cases} \frac{\omega}{\omega_1} M_s & \text{if } \omega < \omega_1, \\ M_s & \text{if } \omega_1 \leq \omega < \omega_a, \\ 1 & \text{if } \omega_a \leq \omega < \infty, \end{cases}$$

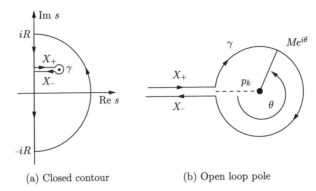

(a) Closed contour   (b) Open loop pole

Figure 14.4: Contour used to prove Bode's theorem. For each right half-plane pole p_k of the loop transfer function $L(s)$, which is also a singularity of $\log S(s)$, we create a path from the imaginary axis that encircles the pole. To avoid clutter only one of the paths is shown.

corresponding to Figure 14.3b. Bode's integral becomes

$$\int_0^\infty \log|S(i\omega)|\,d\omega = \int_0^{\omega_a} \log|S(i\omega)|\,d\omega$$
$$= \int_0^{\omega_1} \log\frac{\omega M_s}{\omega_1}\,d\omega + (\omega_a - \omega_1)\log M_s = \pi p.$$

Integration by parts gives, after some calculation, $-\omega_1 + \omega_a \log M_s = \pi p$ or

$$M_s = e^{(\pi p + \omega_1)/\omega_a}.$$

This formula tells us what the achievable value of M_s will be for the given control specifications. In particular, using $p = 6\,\text{rad/s}$, $\omega_1 = 3\,\text{rad/s}$ and $\omega_a = 40\,\text{rad/s}$, we find that $M_s = 1.75$, which means that in the range of frequencies between ω_1 and ω_a, disturbances at the input to the process dynamics (such as wind) will be amplified by a factor of 1.75 in terms of their effect on the aircraft. With $M_s = 1.75$ we can also obtain an estimate of the phase margin as $\varphi_m \geq 2\arcsin 1/(2M_s) = 33°$ (equation (10.7)), which indicates that the requirement $\varphi_m = 45°$ may not be achievable. ∇

 Derivation of Bode's Integral Formula

Bode's integral formula (Theorem 14.3) can be derived by contour integration. We assume that the loop transfer function has distinct poles at $s = p_k$ in the right half-plane and that $L(s)$ goes to zero faster than $1/s$ for large values of s. Consider the integral of the logarithm of the sensitivity function $S(s) = 1/(1 + L(s))$ along the Nyquist contour Γ shown in Figure 14.4. The contour encloses the right half-plane except for the points $s = p_k$ where the loop transfer function $L(s) = P(s)C(s)$ has poles and the sensitivity function $S(s)$ therefore has singularities (only one p_k is shown in the figure). The direction of the contour is counterclockwise. The integral of the logarithm of the sensitivity function around the contour Γ is given by

$$I = \int_\Gamma \log(S(s))\,ds = I_1 + I_2 + I_3 = 0.$$

The integral I is zero because the function $\log S(s)$ is analytic with no poles or zeros inside the contour. The term I_1 is the integral along the imaginary axis, the term I_2 is the integral along a large semicircle to the right with a radius R that we will make infinitely large. The term I_3 is the integral along two parallel horizontal lines and a small circle enclosing p_k as shown in Figure 14.4.

We now compute each of the terms in the contour integration. We have

$$I_1 = -i \int_{-R}^{R} \log(S(i\omega))d\omega = -2i \int_0^R \log(|S(i\omega)|)d\omega$$

because the real part of $\log S(i\omega)$ is an even function and the imaginary part is an odd function. Furthermore we have

$$I_2 = \int_{\supset} \log(S(s))\,ds = -\int_{\supset} \log(1 + L(s))\,ds \approx -\int_{\supset} L(s)\,ds,$$

where $\supset$ represents the semicircular portion of Γ at radius R. Since $L(s)$ goes to zero faster than $1/s$ for large s, the integral goes to zero when the radius of the semicircle goes to infinity.

Next we consider the integral I_3. We split the contour into three parts: X_+, γ, and X_-, where X_+ and X_- are horizontal lines from the imaginary axis to p_k, and γ is a small circle with radius r around the point p_k (see Figure 14.4b). We can write the contour integral as

$$I_3 = \int_{X_+} \log S(s)\,ds + \int_{\gamma} \log S(s)\,ds + \int_{X_-} \log S(s)\,ds.$$

The point p_k is a pole of $L(s)$ and hence a zero of $S(s)$, which causes $\log S(s)$ to become singular at p_k. The magnitude of the integrand for the middle integral (along γ) is of the order $\log r$ and the length of the path is $2\pi r$, and it can be shown that the magnitude of the integral goes to zero as the radius r goes to zero. At the same time, $S(s) \approx k(s - p_k)$ near p_k, so the argument of $\log S(s)$ decreases by 2π as the contour encircles p_k (in the clockwise direction). On the contours X_+ and X_- we thus have

$$|S_{X_+}| = |S_{X_-}|, \qquad \arg S_{X_-} = \arg S_{X_+} - 2\pi.$$

Hence

$$\log(S_{X_+}) - \log(S_{X_-}) = \log(|S_{X_+}|) + i\,\arg(S_{X_+}) - \log(|S_{X_-}|) - i\,\arg(S_{X_-}) = 2\pi i.$$

Using the fact that the path X_+ is traversed in the opposite direction from X_-, the first and third terms can be combined to give

$$\int_{X_+} \log S(s)\,ds + \int_{X_-} \log S(s)\,ds = \int_{X_+} \left(\log S_{X_+}(s) - \log S_{X_-}(s)\right)\,ds.$$

The length of the path from the imaginary axis to p_k is $\mathrm{Re}\,p_k$ and we get

$$\int_{X_+} \log S(s)\,ds + \int_{X_-} \log S(s)\,ds = 2\pi i \cdot \mathrm{Re}\,p_k.$$

Repeating the argument for all p_k in the right half-plane, and letting the small circles go to zero gives

$$I_1 + I_2 + I_3 = -2i \int_0^\infty \log |S(i\omega)| \, d\omega + i \sum_k 2\pi \operatorname{Re} p_k = 0.$$

Since the p_k's appear as complex conjugate pairs, we have $\sum_k \operatorname{Re} p_k = \sum_k p_k$, which gives Bode's formula (14.5).

14.3   GAIN CROSSOVER FREQUENCY INEQUALITY

We will now investigate the effect of non-minimum phase process dynamics for loop shaping design. The key idea of loop shaping design is to shape the loop transfer function $L(i\omega) = P(i\omega)C(i\omega)$ so that the closed loop system has good performance and robustness. Good performance is obtained by making $|L(i\omega)|$ large for frequencies where we want disturbance attenuation and small for high frequencies where measurement noise dominates. Recall from Figure 12.8 that good robustness is obtained by shaping the loop transfer function around the gain crossover frequency ω_{gc}. The performance limits show up very clearly in the design.

To explore the limits due to right half-plane poles and zeros, we factor the process transfer function as

$$P(s) = P_{\mathrm{mp}}(s) P_{\mathrm{ap}}(s), \tag{14.7}$$

where P_{mp} is the minimum phase factor and P_{ap} is the non-minimum phase factor. We do the factorization so that P_{mp} has all its poles and zeros in the open left half-plane. The factorization is normalized so that $|P_{\mathrm{ap}}(i\omega)| = 1$, and the sign is chosen so that P_{ap} has negative phase. The transfer function P_{ap} is called an *all-pass system* because it has unit gain for all frequencies. For example,

$$P(s) = \frac{s-2}{(s+1)(s-1)} = \frac{s+2}{(s+1)^2} \cdot \frac{(s-2)(s+1)}{(s+2)(s-1)} = P_{\mathrm{mp}}(s) \cdot P_{\mathrm{ap}}(s). \tag{14.8}$$

Since $|P_{\mathrm{ap}}(i\omega)| = 1$, the transfer functions $P(s)$ and $P_{\mathrm{mp}}(s)$ have the same gain curves but the transfer function $P(s)$ has larger phase lag than $P_{\mathrm{mp}}(s)$.

Consider the closed loop system obtained with a controller with the transfer function $C(s)$. Requiring that the phase margin be φ_{m}, we get the inequality

$$\arg L(i\omega_{\mathrm{gc}}) = \arg P_{\mathrm{ap}}(i\omega_{\mathrm{gc}}) + \arg P_{\mathrm{mp}}(i\omega_{\mathrm{gc}}) + \arg C(i\omega_{\mathrm{gc}}) \geq -\pi + \varphi_{\mathrm{m}}, \tag{14.9}$$

where ω_{gc} is the gain crossover frequency. Let n_{gc} be the slope of the gain curve of the loop transfer function $L(s) = P(s)C(s)$ at the crossover frequency. Since $|P_{\mathrm{ap}}(i\omega)| = 1$ it follows that

$$n_{\mathrm{gc}} = \left. \frac{d \log |L(i\omega)|}{d \log \omega} \right|_{\omega = \omega_{\mathrm{gc}}} = \left. \frac{d \log |P_{\mathrm{mp}}(i\omega)C(i\omega)|}{d \log \omega} \right|_{\omega = \omega_{\mathrm{gc}}}.$$

Assuming that the controller $C(s)$ has neither poles nor zeros in the right half-plane, it then follows from Bode's relations (equation (10.9)) that

$$\arg P_{\mathrm{mp}}(i\omega) + \arg C(i\omega) \approx n_{\mathrm{gc}} \frac{\pi}{2}.$$

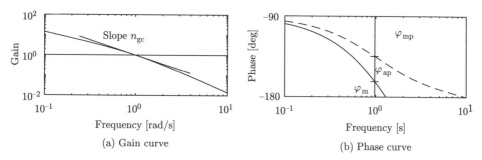

(a) Gain curve   (b) Phase curve

Figure 14.5: Illustration of the gain crossover frequency inequality. (a) Gain curve of the transfer function, with the slope of the curve at the gain crossover frequency n_{gc} marked. (b) Phase of the transfer function (solid) and its minimum phase component (dashed). The phase margin φ_{m}, the phase lags φ_{mp} and φ_{ap} of the minimum phase component, and the all-pass component are shown in the figure.

Combining this with equation (14.9) gives the following inequality for the allowable phase lag of the all-pass part at the gain crossover frequency, which we state as a theorem.

Theorem 14.4 (Gain crossover frequency inequality). *Let* $P(s) = P_{mp}(s)P_{ap}(s)$, *where* P_{ap} *is an all-pass transfer function containing the non-minimum phase portion of* $P(s)$. *If* $C(s)$ *is a stabilizing compensator for the closed loop system with no right half-plane poles and zeros and with phase margin* φ_m, *gain crossover frequency* ω_{gc}, *and gain crossover slope* n_{gc}, *then the allowable phase lag for the all-pass transfer function must satisfy the inequality*

$$\varphi_{ap} := -\arg P_{ap}(i\omega_{gc}) \leq \pi - \varphi_m + n_{gc}\frac{\pi}{2} := \bar{\varphi}_{ap}. \qquad (14.10)$$

The gain crossover frequency inequality is illustrated in Figure 14.5. The condition (14.10) requires that the gain crossover frequency must be chosen so that the phase lag of the all-pass factor is not too large. For systems with high robustness requirements we may choose a phase margin of 60° ($\varphi_{\mathrm{m}} = \pi/3$). To have a reasonable flexibility in choosing the gain crossover frequency we choose $n_{\mathrm{gc}} = -1$, which gives an admissible phase lag $\bar{\varphi}_{\mathrm{ap}} = \pi/6 \approx 0.52$ rad (30°) for the all-pass component. For systems where we can accept a lower robustness we might choose a phase margin of 45° ($\varphi_{\mathrm{m}} = \pi/4$) and the slope $n_{\mathrm{gc}} = -1/2$, which gives an admissible phase lag $\bar{\varphi}_{\mathrm{ap}} = \pi/2 \approx 1.57$ rad (90°).

The gain crossover frequency inequality (14.10) shows that non-minimum phase components impose severe restrictions on possible crossover frequencies and that there are systems that cannot be controlled with sufficient stability margins. We illustrate the limits in a number of commonly encountered situations.

Example 14.3 Crossover frequency limits for a process with a zero in the right half-plane
The non-minimum phase part of the process transfer function for a system with a right half-plane zero is

$$P_{\mathrm{ap}}(s) = \frac{z - s}{z + s},$$

where $z > 0$. Notice that we have $z - s$ in the numerator instead of $s - z$ to satisfy the condition that P_{ap} should have negative phase. The phase lag of the all-pass factor is

$$\varphi_{\mathrm{ap}} = -\arg P_{\mathrm{ap}}(i\omega) = 2 \arctan \frac{\omega}{z}.$$

Let the admissible phase lag of the all-pass factor be $\bar{\varphi}_{\mathrm{ap}}$. The inequality (14.10) then gives the following bound on the crossover frequency:

$$\omega_{\mathrm{gc}} \leq z \tan(\bar{\varphi}_{\mathrm{ap}}/2). \tag{14.11}$$

With $\bar{\varphi}_{\mathrm{ap}} = \pi/3$ we get $\omega_{\mathrm{gc}} < 0.6\,z$. We can thus conclude that a right half-plane zero limits the achievable gain crossover frequency ω_{gc}, and slow right half-plane zeros (z small) give lower crossover frequency than fast right half-plane zeros. $\quad\nabla$

Processes with zeros in the right half-plane are not uncommon, and they are often due to inherent consequences of the physics, as in Exercise 14.5, which models hydroelectric power generation. Another example is the shrink and swell phenomenon in drum level control discussed in Example 3.14. In that example the zero in the right half-plane is associated with the inverse response characteristic, where the step response initially moves in the wrong direction. The effect also appears in product development projects where the cost initially increases during the development phase and then hopefully decreases to give profit when the product appears on the market.

We next consider the case of right half-plane poles.

Example 14.4 Crossover frequency limits for a process with a pole in the right half-plane

The non-minimum phase part of the transfer function for a system with a pole in the right half-plane is

$$P_{\mathrm{ap}}(s) = \frac{s + p}{s - p},$$

where $p > 0$. The sign of P_{ap} is dictated by the condition that it should have negative phase. The phase lag of the non-minimum phase part is

$$\varphi_{\mathrm{ap}} = -\arg P_{\mathrm{ap}}(i\omega) = 2 \arctan \frac{p}{\omega},$$

and the inequality (14.10) gives the following bound on the crossover frequency:

$$\omega_{\mathrm{gc}} \geq \frac{p}{\tan(\bar{\varphi}_{\mathrm{ap}}/2)}, \tag{14.12}$$

where $\bar{\varphi}_{\mathrm{ap}}$ is the maximum admissible phase lag of the all-pass factor P_{ap}. Right half-plane poles thus require that the closed loop system has a sufficiently high gain crossover frequency. With $\bar{\varphi}_{\mathrm{ap}} = \pi/3$ we get $\omega_{\mathrm{gc}} > 1.7p$. Fast right half-plane poles (p large) therefore require a larger gain crossover frequency than slower right half-plane poles. Robust control of unstable systems thus requires that the bandwidths of the process, the actuators, and the sensors are sufficiently high. $\quad\nabla$

Example 14.5 Phase lag for processes with a right half-plane pole/zero pair

Consider a system with a right half-plane zero z and a right half-plane pole p. The transfer function of the process and its all-pass factor are given by

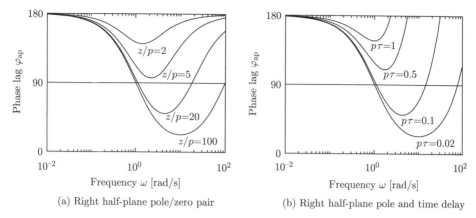

(a) Right half-plane pole/zero pair

(b) Right half-plane pole and time delay

Figure 14.6: Illustration of the gain crossover frequency inequality for systems with a zero and a pole in the right half-plane (a) and systems with a time delay and a right half-plane pole (b). The figures show the phase lag φ_{ap} of the all-pass factor P_{ap} as a function of frequency for the systems using equations (14.14) and (14.15). All systems have a right half-plane pole at $p=1$. The systems in (a) have zeros at $z=2$, 5, 20, and 100, and the systems in (b) have time delays $\tau=0.02$, 0.1, 0.5, and 1.

$$P(s) = \frac{a-z}{s-p}, \qquad P_{\mathrm{ap}}(s) = \frac{(z-s)(s+p)}{(z+s)(s-p)}. \tag{14.13}$$

The all-pass factor has the phase lag

$$\varphi_{\mathrm{ap}} = -\arg P_{\mathrm{ap}}(i\omega) = 2\arctan(\omega/z) + 2\arctan(p/\omega), \tag{14.14}$$

which is plotted in Figure 14.6a for $z/p = 2, 5, 20$, and 100.

We will illustrate with some numerical values. If we require that the phase lag φ_{ap} of the non-minimum phase factor be less than 90°, we must require that the ratio z/p be larger than 6 (from Figure 14.6). The pole and the zero must thus be sufficiently separated (Exercise 14.6). The values of the gain crossover frequency ω_{gc} are also quite restricted.

Notice that we cannot apply Theorem 14.4 if $p > z$ because a stabilizing controller must then have a pole in the right half-plane (see Figure 14.1). ∇

Time delays also impose limits similar to those given by zeros in the right half-plane. For a process with time delay, $P_{\mathrm{ap}}(s) = e^{\tau s}$. Using the gain crossover frequency inequality (14.10) we get $\omega_{\mathrm{gc}}\tau \leq \overline{\varphi}_{\mathrm{ap}}$, where τ is the time delay. Time delays are thus similar to right half-plane zeros because they require that the bandwidth and the crossover frequencies be sufficiently small.

Example 14.6 Phase lag for processes with a right half-plane pole and time delay

Consider a system with all-pass factor and phase lag given by

$$P_{\mathrm{ap}}(s) = \frac{s+p}{s-p} e^{-\tau s}, \qquad \varphi_{\mathrm{ap}} = -\arg P_{p\tau}(i\omega) = \omega\tau + 2\arctan(p/\omega). \tag{14.15}$$

A plot of the phase lag of the all-pass factor is given in Figure 14.6b. The figure shows that the behavior is similar to a system with a right half-plane pole/zero pair. The phase lag φ_{ap} has a minimum $\sqrt{\tau(2 - p\tau)} + 2\arctan\sqrt{p\tau/(2 - p\tau)}$ for $\omega\tau = \sqrt{p\tau(2 - p\tau)}$ (Exercise 14.7). It follows from equation (14.9) that a system with a right half-plane pole p and a time delay τ cannot be stabilized by a controller with no poles and zeros in the right half-plane if $p\tau \geq 2$. ∇

Systems with a pole/zero pair in the right half-plane are not common. In Example 14.2 we encountered the X-29 aircraft (Exercise 14.8). The next example is another illustration.

Example 14.7 Balance system

As an example of a system with both right half-plane poles and zeros, consider the balance system with zero damping introduced in Example 3.2. The transfer functions from force F to output angle θ and position q were derived in Example 9.11:

$$H_{\theta F}(s) = \frac{ml}{(M_{\mathrm{t}}J_{\mathrm{t}} - m^2 l^2)s^2 - mglM_{\mathrm{t}}},$$

$$H_{qF}(s) = \frac{J_{\mathrm{t}}s^2 - mgl}{s^2\big((M_{\mathrm{t}}J_{\mathrm{t}} - m^2 l^2)s^2 - mglM_{\mathrm{t}}\big)}.$$

Assume that we want to stabilize the pendulum by using the cart position as the measured signal. The transfer function H_{qF} from the input force F to the cart position q has poles $\{0, 0, \pm\sqrt{mglM_{\mathrm{t}}/(M_{\mathrm{t}}J_{\mathrm{t}} - m^2 l^2)}\}$ and zeros $\{\pm\sqrt{mgl/J_{\mathrm{t}}}\}$. Using the parameters in Example 7.7, the right half-plane pole is at $p = 2.68$ and the zero is at $z = 2.09$. With the best choice of the gain crossover frequency, it follows from equation (14.14) that the phase lag of the all-pass component P_{ap} is 166°, which implies that it is impossible to obtain a reasonable phase margin. The pole/zero ratio is 1.28, which is far from the value 6 required to control the system robustly. Using Figure 14.6, we see that the amount of achievable phase margin for the system is very small if we desire a bandwidth in the range of 2–4 rad/s.

The right half-plane zero of the system can be eliminated by changing the output of the system. For example, if we choose the output to correspond to a position at a distance r along the pendulum, we have $y = q - r\sin\theta$ and the transfer function for the linearized output becomes

$$H_{yF}(s) = H_{qF}(s) - rH_{\theta F}(s) = \frac{(J_{\mathrm{t}} - mlr)s^2 - mgl^2}{s^2\big((M_{\mathrm{t}}J_{\mathrm{t}} - m^2 l^2)s^2 - mglM_{\mathrm{t}}\big)}.$$

If we choose r sufficiently large, then $mlr - J_{\mathrm{t}} > 0$ and we eliminate the right half-plane zero, obtaining instead two pure imaginary zeros. The gain crossover frequency is determined by the right half-plane pole $p = \sqrt{mglM_{\mathrm{t}}/(M_{\mathrm{t}}J_{\mathrm{t}} - m^2 l^2)}$ (Example 14.4). If our admissible phase lag for the non-minimum phase part is $\varphi_l = 45°$, then our gain crossover must satisfy

$$\omega_{\mathrm{gc}} \geq \frac{p}{\tan(\varphi_l/2)} = 6.48\,\mathrm{rad/s}.$$

If the actuators have sufficiently high bandwidth, e.g., a factor of 10 above ω_{gc} or roughly 10 Hz, then we can provide robust tracking up to this frequency. ∇

14.4   THE MAXIMUM MODULUS PRINCIPLE

Significant insight into the fundamental limits imposed by poles and zeros in the right half-plane and time delays can be obtained with simple calculations by using the *maximum modulus principle*.

Theorem 14.5 (Maximum modulus principle). *Let $\Omega \subset \mathbb{C}$ be a nonempty, bounded, open, and connected set in the complex plane and let $G : \overline{\Omega} \to \mathbb{C}$ be continuous on the closure of Ω and analytic on Ω. Then*

$$\sup_{s \in \overline{\Omega}} |G(s)| = \sup_{s \in \partial \Omega} |G(s)|.$$

This theorem can be used to give bounds on transfer functions, such as the sensitivity functions, by using the Nyquist contour as the boundary of the open right half-plane. We state this result as a corollary.

Corollary. *Let $G(s)$ be a bounded analytic transfer function on the closed, right half-plane. Then $|G(s)|$ assumes its largest value on the imaginary axis:*

$$\max_{\omega \in \mathbb{R}} |G(i\omega)| = \max_{\mathrm{Re}\, s \geq 0} |G(s)|.$$

To see how this result can be applied, consider the transfer functions

$$S(s) = \frac{1}{1 + P(s)C(s)}, \qquad T(s) = \frac{P(s)C(s)}{1 + P(s)C(s)},$$

and note that $S(s) + T(s) = 1$. The zeros of the sensitivity function $S(s)$ are the poles of the process and the controller, and the zeros of the complementary sensitivity function are the zeros of the process and the controller. We find from the above equation that $S(z) = 1$ for zeros z of the process or the controller. Similarly we have $T(p) = 1$ for poles p of the poles of the process or the controller.

We can use the maximum modulus principle to obtain requirements on disturbance attenuation and robustness, formulated as conditions on the sensitivity functions. We will use the following nominal transfer functions to capture our desired sensitivity requirements:

$$S_r(s) = \frac{M_s\, s}{s + a}, \qquad T_r(s) = \frac{M_t\, b}{s + b}. \tag{14.16}$$

Bode plots of the gain curves of the transfer functions $S_r(s)$ and $T_r(s)$ are shown in Figure 14.7. We will consider requirements defined by

$$|S(i\omega)| \leq |S_r(i\omega)|, \qquad |T(i\omega)| \leq |T_r(i\omega)|, \tag{14.17}$$

which guarantee that the maximum sensitivities are less than M_s or M_t. The sensitivity crossover frequencies of the transfer functions (14.16) and the bandwidth are given by

$$\omega_{sc} = \frac{a}{\sqrt{M_s^2 - 1}}, \qquad \omega_{tc} = b\sqrt{M_t^2 - 1}, \qquad \omega_b = b\sqrt{2M_t^2 - 1}. \tag{14.18}$$

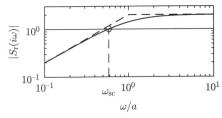 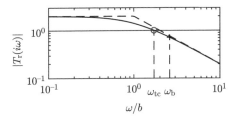

(a) Requirements for sensitivity   (b) Requirements for complementary sensitivity

Figure 14.7: Gain curves for the transfer functions (a) $S_r(s) = M_s\, s/(s+a)$ and (b) $T_r(s) = M_t\, b/(s+b)$, which give requirements for sensitivity and complementary sensitivity. The dashed curves represent the piecewise linear approximations to the first-order sensitivity requirements. The plots are drawn for $M_s = M_t = 2$, the gain crossover frequencies are denoted by $\circ$, and the bandwidth defined by $T(\omega_b) = 1/\sqrt{2}$ by $+$.

We will now use the maximum modulus principle to investigate the effects of poles and zeros in the right half-plane, and to establish limits on achievable performance.

Example 14.8 Sensitivity limits for a system with a zero in the right half-plane

Assume that the process $P(s)$ has a zero $s = z$ in the right half-plane and no other poles and zeros in the right half-plane. The sensitivity function is analytic in the right half-plane for all controllers that stabilize the system, and equation (14.17) implies that

$$\max_\omega \left| \frac{S(i\omega)}{S_r(i\omega)} \right| \leq 1. \tag{14.19}$$

The function $S(s)/S_r(s)$ is analytic in the right half-plane and on the imaginary axis. If the process has a zero $s = z$ in the right half-plane the sensitivity function has the property that $S(z) = 1$. Applying the maximum modulus principle to the function $S(s)/S_r(s)$ then gives

$$\max_\omega \left| \frac{S(i\omega)}{S_r(i\omega)} \right| \geq \left| \frac{S(z)}{S_r(z)} \right| = S(z)\frac{z+a}{M_s z} = \frac{z+a}{M_s z}.$$

This inequality is compatible with equation (14.19) only if $z + a \leq M_s z$, hence

$$a \leq z\,(M_s - 1), \qquad \omega_{sc} \leq z\,\sqrt{\frac{M_s - 1}{M_s + 1}}, \tag{14.20}$$

where the bound on ω_{sc} follows after some algebra. We see that a right half-plane zero z limits the sensitivity crossover frequency ω_{sc} of the closed loop system and thus also the range of frequencies over which the sensitivity can be kept small (compare with Example 14.3). $\qquad\qquad \nabla$

If we make the calculations for a system with complex zeros $s = z_{re} \pm i\,z_{im}$, we obtain the following conditions (Exercise 14.9):

$$a \leq \sqrt{M_s^2 z_{re}^2 + (M_s^2 - 1)z_{im}^2} - z_{re},$$

$$\omega_{\text{sc}} = \frac{a}{\sqrt{M_{\text{s}}^2 - 1}} \leq \frac{\sqrt{M_{\text{s}}^2 z_{\text{re}}^2 + (M_{\text{s}}^2 - 1)z_{\text{im}}^2} - z_{\text{re}}}{\sqrt{M_{\text{s}}^2 - 1}}, \tag{14.21}$$

which are equal to equation (14.20) for $z_{\text{im}} = 0$. Robust control of a process with right half-plane zeros therefore requires that the sensitivity crossover frequency ω_{sc} is not too high (equations (14.20) and (14.21)). If there are several right half-plane zeros the limit is given by the smallest bound.

A similar analysis based on the complementary sensitivity function gives the consequences of right half-plane poles (see Exercise 14.10). We conclude that robust control in the presence of right half-plane poles requires that the complementary sensitivity crossover frequency ω_{tc} and the bandwidth ω_{b} are sufficiently large.

Next we will consider the effect of both poles and zeros in the right half-plane. Since robust control of a process with a right half-plane zero z requires that the sensitivity crossover frequency (or the bandwidth) is sufficiently low and a right half-plane pole requires that the sensitivity crossover frequency is sufficiently high, we may expect that systems with a right half-plane pole/zero pair cannot be controlled robustly if the poles and zeros are close and we may expect that a system cannot be controlled at all if $p > z$. Indeed, it can be shown (Exercise 12.16) that a process cannot be stabilized by a stable controller if $p > z$. We will analyze the situation in the next example.

Example 14.9 Sensitivity limits for processes with poles and zeros in the right half-plane

Consider a process $P(s)$ with right half-plane zeros z_k and right half-plane poles p_k. Introduce the polynomial $n(s)$ with zeros $s = z_k$ and the polynomial $d(s)$ with zeros $s = p_k$. The process transfer function can then be written as

$$P(s) = \frac{n(s)}{d(s)} \tilde{P}(s), \tag{14.22}$$

where $\tilde{P}(s)$ has no poles or zeros in the right half-plane. Furthermore we consider controllers that stabilize the process. The sensitivity function

$$S(s) = \frac{1}{1 + P(s)C(s)} = \frac{d(s)}{d(s) + n(s)\tilde{P}(s)C(s)},$$

has the zeros $s = p_k$ in the right half-plane, and we have $S(z_k) = 1$ for all zeros z_k of the polynomial $n(s)$. Introduce the weighting function

$$W_{\text{p}}(s) = \frac{d(-s)}{d(s)}.$$

The poles and zeros of this function are symmetric with respect to the imaginary axis, which implies that $|W_{\text{p}}(i\omega)| = 1$. The function $W_{\text{p}}(s)S(s)$ is analytic in the right half-plane, since the polynomial $d(s)$ is canceled and $d(-s)$ has all its roots in the left half-plane. Since $S(z_k) = 1$, it follows from the maximum modulus principle that

$$M_{\text{s}} = \max_{\omega} |S(i\omega)| = \max_{\omega} |W_{\text{p}}(i\omega)S(i\omega)| \geq |W_{\text{p}}(z_k)S(z_k)| = \left| \frac{d(-z_k)}{d(z_k)} \right|, \tag{14.23}$$

which implies

$$M_s \geq \max_k \left| \frac{d(-z_k)}{d(z_k)} \right|. \qquad (14.24)$$

For a system with a pole/zero pair in the right half-plane we have $n(s) = s - z$ and $d(s) = s - p$. Since there is only one zero equation (14.24) becomes

$$M_s \geq \left| \frac{z+p}{z-p} \right|, \qquad (14.25)$$

which implies that

$$\frac{z}{p} \geq \frac{M_s+1}{M_s-1} \quad \text{if } z > p \qquad \text{or} \qquad \frac{z}{p} \leq \frac{M_s-1}{M_s+1} \quad \text{if } z < p. \qquad (14.26)$$

$$\nabla$$

To find controllers with a maximum sensitivity less than M_s for a process with a right half-plane pole/zero pair, it follows from equation (14.26) that the pole and zero must be sufficiently separated. The zero/pole ratio must either be smaller than $(M_s - 1)/(M_s + 1)$ or larger than $(M_s + 1)/(M_s - 1)$. For $M_s = 3$ the critical ratios are 0.5 and 2 and for $M_s = 1.4$ they are $1/6$ and 6.

A calculation similar to the one in Example 14.9 for the complementary sensitivity gives (Exercise 14.11)

$$M_t \geq \max_k \left| \frac{n(-p_k)}{n(p_k)} \right|. \qquad (14.27)$$

In the special case of a single pole/zero pair the condition becomes

$$M_t \geq \left| \frac{z+p}{z-p} \right| \qquad \Longrightarrow \qquad \frac{z}{p} \geq \frac{M_t+1}{M_t-1} \quad \text{or} \quad \frac{z}{p} \leq \frac{M_t-1}{M_t+1}. \qquad (14.28)$$

We illustrate the results with an example.

Example 14.10 Bicycle with rear-wheel steering
Figure 14.8 shows two bicycles with rear wheel steering. Bicycle dynamics were discussed in Section 4.2, where the following model was obtained:

$$J \frac{d^2\varphi}{dt^2} - \frac{Dv_0}{b} \frac{d\delta}{dt} = mgh \sin\varphi + \frac{mv_0^2 h}{b} \delta.$$

The wheelbase is b, the mass of the bicycle and the driver is m, and the distance from the center of mass to ground is h. Furthermore, J is the moment of inertia with respect to the line through the contact points of the wheels with the ground and D is the inertia product. We have $J \approx mh^2$ and $D \approx mah$, where a is the distance between the projection of the center of mass on the ground and the contact point of the driving wheel. The model for a bicycle with rear wheel steering is obtained simply by reversing the sign of the velocity and we get

$$mh^2 \frac{d^2\varphi}{dt^2} + \frac{mhav_0}{b} \frac{d\delta}{dt} = mgh \sin\varphi + \frac{mv_0^2 h}{b} \delta.$$

The transfer function from steering angle δ to tilt angle φ is

$$P_{\varphi\delta} = \frac{-av_0 s + v_0^2}{b(hs^2 - g)} = \frac{av_0}{bh} \frac{-s + v_0/a}{s^2 - g/h}.$$

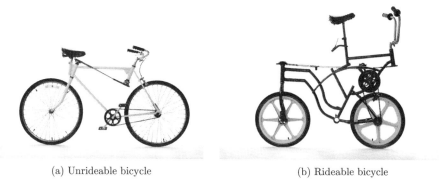

(a) Unrideable bicycle   (b) Rideable bicycle

Figure 14.8: Two bicycles with rear wheel steering: (a) is unrideable and (b) is rideable. (Figures courtesy of Richard Klein [20].)

The transfer function has a right half-plane pole $p = \sqrt{g/h}$ and a right half-plane zero at $z = v_0/a$. The condition (14.26) then gives

$$\frac{z}{p} = \frac{v_0}{a} \sqrt{\frac{h}{g}} \geq \frac{M_s + 1}{M_s - 1} \qquad \Longrightarrow \qquad v_0 \geq a \sqrt{\frac{g}{h}} \, \frac{M_s + 1}{M_s - 1}.$$

The unstable pole $p = \sqrt{g/h}$ does not depend on the velocity but the right half-plane zero $z = v_0/a$ is proportional to the velocity. To ride the bicycle comfortably the velocity must therefore be sufficiently large. Evaluating the parameters for the bicycles in Figure 14.8 with $M_s = 2$ we find $v_0 \geq 9.4\,\mathrm{m/s}$ ($34\,\mathrm{km/h}$) for the bicycle in Figure 14.8a and $v_0 \geq 1.2\,\mathrm{m/s}$ ($3.8\,\mathrm{km/h}$) for the bicycle in Figure 14.8b. The bicycle in Figure 14.8a has indeed proven to be unrideable, while the bicycle in Figure 14.8b is rideable [148]. $\qquad\qquad \nabla$

In view of the robustness results for systems with a single right half-plane pole or single right half-plane zero, it is perhaps surprising that processes with $p > z$ can actually be controlled robustly. This is in fact possible, though it requires more clever design techniques. A detailed discussion of stabilizability is given by Youla [260], where it is proven that a system with right half-plane poles and zeros can be stabilized with a stable controller if and only if the number of poles between every pair of right half-plane zeros is even (Theorem 14.2).

We have focused here on the effects of right half-plane poles and zeros. Another common source of limits is the existence of time delays. The limits imposed by a time delay and a right half-plane pole are similar to the limits by a right half-plane pole/zero pair. A list of various limits are summarized in Table 14.1.

14.5   ROBUST POLE PLACEMENT

When using any design method that does not include requirements on robustness it is necessary to check the robustness of the design. In Section 7.2 we used state feedback to assign the eigenvalues of the closed loop system and showed that if a system is reachable then the eigenvalues of the closed loop system can be set to

Table 14.1: Summary of limits by time delays and right half-plane (RHP) poles and zeros; ω_{sc} and ω_{tc} are the crossover frequencies for the sensitivity function and the complementary sensitivity function.

Process feature	Limits
Real RHP zero z	$\omega_{sc} \leq z\sqrt{\dfrac{M_s - 1}{M_s + 1}}$
Complex RHP zeros $z = z_{re} \pm iz_{im}$	$\omega_{sc} \leq \dfrac{\sqrt{M_s^2 z_{re}^2 + (M_s^2 - 1)z_{im}^2} - z_{re}}{\sqrt{M_s^2 - 1}}$
Real RHP pole p	$\omega_{tc} \geq p\sqrt{\dfrac{M_t + 1}{M_t - 1}}$
Complex RHP poles $p = p_{re} \pm ip_{im}$	$\omega_{tc} \geq \dfrac{\sqrt{M_t^2 p_{re}^2 + (M_t^2 - 1)p_{im}^2} + p_{re}}{\sqrt{M_t^2 - 1}}$
RHP pole/zero pair p, z	$M_s \geq \left|\dfrac{p+z}{p-z}\right|, \; M_t \geq \left|\dfrac{p+z}{p-z}\right|$
RHP poles and zeros $d(s), n(s)$	$M_s \geq \max_k \left|\dfrac{d(-z_k)}{d(z_k)}\right|, \; M_t \geq \max_k \left|\dfrac{n(-p_k)}{n(p_k)}\right|$
RHP pole p and time delay τ	$M_s \geq e^{p\tau}, \; M_t \geq e^{p\tau} - 1$

arbitrary values. This design technique is also called "pole placement" and in this section we will show that the insights into the roles of poles and zeros can give us a deeper understanding of how to design such controllers. In particular we will show that it is necessary to take the process zeros into account when choosing the desired closed loop poles. We will first analyze examples where seemingly reasonable designs lead to closed loop systems that are not robust. We will then present design rules for pole (eigenvalue) placement that guarantee that the closed loop system is robust.

Fast Stable Process Poles

A pole is stable if it is in the left half-plane and unstable if it is in the right half-plane. We call it "fast" if its magnitude is larger than the intended closed loop bandwidth. We will explore the effects of fast stable process poles on pole placement design through a simple example that illustrates the basic design rule.

Example 14.11 Robust pole placement for fast process poles
Consider a PI controller for a first-order system, where the process and the controller have the transfer functions $P(s) = b/(s+a)$, with $a > 0$, and $C(s) = k_p + k_i/s$. The loop transfer function is

$$L(s) = \frac{b(k_p s + k_i)}{s(s+a)},$$

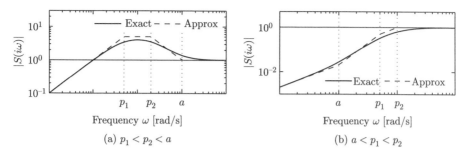

(a) $p_1 < p_2 < a$ (b) $a < p_1 < p_2$

Figure 14.9: Gain curves of the sensitivity function S for designs in Example 14.11. The solid lines are the true sensitivities, and the dashed lines are the asymptotes. Notice the high peak of the sensitivity function in (a) and that there is no peak in (b).

and the closed loop characteristic polynomial is

$$s(s+a) + b(k_p s + k_i) = s^2 + (a + bk_p)s + k_i b.$$

If we specify that the desired closed loop poles should be $-p_1$ and $-p_2$, we find that the controller parameters are given by

$$k_p = \frac{p_1 + p_2 - a}{b}, \qquad k_i = \frac{p_1 p_2}{b}.$$

The sensitivity functions are then

$$S(s) = \frac{s(s+a)}{(s+p_1)(s+p_2)}, \qquad T(s) = \frac{(p_1 + p_2 - a)s + p_1 p_2}{(s+p_1)(s+p_2)}.$$

Assume that the process pole a is faster than the closed loop poles $p_1 < p_2 < a$. The proportional gain k_p is then negative and the controller has a zero in the right half-plane, an indication that the system may have bad properties. Consider the gain $|S(i\omega)|$ of the sensitivity function plotted in Figure 14.9a for $a = b = 1$, $p_1 = 0.05$, and $p_2 = 0.2$. We have $S(i\omega) \approx 1$ for high frequencies. Moving backwards in frequency we find that the sensitivity increases around $\omega = a$ corresponding to the fast process pole. The sensitivity continues to increase with decreasing frequency and it does not decrease until the frequency is below the closed loop pole p_2. The net effect is a large sensitivity peak, approximately $\omega = a/\sqrt{p_1 p_2} \approx 10$.

The problem with poor robustness can be avoided by choosing one closed loop pole equal to the process pole, i.e., $p_2 = a$. The controller gains then become

$$k_p = \frac{p_1}{b}, \qquad k_i = \frac{a p_1}{b},$$

which means that the fast process pole is canceled by a controller zero at $s = -a$. The loop transfer function and the sensitivity functions are

$$L(s) = \frac{bk_p}{s}, \qquad S(s) = \frac{s}{s + bk_p}, \qquad T(s) = \frac{bk_p}{s + bk_p}.$$

Figure 14.9b shows the gain curve of the sensitivity function for the case when the closed loop poles ($p_1 = 5$, $p_2 = 20$) are faster than the process pole ($a = 1$). There is no peak of the sensitivity function in this case.   ▽

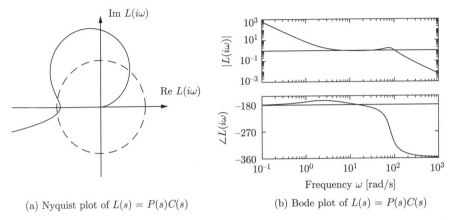

(a) Nyquist plot of $L(s) = P(s)C(s)$ (b) Bode plot of $L(s) = P(s)C(s)$

Figure 14.10: Observer-based control of vehicle steering. Nyquist and Bode plots of the loop transfer function for vehicle steering with a controller based on state feedback and an observer. The controller provides stable operation, but with very poor robustness.

Slow Stable Process Zeros

We call a zero "stable" if it is in the left half-plane and "unstable" if it is in the right half-plane. Furthermore a zero is said to be "slow" if its magnitude is smaller than the intended closed loop bandwidth. We will explore the effects of slow stable process zeros in pole placement design, and we begin with a simple example.

Example 14.12 Vehicle steering

Consider the model for vehicle steering in Example 9.10, where the transfer function from steering angle to lateral position is

$$P(s) = \frac{\gamma s + 1}{s^2} = \gamma \frac{s + 1/\gamma}{s^2}.$$

A controller based on state feedback was designed in Example 7.4, and state feedback was combined with an observer in Example 8.4. The system simulated in Figure 8.8 has closed loop poles specified by $\omega_c = 0.7$, $\zeta_c = 0.707$, $\omega_o = 1$, and $\zeta_o = 0.707$. Assume that we want a faster closed loop system and choose $\omega_c = 10$, $\zeta_c = 0.707$, $\omega_o = 20$, and $\zeta_o = 0.707$. Using the state representation in Example 8.3, a pole placement design gives state feedback gains $k_1 = 100$ and $k_2 = -35.86$ and observer gains $l_1 = 28.28$ and $l_2 = 400$. The controller transfer function is

$$C(s) = \frac{-11516s + 40000}{s^2 + 42.4s + 6657.9}. \tag{14.29}$$

Figure 14.10 shows Nyquist and Bode plots of the loop transfer function.

    The Nyquist plot indicates that the robustness is poor since the loop transfer function is very close to the critical point -1. The phase margin is $7°$ and the gain margin is $g_m = 1.08$, which means that the system becomes unstable if the gain is increased by 8%. The poor robustness also shows up in the Bode plot, where the gain curve hovers around the value 1 while the phase curve is close to $-180°$ for a wide frequency range (3-40 rad/s). Additional insight is obtained by analyzing the

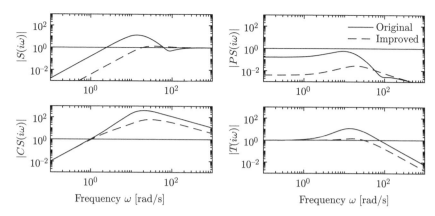

Figure 14.11: Gain curves of the sensitivity functions for systems with observer-based control of vehicle steering. The original controller with $\omega_c = 10$, $\zeta_c = 0.707$, $\omega_o = 20$, $\zeta_o = 0.707$ is shown as solid lines and the improved controller with $\omega_c = 10$, $\zeta_c = 2.6$ is shown as dashed lines.

sensitivity functions, shown as solid lines in Figure 14.11. The maximum sensitivities are $M_s = 13$ and $M_t = 12$.

It is surprising that the closed loop is so sensitive to process variations when we have designed a controller so that the closed loop system has well-damped closed loop poles. We have an indication that something is unusual because the design gives a controller that has a zero in the right half-plane at $s = 3.5$, while the observer and controller have complex poles with $\omega_c = 10$ and $\omega_o = 20$. Recall the results from Example 14.3, which indicate that robust control of a process with a zero at $s = 3.5$ cannot have a gain crossover frequency larger than $\omega_{gc} = 2$.

To understand what happens, we will investigate the reason for the peaks of the sensitivity functions. Let the transfer functions of the process and the controller be

$$P(s) = \frac{n_p(s)}{d_p(s)}, \qquad C(s) = \frac{n_c(s)}{d_c(s)}, \tag{14.30}$$

where $n_p(s)$, $n_c(s)$, $d_p(s)$, and $d_c(s)$ are the numerator and denominator polynomials. The complementary sensitivity function is

$$T(s) = \frac{P(s)C(s)}{1 + P(s)C(s)} = \frac{n_p(s)n_c(s)}{d_p(s)d_c(s) + n_p(s)n_c(s)}.$$

The poles of $T(s)$ are the poles of the closed loop system and the zeros of $T(s)$ are the zeros of the process and the controller transfer functions. A plot of the gain curve of $T(s)$ for the original controller is shown as the solid line in the lower right plot in Figure 14.11. We have $T(0) = 1$, because $L(0) = P(0)C(0) = \infty$ due to the double integrator of P. The gain $|T(i\omega)|$ increases for increasing ω due to the process zero at $\omega = 2$. It increases further at the controller zero at $\omega = 3.5$, and it does not start to decrease until the closed loop poles appear at $\omega = 10$ and $\omega = 20$. The net result is a high peak of the gain of the complementary sensitivity function.

The peak in the complementary sensitivity function can be avoided by assigning a closed loop pole at the slow process zero or close to it. We can achieve this by choosing $\omega_c = 10$ and $\zeta_c = 2.6$, which gives closed loop poles at $s = -2$ and $s = -50$.

The controller transfer function then becomes

$$C(s) = \frac{3628s + 40000}{s^2 + 80.28s + 156.56} = 3628 \frac{s + 11.02}{(s+2)(s+78.28)}. \tag{14.31}$$

Notice that the new controller has a pole at $s = -2$ that cancels the process zero. Also notice the large differences in the zero frequency gains of the controllers $C(0) = 6.0$ for the controller (14.29) and $C(0) = 255$ for the controller (14.31). Cancellation of the slow zero gives a dramatic increase of the low-frequency gain of the controller. The gain curves for the sensitivity function of the improved controller are shown with dashed lines in Figure 14.11. The closed loop system has the maximum sensitivities $M_s = 1.34$ and $M_t = 1.41$, which indicate good robustness.

This example shows that a robust design can be obtained by first canceling the slow stable process zero, designing the controller for the system without the zero, and then adding the pole to the controller. Notice that the plot of $|PS(i\omega)|$ shows that the improved system has much better disturbance attenuation and the plot of $|CS(i\omega)|$ shows that it is not as sensitive to measurement noise. The large differences in low-frequency gains of the controllers are clearly visible in the gain curves for S and PS.   ∇

We can learn several things from this example. First, it is essential to evaluate the closed loop system carefully, for example by plotting the gain curves of the Gang of Four. We have also seen that seemingly reasonable design methods do not necessarily give robust closed loop systems. For designs based on pole placement it is necessary to consider the open loop poles and zeros when specifying the desired closed loop dynamics, and in particular robustness requires that there must be closed loop poles that are equal to or close to slow stable process zeros. Another lesson is that slow unstable process zeros impose limits on the achievable bandwidth, as already noted in Section 14.4.

One potential issue with the choice of controller poles and zeros that exactly cancel the open loop poles and zeros is that they may lead to undesirable dynamics or lack of robustness (if there are model uncertainties). We address this important issue in more detail below.

Design Rules for Robust Pole Placement

Based on the insight gained from the previous examples, we can now formulate design rules that give controllers with good robustness for pole placement design. Consider the expression (13.12) for maximum complementary sensitivity, repeated here:

$$M_t = \sup_\omega |T(i\omega)| = \left\| \frac{PC}{1 + PC} \right\|_\infty.$$

Let ω_{gc} be the desired gain crossover frequency, and assume that the process has zeros that are slower than ω_{gc}. The complementary sensitivity function is 1 for low frequencies, and it increases for frequencies close to the process zeros unless there is a closed loop pole in the neighborhood (as seen, for instance, in Figure 14.11 of the previous example). To avoid large values of the complementary sensitivity function we find that the closed loop system should therefore have poles close to or equal to the slow stable zeros. This means that slow stable zeros should be canceled by controller poles. Since unstable zeros cannot be canceled, the presence of slow *unstable* zeros means that achievable gain crossover frequency must be smaller than the slowest unstable process zero.

Now consider process poles that are faster than the desired gain crossover frequency. Consider the expression for the maximum of the sensitivity function:

$$M_s = \sup_\omega |S(i\omega)| = \left\| \frac{1}{1+PC} \right\|_\infty.$$

The sensitivity function is 1 for high frequencies. Moving from high to low frequencies, the sensitivity function increases at the fast process poles. The sensitivity function will have large peaks unless there are closed loop poles that are close to the fast process poles. To avoid large peaks in the sensitivity, the closed loop system should therefore have poles close the fast process poles. One way to achieve this is to have controller zeros close to the fast process pole. Since unstable modes cannot be canceled, the presence of a fast unstable pole implies that the gain crossover frequency must be sufficiently large, as was discussed in Section 14.3 (Example 14.4).

To summarize, we obtain the following simple rules for choosing closed loop poles: slow stable process zeros should be matched by slow closed loop poles, and fast stable process poles should be matched by fast closed loop poles. Slow unstable process zeros and fast unstable process poles impose severe limits.

14.6 NONLINEAR EFFECTS

Although we focus primarily on linear systems in this chapter, there are some nonlinearities that must be considered when designing a control system. Limits on actuation power set bounds on response speed. Nonlinearities due to friction, round-off error in A/D and D/A converters, and numerical representations in computation bound the precision that can be obtained in regulation and tracking. We briefly describe some of the effects of these limits here, illustrated primarily through examples.

Actuation Limits

Many limits are associated with constraints on how large signals and variables can be. Motors have limited torque, amplifiers have limits on currents, and pumps have limited flow. There are also limits due to equipment protection: the temperature of a component must not be too high and compressor stall must be avoided, for example. Limits may appear as restrictions on the amplitude and the rate of change of the control signal. There may also be restrictions on internal process variables and their rates.

A real-world example of the consequences of actuator limits is the grounding of a Swedish passenger ferry in 2004. The ferry was grounded while entering the port of Umeå due to high winds (20 m/s). The incident analysis revealed that the wind forces of 600 kN and higher were much larger than the forces generated by the ship's propellers and rudder, and even assistance from a tugboat capable of applying 260 kN of thrust could not have helped. In the setting of control systems, this example illustrates a situation where actuators do not have the sufficient power to counteract the load disturbances.

The following simple analytical example demonstrates how these types of considerations can be taken into account in the design stage of a project.

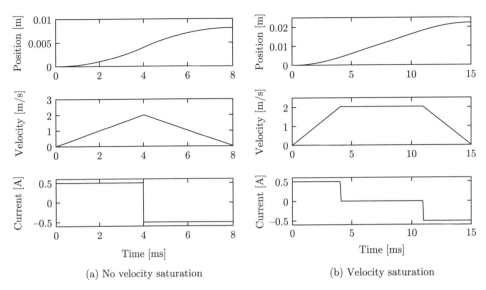

(a) No velocity saturation   (b) Velocity saturation

Figure 14.12: Minimum time transition for a servo system. (a) The case of short movements when the velocity does not reach the saturation limit. The control is of the "bang-bang" type where maximum current is applied to accelerate or brake. (b) Illustration of what happens for large motions. Full acceleration $a_{max} = 500\,\text{m/s}^2$ is applied until $t = 5\,\text{ms}$ when maximum velocity $v_{max} = 2\,\text{m/s}$ is reached and the drive circuit saturates. The current is then zero until time $t = 10\,\text{ms}$ when full braking current is applied. The parameter values are $m = 2.5 \times 10^{-3}\,\text{kg}$, $k_I = 2.5\,\text{N/A} = 2.5\,\text{Vs/m}$, $I_{max} = 0.5\,\text{A}$, and $V_{max} = 5\,\text{V}$.

Example 14.13 Current limits in servo systems

Response time is a common requirement for motor drives. The achievable response time depends critically on actuation power and physical limits of the process. To determine the response time we can compute the minimum time to make transitions from one state to the other, subject to the physical constraints on the process and the actuator.

Consider a simple servo system where the actuator is a current-driven voice coil. The system can be modeled by

$$m\frac{d^2x}{dt^2} = F = k_I I, \tag{14.32}$$

where m is the mass of the system, x is the position of the mass, F is the force, I is the current through the voice coil, and k_I is the motor constant. The maximum acceleration $a_{max} = F_{max}/m = k_I I_{max}/m$ is given by the maximum current I_{max}. There is also a limit on the maximum velocity: for a voice coil drive the maximum velocity is $v_{max} = V_{max}/k_I$, where V_{max} is the largest supply voltage.

If there is no limit on the velocity, the problem of moving the mass from one position to another in minimum time is simply to apply maximum acceleration until the mid position is reached and then apply maximum deceleration, so-called "bang-bang" control. If there is a velocity limit, the maximum acceleration is only applied until the maximum velocity is reached. The minimum time solutions are illustrated in Figure 14.12. When the acceleration a is constant, the velocity increases as

$v(t) = at$ and the position is $x(t) = at^2/2 = v^2(t)/(2a)$. A straightforward calculation shows that the minimum time for a transition over a distance ℓ with zero velocity at start and end is

$$t = \begin{cases} 2\sqrt{\ell/a_{\max}} & \text{if } \ell \leq v_{\max}^2/a_{\max}, \\ \ell/v_{\max} + v_{\max}/a_{\max} & \text{if } \ell > v_{\max}^2/a_{\max}. \end{cases} \tag{14.33}$$

We can derive requirements on the actuator from this equation. $\triangledown$

This simple example can be solved analytically. Software for computing minimum time control is readily available for more complex systems.

Saturation limits can also affect the stability of a feedback system. We saw in Section 10.5 two different methods for reasoning about the effects of (static) non-linearities in a feedback system: the circle criterion and describing functions. Both of these techniques use the Nyquist plot as a means of analyzing the effects of the nonlinearity on closed loop stability. In the particular case of actuation limits, the circle criterion allows the saturation to be modeled as a sector-bounded nonlinearity with $k_{\text{low}} = 0$ and $k_{\text{high}} = 1$, which implies that the system is stable if the Nyquist curve for the linear dynamics has $\operatorname{Re} H(s) > -1$. The describing function method is slightly less constraining, since the image of the describing function for a saturation nonlinearity is given by the negative real axis from $-\infty$ to -1, and hence the Nyquist curve for $H(s)$ should not cross the negative real axis at a gain greater than one. (Note that the describing function method is only an approximation, although it is often a very useful for preliminary design.)

Measurement Noise and Friction

There are many sources of measurement noise: the physics of the sensor, the electronics, the transmission equipment, and the A/D and D/A converters. The controller in a closed loop system feeds measurement noise into the system, creating fluctuations in all variables. Fluctuations in the output limits regulation and tracking performance. Fluctuations in the control signal causes wear or even saturation of the actuator, and cannot be permitted to be too large. Since measurement noise is typically dominated by high frequencies, it limits the high-frequency gain of the controller, the bandwidth, and thus the response time of the closed loop system.

The effects of measurement noise and quantization can be estimated using linear methods by calculating the transfer function from the noise sources to the control signal and the process variables, and they can be alleviated by filtering and a controller with high-frequency roll-off. Quantization can be approximated as noise with a variance of $\delta^2/12$, where δ is the quantization level.

Friction typically generates oscillations that limit regulation and tracking performance. Similar oscillations can be caused by quantization. Oscillations can be reduced by nonlinear friction compensation. Friction is inherently a nonlinear phenomenon, and accurate analysis requires nonlinear methods. Some insight can be obtained using the describing function method discussed in Section 10.5. We illustrate with an example.

Example 14.14 Effect of friction in a cart–pendulum system
The cart-pendulum or balance system was introduced in Example 3.2 and we designed a state feedback for it in Example 7.7. Experiments with cart–pendulum

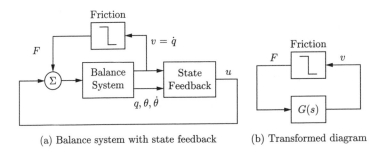

(a) Balance system with state feedback   (b) Transformed diagram

Figure 14.13: Block diagrams of a balance system with state feedback and friction. (a) Detailed block diagram showing the balance system with inputs u and F and outputs q, θ, $v = \dot{q}$, $\dot{\theta}$. (b) Block diagram obtained after transformations. It has two blocks: the nonlinear friction block a linear block with the transfer function $G(s)$ from friction force F to velocity v.

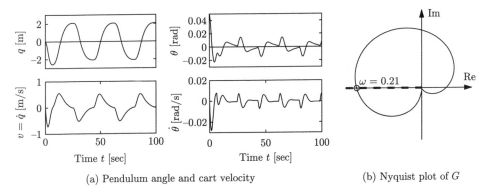

(a) Pendulum angle and cart velocity   (b) Nyquist plot of G

Figure 14.14: Time and frequency responses of the cart-pendulum system. (a) Time responses when the pendulum has an initial misalignment. (b) Frequency response of the transfer function $G(s)$ (solid line), given by (14.36), and the locus of the negative inverse $-1/N(a)$ (dashed line) of the describing function $N(a)$ for friction, given by equation (14.35).

systems have shown that friction on the cart creates oscillations. To explore this we will investigate the effects of friction by simulation and analysis.

A block diagram of a balance system with friction is shown in Figure 14.13a. To simulate the system we use Coulomb's model for friction, where the friction force is F is given by

$$F = -\mu_f M_t g \, \text{sgn}(v), \tag{14.34}$$

where $\mu_f = 0.001$ is the coefficient for rolling friction, M_t is the total mass, g is the acceleration due to gravity, and v is the cart velocity. We use the parameter values from Example 3.2, and the controller is the state space feedback in Example 7.7 with the slower closed loop poles. Results of a simulation of the system are shown in Figure 14.14a. The upper plots in the figure show the cart position q (left) and the pendulum angle θ (right), and the lower plots show the cart velocity $v = \dot{q}$ (left) and the angular velocity of the pendulum $\dot{\theta}$ (right). The plots show clearly that there are oscillations with period $T_p = 37$ s. The oscillation of the cart velocity has amplitude $A \approx 0.52$ m/s. The waveforms of the oscillations are far from sinusoidal, as can be seen in the plots on the right in Figure 14.14a.

We can make a simple physical argument to understand how friction may cause oscillation. The pendulum is unstable and will start to fall for any perturbation. The control law then attempts to stabilize the system by applying a force to the cart, but the cart will remain stationary until the pendulum has fallen so much that the control signal is large enough to generate a force that is larger than the friction force. The cart then moves, causing the pendulum to move towards the upright position. The process will repeat itself creating an oscillation.

We will now use the describing function method, introduced in Section 10.5, to understand the behavior of the system. To do this we first use block diagram algebra to reduce Figure 14.13a to the two-block system in Figure 14.13b. One block represents the nonlinear friction model (14.34), which has the describing function

$$N(a) = \frac{4\mu_f M_t g}{a\pi}, \tag{14.35}$$

where a is the amplitude of the input (cart velocity). The other block in Figure 14.13b represents the linear closed loop dynamics from friction force F to cart velocity v, when friction is not present. The transfer function can be computed from the state space representation of the closed loop dynamics

$$\frac{d}{dt}x = (A - BK)x + BF, \qquad v = \begin{pmatrix} 0 & 0 & 1 & 0 \end{pmatrix} x,$$

where $x = (q, \theta, \dot{q}, \dot{\theta})$, A, B, and K are given in Example 7.7. The resulting transfer function is given by

$$G(s) = \frac{0.01837s^3 - 0.08s}{s^4 + 1.046s^3 + 0.9109s^2 + 0.2552s + 0.03781}, \tag{14.36}$$

where the numerical values are based on the parameter values from Example 7.7.

Figure 14.14b shows a Nyquist plot of the transfer function G (solid line) and the negative inverse $-1/N(a)$ of the describing function (dashed line). Recall that the condition for oscillation is $G(i\omega)N(a) = -1$, which corresponds to an intersection of the solid and dashed lines in the figure. The intersection occurs for $\omega = 0.21$, and $1/N(a) = 0.39$. The describing function method then indicates that there may be an oscillation with period $T_p = 2\pi/0.21 = 30$ s and amplitude $a = 4 \times 0.39 \, \mu_f M_t g/\pi = 0.43$ m/s. Notice that the describing function method assumes that the velocity variation is sinusoidal, which explains the difference from the values $T = 37$ s and $a = 0.52$ m/s obtained by simulation. $\triangledown$

14.7 FURTHER READING

The limitations caused by right half-plane poles and zeros were well known by Bode, who coined the term non-minimum phase to emphasize that such systems had much more phase lag than the equivalent minimum phase systems [51]. The paper [229], which is based on the inaugural IEEE Bode Lecture gives important insights into the effects of unstable poles and is strongly recommended. Horowitz [121] also discussed the limits caused by poles and zeros in the right half-plane. The section on the maximum modulus theorem is based on [206]; more details are found in [107, 223]. The section on loop shaping design is based on [16]. The design rules for pole placement are not widely known. The effects of actuator limits are conveniently explored using optimal control theory [24, 58], which permits solution of problems that are much more complicated than the one in Figure 14.12.

EXERCISES

14.1 (Right half-plane pole/zero pair PI control) Consider a process with the transfer function

$$P(s) = \frac{s - z}{s - p}.$$

a) Show that the system can be controlled by a PI controller and design a PI controller that gives a closed loop system with poles at $s = -\zeta\omega_0 \pm \omega_0\sqrt{1 - \zeta^2}$.

b) Calculate the maximum sensitivity of the closed loop system as a function of ω_0 and compare with the bound imposed by the right half-plane poles and zeros of the system. Discuss the differences between the cases $z > p$ and $z < p$.

c) Plot the root locus of the process with the PI controller and qualitatively describe how it changes with the process pole and the process zero. Use the numerical values $\omega_0 = 1, \zeta = 1$; $p = 1$, $z = 5$; and $p = 5$, $z = 1$.

14.2 (Effect of roll-off) Consider a closed loop system consisting of a first-order process and a proportional controller. Let the loop transfer function be

$$L(s) = P(s)C(s) = \frac{k}{s + 1},$$

where parameter $k > 0$ is the controller gain. Show that the magnitude of the sensitivity function is bounded above by 1 and can be made arbitrarily small up to any frequency ω.

14.3 (Bode's integral formula) In Theorem 14.3 it was assumed that $sL(s)$ goes to zero as $s \to \infty$. Assume instead that $\lim sL(s) = a$ and show that

$$\int_0^\infty \log |S(i\omega)| \, d\omega = \int_0^\infty \log \frac{1}{|1 + L(i\omega)|} \, d\omega = \pi \sum p_k - a\frac{\pi}{2},$$

where p_k are the poles of the loop transfer function $L(s)$ in the right half-plane.

 14.4 (Integral formula for complementary sensitivity) Prove the formula (14.6) for the complementary sensitivity.

14.5 (Water turbine dynamics) Consider the problem of power generation in a hydroelectric power station. Let the control signal be the opening area a at the turbine entrance and ℓ be the length of the tube, which has area A. Formulate a mathematical model for the system, then linearize the model around a nominal valve opening $u_0 = a/A$ and a nominal power P_0. Show that the linearization is non-minimum phase, with transfer function

$$G(s) = \frac{P_0}{a_0} \frac{1 - 2u_0 s\tau}{1 + u_0 s\tau},$$

where $\tau = \ell/\sqrt{2gh}$ and g is the acceleration due to gravity.

14.6 (The pole/zero ratio) Consider a process with the loop transfer function

$$L(s) = k\frac{z - s}{s - p},$$

with positive z and p. Show that the system is stable if $p/z < k < 1$ or $1 < k < p/z$ and that the largest stability margin is $s_m = |p - z|/(p + z)$, which is obtained for $k = 2p/(p + z)$. Determine the pole/zero ratios that give the stability margin $s_m = 2/3$.

14.7 (Phase lag of systems with right half-plane pole/zero pair and delay and right half-plane pole) Consider the transfer functions for a process with a right half-plane pole and right half-plane zero as in Example 14.5 and a right half-plane pole and a time delay as in Example 14.6. The phase lags of their all-pass factors are given in equations (14.14) and (14.15). Show that the largest phase lags are

$$\varphi_{\mathrm{ap1}} = -\arg P_{pz}(i\omega) \leq 2\arctan\left(2\sqrt{pz}/|z - p|\right),$$

$$\varphi_{\mathrm{ap2}} = -\arg P_{p\tau}(i\omega) \leq \sqrt{p\tau(2 - p\tau)} + 2\arctan\sqrt{p\tau/(2p - p\tau)}$$

and that they occur for $\omega_1 = \sqrt{pz}$ and $\omega_2 = \sqrt{2p/\tau - p^2}$ respectively.

14.8 (X-29) A simplified model of the X-29 aircraft in a certain flight condition has a right-hand pole/zero pair with $p = 6\,\mathrm{rad/s}$ and $z = 26\,\mathrm{rad/s}$. Estimate the achievable stability margins and compare with the results in Example 14.2.

14.9 (Sensitivity inequalities) Prove the inequalities given by equation (14.21). (Hint: Use the maximum modulus theorem.)

14.10 (Sensitivity limits due to poles in the right half-plane) Let $T_r = M_t b/(s + b)$ represent an upper bound on the desired sensitivity and let ω_{tc} represent the complementary sensitivity crossover frequency. Show that for a process $P(s)$ with a right half-plane pole $s = p$ but no other singularities in the right half-plane, the following inequalities hold:

$$b \geq \frac{p_{\mathrm{re}} + \sqrt{M_t^2 p_{\mathrm{re}}^2 + (M_t^2 - 1)p_{\mathrm{im}}^2}}{M_t^2 - 1}, \qquad \omega_{\mathrm{tc}} \leq \frac{p_{\mathrm{re}} + \sqrt{M_t^2 p_{\mathrm{re}}^2 + (M_t^2 - 1)p_{\mathrm{im}}^2}}{\sqrt{M_t^2 - 1}},$$

$$(14.37)$$

where $p = p_{\mathrm{re}} + i p_{\mathrm{im}}$.

14.11 [Maximum complementary sensitivity for multiple right half-plane poles and zeros] Consider a process $P(s)$ with the right half-plane zeros z_k and right half-plane poles p_k. Introduce the polynomial $n(s)$ with zeros $s = z_k$ and the polynomial $d(s)$ with zeros $s = p_k$. Show that the complementary sensitivity function has the property

$$M_t \geq \max_k \left| \frac{n(-p_k)}{n(p_k)} \right|.$$

Also show that the equations (14.28) hold.

14.12 (Vehicle steering) Consider the Nyquist curve in Figure 14.10. Explain why part of the curve is approximately a circle. Derive a formula for the center and the radius and compare with the actual Nyquist curve.

14.13 Consider a process with the transfer function

$$P(s) = \frac{(s+3)(s+200)}{(s+1)(s^2+10s+40)(s+40)}.$$

Discuss suitable choices of closed loop poles for a design that gives dominant poles with undamped natural frequency 1 and 10.

14.14 (Large signals) Verify Figure 14.12 by hand calculation.

14.15 (Noise limits bandwidth) Consider PI control of an integrator, where the transfer functions of the process and the controller are

$$P(s) = \frac{1}{s}, \qquad C(s) = k_p + \frac{k_i}{s},$$

with $k_p = 2\zeta\omega_0$, $k_i = \omega_0^2$, and $\zeta = 0.707$. Assume that the inputs and outputs range from 0 to 10, that there is measurement noise with a standard deviation of 0.01, and that the largest permissible variation in the control signal due to noise is 2. Show that the bandwidth, defined as $\omega_{bw} = 2\omega_0$, cannot be larger than 283.

Bibliography

[1] M. A. Abkowitz. *Stability and Motion Control of Ocean Vehicles*. MIT Press, Cambridge, MA, 1969.

[2] R. H. Abraham and C. D. Shaw. *Dynamics—The Geometry of Behavior, Part 1: Periodic Behavior*. Aerial Press, Santa Cruz, CA, 1982.

[3] J. Ackermann. Der Entwurf linearer Regelungssysteme im Zustandsraum. *Regelungstechnik und Prozessdatenverarbeitung*, 7:297–300, 1972.

[4] J. Ackermann. *Sampled-Data Control Systems*. Springer, Berlin, 1985.

[5] C. E. Agnew. Dynamic modeling and control of congestion-prone systems. *Operations Research*, 24(3):400–419, 1976.

[6] L. V. Ahlfors. *Complex Analysis*. McGraw-Hill, New York, 1966.

[7] P. Albertos and I. Mareels. *Feedback and Control for Everyone*. Springer, 2010.

[8] R. Alur. *Principles of Cyber-Physical Systems*. MIT Press, 2015.

[9] B. D. O. Anderson and J. B. Moore. *Optimal Control Linear Quadratic Methods*. Prentice Hall, Englewood Cliffs, NJ, 1990. Republished by Dover Publications, 2007.

[10] A. A. Andronov, A. A. Vitt, and S. E. Khaikin. *Theory of Oscillators*. Dover, New York, 1987.

[11] T. M. Apostol. *Calculus, Vol. II: Multi-Variable Calculus and Linear Algebra with Applications*. Wiley, New York, 1967.

[12] T. M. Apostol. *Calculus, Vol. I: One-Variable Calculus with an Introduction to Linear Algebra*. Wiley, New York, 1969.

[13] R. Aris. *Mathematical Modeling Techniques*. Dover, New York, 1994. Originally published by Pitman, 1978.

[14] V. I. Arnold. *Mathematical Methods in Classical Mechanics*. Springer, New York, 1978.

[15] V. I. Arnold. *Ordinary Differential Equations*. MIT Press, Cambridge, MA, 1987. 10th printing 1998.

[16] K. J. Åström. Limitations on control system performance. *European Journal on Control*, 6(1):2–20, 2000.

[17] K. J. Åström. *Introduction to Stochastic Control Theory*. Dover, New York, 2006. Originally published by Academic Press, New York, 1970.

[18] K. J. Åström and R. D. Bell. Drum-boiler dynamics. *Automatica*, 36:363–378, 2000.

[19] K. J. Åström and T. Hägglund. *Advanced PID Control*. ISA—The Instrumentation, Systems, and Automation Society, Research Triangle Park, NC, 2006.

[20] K. J. Åström, R. E. Klein, and A. Lennartsson. Bicycle dynamics and control. *IEEE Control Systems Magazine*, 25(4):26–47, 2005.

[21] K. J. Åström and P. R. Kumar. Control: A perspective. *Automatica*, 50:3–43, 2014.

[22] K. J. Åström and B. Wittenmark. *Computer-Control Systems: Theory and Design*, 3rd ed. Prentice Hall, Englewood Cliffs, NJ, 1997.

[23] K. J. Åström and B. Wittenmark. *Adaptive Control*, 2nd ed. Dover, New York, 2008. Originally published by Addison Wesley, 1995.

[24] M. Athans and P. Falb. *Optimal Control*. McGraw-Hill, New York, 1966. Dover Reprint 2007.

[25] D. P. Atherton. *Nonlinear Control Engineering*. Van Nostrand, New York, 1975.

[26] M. Atkinson, M. Savageau, J. Myers, and A. Ninfa. Development of genetic circuitry exhibiting toggle switch or oscillatory behavior in *Escherichia coli*. *Cell*, 113(5):597–607, 2003.

[27] M. B. Barron and W. F. Powers. The role of electronic controls for future automotive mechatronic systems. *IEEE Transactions on Mechatronics*, 1(1): 80–89, 1996.

[28] T. Basar (editor). *Control Theory: Twenty-five Seminal Papers (1932–1981)*. IEEE Press, New York, 2001.

[29] T. Basar and P. Bernhard. H^∞-*Optimal Control and Related Minimax Design Problems: A Dynamic Game Approach*. Birkhauser, Boston, 1991.

[30] J. Bechhoefer. Feedback for physicists: A tutorial essay on control. *Reviews of Modern Physics*, 77:783–836, 2005.

[31] J. Bechhoefer. *Control Theory for Physicists*. Cambridge University Press, 2020. In press.

[32] R. Bellman and K. J. Åström. On structural identifiability. *Mathematical Biosciences*, 7:329–339, 1970.

[33] R. E. Bellman and R. Kalaba. *Selected Papers on Mathematical Trends in Control Theory*. Dover, New York, 1964.

[34] S. Bennett. *A History of Control Engineering: 1800–1930*. Peter Peregrinus, Stevenage, UK, 1979.

[35] S. Bennett. *A History of Control Engineering: 1930–1955*. Peter Peregrinus, Stevenage, UK, 1993.

[36] B. W. Bequette. Challenges and recent progress in the development of a closed-loop artificial pancreas. *Annual Reviews in Control*, 36:255–268, 2012.

[37] B. W. Bequette. Algorithms for a closed-loop artificial pancreas: The case for model predictive control. *Journal of Diabetes Science and Technology*, 7(6):1632–1643, 2013.

[38] L. L. Beranek. *Acoustics*. McGraw-Hill, New York, 1954.

[39] R. N. Bergman. Toward physiological understanding of glucose tolerance: Minimal model approach. *Diabetes*, 38:1512–1527, 1989.

[40] R. N. Bergman. The minimal model of glucose regulation: A biography. In J. Novotny, M. Green, and R. Boston (editors), *Mathematical Modeling in Nutrition and Health*. Kluwer Academic/Plenum, New York, 2001.

[41] J. Berner, T. Hägglund, and K. J. Åström. Asymmetric relay autotuning— Practical features for industrial use. *Control Engineering Practice*, 54: 231–245, 2016.

[42] J. Berner, K. Soltesz, K. J. Åström, and T. Hägglund. Practical evaluation of a novel multivariable relay autotuner with short and efficient excitation. *IEEE Conference on Control Technology and Applications*, 2017.

[43] D. Bertsekas and R. Gallager. *Data Networks*. Prentice Hall, Englewood Cliffs, NJ, 1987.

[44] B. Bialkowski. Process control sample problems. In N. J. Sell (editor), *Process Control Fundamentals for the Pulp & Paper Industry*. Tappi Press, Norcross, GA, 1995.

[45] H. S. Black. Stabilized feedback amplifiers. *Bell System Technical Journal*, 13:1–2, 1934.

[46] H. S. Black. Inventing the negative feedback amplifier. *IEEE Spectrum*, 14(12):55–60, 1977.

[47] J. F. Blackburn, G. Reethof, and J. L. Shearer. *Fluid Power Control*. MIT Press, Cambridge, MA, 1960.

[48] J. H. Blakelock. *Automatic Control of Aircraft and Missiles*, 2nd ed. Addison-Wesley, Cambridge, MA, 1991.

[49] G. Blickley. Modern control started with Ziegler-Nichols tuning. *Control Engineering*, 37:72–75, 1990.

[50] H. W. Bode. Relations between attenuation and phase in feedback amplifier design. *Bell System Technical Journal*, 19:421–454, 1940.

[51] H. W. Bode. *Network Analaysis and Feedback Amplifier Design*. Van Nostrand, New York, 1945.

[52] H. W. Bode. Feedback—The history of an idea. *Symposium on Active Networks and Feedback Systems*. Polytechnic Institute of Brooklyn, New York, 1960. Reprinted in [33].

[53] W. E. Boyce and R. C. DiPrima. *Elementary Differential Equations*. Wiley, New York, 2004.

[54] B. Brawn and F. Gustavson. Program behavior in a paging environment. *Proceedings of the AFIPS Fall Joint Computer Conference*, pp. 1019–1032, 1968.

[55] R. W. Brockett. *Finite Dimensional Linear Systems*. Wiley, New York, 1970.

[56] R. W. Brockett. New issues in the mathematics of control. In B. Engquist and W. Schmid (editors), *Mathematics Unlimited—2001 and Beyond*, pp. 189–220. Springer, Berlin, 2000.

[57] G. S. Brown and D. P. Campbell. *Principles of Servomechanims*. Wiley, New York, 1948.

[58] A. E. Bryson, Jr. and Y.-C. Ho. *Applied Optimal Control: Optimization, Estimation, and Control*. Wiley, New York, 1975.

[59] F. M. Callier and C. A. Desoer. *Linear System Theory*. Springer, London, 1991.

[60] R. H. Cannon. *Dynamics of Physical Systems*. Dover, New York, 2003. Originally published by McGraw-Hill, 1967.

[61] H. S. Carslaw and J. C. Jaeger. *Conduction of Heat in Solids*, 2nd ed. Clarendon Press, Oxford, UK, 1959.

[62] H. Chestnut and R. W. Mayer. *Servomechanisms and Regulating System Design, Vol. 1*. Wiley, New York, 1951.

[63] C. Cobelli, E. Renard, and B. Kovatchev. Artificial pancreas: Past, present, future. *Diabetes,* 68(11):2672–2682, 2011.

[64] J. Cortés. Distributed algorithms for reaching consensus on general functions. *Automatica,* 44(3):726–737, March 2008.

[65] R. F. Coughlin and F. F. Driscoll. *Operational Amplifiers and Linear Integrated Circuits*, 6th ed. Prentice Hall, Englewood Cliffs, NJ, 1975.

[66] L. B. Cremean, T. B. Foote, J. H. Gillula, G. H. Hines, D. Kogan, K. L. Kriechbaum, J. C. Lamb, J. Leibs, L. Lindzey, C. E. Rasmussen, A. D. Stewart, J. W. Burdick, and R. M. Murray. Alice: An information-rich autonomous vehicle for high-speed desert navigation. *Journal of Field Robotics*, 23(9): 777–810, 2006.

[67] Crocus. *Systemes d'Exploitation des Ordinateurs*. Dunod, Paris, 1993.

[68] W. J. Culver. On the existence and uniqueness of the real logarithm of a matrix. *Proc. American Mathematical Society*, 17(5):1146–1151, 1966.

[69] C. Dalla Man, R. A. Rizza, and C. Cobelli. Meal simulation model of the glucose-insulin system. *IEEE Transactions on Biomedical Engineeing*, 54(10):1740–1749, 2007.

[70] D. Del Vecchio and R. M. Murray. *Biomolecular Feedback Systems*. Princeton University Press, Princeton, NJ, 2014.

[71] L. Desborough and R. Miller. Increasing customer value of industrial control performance monitoring—Honeywell's experience. *Sixth International Conference on Chemical Process Control*. AIChE Symposium Series Number 326 (Vol. 98), 2002.

[72] Y. Diao, N. Gandhi, J. L. Hellerstein, S. Parekh, and D. M. Tilbury. Using MIMO feedback control to enforce policies for interrelated metrics with application to the Apache web server. *Proceedings of the IEEE/IFIP Network Operations and Management Symposium*, pp. 219–234, 2002.

[73] R. C. Dorf and R. H. Bishop. *Modern Control Systems*, 10th ed. Prentice Hall, Upper Saddle River, NJ, 2004.

[74] F. H. Dost. *Grundlagen der Pharmakokinetik*. Thieme Verlag, Stuttgart, 1968.

[75] J. C. Doyle. Guaranteed margins for LQG regulators. *IEEE Transactions on Automatic Control*, 23(4):756–757, 1978.

[76] J. C. Doyle, B. A. Francis, and A. R. Tannenbaum. *Feedback Control Theory*. Macmillan, New York, 1992.

[77] J. C. Doyle, K. Glover, P. P. Khargonekar, and B. A. Francis. State-space solutions to standard H_2 and H_∞ control problems. *IEEE Transactions on Automatic Control*, 34(8):831–847, 1989.

[78] C. S. Draper. Flight control. *Journal Royal Aeronautical Society*, 59(July):451–477, 1955. 45th Wilber Wright Memorial Lecture.

[79] L. E. Dubins. On curves of minimal length with a constraint on average curvature, and with prescribed initial and terminal positions and tangents. *American Journal of Mathematics*, 79:497–516, 1957.

[80] F. Dyson. A meeting with Enrico Fermi. *Nature*, 427(6972):297, 2004.

[81] H. El-Samad, J. P. Goff, and M. Khammash. Calcium homeostasis and parturient hypocalcemia: An integral feedback perspective. *Journal of Theoretical Biology*, 214:17–29, 2002.

[82] J. R. Ellis. *Vehicle Handling Dynamics*. Mechanical Engineering Publications, London, 1994.

[83] S. P. Ellner and J. Guckenheimer. *Dynamic Models in Biology*. Princeton University Press, Princeton, NJ, 2005.

[84] E. N. Elnozahy, M. Kistler, and R. Rajamony. Energy-efficient server clusters. *Power-Aware Computer Systems*, pp. 179–197. Springer, Berlin, 2003.

[85] M. B. Elowitz and S. Leibler. A synthetic oscillatory network of transcriptional regulators. *Nature*, 403(6767):335–338, 2000.

[86] P. G. Fabietti, V. Canonico, M. O. Federici, M. Benedetti, and E. Sarti. Control oriented model of insulin and glucose dynamics in type 1 diabetes. *Medical and Biological Engineering and Computing*, 44:66–78, 2006.

[87] M. Fliess, J. Levine, P. Martin, and P. Rouchon. On differentially flat nonlinear systems. *Comptes Rendus des Séances de l'Académie des Sciences,* Serie I, 315:619–624, 1992.

[88] M. Fliess, J. Levine, P. Martin, and P. Rouchon. Flatness and defect of nonlinear systems: Introductory theory and examples. *International Journal of Control*, 61(6):1327–1361, 1995.

[89] J. W. Forrester. *Industrial Dynamics*. MIT Press, Cambridge, MA, 1961.

[90] J. B. J. Fourier. On the propagation of heat in solid bodies. Memoir, read before the Class of the Institut de France, 1807.

[91] A. Fradkov. *Cybernetical Physics: From Control of Chaos to Quantum Control*. Springer, Berlin, 2007.

[92] B. A. Francis. *A Course in $\mathcal{H}_\infty$ Control*. Springer, Berlin, 1987.

[93] G. F. Franklin, J. D. Powell, and A. Emami-Naeini. *Feedback Control of Dynamic Systems*, 5th ed. Prentice Hall, Upper Saddle River, NJ, 2005.

[94] B. Friedland. *Control System Design: An Introduction to State Space Methods*. Dover, New York, 2004.

[95] P. Fritzson. *Principles of Object-Oriented Modeling and Simulation with Modelica 3.3: A Cyber-Physical Approach*, 2 ed. IEEE Press. Wiley, 2015.

[96] A. De Gaetano, D. Di Martino, A. Germani, and C. Manes. Mathematical models and state observation of the glucose-insulin homeostasis. In J. Cagnol and J.-P. Zolesio (editors), *System Modeling and Optimization – Proceedings of the 21st IFIP TC7 Conference*, pp. 281–294. Springer, 2005.

[97] F. R. Gantmacher. *The Theory of Matrices, Vol. 1 and 2*. Chelsea Publishing Company, New York, 1960.

[98] M. A. Gardner and J. L. Barnes. *Transients in Linear Systems*. Wiley, New York, 1942.

[99] T. T. Georgiou and A. Lindquist. The separation principle in stochastic control, redux. *IEEE Transactions on Automatic Control*, 58(10):2481–2494, 2013.

[100] E. Gilbert. Controllability and observability in multivariable control systems. *SIAM Journal of Control*, 1(1):128–151, 1963.

[101] J. C. Gille, M. J. Pelegrin, and P. Decaulne. *Feedback Control Systems; Analysis, Synthesis, and Design*. McGraw-Hill, New York, 1959.

[102] M. Giobaldi and D. Perrier. *Pharmacokinetics*, 2nd ed. Marcel Dekker, New York, 1982.

[103] K. Godfrey. *Compartment Models and Their Application*. Academic Press, New York, 1983.

[104] R. Goebel, R. G. Sanfelice, and A. R. Teel. *Hybrid Dynamical Systems: Modeling, Stability, and Robustness*. Princeton University Press, Princeton, NJ, 2012.

[105] H. Goldstein. *Classical Mechanics*. Addison-Wesley, Cambridge, MA, 1953.

[106] S. W. Golomb. Mathematical models—Uses and limitations. *Simulation*, 4(14):197–198, 1970.

[107] G. C. Goodwin, S. F. Graebe, and M. E. Salgado. *Control System Design*. Prentice Hall, Upper Saddle River, NJ, 2001.

[108] D. Graham and D. McRuer. *Analysis of Nonlinear Control Systems*. Wiley, New York, 1961.

[109] M. Green and D. J. N. Limebeer. *Linear Robust Control*. Prentice Hall, Englewood Cliffs, NJ, 1995.

[110] J. Guckenheimer and P. Holmes. *Nonlinear Oscillations, Dynamical Systems, and Bifurcations of Vector Fields*. Springer, Berlin, 1983.

[111] E. A. Guillemin. *Theory of Linear Physical Systems*. MIT Press, Cambridge, MA, 1963.

[112] L. Gunkel and G. F. Franklin. A general solution for linear sampled data systems. *IEEE Transactions on Automatic Control*, AC-16:767–775, 1971.

[113] W. Hahn. *Stability of Motion*. Springer, Berlin, 1967.

[114] D. Hanahan and R. A. Weinberg. The hallmarks of cancer. *Cell*, 100:57–70, 2000.

[115] R. J. Hawkins, J. K. Speakes, and D. E. Hamilton. Monetary policy and PID control. *Journal of Economic Interaction and Coordination*, 10(1):183–197, 2015.

[116] J. K. Hedrick and T. Batsuen. Invariant properties of automobile suspensions. *Proceedings of the Institution of Mechanical Engineers*, Vol. 204, pp. 21–27, London, 1990.

[117] J. L. Hellerstein, Y. Diao, S. Parekh, and D. M. Tilbury. *Feedback Control of Computing Systems*. Wiley, New York, 2004.

[118] D. V. Herlihy. *Bicycle—The History*. Yale University Press, New Haven, CT, 2004.

[119] M. B. Hoagland and B. Dodson. *The Way Life Works*. Times Books, New York, 1995.

[120] A. L. Hodgkin and A. F. Huxley. A quantitative description of membrane current and its application to conduction and excitation in nerve. *Journal of Physiology*, 117:500–544, 1952.

[121] I. M. Horowitz. *Synthesis of Feedback Systems*. Academic Press, New York, 1963.

[122] I. M. Horowitz. Superiority of transfer function over state-variable methods in linear, time-invariant feedback system design. *IEEE Transactions on Automatic Control*, AC-20(1):84–97, 1975.

[123] I. M. Horowitz. Survey of quantitative feedback theory. *International Journal of Control*, 53:255–291, 1991.

[124] I. M. Horowitz. *Quantitative Feedback Design Theory (QFT)*. QFT Publications, Boulder, CO, 1993.

[125] T. P. Hughes. *Elmer Sperry: Inventor and Engineer*. John Hopkins University Press, Baltimore, MD, 1993.

[126] A. Isidori. *Nonlinear Control Systems*, 3rd ed. Springer, Berlin, 1995.

[127] M. Ito. Neurophysiological aspects of the cerebellar motor system. *International Journal of Neurology*, 7:162–178, 1970.

[128] V. Jacobson. Congestion avoidance and control. *ACM SIGCOMM Computer Communication Review*, 25:157–173, 1995.

[129] J. A. Jacquez. *Compartment Analysis in Biology and Medicine*. Elsevier, Amsterdam, 1972.

[130] H. James, N. Nichols, and R. Phillips. *Theory of Servomechanisms*. McGraw-Hill, New York, 1947.

[131] P. K. Janert. *Feedback Control for Computer Systems*. O'Reilly Media, Sebastopol, CA, 2014.

[132] P. D. Joseph and J. T. Tou. On linear control theory. *Transactions of the AIEE*, 80(18), 1961.

[133] W. G. Jung (editor). *Op Amp Applications*. Analog Devices, Norwood, MA, 2002.

[134] R. E. Kalman. Contributions to the theory of optimal control. *Boletin de la Sociedad Matématica Mexicana*, 5:102–119, 1960.

[135] R. E. Kalman. New methods and results in linear prediction and filtering theory. Technical Report 61-1. Research Institute for Advanced Studies (RIAS), Baltimore, MD, February 1961.

[136] R. E. Kalman. On the general theory of control systems. *Proceedings of the First IFAC Congress on Automatic Control, Moscow, 1960*, Vol. 1, pp. 481–492. Butterworths, London, 1961.

[137] R. E. Kalman and R. S. Bucy. New results in linear filtering and prediction theory. *Transactions of the ASME (Journal of Basic Engineering)*, 83 D:95–108, 1961.

[138] R. E. Kalman, P. L. Falb, and M. A. Arbib. *Topics in Mathematical System Theory*. McGraw-Hill, New York, 1969.

[139] R. E. Kalman, Y. Ho, and K. S. Narendra. *Controllability of Linear Dynamical Systems*, Vol. 1 of *Contributions to Differential Equations*. Wiley, New York, 1963.

[140] J. Keener and J. Sneyd. *Mathematical Physiology I: Cellular Physiology*, 2nd ed. Springer, New York, 2008.

[141] J. Keener and J. Sneyd. *Mathematical Physiology II: Systems Physiology*, 2nd ed. Springer, New York, 2009.

[142] F. P. Kelly. Stochastic models of computer communication. *Journal of the Royal Statistical Society*, B47(3):379–395, 1985.

[143] K. Kelly. *Out of Control*. Addison-Wesley, Reading, MA, 1994. Available at http://www.kk.org/outofcontrol.

[144] H. K. Khalil. *Nonlinear Systems*, 3rd ed. Macmillan, New York, 2001.

[145] U. Kiencke and L. Nielsen. *Automotive Control Systems: For Engine, Driveline, and Vehicle*. Springer, Berlin, 2000.

[146] H. Kitano. Biological robustness. *Nature Reviews Genetics*, 5(11):826–837, 2004.

[147] C. Kittel. *Introduction to Solid State Physics*. Wiley, New York, 1995.

[148] R. E. Klein. Using bicycles to teach dynamics. *Control Systems Magazine*, 9(3):4–8, 1989.

[149] L. Kleinrock. *Queuing Systems, Vols. I and II*, 2nd ed. Wiley-Interscience, New York, 1975.

[150] A. J. Kowalski. Can we really close the loop and how soon? Accelerating the availability of an artificial pancreas: A roadmap to better diabetes outcomes. *Diabetes Technology & Therapeutics*, 11, Supplement 1:113–119, 2009.

[151] N. N. Krasovski. *Stability of Motion*. Stanford University Press, Stanford, CA, 1963.

[152] M. Krstić, I. Kanellakopoulos, and P. Kokotović. *Nonlinear and Adaptive Control Design*. Wiley-Interscience, New York, 1995.

[153] Paul Krugman. *The Return of Depression Economics and the Crisis of 2008*. W. W. Norton & Company, New York, 2009.

[154] P. R. Kumar. New technological vistas for systems and control: The example of wireless networks. *Control Systems Magazine*, 21(1):24–37, 2001.

[155] P. R. Kumar and P. Varaiya. *Stochastic Systems: Estimation, Identification, and Adaptive Control*. Prentice Hall, Englewood Cliffs, NJ, 1986.

[156] P. Kundur. *Power System Stability and Control*. McGraw-Hill, New York, 1993.

[157] B. C. Kuo and F. Golnaraghi. *Automatic Control Systems*, 8th ed. Wiley, New York, 2002.

[158] F. Lamnabhi-Lagarrigue, A. Annaswamy, S. Engell, A. Isaksson, P. Khargonekar, R. M. Murray, H. Nijmeijer, T. Samad, D. Tilbury, and P. Van den Hof. Systems & control for the future of humanity, research agenda: Current and future roles, impact and grand challenges. *Annual Reviews in Control*, 43:1–64, 2017.

[159] J. P. LaSalle. Some extensions of Lyapunov's second method. *IRE Transactions on Circuit Theory*, CT-7(4):520–527, 1960.

[160] S. Laxminaryan, J. Reifman, and G. M. Steil. Use of a food and drug administration-approved type 1 diabetes mellitus simulator to evaluate and optimize a proportional-integral-derivative controller. *Journal of Diabetes Science and Technology*, 6:1401–1409, 2012.

[161] E. A. Lee and S. A. Seshia. *Introduction to Embedded Systems, A Cyber-Physical Systems Approach*. http://LeeSeshia.org, 2015. ISBN 978-1-312-42740-2.

[162] E. A. Lee and P. Varaiya. *Structure and Interpretation of Signals and Systems*. LeeVaraiya.org, 2011. Available online at http://leevaraiya.org.

[163] A. D. Lewis. A mathematical approach to classical control. Technical report. Queens University, Kingston, Ontario, 2003.

[164] D. J. N. Limebeer and R. S. Sharp. Bicycles, motorcycles and models. *Control Systems Magazine*, 26(5):34–61, 2006.

[165] A. Lindquist and G. Picci. *Linear Stochastic Systems: A Geometric Approach to Modeling, Estimation and Identification*. Springer, Berlin, Heidelberg, 2015.

[166] L. Ljung. *System Indentification – Theory for the User*, 2nd ed. Prentice Hall, Upper Saddle River, NJ, 1999.

[167] S. H. Low. *Analytical Methods for Network Congestion Control*. Morgan and Claypool, San Rafael, CA, 2017.

[168] S. H. Low, F. Paganini, and J. C. Doyle. Internet congestion control. *IEEE Control Systems Magazine*, pp. 28–43, February 2002.

[169] S. H. Low, F. Paganini, J. Wang, S. Adlakha, and J. C. Doyle. Dynamics of TCP/RED and a scalable control. *Proceedings of IEEE Infocom*, pp. 239–248, 2002.

[170] K. H. Lundberg. History of analog computing. *IEEE Control Systems Magazine*, pp. 22–28, March 2005.

[171] L. A. MacColl. *Fundamental Theory of Servomechanisms*. Van Nostrand, Princeton, NJ, 1945. Dover reprint 1968.

[172] J. M. Maciejowski. *Multivariable Feedback Design*. Addison Wesley, Reading, MA, 1989.

[173] D. A. MacLulich. *Fluctuations in the Numbers of the Varying Hare (Lepus americanus)*. University of Toronto Press, 1937.

[174] A. Makroglou, J. Li, and Y. Kuang. Mathematical models and software tools for the glucose-insulin regulatory system and diabetes: An overview. *Applied Numerical Mathematics*, 56:559–573, 2006.

[175] J. G. Malkin. *Theorie der Stabilität einer Bewegung*. Oldenbourg, München, 1959.

[176] R. Mancini. *Op Amps for Everyone*. Texas Instruments, Houston. TX, 2002.

[177] J. E. Marsden and M. J. Hoffmann. *Basic Complex Analysis*. W. H. Freeman, New York, 1998.

[178] J. E. Marsden and T. S. Ratiu. *Introduction to Mechanics and Symmetry*. Springer, New York, 1994.

[179] O. Mayr. *The Origins of Feedback Control*. MIT Press, Cambridge, MA, 1970.

[180] M. W. McFarland (editor). *The Papers of Wilbur and Orville Wright*. McGraw-Hill, New York, 1953.

[181] D. C. McFarlane and K. Glover. *Robust Controller Design Using Normalized Coprime Factor Plant Descriptions*. Springer, New York, 1990.

[182] H. T. Milhorn. *The Application of Control Theory to Physiological Systems*. Saunders, Philadelphia, 1966.

[183] D. A. Mindel. *Between Human and Machine: Feedback, Control, and Computing Before Cybernetics*. Johns Hopkins University Press, Baltimore, MD, 2002.

[184] D. A. Mindel. *Digital Apollo: Human and Machine in Spaceflight*. The MIT Press, Cambridge, MA, 2008.

[185] C. A. Monje, Y. Q. Chen, B. M. Vinagre, D. Xue, and V. Feliu. *Fractional-order Systems and Controls: Fundamentals and Applications*. Springer, 2010.

[186] J. D. Murray. *Mathematical Biology, Vols. I and II*, 3rd ed. Springer, New York, 2004.

[187] R. M. Murray (editor). *Control in an Information Rich World: Report of the Panel on Future Directions in Control, Dynamics and Systems*. SIAM, Philadelphia, 2003.

[188] R. M. Murray, K. J. Åström, S. P. Boyd, R. W. Brockett, and G. Stein. Future directions in control in an information-rich world. *Control Systems Magazine*, April 2003.

[189] R. M. Murray, Z. Li, and S. S. Sastry. *A Mathematical Introduction to Robotic Manipulation*. CRC Press, Boca Raton, FL, 1994.

[190] P. J. Nahin. *Oliver Heaviside: Sage in Solitude: The Life, Work and Times of an Electrical Genius of the Victorian Age*. IEEE Press, New York, 1988.

[191] H. Nijmeijer and J. M. Schumacher. Four decades of mathematical system theory. In J. W. Polderman and H. L. Trentelman (editors), *The Mathematics of Systems and Control: From Intelligent Control to Behavioral Systems*, pp. 73–83. University of Groningen, Groningen, NL, 1999.

[192] H. Nyquist. Regeneration theory. *Bell System Technical Journal*, 11:126–147, 1932.

[193] H. Nyquist. The regeneration theory. In R. Oldenburger (editor), *Frequency Response*, p. 3. MacMillan, New York, 1956.

[194] A. O'Dwyer. *Handbook of PI and PID Controller Tuning Rules*, 3rd ed. Imperial College Press, 2006.

[195] K. Ogata. *Modern Control Engineering*, 4th ed. Prentice Hall, Upper Saddle River, NJ, 2001.

[196] R. Oldenburger (editor). *Frequency Response*. MacMillan, New York, 1956.

[197] R. Olfati-Saber, J. A. Fax, and R. M. Murray. Consensus and cooperation in networked multi-agent systems. *Proceedings of the IEEE*, 95(1):215–233, 2007.

[198] A. V. Oppenheim, A. S. Willsky, and S. H. Nawab. *Signals and Systems*, 2nd ed. Prentice-Hall, Saddle River, NJ, 1996.

[199] G. Pacini and R. N. Bergman. A computer program to calculate insulin sensitivity and pancreatic responsivity from the frequently sampled intravenous glucose tolerance test. *Computer Methods and Programs in Biomedicine*, 23:113–122, 1986.

[200] D. Packard. *The HP Way: How Bill Hewlett and I Built Our Company*. Harper Collins, New York, 2013.

[201] G. A. Philbrick. Designing industrial controllers by analog. *Electronics*, 21(6):108–111, 1948.

[202] T. Van Pottelbergh, G. Deion, and R. Sepulchre. Robust modulation of integrate–and–fire models. *Neural Computing*, 30:987–1011, 2018.

[203] W. F. Powers and P. R. Nicastri. Automotive vehicle control challenges in the 21st century. *Control Engineering Practice*, 8:605–618, 2000.

[204] S. Prajna, A. Papachristodoulou, and P. A. Parrilo. SOSTOOLS: Sum of squares optimization toolbox for MATLAB, 2002. Available from http://www.cds.caltech.edu/sostools.

[205] C. Ptolemaeus (editor). *System Design, Modeling, and Simulation using Ptolemy II*. Ptolemy.org, 2014.

[206] A. Rantzer and K. J. Åström. Control theory. In N. J. Higham (editor), *The Princeton Companion to Applied Mathematics*. Princeton University Press, Princeton and Oxford, 2015.

[207] M. B. Reiser, J. S. Humbert, M. J. Dunlop, D. Del Vecchio, R. M. Murray, and M. H. Dickinson. Vision as a compensatory mechanism for disturbance rejection in upwind flight. *Proc. American Control Conference*, Vol. 1, pp. 311–316, 2004.

[208] D. S. Riggs. *The Mathematical Approach to Physiological Problems*. MIT Press, Cambridge, MA, 1963.

[209] J. Rissanen. Control system synthesis by analogue computer based on the generalized feedback concept. In Robert Vichnevetsky (editor), *Proceedings of the Symposium on Analogue Computation Applied to the Study of Chemical Processes*, pp. 1–13, Gordon & Breach. New York, 1960.

[210] H. H. Rosenbrock and P. D. Moran. Good, bad or optimal? *IEEE Transactions on Automatic Control*, AC-16(6):552–554, 1971.

[211] W. J. Rugh. *Linear System Theory*, 2nd ed. Prentice Hall, Englewood Cliffs, NJ, 1995.

[212] E. B. Saf and A. D. Snider. *Fundamentals of Complex Analysis with Applications to Engineering, Science and Mathematics*. Prentice Hall, Englewood Cliffs, NJ, 2002.

[213] T. Samad and A. M. Annaswamy (editors). *The Impact of Control Technology*, 2nd ed. IEEE Control Systems Society, 2014. Available at www.ieeecss.org.

[214] D. Sarid. *Atomic Force Microscopy*. Oxford University Press, Oxford, UK, 1991.

[215] S. Sastry. *Nonlinear Systems*. Springer, New York, 1999.

[216] G. Schitter. High performance feedback for fast scanning atomic force microscopes. *Review of Scientific Instruments*, 72(8):3320–3327, 2001.

[217] G. Schitter, K. J. Åström, B. DeMartini, P. J. Thurner, K. L. Turner, and P. K. Hansma. Design and modeling of a high-speed AFM-scanner. *IEEE Transactions on Control System Technology*, 15(5):906–915, 2007.

[218] M. Schwartz. *Telecommunication Networks*. Addison Wesley, Reading, MA, 1987.

[219] D. E. Seborg, T. F. Edgar, and D. A. Mellichamp. *Process Dynamics and Control*, 2nd ed. Wiley, Hoboken, NJ, 2004.

[220] S. D. Senturia. *Microsystem Design*. Kluwer, Boston, MA, 2001.

[221] R. Sepulchre, G. Drion, and A. Franci. Excitable behaviors. In R. Tempo, S. Yurkovich, and P. Misra (editors), *Emerging Applications of Control and Systems Theory. Lecture Notes in Control and Information Sciences— Proceedings*, pp. 269–280. Springer, 2018.

[222] F. G. Shinskey. *Process-Control Systems. Application, Design, and Tuning*, 4th ed. McGraw-Hill, New York, 1996.

[223] S. Skogestad and I. Postlethwaite. *Multivariable Feedback Control*, 2nd ed. Wiley, Hoboken, NJ, 2005.

[224] J. M. Smith. The importance of the nervous sytem in the evolution of animal flight. *Evolution*, 6:127–129, 1952.

[225] E. D. Sontag. *Mathematical Control Theory: Deterministic Finite Dimensional Systems*, 2nd ed. Springer, New York, 1998.

[226] M. W. Spong and M. Vidyasagar. *Robot Dynamics and Control*. John Wiley, New York, 1989.

[227] L. Stark. *Neurological Control Systems—Studies in Bioengineering*. Plenum Press, New York, 1968.

[228] G. M. Steil. Algorithms for a closed-loop artificial pancreas: The case for proportional-integral-derivative control. *Journal of Diabetes Science and Technology*, 7(6):1621–1631, 2013.

[229] G. Stein. Respect the unstable. *Control Systems Magazine*, 23(4):12–25, 2003.

[230] J. Stelling, U. Sauer, Z. Szallasi, F. J. Doyle III, and J. Doyle. Robustness of cellular functions. *Cell*, 118(6):675–685, 2004.

[231] J. Sternby, K. J. Åström, and P. Hagander. Zeros of sampled systems. *Automatica*, 20(1):31–38, 1984.

[232] J. Stewart. *Calculus: Early Transcendentals*. Brooks Cole, Pacific Grove, CA, 2002.

[233] G. Strang. *Linear Algebra and Its Applications*, 3rd ed. Harcourt Brace Jovanovich, San Diego, CA, 1988.

[234] S. H. Strogatz. *Nonlinear Dynamics and Chaos, with Applications to Physics, Biology, Chemistry, and Engineering*. Addison-Wesley, Reading, MA, 1994.

[235] L. Sun, D. Li, and K. Y. Lee. Optimal disturbance rejection for PI controller with constraints on relative delay margin. *ISA Transactions*, 63: 103–111, 2016.

[236] A. S. Tannenbaum. *Computer Networks*, 3rd ed. Prentice Hall, Upper Saddle River, NJ, 1996.

[237] T. Teorell. Kinetics of distribution of substances administered to the body, I and II. *Archives Internationales de Pharmacodynamie et de Therapie*, 57: 205–240, 1937.

[238] G. T. Thaler. *Automatic Control Systems*. West Publishing, St. Paul, MN, 1989.

[239] M. Tiller. *Introduction to Physical Modeling with Modelica*. Springer, Berlin, 2001.

[240] D. Tipper and M. K. Sundareshan. Numerical methods for modeling computer networks under nonstationary conditions. *IEEE Journal of Selected Areas in Communications*, 8(9):1682–1695, 1990.

[241] J. G. Truxal. *Automatic Feedback Control System Synthesis*. McGraw-Hill, New York, 1955.

[242] H. S. Tsien. *Engineering Cybernetics*. McGraw-Hill, New York, 1954.

[243] H. Tullberg, M. Fallgren, K. Kusume, and Andreas Höglund. 5G use cases and system concept. In A. Osseiran, J. F. Monserrat, and P. Marsch (editors), *5G Mobile and Wireless Communications Technology*, chapter 2. Cambridge University Press, 2016.

[244] A. Tustin. Feedback. *Scientific American*, 187:48–54, 1952.

[245] M. Vidyasagar. The graph metric for unstable plants and robustness estimates for feedback stability. *IEEE Transactions on Automatic Control*, 29(5): 403–418, 1984.

[246] M. Vidyasagar. *Control Systems Synthesis*. MIT Press, Cambridge, MA, 1985.

[247] G. Vinnicombe. Frequency domain uncertainty and the graph topology. *IEEE Transactions on Automatic Control*, 38(9):1371–1383, 1993.

[248] G. Vinnicombe. *Uncertainty and Feedback: $\mathcal{H}_\infty$ Loop-Shaping and the ν-Gap Metric*. Imperial College Press, London, 2001.

[249] F. J. W. Whipple. The stability of the motion of a bicycle. *Quarterly Journal of Pure and Applied Mathematics*, 30:312–348, 1899.

[250] F. R. Whitt and D. G. Wilson. *Bicycling Science*. MIT Press, 1997.

[251] D. V. Widder. *Laplace Transforms*. Princeton University Press, Princeton, NJ, 1941.

[252] E. P. M. Widmark and J. Tandberg. Über die Bedingungen für die Akkumulation indifferenter Narkotika. *Biochemische Zeitung*, 148:358–389, 1924.

[253] N. Wiener. *Cybernetics: Or Control and Communication in the Animal and the Machine*. Wiley, New York, 1948.

[254] S. Wiggins. *Introduction to Applied Nonlinear Dynamical Systems and Chaos*. Springer, Berlin, 1990.

[255] D. G. Wilson. *Bicycling Science*, 3rd ed. MIT Press, Cambridge, MA, 2004. With contributions by Jim Papadopoulos.

[256] H. R. Wilson. *Spikes, Decisions, and Actions: The Dynamical Foundations of Neuroscience*. Oxford University Press, Oxford, UK, 1999.

[257] K. A. Wise. Guidance and control for military systems: Future challenges. *AIAA Conference on Guidance, Navigation, and Control*, 2007. AIAA Paper 2007-6867.

[258] S. Yamamoto and I. Hashimoto. Present status and future needs: The view from Japanese industry. In Y. Arkun and W. H. Ray (editors), *Chemical Process Control—CPC IV*, 1991.

[259] T. M. Yi, Y. Huang, M. I. Simon, and J. C. Doyle. Robust perfect adaptation in bacterial chemotaxis through integral feedback control. *Proceedings of the National Academy of Sciences*, 97(9):4649–4653, 2000.

[260] D. C. Youla, J. J. Bongiorno, Jr., and C. N. Lu. Single-loop feedback stabilization of linear multivariable plants. *Automatica*, 10(2):159–173, 1974.

[261] L. Zaccariand and A. R. Teel. *Modern Anti-windup Synthesis: Control Augmentation for Actuator Saturation*. Princeton University Press, Princeton, NJ, 2011.

[262] L. A. Zadeh and C. A. Desoer. *Linear System Theory: the State Space Approach*. McGraw-Hill, New York, 1963.

[263] G. Zames. Feedback and optimal sensitivity: Model reference transformations, multiplicative seminorms, and approximative inverse. *IEEE Transactions on Automatic Control*, AC-26(2):301–320, 1981.

[264] G. Zames and A. K. El-Sakkary. Unstable systems and feedback: The gap metric. *Proc. Allerton Conference*, pp. 380–385, 1980.

[265] J. C. Zhou, J. C. Doyle, and K. Glover. *Robust and Optimal Control*. Prentice Hall, Englewood Cliffs, NJ, 1996.

[266] J. G. Ziegler and N. B. Nichols. Optimum settings for automatic controllers. *Transactions of the ASME*, 64:759–768, 1942.

Index

acausal modeling, 67

access control, *see* admission control

acknowledgment (ack) packet, 120–122

activator, 13, 100, *see also* biological circuits

active filter, 197, *see also* operational amplifier

actuators, 5, 6, 65, 89, 108, 124, 277, 343, 375, 389, 446, 449, 454, 456
    effect on zeros, 343, 446
    in computing systems, 118
    saturation, 68, 278, 363, 369–371, 375, 389, 467–469

A/D converters, *see* analog-to-digital converters

adaptation, 360

adaptive control, 439, 440

additive uncertainty, 415, 421, 424

adjacency matrix, 97

admission control, 94, 106, 120, 122, 332

aerospace systems, 8, 15, 449, *see also* vectored thrust aircraft; X-29 aircraft

AFM, *see* atomic force microscopes

air-fuel ratio control, 23

aircraft, *see* flight control

alcohol, metabolism of, 136

algebraic loops, 85–86, 302–303

aliasing, 278

all-pass transfer function, 452

alternating current (AC), 199

amplifier, *see* operational amplifier

amplitude ratio, *see* gain

analog computing, 69, 86, 114, 303, 374

analog implementation, of controllers, 117, 320, 374

analog-to-digital converters, 5, 6, 125, 277, 375

angle, of frequency response, *see* phase

anticipation, in controllers, 20, 359, *see also* derivative action

antiresonance, 199, 200

anti-windup compensation, 20, 370–372, 375, 377, 379
    stability analysis, 379

Apache web server, 118, *see also* web server control

Arbib, M. A., 213

architectures, for control systems, 17, 23–27, 49, 270

argument, of a complex number, 309

arrival rate (queuing systems), 95

artificial pancreas, 131

asymptotes, in Bode plot, 311, 312

asymptotic stability, 81, 145, 146, 148, 149, 154, 156, 158, 162, 163, 165, 184
    discrete-time systems, 210

atmospheric dynamics, *see* environmental science

atomic force microscopes, 3, 89, 124–127
    contact mode, 124, 199
    horizontal positioning, 341, 431
    system identification, 317–318
    tapping mode, 124, 351, 362, 367, 395–396
    with preloading, 136

attractor (equilibrium point), 146

automatic reset, in PID control, 358

automatic tuning, 368, 439

automotive control systems, 18, 89, 112, *see also* cruise control; vehicle steering

autonomous differential equation, 63, *see also* time-invariant systems

autonomous vehicles, 9, 10, 26–27, *see also* robotics